Lecture Notes in Control and Information Sciences

Edited by M. Thoma and A. Wyner

Lecture Notes in Control and Information Sciences

Edited by M. Thoma and A. Wyner

139

V. Hayward, O. Khatib (Eds.)

Experimental Robotics I

The First International Symposium
Montreal, June 19-21, 1989

Springer-Verlag
Berlin Heidelberg GmbH

Editors
Prof. Vincent Hayward
Dept. of Electrical Engineering
McGill University
Montreal, Quebec, H3A 2A7
Canada

Prof. Oussama Khatib
Dept. of Computer Science
Stanford University
Stanford, CA 94305
USA

ISBN 978-3-540-52182-2   ISBN 978-3-540-46917-9 (eBook)
DOI 10.1007/978-3-540-46917-9

First International Symposium On

Experimental Robotics

Montréal, Canada
June 19-21, 1989

Design and Photography: E. M. Fenton

About the Editors

Vincent Hayward received the degree of "Docteur-Ingénieur en Informatique" from "Université de Paris XI" at Orsay in 1981.

He held research positions at Purdue University, Indiana, USA, and at CNRS, France, and is now Assistant Professor with the Department of Electrical Engineering at McGill University. His research interests include sensor-based robot design, programming and control, spatial reasoning and perception, and computational architectures.

Oussama Khatib received the degree of "Docteur-Ingénieur en Automatique et Systèmes" in 1980 from "l'École Nationale Supérieure de l'Espace," Toulouse, France.

He is an Associate Professor with the department of Computer Science at Stanford University, California, USA, and leads the *Manipulation Project* at the Robotics Laboratory of this department. His research interests include motion and force control, motion primitives and sensor-based strategies for real-time collision avoidance and part-mating operations, as well as the design of high-performance force-controlled manipulator and micro-manipulators systems.

Preface

Experimental Robotics 1 — The First International Symposium is the first attempt at collecting works in Robotics from the point of view of experimental research. The meeting at which these contributions were presented took place in Montréal in June 1989. It is the first of a series to be organized in a circular fashion around North America, Europe, and Asia.

The series of events that led to this meeting can be traced back to Spring 1987 in Albany, New York. Several members of the organizing committee, who were attending a SIAM conference on Applied Geometry, began to discuss the major trends that would underly Robotics Research toward the rapidly looming end of this Century. We all agreed that experiments were called to play a larger role in the very way Robotics Research will be approached. Of course we did not mean that theoretical developments are not important, we simply felt that their importance can only be assessed through experimentation, that is, synthesis.

We proposed that the presentations be centered on theories and principles, as applied to robotics, that are validated by experiments. One of the conclusions of the meeting was thus to draw a clear distinction between "demonstrations" and "experiments". An increasing amount of researchers in the field of robotics no longer feel satisfied with hypothetical developments. They strive to distinguish what *can* be done from what *could* be done, and are willing to submit their theories to the test of physical implementation. The content of this collection of contributions reflects a cross-section of the current state of robotic research from one particular aspect: experimental work, and how it affects the theoretical basis of subsequent developments.

Of course, the selection of topics: the study of friction, calibration issues, and design of manipulators, for example, reflects this theme. Some of these problems may have been considered in the past as secondary ones, but they are now revealed as being critical, and we conjecture that progress toward their solution will have a significant impact on most aspects of robotics research. A cursory examination of the contents of the papers shows that, in Robotics, once again, practice is sometimes ahead of theory: successes and failures are rarely fully explained by the theory and investigators are left with conjectures. This is a further justification for the choice of the experimental theme.

The international program committee was composed of the following individuals: V. Hayward, *McGill University*, Canada; O. Khatib, *Stanford University*, USA; J. Angeles, *McGill University*, Canada; R. Chatila, LAAS/CNRS, France; J. Craig, *Silma*, USA; P. Dario, *University of Pisa*, Italy; B. Espiau, IRISA/INRIA, France; G. Hirzinger, DFVLR, FRG; K. Salisbury, MIT, USA; and T. Yoshikawa, *Kyoto University*, Japan. A total of 35 papers have been included in the program, representing Australia, Belgium, Canada, England, France, Germany, Italy, Japan, USA, and Yugoslavia. A keynote lecture given by Prof. Lozano-Perez from MIT on the topic "Tasks, Experiments, and Strategies" kicked off the meeting with the delivery of numerous inspiring ideas. Due to the small size of this meeting and the quality of the attendees, a large amount of interaction took place during the three days of this meeting.

The international committee has proposed that the next meeting should take place in Toulouse, France, in the Spring 1991. The committee has asked Dr. Raja Chatilla from LAAS-CNRS, Toulouse, France, and Dr. Gerd Hirzinger from from DLR, German Aerospace Research Establishment, Institute for Flight Systems, to co-chair the next meeting.

The McGill Research Center for Intelligent Machines (McRCIM) hosted this meeting on the McGill University campus. All the persons involved in this center must be thanked for helping to make this event possible. In particular, Prof. Martin Levine, director of McRCIM, gave a warm welcome address which created an informal, yet productive atmosphere. Dean Bélanger, forefather of McRCIM, then discussed the important connection between theory and experiments in Robotics. He must be thanked for his contribution and support. Prof. Lozano-Perez's keynote lecture alone made the effort of putting this meeting together worthwhile. The members of the program committee must be gratefully acknowledged for accepting to shoulder the difficult task of selecting the papers.

This meeting could not have taken place without the backing of many individuals, institutions and companies: Norman Kaplan from the National Science Foundation, Washington D.C.; Christine Quérido from "Les Fonds pour la Formation des Chercheurs et l'Aide à la Recherche (FCAR), Sainte Foy, Québec"; Pierre Girard and René Blais from "L'Institut de Recherche d'Hydro-Québec (IREQ), Varennes, Québec"; Fred Christie and Pierre Maltais from The Canadian Space Agency, Ottawa, Ontario; Len Allen and Roy Hoffman from CAE Electronics Limited, Saint Laurent, Québec; Samad Hayati from the Jet Propulsion Laboratory, Pasadena, California; Ian Rowe and Ravi Ravindran from SPAR Aerospace Limited, Weston, Ontario.

In the end, the credits must go to the authors for the quality of their contributions and their availability during the conference. The final kudos go to Margaret Dalziel, Manager of McRCIM, who skillfully engineered the organization of this meeting and whose talent was very much appreciated.

Finally, we hope that the video document which accompanies this collection of contributions will prove to be a useful illustration of the reported research.

Montréal, November 1989

Vincent Hayward (McGill University)
Oussama Khatib (Stanford University)

Introduction

The classification adopted in this collection of contributions proceeds along the lines of a nearly traditional division of topics in robotics. The reader will notice numerous cross-correlations between sections. As with most classifications, it is perhaps artificial and somewhat unsatisfactory.

Section 1 covers the control of flexible limbs, intermittent tasks, the control of cooperating robots, force control, and adaptive control. The second section deals with design issues: control of friction at low velocities, actuators, joints, and manipulators. Section 3 investigates questions in perception in terms of model construction, task guidance, visual servoing, and tactile feedback. Section 4 tackles kinematic problems such as inversion and calibration. Finally, section 5 deals with problems in motion planning.

Most of the papers share three overridding concerns that we might wish to use as guidelines for future work in Experimental Robotics.

Dealing with uncertainty, is one of these concerns. It is certainly not a new issue in engineering, but it is treated in various and particular ways by the robotics approach. Experimental robotics research suggests that there might be dual perspectives to reducing uncertainty. Conventional wisdom tells us that we need to "see to act" (as in control), whereas the dialectic reversal of this proposition requires us to "act to see" (as in perception). Dealing with uncertainty is an attribute of intelligence and autonomy. Whether it lies in the effectors, the sensors or elsewhere is still an open question; what is sure is that robots need a lot of it.

The second general theme which stems from these contributions is the *extension of the task repertoire.* This certainly indicates a major trend in robotics research. Clearly, no systematic methodology is proposed; however, biological systems seem to provide the largest source of inspiration (*i.e.* running, juggling, etc...). It is of course an exercise in synthesis, for which the tools are difficult to find, whereas analytical tools are already abundantly available.

The third general direction suggested by the reading of the papers is *redundancy and co-operation.* Advanced robots will undoubtly be endowed with large amount of redundancy from the viewpoints of action and perception. This redundancy must be orchestrated to achieve cooperation. This is the running theme of a large amount of current research. The necessity of multi-sensory perception systems is generaly agreed upon. Similarly, multi-actuator action systems are also coming into focus. Cooperative action, for lack of a better word, should become the center of focus for understanding advanced robotic systems.

Each of these three themes encompasses the others and it can be observed that most of the papers incorporate elements of all three.

Section 1
Control

This first set of papers is concerned with feedback control of manipulators. The papers by Oakley and Cannon and by Sweevers, Adams, De Schutter, Van Brussel, and Thielemans are concerned with the accurate control of manipulators with flexible links. Both papers insist on the importance of accurate structural modeling to achieve the control of manipulators with flexible links. Modeling goes hand-in-hand with identification. This issue is treated in depth by Sweevers and co-workers.

In the past few years, researchers have become increasingly interested in augmenting the vocabulary of tasks performed by manipulators. The class of "intermittent tasks" should obviously belong to the vocabulary of advanced manipulators. Useful intermittent tasks are observed frequently in everyday life: walking, tossing, tapping, and so-on. The papers by Bühler, Koditschek and Kindlmann and by Thompson and Raibert treat this question from radically different viewpoints. Bühler and co-workers offer an in depth analysis of a very simple task: "the vertical juggle," and propose a non-linear control approach to regulate it. Thompson and Raibert consider the inherently complex task of running and explore the structure of a passive mechanical system to accomplish it without explicit feedback or supply of energy.

It has been often observed that the cooperation of multiple manipulators could also significantly augment the class of achievable tasks, for example in grasping and sharing loads. The paper by Miyazaki, Sonoyama, Manabe, and Manabe explores the issue of cooperation through a series of examples: antagonist "muscle-like" pneumatic actuators and a multiple finger hand. A collection of task-dependent control strategies are evaluated for these examples. An adaptive load-sharing algorithm detailed by Uchiyama and Yamashita is demonstrated to provide the basis of a remarkable collection of dual-arm cooperative tasks.

The control of forces applied by manipulators to their environment has been for a long time a central problem in robotics. The paper by Yoshikawa and Sudou sets the problem in a larger framework than that usually encountered in the literature. The constraints dictated by the manipulator (dynamics) and by the task (shape and surface orientation estimation) are included in the problem framework. Hannaford and Lee propose a stochastic model of force histories recorded during the successful execution of tasks performed under telemanipulation. This model can be utilized to reliably segment tasks into distinct phases.

The paper by Daneshmend, Hayward and Pelletier reports on an adaptive algorithm applied to stabilize damping control (a form of force control) in the presence of uncertainty about the knowledge of environmental stiffness properties. Niemeyer and Slotine apply adaptive control theory to combine high-speed and high-precision robotic "whole-arm" manipulation.

Section 2
Design

The underlying properties of manipulator mechanisms have a major impact on the performance of the control strategies that are applied to them. Hence, a greater emphasis has been recently placed on the design of manipulators.

Papers on the analysis and control of friction have been included in the design section because the control of friction disturbances has a major impact on the design of manipulators. The paper by Armstrong is concerned with the modeling and control of Stribeck friction which is at the root of the stick-slip behavior of machines. The paper by Canudas de Wit tackles a similar problem and proposes an adaptive control algorithm.

The design of actuators is currently undergoing a tremendous evolution as testified by the two contributions in this area. The first paper by Bobrow and Desai describes a hybrid actuator design in which an electric source of energy powers a joint via hydraulic velocity reduction, thereby optimizing a number of design parameters. A second paper by Tsuda, Higuchi and Nakamura describes the high-speed digital control of a magnetically levitated robot wrist.

The design of a passive poly-articulated joint is tackled by Xu, Paul and Corke. They propose an instrumented six d-o-f compliant wrist assembly to implement a robust hybrid position/force control law. Vischer and Khatib present the development and testing of a low-geared torque-controlled joint and describe a new design for an inductive contactless transducer torque sensor. The performance of this joint, optimized for the ARTISAN manipulator, is checked against an earlier design implemented on one joint of a PUMA manipulator.

The section on design concludes with two papers on manipulators. Manipulators mounted on a compliant base, such as in the cases of vehicle support or outer space, are investigated by West, Hootsmans, Dubowsky, and Stelman. A "vehicle emulator" has been designed and built to carry out experiments regarding the control of manipulators placed under such conditions. Dietrich, Hirzinger, Gombert, and Schott discuss several concepts oriented towards the development of light-weight manipulators. The use of advanced materials such as carbon fibers is discussed, as well as issues in sensor integration.

Section 3
Perception

The issue of perception overshadows many advanced robotic application. The paper by Moutarlier and Chatila deals with combined sensing and motion inaccuracies in the context of mobile robots. They propose two approaches. The first one is based on the estimation of the robot state. The second approach relies on actively re-positioning the robot to reduce uncertainty before fusing sensory information into a global environmental model. The problems discussed in the paper by Even, Marcé, Morillon and Fournier differ from those of the previous paper because the presence of a human operator is assumed and the goal is to provide an accurate synthetic feedback.

The paper by Steer focuses on the issue of autonomously navigating a mobile robot with multi-sensory feedback given a known environment. Thus, it differs fundamentally from the two previous papers. The paper by Ijel, Laugier and Troccaz considers the problem of automatically grasping polyhedral objects, given a partial knowledge of the geometry of the environment. Their method relies on a local model refinement method combined with automatic sensor placement, thus combining problems addressed in the three previous papers.

The ability of to use visual perception at high rates is an exciting prospect for a truly sensor based robot. Rives, Chaumette, and Espiau suggest the possibility of closing the loop directly in the sensor frame, thus offering the ability for a manipulator equipped with a wrist mounted camera to overcome problems in sensor and kinematic calibration. Robust control theory is then used as a theoretical framework. Corke and Paul discuss the design of a robot position control loop via visual feedback at video-rate.

Tactile sensing is addressed by Sabatini, Dario, and Bergamasco in the context of tactile contact with soft material. Once again, their work is carried out with a view to extending the task repertoire of robots toward advanced applications such as automated palpation in medicine and other fields, agriculture for example. Eberman and Salisbury address the problem of minimizing the number of sensors to resolve contact information. They show under which conditions contact information can be resolved solely from joint torque measurements. Berger and Khosla discuss the application of tactile sensing to edge tracking, as part of a dynamic exploratory procedure.

Section 4
Kinematics

The efficient kinematic inversion of manipulators with arbitrary architectures still remains an open problem. Lenarčič and Košutnik consider in their paper the intriguing idea of deriving approximate solutions of the inverse kinematic problem, having observed that in many situations an exact solution is, in fact, not necessary. They propose a collection of methods to efficiently compute reliable approximate solutions, particularly when the accuracy of the end-effector orientation is of lesser importance than its position. Charentus and Renaud discuss the kinematic modeling of a modular redundant manipulator (up to 24 actuators) in view of its control. The control of a manipulator with an architecture which is both serial and parallel, as well as heavily redundant is thus discussed.

Khalil, Caenen, and Enguehard present a method to identify the Denavit-Hartenberg parameters of serial manipulators from theodolite measurements. Their method is based on the identification of differential changes of the geometric parameters. Bennett and Hollerbach propose to sweep the kinematic null-space of a redundant closed kinematic chain to sufficiently over-constrain the problem for self-calibration, thus eliminating the need for external instrumentation. They apply this method to pairs of fingers of a mechanical hand. Ioannides, Angeles, Flanagan, and Ostry suggest the use of polar-decomposition filtering to improve the performance of a least-square estimation procedure in the presence of noisy sensor data.

Section 5
Motion Planning

Mason discusses manipulation strategies arising in the context of sliding blocks along walls which suggest a possible unification of the methods available for the analysis of fine motions and those available for synthesis.

Bodduluri, McCarthy and Bobrow investigate motion planning for two manipulators sharing a common load. A general path-planning algorithm developed for multi-dimensional chains produces collision-free motions for two cooperating arms with astonishing ease. Finally, Flanagan and Ostry investigate the properties of multi-joint human motions and propose a model combining staggered joint interpolation and minimum jerk joint trajectories.

Table of Contents

Section 1: Control

Flexible Limbs

Intermittent Tasks

Cooperating Robots

Force

Section 2: Design

Section 3: Perception

Section 4: Kinematics

Section 5: Motion Planning

Theory and Experiments in Selecting Mode Shapes for Two-Link Flexible Manipulators

Celia M. Oakley*   Robert H. Cannon, Jr.[†]

Stanford University Aerospace Robotics Laboratory (ARL)
Stanford, California 94305, USA

Abstract

The design of a control system is typically based on a model of the actual plant. The achievable performance is thus intimately related to the modelling accuracy. A popular method for modelling flexible manipulators is the assumed-modes method. However, in order to generate an accurate model with a minimal number of modes, appropriate component mode shapes must be selected. This paper explores the selection of component mode shapes to be used in models for two-link flexible manipulators. The theoretical natural frequencies and system mode shapes predicted by these models are compared to those of the experimental Stanford Multi-Link Flexible Manipulator. Strobe photographs taken to capture the experimental system mode shapes are included.

1 Introduction

There are many robotic manipulators that possess significant structural link flexibility. These manipulators, used in both space and industrial environments, are utilized for dangerous situations, handling large payloads, and executing precise maneuvers. The speed and accuracy with which these tasks are performed depends intimately on the control system implemented. Since the design of the control system is based on a model of the actual manipulator, the achievable performance is limited by the modelling accuracy. For example, if the model is simplified by ignoring link flexibility, the control bandwidth must be significantly lower than the first natural frequency of the manipulator [1].

If, however, link flexibility is suitably modelled and advanced control strategies are formulated from these models, the performance of existing flexible manipulators can be improved and certain characteristics of link flexibility can be advantageously used in future designs. For instance, flexible manipulators offer high payload-to-mass ratios in addition to light-weight and power-efficient configurations.

Single-link flexible manipulators can often be adequately described by a linear model [2]. That is, poles and zeros of the transfer function of a single-link manipulator can be accurately identified experimentally, and a control system can be designed. The more useful configuration of a two-link manipulator, however, must be described by a nonlinear model. In this case, the physical manipulator parameters must be identified and a model suitably developed so that the control system can be designed using accurate equations of motion.

---

*NSF Fellow, Department of Mechanical Engineering.
[†]Charles Lee Powell Professor and Chairman, Department of Aeronautics and Astronautics.

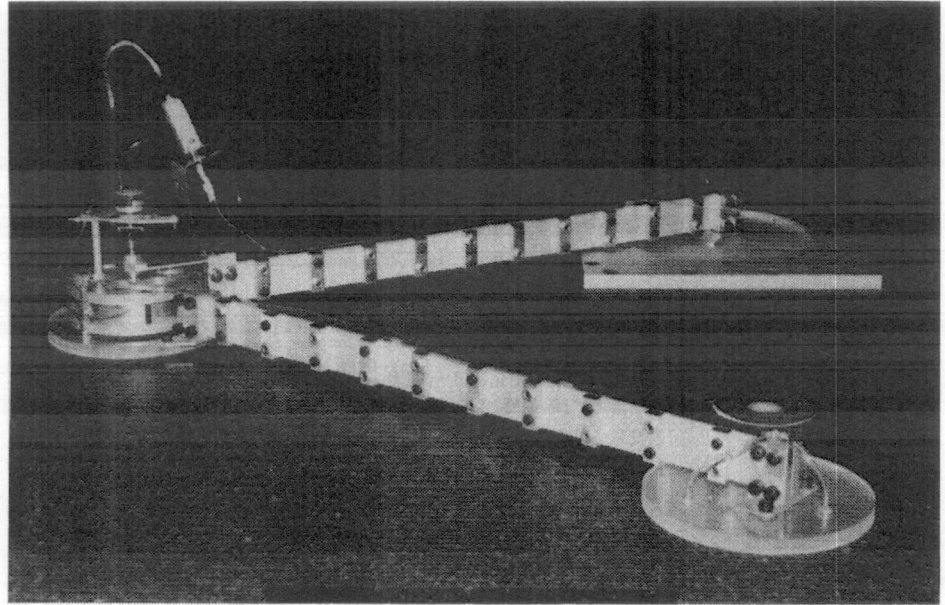

Figure 1: Photograph of the Stanford Multi-Link Flexible Manipulator

The motivation to pursue more appropriate mode shapes arose when a rather large number of cantilever modes was required to sufficiently match the model of an experimental manipulator to a sine-sweep identification [5]. (The experimental two-link manipulator was configured with a rigid upper arm and a very flexible forearm.) Proper mode shape selection was even more critical for controller design. As a result of limited computing power with which to control the experimental manipulator, a model consisting of a small number of cantilever modes was first employed to design a high-performance optimal end-point controller. The resulting controller worked fine in simulation, but when implemented on the experimental manipulator, was easily be driven unstable, indicating that refinement of the theoretical model (or mode shapes) was required [6].

We consequently set out to explore both theoretically and experimentally the use of different component mode shapes for an assumed-modes model of a two-link manipulator with flexibility in both the links. (The word "component" is used in the sense that these mode shapes are "building-blocks" for predicting the true system mode shapes.) Of particular importance is the number of states (or order) necessary to accurately model the physical manipulator, as this number will influence the complexity and implementability of the control system. This paper presents some of our results and identifies how revisions to the model were guided by the experimental facility.

2 The Experimental Manipulator

The experimental Stanford Multi-Link Flexible Manipulator has been constructed to serve as a test bed for validating new theories and to suggest new directions for basic research. A photograph of the manipulator is shown in Figure 1. The manipulator operates in the horizontal plane on top of a flat 4 ft by 8 ft granite table. The shoulder motor is mounted on the side of the table, while both the elbow motor and tip pad are supported by air cushions on the table. Located on each of the motor shafts are rotary variable differential transformers (RVDT's) to measure joint angles. Angular rate signals are generated by passing the RVDT signals through analog band-pass filters. End-point sensing is achieved using a technique developed in the ARL by Schneider [7]. A CCD television camera tracks a special variable reflectivity target located at the manipulator end-point as seen in the photograph of Figure 1.

The experimental apparatus is designed to permit a variety of configurations. Rigid or flexible links of different cross-sections, lengths, or mass distributions can be interchanged to study the effects of structural flexibility in either or both links. The configuration studied in this paper, as seen in the photograph, consists of a flexible upper arm and a flexible forearm. "Mass Intensifiers" [6], which are discrete masses evenly spaced along the length of the flexible links, have been added to the flexible links in order to increase the link inertia without changing the link flexibility. A complete set of the geometric and mass properties of the experimental two-link flexible manipulator is given in [8].

3 Modelling

We are interested in developing models of the two-link flexible manipulator that contain the significant dynamics of the system and will be suitable for control system design and implementation. Other models are also developed that are useful for verifying the models for control.

In the models that follow, each link of the manipulator has a hub and a tip, each with accountable mass, inertia, and mechanical offsets. The notation is easily extensible to configurations with more than two links.

3.1 Assumed-Modes Model

For the purpose of control system design, we have used the assumed-modes method to derive the equations of motion of a two-link flexible manipulator. Using this modelling technique we are able to obtain a finite set of ordinary differential equations (useful for controller design). We also may retain the nonlinear terms in the equations of motion so that the model will be accurate when the manipulator undergoes rapid, large-angle slews. The assumed-modes method has the potential to properly model the manipulator in a relatively small number of states. However, the choice of appropriate mode shapes is crucial.

The manipulator schematic diagram used to develop the assumed-modes model is given in Figure 2. The $\theta_{(.)}$'s, $w_{(.)}$'s, $x_{(.)}$'s, and $s_{(.)}$'s are the rigid-body rotations, deflections of the flexible links, distances along the undeflected axes of the links, and arc lengths along the links, respectively. The subscript 1 designates the upper arm and the 2 the forearm. T_1 and T_2 are the shoulder and elbow torques.

A proper analytical description of the motion of the manipulator can be obtained from the shoulder and elbow angles θ_1 and θ_2 and from the lateral displacements of the flexible links $w_1(x_1,t)$ and $w_2(x_2,t)$ and their first and second time derivatives. The assumed-modes method is then used to develop ordinary differential equations by making the approximation that the deflections of the flexible links may be expressed as a weighted superposition of a finite set of spatial mode shapes ($\phi_{(.)}(x_{(.)})$'s) that satisfy the geometric boundary conditions of the link.

The lateral displacements of the flexible links are then written as

$$w_1(x_1,t) = \sum_{j_1=1}^{\nu_1} \phi_{1j_1}(x_1)\, q_{1j_1}(t) \tag{1}$$

$$w_2(x_2, t) = \sum_{j_2=1}^{\nu_2} \phi_{2j_2}(x_2) \, q_{2j_2}(t) \tag{2}$$

where $\nu_{(.)}$ is the number of modes, or degrees of freedom, used to describe the deflection of a link and the $q_{(.)}(t)$'s are the time dependent modal amplitudes or weights.

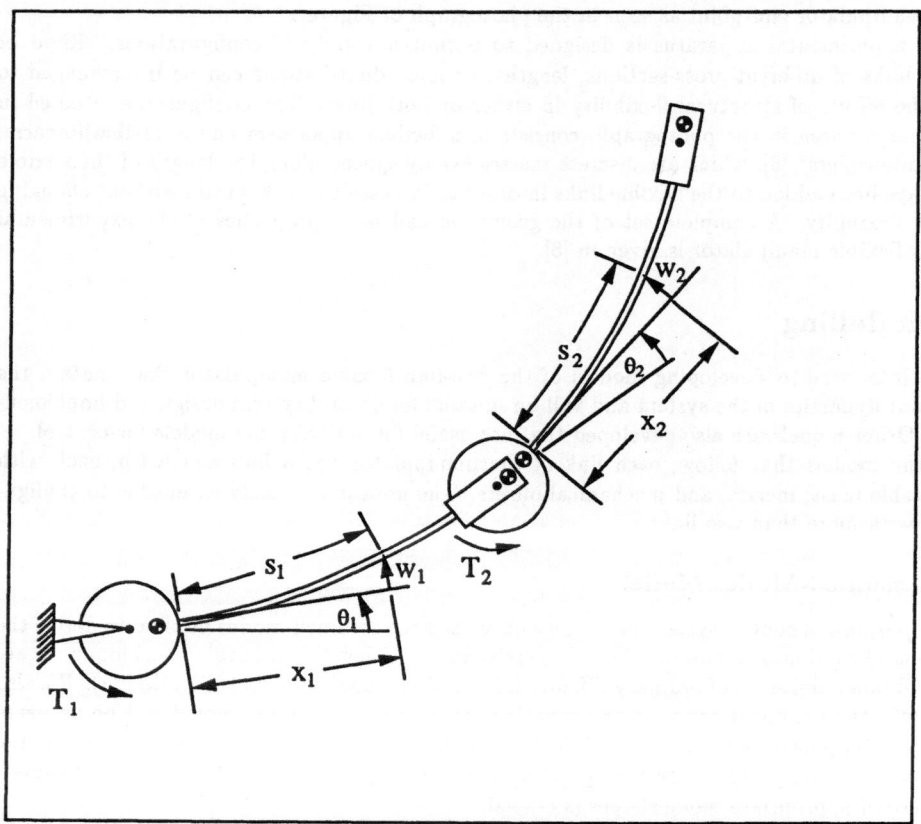

Figure 2: Assumed-Modes Manipulator Model

Link Foreshortening When using an assumed-modes model, link foreshortening may introduce a significant kinematical effect. In particular, if we excite the experimental manipulator with a sufficiently large amplitude sinusoidal disturbance (to observe a natural shape of the manipulator), link foreshortening will have a significant effect and should be included in the model.

Accounting for the link foreshortening can be simplified if we assume that the mode shapes depend on the arc length of the deflected beam [9], rather than on the distance along the undeflected axis. In this case, the projection of the flexible link on the undeformed axis can be calculated from

$$x_i = \int_0^{s_i} \sqrt{1 - \left(\frac{\delta w_i(\sigma, t)}{\delta \sigma}\right)^2} \, d\sigma \ . \tag{3}$$

An example of how this effects the spatial representation of a "generic" mode shape is presented in Figure 3. The figure shows how a *plot* of the mode shape at various amplitudes is affected by

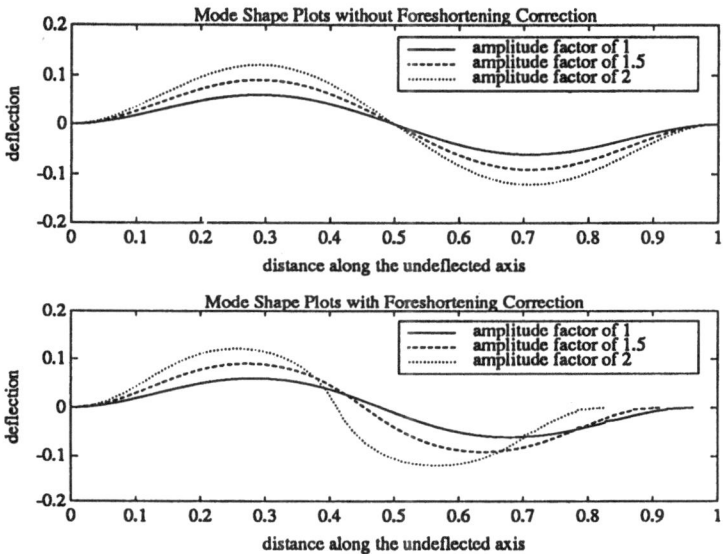

Figure 3: How the Plot of a Generic Modeshape at Various Amplitudes Changes with and without Link Foreshortening Correction

the absence and presence of the foreshortening correction.

The link foreshortening correction is used in the remainer of this paper.

3.2 Lumped Spring-Mass Model

For the purpose of verifying and further developing the assumed-modes model, we've formulated an additional model that treats each of the flexible links as a series of rigid link segments interconnected by torsional springs. This model is of particular value since the equations of motion, although high in order for a large number of segments (difficult for control), are independent of the mode shapes that must be chosen for the assumed-modes method. Note, however, that the torsional spring stiffnesses must be appropriately selected.

The manipulator schematic diagram used to develop the lumped spring-mass model is given in Figure 4. The hub and tip inertias are the same as in the assumed-modes model, and the segment geometric and mass parameters are chosen to match those of the flexible link in the assumed-modes model. The spring stiffnesses, $k_{(\cdot)}$, are assumed to be constant for each link. n_1 is the number of segments in the upper arm, and n_2 is the number of segments in the forearm.

Selecting the Torsional Spring Stiffness The torsional spring stiffnesses are found by matching the end-point deflection of a lumped spring-mass model of a cantilevered beam, shown in Figure 5, to that of a cantilevered continuous flexible beam. A force F is applied to the end of the spring-mass model and the spring stiffness k is selected so that the deflection y at the end-point is the same as that of a continuously flexible beam. Assuming small deflections (i.e. neglecting higher than first order terms in the $q_{(\cdot)}$'s) the appropriate spring stiffness can be

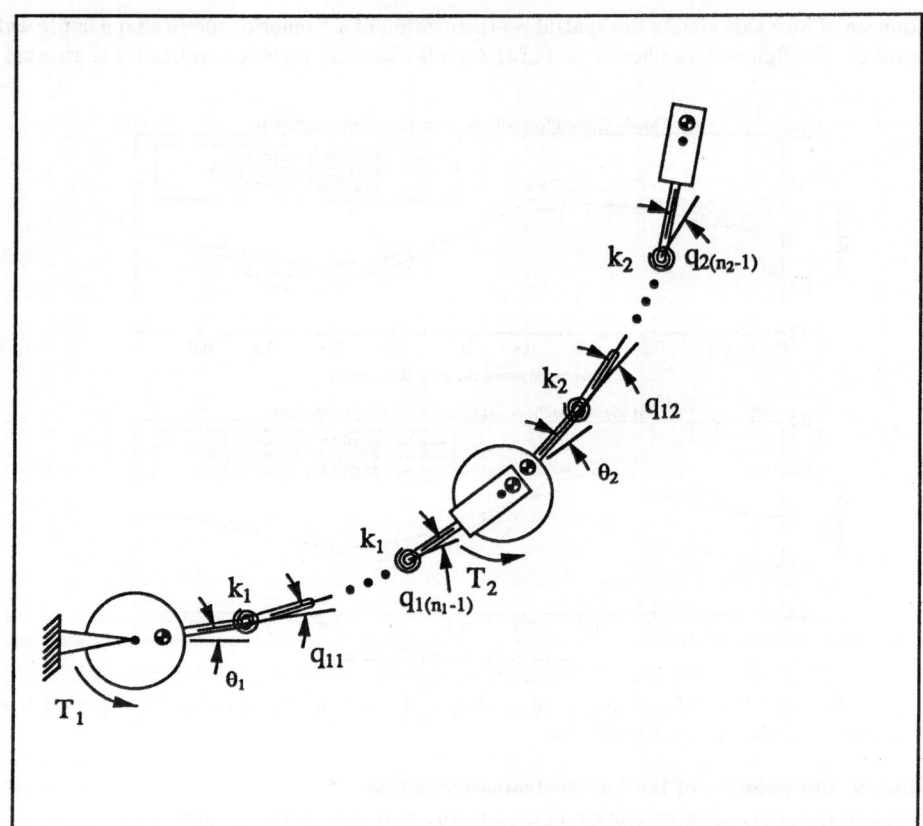

Figure 4: Lumped Spring-Mass Manipulator Model

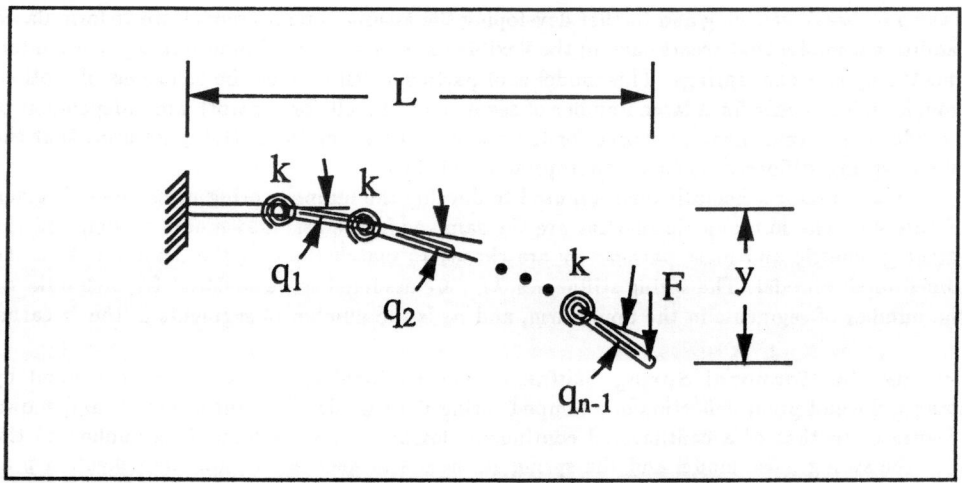

Figure 5: Lumped Spring-Mass Model of a Cantilevered Beam

found from

$$k = \frac{3EI}{n^2 L} \left[\frac{n(n-1)(2n-1)}{6} \right] \qquad (4)$$

where n is the number of segments in the spring-mass model and EI and L are the flexural rigidity and length of the continuous model, respectively.

3.3 Equations of Motion

The dynamical equations of motion of both the assumed-modes and lumped spring-mass models of the two-link flexible manipulator, linearized about an elbow angle θ_{20} with the other states zero, were derived using Kane's Method [10, 11] and can be expressed in matrix form as

$$\mathbf{M}(\theta_{20})\ddot{\mathbf{x}} + \mathbf{K}\mathbf{x} = \mathbf{T} \qquad (5)$$

where for the assumed-modes model

$$\mathbf{x} = [\theta_1 \quad \theta_2 \quad q_{11} \quad q_{12} \quad \cdots \quad q_{1\nu_1} \quad q_{21} \quad q_{22} \quad \cdots \quad q_{2\nu_2}]^T \qquad (6)$$

and for the lumped spring-mass model

$$\mathbf{x} = [\theta_1 \quad q_{11} \quad q_{12} \quad \cdots \quad q_{1n_1} \quad \theta_2 \quad q_{21} \quad q_{22} \quad \cdots \quad q_{2n_2}]^T \ . \qquad (7)$$

The size of the mass matrix $\mathbf{M}$, stiffness matrix $\mathbf{K}$, and torque matrix $\mathbf{T}$ depend upon the number of modes retained for the assumed-modes model and the number of segments per link for the lumped spring-mass model.

To model systems that are degenerates of a two-link flexible manipulator, the $\mathbf{M}$, $\mathbf{K}$, and $\mathbf{T}$ matrices and the $\ddot{\mathbf{x}}$ and $\mathbf{x}$ vectors can be simply partitioned and reduced to obtain the equations of motion. For example, the equations of motion of a single cantilever beam with a rotating tip mass can be obtained by partitioning all the rows and columns associated with the flexibility coordinates of the second link and setting all the parameters associated with the second link equal to zero. Similar techniques were used to obtain models for all the degenerate systems used in this paper.

The elements of the $\mathbf{M}$, $\mathbf{K}$, and $\mathbf{T}$ matrices for the assumed-modes model can be found in [8].

4 Selection of Component Mode Shapes

The assumed-modes method has the potential to yield an accurate low-order model for the purposes of simulation and controller design. However, spatial shapes used to describe the deflections of each of the flexible links must be selected. These spatial shapes are termed "component mode shapes" (CMS) and are the "building-blocks" used to determine the true system mode shapes. They are selected for each link by considering the geometric and mass properties of the link as well as the boundary conditions. The equations of motion, developed by describing the link deflections as time-weighted sums of component mode shapes, provide eigenvectors associated with the natural frequencies of the system from which "system mode shapes" (SMS) are determined. The distinction between "component mode shapes" and "system mode shapes" is illustrated in Figure 6. The selection and number of component mode shapes used to describe each flexible link determines how well the predicted system mode shapes will match those of the experimental manipulator. In this section, we consider the selection of a suitable set of component mode shapes for each link of the manipulator.

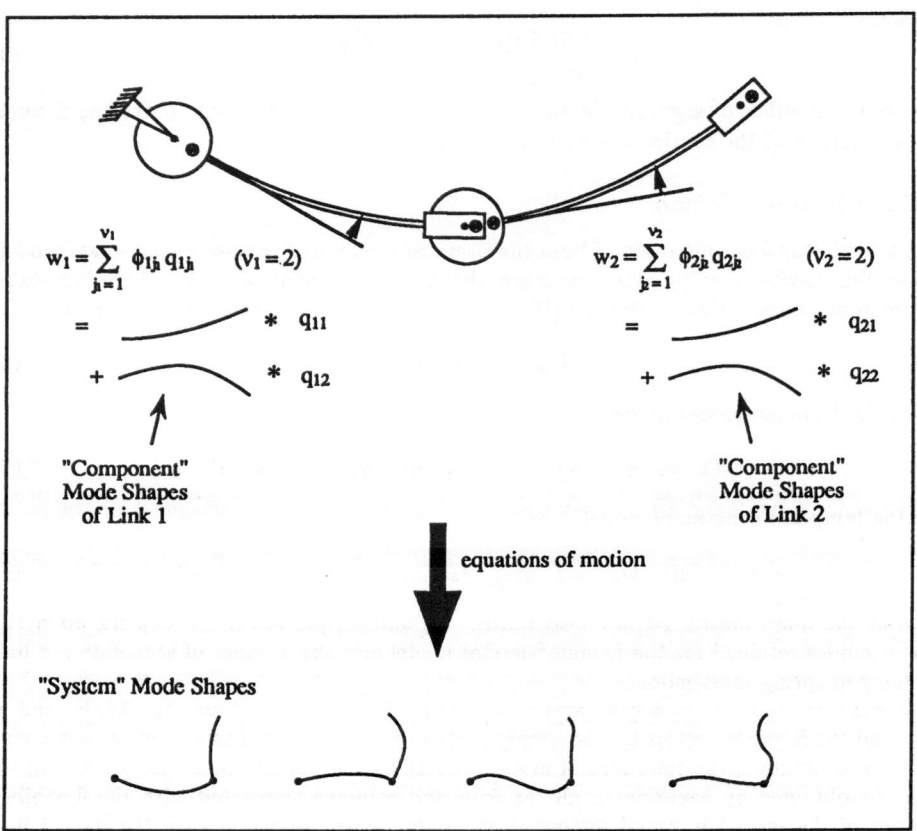

Figure 6: Illustration of The Difference Between Component Mode Shapes and System Mode Shapes

4.1 Cantilever Mode Shapes

Cantilever component mode shapes are often used [4] with the assumed-modes method to develop equations of motion for multi-link flexible manipulators. The equations describing the mode shapes of a cantilever beam are relatively simple and are independent of the mass properties of the beam [3]. For the *simulation* of systems composed of interconnected flexible beams, a large number of cantilever mode shapes may be used to predict accurately the system motion. We are interested, however, in simplifying the *implementation* and *design* of high-performance controllers by selecting component mode shapes that will allow a smaller, more tractable number of modes to be used.

When relatively large inertias are located at the joint and/or tip of each flexible link, cantilever component mode shapes may not result in an accurate description of the flexible manipulator in a minimal number of modes. This is typical of most common flexible manipulators, including our experimental manipulator. How well a particular number of cantilever component mode shapes predicts the system mode shapes of a theoretical model of a cantilever beam with a tip inertia is illustrated in Figure 7. In the figure, the solid line represents the actual system mode shape, determined theoretically. The dashed, dotted, and dash-dotted lines represent the predicted system mode shapes using 2, 3, and 5 cantilever component mode shapes, respectively. The system mode shapes are drawn for 0.25, 1, and 4 times the tip mass and inertia of the forearm of the experimental manipulator. As can be seen, the first system mode shape is matched quite well using cantilever component mode shapes, but the match gets worse for higher system mode shapes and increasing tip inertia.

4.2 Forearm Link

Since a relatively large inertia is located at the end-point of the manipulator, we have investigated the use of mode shapes derived from a cantilever beam with an inertia located at the tip as the component mode shapes to describe the deflection of the forearm link. This beam configuration is designated by the acronym CLTI (cantilever with tip inertia), and an explicit form of the corresponding mode shape and eigenvalue equations can be found in [6]. A schematic diagram of a CLTI beam and plots of the first three mode shapes (derived using the geometric and mass properties of the experimental manipulator's forearm) are given in Figure 8. For comparison, the mode shapes for a cantilever beam with zero tip inertia are located below the CLTI mode shapes in the figure. As can be seen in the figure, the deflection at the end of the CLTI beam is closer to the undeformed axis (zero deflection) than that of the cantilever beam. In fact, as is physically intuitive, the CLTI mode shapes approach the mode shapes of a clamped-clamped beam as the tip inertia tends to infinity.

4.3 Upper Arm Link

While the mode shapes of a CLTI beam are appropriate to use as component mode shapes to describe the deflection of the forearm link, they are not appropriate for the upper arm link. The tip inertia of the forearm link is fixed (clamped) to the end of the link, whereas a significant portion of the effective tip mass (namely the forearm link) of the upper arm link is free to rotate (pinned). Rather than derive closed-form analytical expressions of the mode shapes for a cantilever beam with a fixed tip inertia and a rotating tip inertia, we have generated polynomials in the spatial coordinate x that fit the eigenvector shapes resulting from a lumped spring-mass model of such a beam. This beam configuration is designated by the acronym LSMCR (lumped spring-mass cantilevered at the hub and rotating tip). These polynomials are then used as the component mode shapes of the upper arm. Note that the polynomial fit was performed

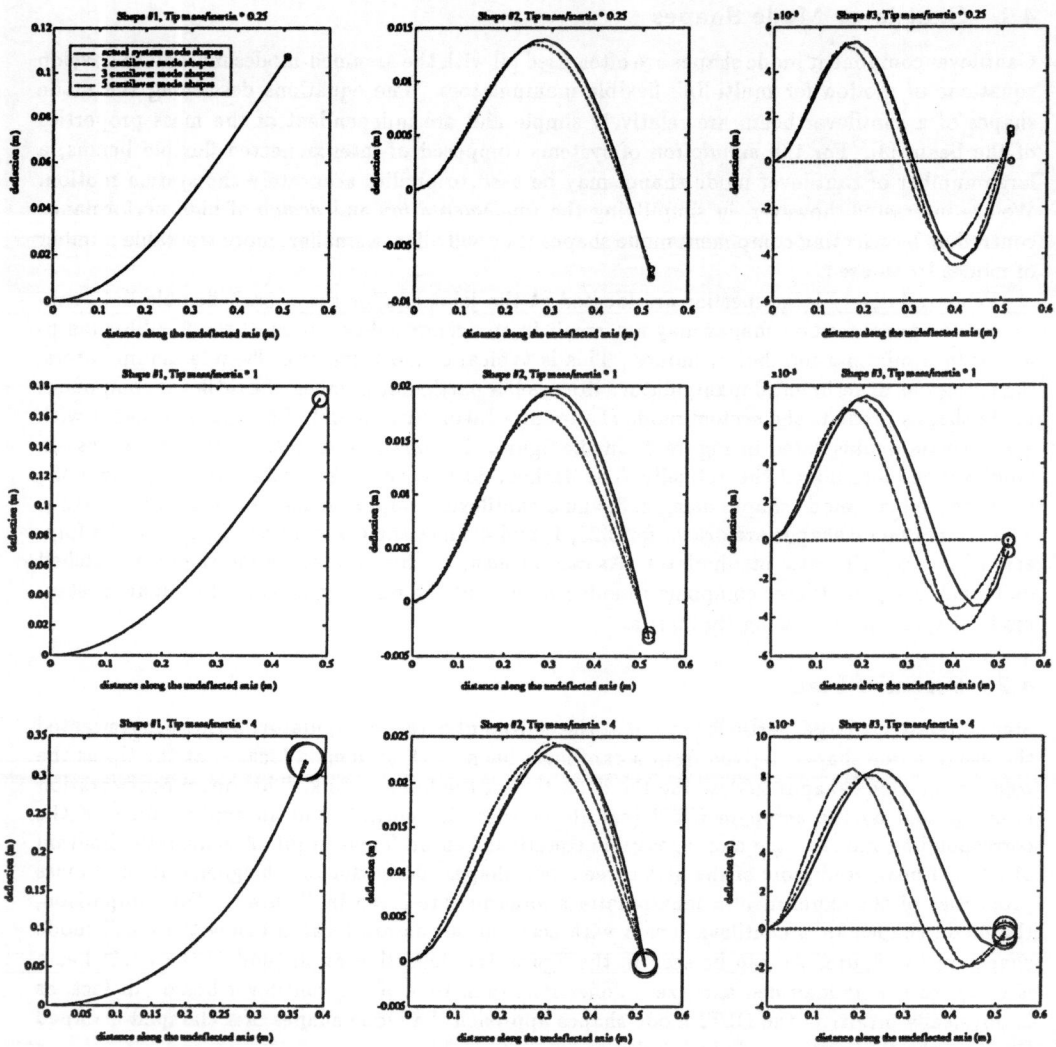

Figure 7: Accuracy of Employing Cantilever Component Mode Shapes to Predict the System Mode Shapes of a Cantilever Beam with a Tip Inertia

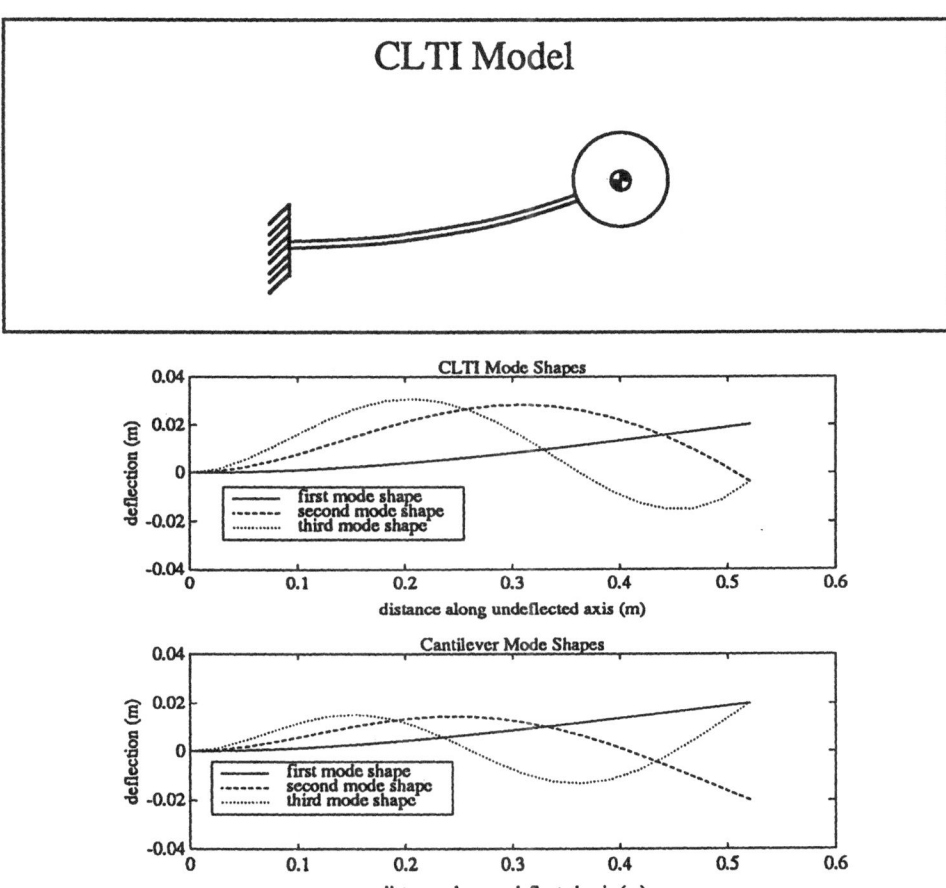

Figure 8: Cantilever with Tip Inertia Model and Corresponding Mode Shapes (cantilever mode shapes included for comparison)

such that the coefficients of the zeroth and first order terms were zero to satisfy the geometric boundary conditions at the hub of the beam (i.e. the polynomials are admissible functions [3]). A schematic diagram of the LSMCR model and the associated mode shapes (derived from the model linearized about a relative angle of 90° between the fixed and rotating tip inertias) are given in Figure 9. Additionally, for comparison, the mode shapes for a cantilever beam with zero tip mass and inertia are located below the LSMCR mode shapes in the figure. Similar to the CLTI mode shapes, the LSMCR mode shapes are more representative of the physical system (they approach pinned-clamped mode shapes as the rotating portion of the tip inertia tends to infinity, and clamped-clamped mode shapes as the fixed portion of the tip inertia tends to infinity), than the cantilever mode shapes.

5 Frequency Response and System Mode Shapes

In the previous section we developed *component* mode shapes to be used in the model to describe the deflections in the upper arm and forearm links. In this section we look at the *system* mode shapes derived theoretically from the model employing the component mode shapes and compare them to those captured using strobe photography on the experimental manipulator.

The natural frequencies and system mode shapes of a linear theoretical model of the two-link flexible manipulator can be found from the eigenvalues and eigenvectors of the equations of motion. The natural frequencies and system mode shapes of the experimental manipulator can be found by injecting sinusoidal signals into the manipulator and determining the frequencies and shapes of the manipulator when the amplitude of the sensor of interest is a maximum. Comparing the theoretical natural frequencies to those obtained experimentally gives an indication of how well the natural frequencies of the various theoretical models match those of the experimental manipulator. A similar comparison can be made for the system mode shapes.

The two theoretical models being compared to each other and to the experimental manipulator were derived using the assumed-modes method, each linearized about an elbow angle of 90°. The first model incorporates two cantilever component mode shapes for each of the flexible links and will be designated as the "CANT/CANT" model. The second model incorporates two LSMCR component mode shapes for the upper arm link and two CLTI component mode shapes for the forearm link, and will be designated as the "LSMCR/CLTI" model.

5.1 Natural Frequencies

Frequency responses in the form of magnitude and phase plots for the shoulder torque to shoulder angle and elbow torque to elbow angle are given in Figures 10 and 11, respectively.   The experimental results were obtained with a nominal elbow angle of 90°, and the theoretical results are based on models linearized about an elbow angle of 90°. The alternating pole and zero patterns appear as expected for the collocated actuators and sensors [12]. All of the natural frequencies (in the frequency band of interest) of the flexible manipulator appear as peaks in the magnitude plot of the elbow collocated transfer function (Figure 11). The first frequency of each of the double peaks corresponds to a natural frequency of the shoulder collocated transfer function. This indicates that as the input frequency to the elbow motor is increased, we will first observe significant excitation in the upper arm link as we approach the first peak of a double peak and then we will observe significant excitation in the forearm link when we reach the second peak of a double peak. This phenomenon was observed and verified on the experimental manipulator.

From the magnitude plot of Figure 11, we can see that the CANT/CANT model predicts the second natural frequency quite well, but misses the first, third, and fourth frequencies.

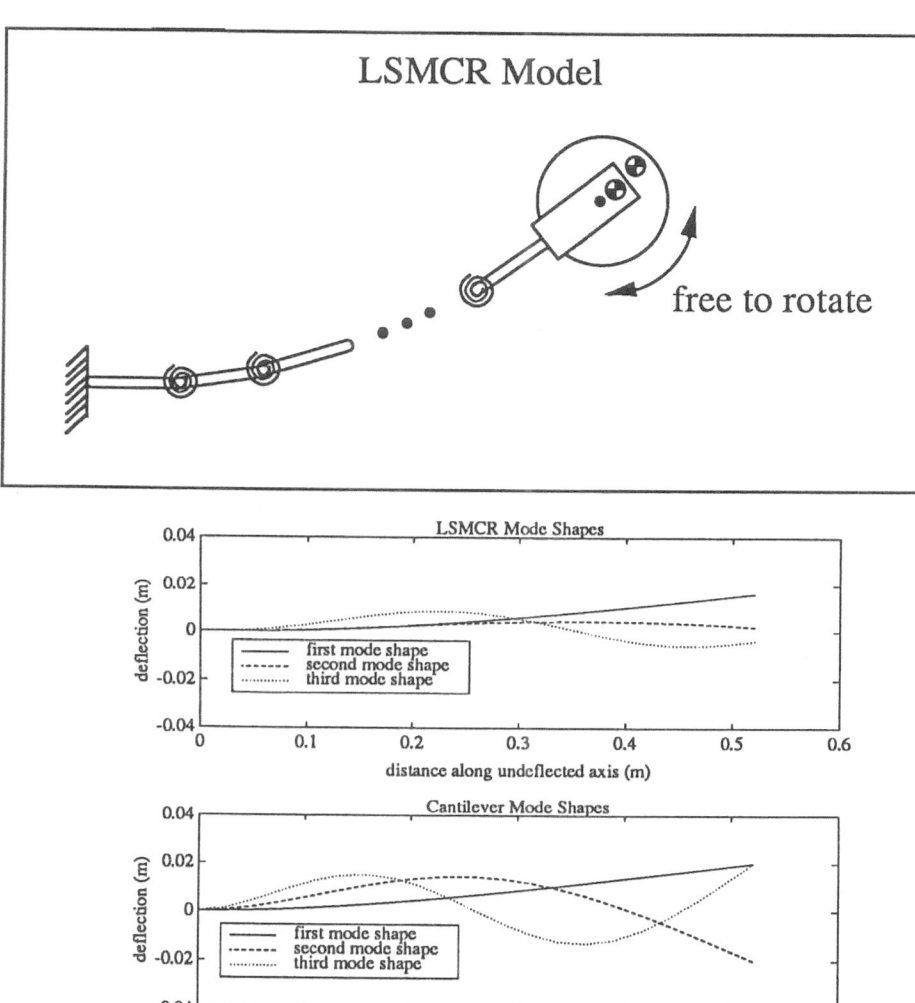

Figure 9: Lumped Spring-Mass Model of a Flexible Beam and a Rotating Tip Inertia and Corresponding Mode Shapes (cantilever mode shapes included for comparison)

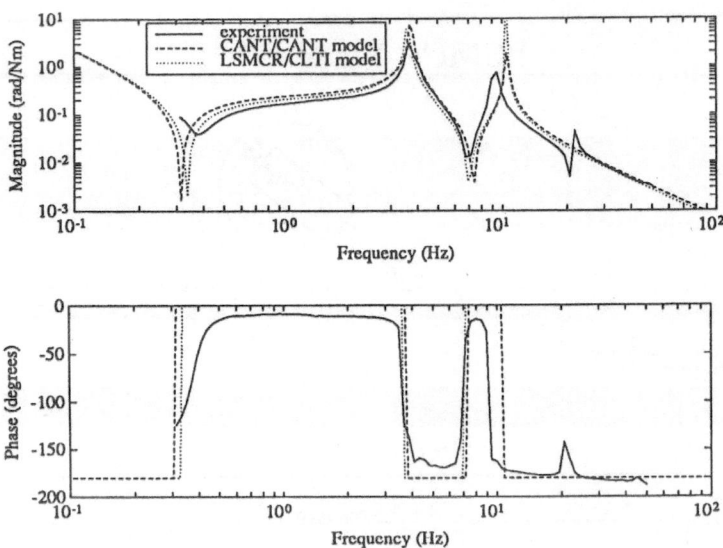

Figure 10: Frequency Response from Shoulder Torque T_1 to Shoulder Angle θ_1

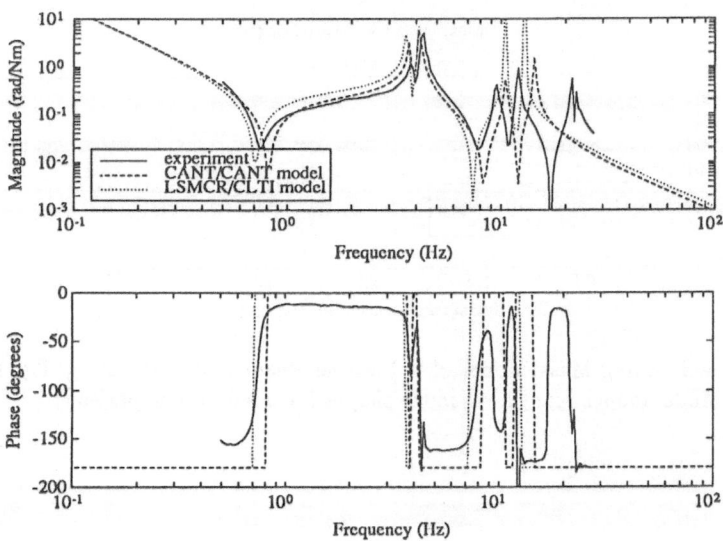

Figure 11: Frequency Response from Elbow Torque T_2 to Elbow Angle θ_2

The mismatch in the fourth frequency is significant enough to limit the performance of a Linear Quadratic Gaussian controller with a nonlinear estimator implemented on the experimental flexible manipulator configured with a rigid upper arm and flexible forearm [6]. The LSMCR/CLTI model, on the other hand, does much better at predicting the first and fourth natural frequencies of the experimental manipulator.

5.2 Photographs of The Experimental Manipulator

The system mode shapes of the experimental manipulator were determined by taking a series of photographs, some of which are shown in Figure 12. Special software was developed to command the joint motors and to synchronize the operation of a camera flash so that these "freeze-frame" photographs of the system mode shapes of the experimental manipulator could be obtained. With the room dark, the camera shutter open, and the computer generating a sinusoidal input at one of the natural frequencies of the manipulator into one of the motors, the flash was triggered at one of the peaks and one of the valleys of the sinusoidal input to the motors, resulting in the "double-exposure" effect in the photographs. The sinusoidal signal was input into the shoulder to photograph system mode shapes I and III, and into the elbow for system mode shapes II and IV. A very low gain proportional (P) controller was placed around each of the joints to prevent drifting.

5.3 System Mode Shape Comparison

Presented in Figure 13 is a comparison of the system mode shapes determined from the two models and from the experimental manipulator. Within the resolution of the plots we see that the first 4 experimental system mode shapes are predicted reasonably well by both of the models, with a slightly better prediction with the LSMCR/CLTI model.

From these results we conclude that although the selection of the component mode shapes influences the predicted system mode shapes, a more accurate comparison of different sets of component mode shapes can be derived from the frequency response. It is important, though, that we were able to reasonably predict the system mode shapes given by the strobe photographs; the optimal placement of sensors (such as strain gages) along the length of the manipulator links depends on an accurate prediction of the nodal and anti-nodal locations.

All of the results and discussion presented in this paper so far have been based on the manipulator in a configuration of a relative elbow angle of 90°. In Figure 14, we show how the system mode shapes for the LSMCR/CLTI model change as a function of the elbow angle. For our experimental system and within the resolution of the plots, a significant change in the shapes is not noticeable as the elbow angle is varied. This is not necessarily true for all two-link flexible manipulators and depends on the geometric and mass properties of both the flexible links and the hub and tip inertias.

6 Concluding Remarks

Besides developing sets of theoretical component mode shapes that permit a low-order model to predict accurately the natural frequencies and system mode shapes of an experimental two-link flexible manipulator, we have also developed a more fundamental understanding of the dynamics of such systems. Such an understanding is beneficial and necessary for designing high-performance control systems and interpreting experimental results.

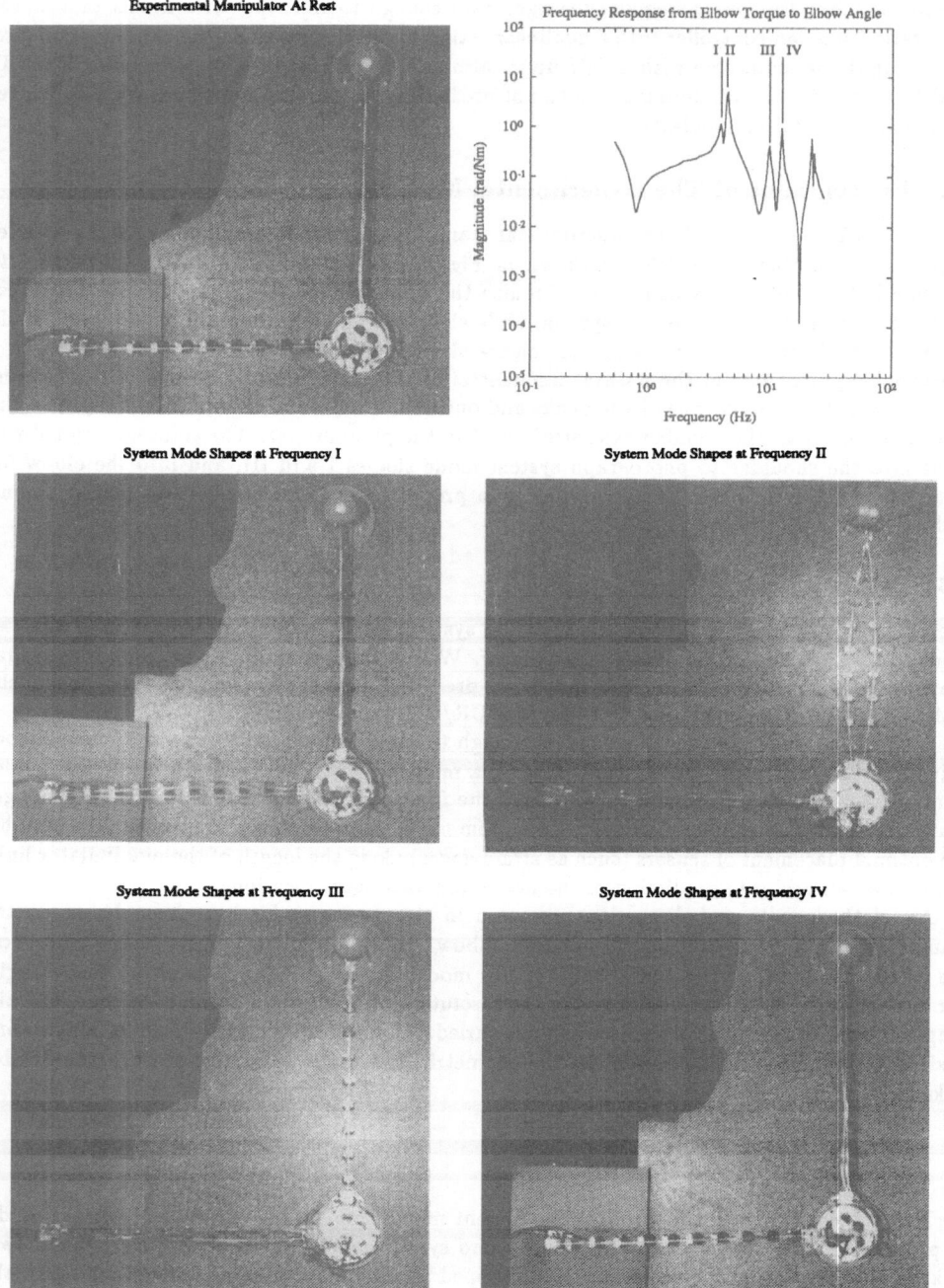

Figure 12: Photographs of the Experimental Manipulator System Mode Shapes

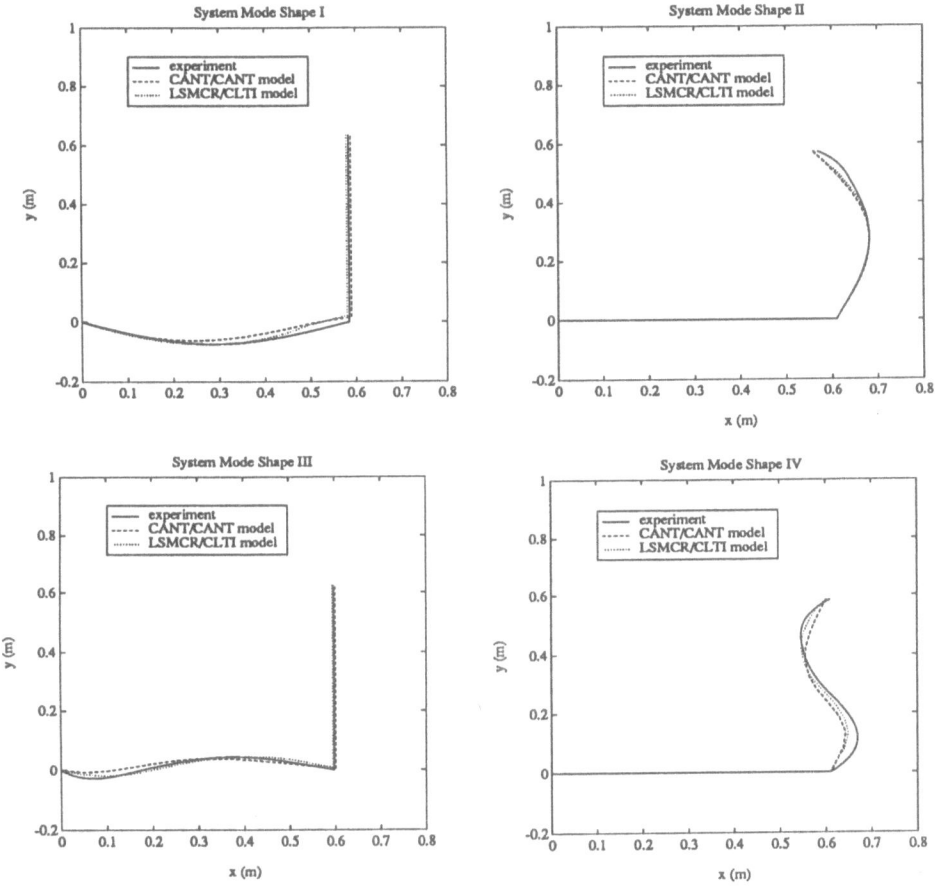

Figure 13: Comparison of the System Mode Shapes of the Experimental Manipulator to those Derived from Various Models

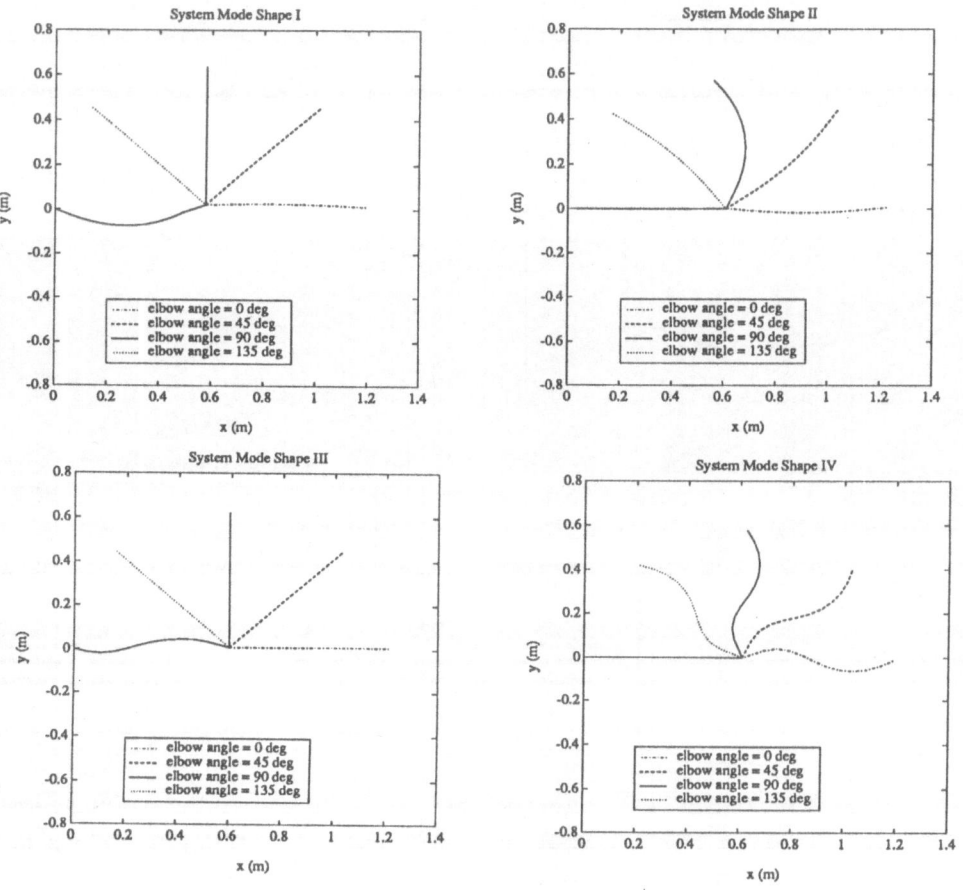

Figure 14: System Mode Shapes of the Assumed-Modes Manipulator Model Employing LSMCR/CLTI Mode Shapes Linearized about Various Elbow Angles

Acknowledgements

The research reported here was sponsored by AFOSR and the Informational Sciences Division of the NASA Ames Research Center. The authors gratefully acknowledge the valuable assistance of Dr. Harold Alexander, Christopher Azzano, Craig Barratt, Yosi Drucker, Professor Gene Franklin, Rebecca Jackson, Stanley Schneider, Gad Shelef, Marc Ullman, Roberto Zanutta, and Godwin Zhang in the design and construction of the experimental apparatus, software development, photography sessions, and technical discussions.

References

[1] R. P. Paul, *Robot Manipulators—Mathematics, Programming, and Control*. The MIT Press Series in Artificial Intelligence, Cambridge, MA: The MIT Press, 1981. Second Printing, 1982.

[2] R. H. Cannon, Jr. and E. Schmitz, "Initial experiments on the end-point control of a flexible one-link robot," *International Journal of Robotics Research*, vol. 3, pp. 62–75, Fall 1984.

[3] L. Meirovitch, *Analytical Methods in Vibrations*. Macmillan Series in Applied Mechanics, New York, NY: Macmillan Company, 1967.

[4] W. J. Book, O. Maizza-Neto, and D. E. Whitney, "Feedback control of two-beam, two-joint systems with distributed flexibility," *ASME Journal of Dynamic Systems, Measurement, and Control*, vol. 97, pp. 424–431, December 1975.

[5] C. M. Oakley and R. H. Cannon, Jr., "Initial experiments on the control of a two-link manipulator with a very flexible forearm," in *Proceedings of the American Control Conference*, (Atlanta, GA), June 1988.

[6] C. M. Oakley and R. H. Cannon, Jr., "End-point control of a two-link manipulator with a very flexible forearm: Issues and experiments," in *Proceedings of the American Control Conference*, (Pittsburgh, PA), June 1989.

[7] S. Schneider, *Experiments in the Dynamic and Strategic Control of Cooperating Manipulators*. PhD thesis, Stanford University, Stanford, CA 94305, May 1989.

[8] C. M. Oakley and R. H. Cannon, Jr., "Equations of motion for an experimental planar two-link flexible manipulator," in *Proceedings of the ASME Winter Annual Meeting*, (San Francisco, CA), December 1989.

[9] K. W. Buffinton, *Dynamics of Beams Moving Over Supports*. PhD thesis, Stanford University, Department of Mechanical Engineering, Stanford, CA 94305, July 1985.

[10] T. R. Kane, P. W. Likins, and D. A. Levinson, *Spacecraft Dynamics*. New York, NY: McGraw-Hill, 1983.

[11] T. R. Kane and D. A. Levinson, *Dynamics: Theory and Application*. McGraw-Hill Series in Mechanical Engineering, New York, NY: McGraw-Hill, 1985.

[12] W. B. Gevarter, "Basic relations for control of flexible vehicles," *AIAA Journal*, vol. 8, pp. 666–672, April 1970.

Limitations of Linear Identification and Control Techniques for Flexible Robots with Nonlinear Joint Friction

J. Swevers, M. Adams, J. De Schutter,
H. Van Brussel, H. Thielemans.
Katholieke Universiteit Leuven
Department of Mechanical Engineering
Celestijnenlaan 300 B
B - 3030 Leuven   Belgium

Abstract

A state feedback controller with acceleration feedforward has been developed for accurate tracking control of a flexible one-link robot. The model, on which the controller is based, is the result of a least squares parameter estimation algorithm that has been developed especially for the test setup. The algorithm divides the total model into two submodels in series, and estimates their parameters separately. The controller does not use a direct end point position measurement, but estimates this position from the measured motor angle and strain gauge signals. Acceleration feedforward is introduced in the state feedback controller to reduce tracking errors to negligible values. Integral control eliminates positioning errors caused by nonlinear motor friction.

1 Introduction

Identification and control of a flexible one-link robot is a first step in the development of a controller for a more complex configuration. Linear strategies, applied to the single degree of freedom, must be extended to nonlinear or adaptive techniques for the more complex case. To avoid unexpected problems, the characteristics of the one-link robot must be chosen as close as possible to those of a real robot. These characteristics include nonlinear joint friction and low amplitude oscillations. A direct measurement of the end point position is also unrealistic in an industrial environment. This paper reports on experimental experience with identification and control of a one-link flexible robot which exhibits substantial nonlinear joint friction. This research is carried out in the framework of Esprit project 1561 (SACODY).

Nonlinear friction causes limit cycles if the feedback gains are too high. In the case of a flexible link, these oscillations have a frequency which is different from the structural resonance frequency. Accurate positioning becomes impossible then. Therefore the closed loop bandwidth must be restricted. On the other hand high feedback gains are not sufficient to guarantee fast and accurate operation since they do not eliminate tracking and positioning errors. Acceleration feedforward and integral control have to be included to this end.

A filtered least squares identification algorithm has been developed for the test setup. It derives a more accurate model, with less computational effort, than any other least squares identification algorithm found in literature [1]-[3], [12], [19], [20]. The algorithm divides the total model into two submodels in series, which are calculated separately. The first model relates the input to the beam deflection. The second model relates the beam deflection to the angular motor position.

Control spill-over is avoided by filtering the output signals using a second order digital filter. These filters extend the model order. The controller is an integral state feedback controller with acceleration feedforward. The feedback gains are calculated using a pole placement algorithm. The calculation of the acceleration feedforward is based on the steady state behaviour of the system.

The best control results are obtained for a closed loop bandwidth which is a little higher than the first resonance frequency of the structure. This controller allows accurate tracking and positioning with very limited overshoot, oscillation or tracking errors. Higher bandwidths result in limit cycles, which inhibit accurate control.

Section 2 gives a description of the test setup along with some dynamic characteristics. Section 3 describes the least squares parameter identification algorithm developed for the test setup. Section 4 describes the control design, and discusses the control results.

2 Description of the flexible one-link robot

Figure 1 gives a schematic representation of the test setup. It consists of a flexible beam with a payload, connected to a flexible torsional beam, which itself is connected to a direct drive motor (maximum motor torque = $200\,Nm$). The control input to the system is a voltage between -10 Volt and $+10$ Volt, which is converted by the power supply of the motor into a current. This conversion is linear. A built-in encoder measures the angular motor position. It has a resolution of 1024000 pulses per revolution. One encoder pulse corresponds to an angle of $6.1359\,10^{-6}$ radians or 1.2656 arcseconds. Strain gauges on the flexible beam near to the axis of rotation, and in the middle of the torsional beam measure the beam deflections. The end point position is not measured directly, but can be determined from the strain gauge and encoder signals. The first resonance frequency is a free-free mode in the neighbourhood of $5Hz$. The second resonance frequency is at $38Hz$. This mode and the higher ones have been neglected in the identification and control. An anti-resonance exists at approximately $3Hz$, which corresponds to a low damped clamped-free mode. This mode is excited when, in the neighbourhood of the destination point, the motor velocity suddenly drops to zero due to dry friction (static friction torque $\simeq 6\,Nm$, i.e. 3% of the maximum torque), or when external forces are applied to the beam.

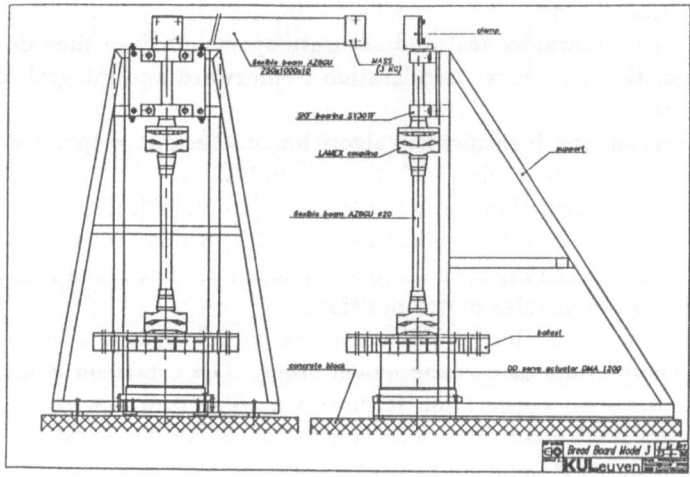

Figure 1: Schematic representation of the test setup.

3 Least squares identification of a flexible one-link robot

The identification of the total multi variable state space model is based on the identification of two transfer functions and on the determination of the relation between the end point position, the motor angle and the beam deflections. The complexity of the model is determined from a physical analysis of the system, based on a mass-spring equivalent. This is allowed because only one eigenfrequency is considered [6]. A least squares parameter estimation algorithm calculates the unknown parameters of the transfer functions. These estimations are based on measured input-output signals which result from an excitation with a band limited random signal. The use of a band limited random signal reduces the unnecessary excitation of the higher modes during the measurement. The cut-off frequency of the input signal spectrum is $20Hz$. The relation between the encoder signal, the strain gauge signals and the end point position is linear if only the first mode is considered. The coefficients of this relation are determined using a static measurement.

3.1 Mathematical formulation of the model

The theoretical model form can be derived from a physical analysis of the system. The system is compared with a mass-spring equivalent, shown by figure 2. Mass' correspond to inertias, and springs and dampers correspond to the flexible parts of the test setup in this equivalent. The two flexible parts of the test setup can be modelled as one spring and damper, because the inertia of the torsional beam is negligible compared with the inertias of the motor and flexible beam with payload. This simplification has been verified experimentally. The strain gauge signal measured on the torsional beam

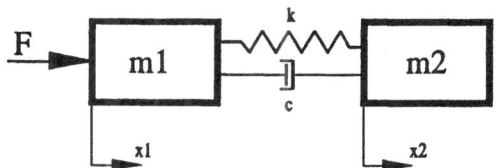

F	:	corresponds to the motor torque, which is proportional to the input of the system.
m_1	:	corresponds to the inertia of the motor.
m_2	:	corresponds to the total inertia of the flexible part of the test setup.
k, c	:	correspond to the stiffness and damping of the flexible part of the bread board model.
x_1	:	corresponds to the displacement of the motor.
x_2	:	corresponds to the displacement of the center of gravity of the flexible part of the bread board model.
$x_1 - x_2$	:	corresponds to the deformation of the flexible part of the bread board. This signal is proportional to the strain gauge signal.

Figure 2: Schematic representation of the mass-spring equivalent.

contains the same information as the signal of the flexible beam strain gauges. Only one of these signals has to be measured for the identification and control. The strain gauge signals of the flexible beam are chosen to represent the beam deflections. This choice was totally arbitrary. The linear motor friction is also neglected in the mass-spring equivalent and in the models. Experiments have shown the correctness of this simplification.

Some interesting transfer functions can be derived from this mass-spring equivalent:

$$
\begin{aligned}
H_1(p) &= \frac{\frac{1}{m_1}}{p^2 + c(\frac{1}{m_1} + \frac{1}{m_2})p + k(\frac{1}{m_1} + \frac{1}{m_2})} \\
H_2(p) &= \frac{p^2 + \frac{c}{m_2}p + \frac{k}{m_2}}{p^2} \\
H_3(p) &= \left(\frac{\frac{1}{m_1}}{p^2 + c(\frac{1}{m_1} + \frac{1}{m_2})p + k(\frac{1}{m_1} + \frac{1}{m_2})} \right) \left(\frac{p^2 + \frac{c}{m_2}p + \frac{k}{m_2}}{p^2} \right)
\end{aligned}
\tag{1}
$$

$H_1(p)$ relates the input force F to the deformation of the spring $(x_1 - x_2)$. $H_2(p)$ relates the deformation of the spring $(x_1 - x_2)$ to the absolute displacement x_1 of mass m_1. $H_3(p)$ relates the input force F to the absolute displacement x_1 of mass m_1. The discrete time transfer functions that model the test setup are the z-transform equivalents of these continuous time transfer functions.

$$
\begin{aligned}
H_1(z^{-1}) &= \frac{b_{11}z^{-1}}{1 + a_{11}z^{-1} + a_{12}z^{-2}} \\
H_2(z^{-1}) &= \frac{b_{20} + b_{21}z^{-1} + b_{22}z^{-2}}{1 - 2z^{-1} + z^{-2}} \\
H_3(z^{-1}) &= H_1(z^{-1})H_2(z^{-1})
\end{aligned}
\tag{2}
$$

$H_1(z^{-1})$ relates the input signal $u(k)$ to the beam deflection or the strain gauge signal $\varepsilon(k)$. $H_2(z^{-1})$ relates the strain gauge signal $\varepsilon(k)$ to the angular motor position $\alpha(k)$.

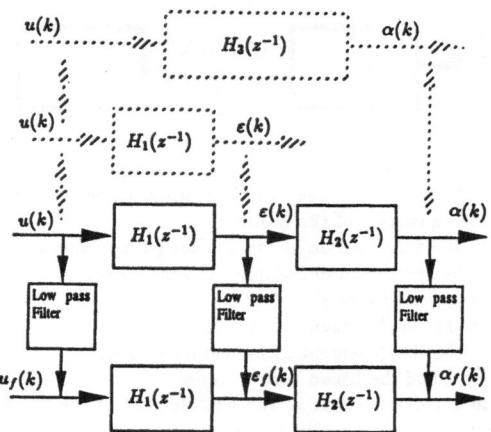

Figure 3: Schematic representation of the relation between $H_1(z^{-1})$, $H_2(z^{-1})$, and $H_3(z^{-1})$, used for the time domain parameter estimation.

$H_3(z^{-1})$ relates the input $u(k)$ to the angular motor position $\alpha(k)$. $H_3(z^{-1})$ is not equivalent to the z-transform of $H_3(p)$. The difference between both discrete transfer functions is of second order. It has been proven experimentally that the model form $H_3(z^{-1})$ allows an accurate fitting of the data. The denominator coefficients -2 and 1 in $H_2(z^{-1})$ define a double pole at $z = 1$.

The unknown parameters $a_{11}, a_{12}, \ldots$, are calculated with a least squares time domain parameter estimation algorithm, which is explained in the next paragraph.

3.2 Least squares time domain parameter estimation

A least squares parameter estimation algorithm fits a discrete domain transfer function to a given set of input-output data. It is well known [12] that least squares methods lead to biased solutions if the input-output measurements are corrupted with measurement noise, high order oscillations and harmonics caused by nonlinearities, etc .... The solution to this problem is low pass filtering of the data. A second order digital low pass Butterworth filter with a cut-off frequency at $10Hz$ has been used for this purpose.

The total model of the test setup is based on three transfer functions, $H_1(z^{-1})$, $H_2(z^{-1})$ and $H_3(z^{-1})$. Identification of $H_1(z^{-1})$, based on the measured input and strain gauge signals, and identification of $H_3(z^{-1})$, based on the measured input and encoder signals, is the most straightforward approach, but certainly not optimal. Identification of the two second order submodels $H_1(z^{-1})$ and $H_2(z^{-1})$ is a better solution. Figure 3 gives the schematic representation of the division into two submodels. It also shows the low pass filtering necessary for a reliable least squares parameter estimation.

The input-output signals used for the identification are measured with a sample frequency of $200Hz$. (The same sample frequency is used for the control). The identification of the parameters of $H_1(z^{-1})$ is based on the filtered input signal $u_f(k)$ and the filtered strain gauge signal $\varepsilon_f(k)$. The identification of the parameters of $H_2(z^{-1})$

is based on the filtered strain gauge signal $\varepsilon_f(k)$ and the filtered encoder signal $\alpha_f(k)$. Only two times three parameters have to be estimated. b_{11}, a_{11} and a_{12} for $H_1(z^{-1})$ and b_{20}, b_{21} and b_{22} for $H_2(z^{-1})$. The a priori information of the denominator coefficients -2 and 1 of $H_2(z^{-1})$ is explicitely taken into account during the parameter estimation of the numinator coefficients of $H_2(z^{-1})$ by considering the values $\alpha_f(k) - 2\alpha_f(k-1) + \alpha_f(k-2)$ instead of $\alpha_f(k)$ as the output signal.

The obtained parameters of $H_1(z^{-1})$ and $H_2(z^{-1})$ are:

$$
\begin{aligned}
H_1(z^{-1}) &= \frac{-3.80z^{-1}}{1 - 1.97z^{-1} + 0.990z^{-2}} \\
H_2(z^{-1}) &= \frac{-1.42\,10^{-4} + 2.82\,10^{-4}z^{-1} - 1.42\,10^{-4}z^{-2}}{1 - 2z^{-1} + z^{-2}}
\end{aligned}
\tag{3}
$$

$H_3(z^{-1})$, which is a series combination of $H_1(z^{-1})$ and $H_2(z^{-1})$, is:

$$
H_3(z^{-1}) = \frac{5.40\,10^{-4}z^{-1} - 1.07\,10^{-3}z^{-2} + 5.39\,10^{-4}z^{-3}}{1 - 3.97z^{-1} + 5.92z^{-2} - 3.95z^{-3} + 0.990z^{-4}}
\tag{4}
$$

The eigenfrequency is estimated in $H_1(z^{-1})$. Its value is $4.89Hz$ with a damping ratio of 0.03 .The anti-resonance frequency is estimated in $H_2(z^{-1})$. Its value is $2.87Hz$ with a damping ratio of 0.008.

The results of the time domain identification method are compared with a non-parametric reference model. This reference model is identified using a stepped sine identification method [21]. This identification technique calculates the frequency response of the system by exciting it with constant-frequency sinusoids. Figure 4 shows the amplitude and the phase of models $H_1(z^{-1})$ and $H_3(z^{-1})$ along with the diagrams of the reference model. There is only a small difference between the diagrams in the neighbourhood of the resonance and anti-resonance frequency. This difference is caused by nonlinear motor friction [13].

Less calculation time and more accurate results are two important advantages of this method with respect to the direct estimation of the parameters of $H_3(z^{-1})$. The new approach also guarantees a minimal multi variable model. It estimates every pole and zero of the transfer function only once.

3.3 Spill-over

A one-link flexible robot has an infinite number of eigenfrequencies and anti-resonance frequencies. The identified model and the controller considers only the first of both. The unmodelled modes influence the reduced order controller and can even cause unstable behaviour [7]. Output filtering with a digital low pass filter reduces these spill-over problems. These filters remove higher order oscillations from the measured output signals. They extend the continuous time physical system and the model with their discrete transfer function. The filter is the same as used for the identification. It has the following discrete transfer function:

$$
H_q(z^{-1}) = \frac{2.01\,10^{-2} + 4.02\,10^{-2}z^{-1} + 2.01\,10^{-2}z^{-2}}{1 - 1.56z^{-1} + 6.41z^{-2}}
\tag{5}
$$

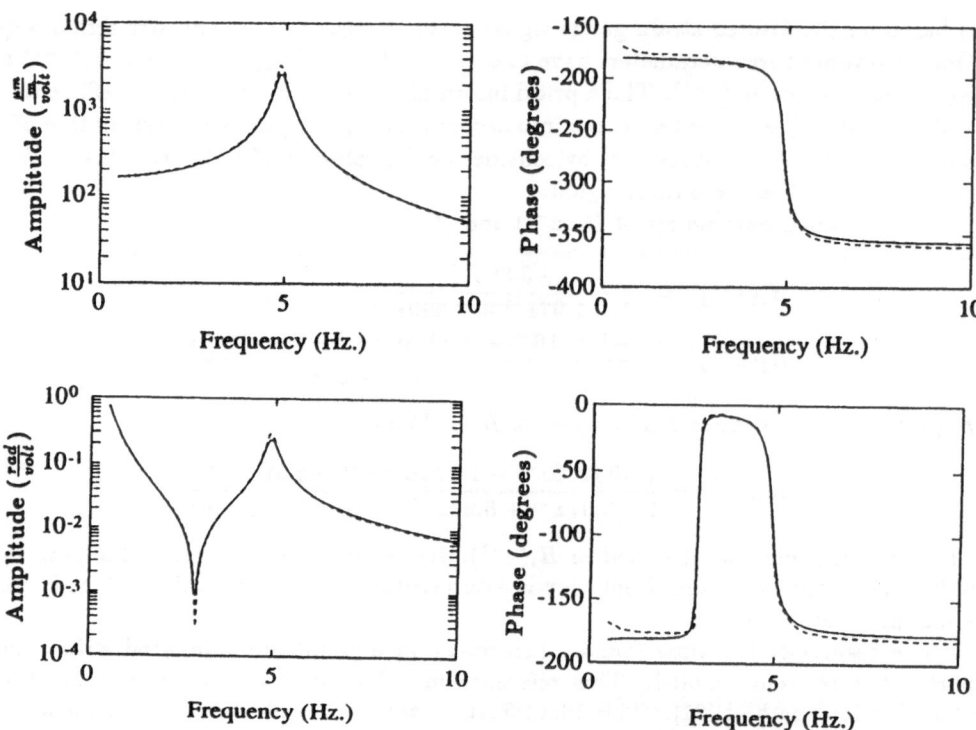

Figure 4: Comparison between the reference models (dotted line) and the identified transfer functions $H_1(z^{-1})$ (upper figures) and $H_3(z^{-1})$ (bottom figures) (solid line).

The discrete transfer function that relates the input signal $u(k)$ to the filtered strain gauge signal $\varepsilon_f(k)$ is a series combination of $H_1(z^{-1})$ and $H_q(z^{-1})$. The discrete transfer function that relates the input signal $u(k)$ to the filtered encoder signal $\alpha_f(k)$ is a series combination of $H_3(z^{-1})$ and $H_q(z^{-1})$.

3.4 The identification of the static relation between the end point position, the strain gauge signal and the motor angle

The position $y(k)$ (in mm) of the end point is a linear combination of the strain gauge signal $\varepsilon(k)$ and the motor angle $\alpha(k)$ if only the first mode is considered.

$$y(k) = a_{p_1}\varepsilon(k) + a_{p_2}\alpha(k) \tag{6}$$

a_{p_2} is equal to the length of the flexible beam ($840\,mm$). a_{p_1} has been determined experimentally. During the experiment the end point position is held constant, while the motor angle and the beam deflection are increased. The values of the strain gauge and encoder signals are measured for different beam deflections, and used to calculate a_{p_1}. The result of this calculation is $a_{p_1} = 0.1234$. The same coefficients are valid for the relation between the filtered values of these signals.

3.5 The total state space model

State feedback control is based on a state space model of the system. The state feedback controller for the test setup uses a state space model which is a combination of a canonical model and a state space model with measurable state variables. This combines the advantages of a minimum number of model parameters and the possibility to use a reduced order state estimator. The total state space model is presented in the following set of equations:

$$
\begin{aligned}
X(k+1) &= [A]\,X(k) + [B]\,u(k) \\
Z(k) &= [C]\,X(k)
\end{aligned}
\tag{7}
$$

with

$$
X(k) = [\ x_1(k)\quad x_2(k)\quad x_3(k)\quad x_4(k)\quad x_5(k)\quad x_6(k)\ ]^{\mathrm{T}}
$$

$$
[A] =
\begin{bmatrix}
3.53 & -4.70 & 2.81 & -0.63 & 0 & 0 \\
1 & 0 & 0 & 0 & 0 & 0 \\
0 & 1 & 0 & 0 & 0 & 0 \\
0 & 0 & 1 & 0 & 0 & 0 \\
3.83\,10^{-5} & -7.24\,10^{-5} & 3.05\,10^{-5} & 3.94\,10^{-6} & 2 & -1 \\
0 & 0 & 0 & 0 & 1 & 0
\end{bmatrix}
$$

$$
[B] = [\ 1\quad 0\quad 0\quad 0\quad 1.08\,10^{-5}\quad 0\ ]^{\mathrm{T}}
$$

$$
Z(k) = [\ \varepsilon_f(k)\quad \alpha_f(k)\quad y_f(k)\ ]^{\mathrm{T}}
$$

$$
[C] =
\begin{bmatrix}
C_1 \\ C_2 \\ C_3
\end{bmatrix}
=
\begin{bmatrix}
-7.63\,10^{-2} & -1.53\,10^{-1} & -7.63\,10^{-2} & 0 & 0 & 0 \\
0 & 0 & 0 & 0 & 1 & 0 \\
-9.41\,10^{-3} & -1.88\,10^{-2} & -9.41\,10^{-3} & 0 & 8.40\,10^{2} & 0
\end{bmatrix}
$$

The input-to-filtered strain gauge signal submodel has a controller canonical form. It is derived from the discrete transfer function that relates these two signals, and depends only on the first 4 state variables. The second submodel contains $x_5(k)$ and $x_6(k)$. $x_5(k)$ is the filtered motor position at the current time instant k, and $x_6(k)$ is the filtered motor position at the previous time instant $k-1$. The numerator coefficients of $H_2(z^{-1})$ determine the link between both submodels.

4 State feedback control

4.1 The state estimator

A reduced order current state estimator has been implemented for the test setup. It estimates the state variables more accurate and it reacts faster on external unmeasurable disturbances than a predictive estimator. State variables $x_1(k), x_2(k), x_3(k)$, and $x_4(k)$ are calculated with a closed loop estimator. The estimator is based on the canonical submodel that contains only these four state variables. It feeds back the filtered strain gauge signal $\varepsilon_f(k)$ because this is the only output signal that detects both the resonance and anti-resonance frequency. $x_5(k)$ and $x_6(k)$ are the filtered values of the measured angular motor position. The feedback gains of the state estimator are calculated using a pole placement algorithm. The choice of the estimator poles is not so critical. The control results shown in this paper are obtained with an estimator bandwidth of $20Hz$.

4.2 The control design

The output of a state feedback controller during a tracking operation consists of an acceleration feedforward and a feedback of the difference between the measured and the desired state variables. For the sixth order system this corresponds to:

$$u(k) = u_{ff}(k) - \sum_{i=1}^{6} K_i(x_i(k) - x_{id}(k)) \tag{8}$$

The control design consists of the calculation of: 1) the feedback gains K_i, 2) the feedforward term $u_{ff}(k)$, and 3) the desired state variables.

1) Several criteria can be handled to calculate the state feedback gains. Pole placement seems to be the most straightforward approach for flexible robots [19]. It allows a fine tuning of the closed loop dynamics.

2) The feedforward term is determined as being the input needed to obtain exact correspondence between output, y, and desired trajectory, y_d. This involves calculation of the inverse model (7). Because this calculation gives stability problems, only the steady state behaviour of the system is considered. The relationship between input and end point acceleration, $\ddot{y}$ or $\ddot{y}_f$, in steady state is given by:

$$\frac{\ddot{y}}{u_{ff}} = \frac{\ddot{y}_f}{u_{ff}} = \lim_{z \to 1} (1 - z^{-1})^2 \, C_3(zI - A)^{-1} B = 0.1577 \tag{9}$$

Therefore, with $\ddot{y}_d$ the desired acceleration, the feedforward input is obtained as:

$$u_{ff} = \frac{\ddot{y}_d}{0.1577} \tag{10}$$

3) The desired state variables x_{1d} to x_{4d} can be calculated by simulating the identified model (7) with input u_{ff}:

$$
\begin{aligned}
x_{1d}(k) &= 3.53\, x_{1d}(k-1) - 4.70\, x_{1d}(k-2) + 2.81\, x_{1d}(k-3) - 0.63\, x_{1d}(k-4) \\
&\quad + u_{ff}(k-1) \\
x_{2d}(k) &= x_{1d}(k-1) \\
x_{3d}(k) &= x_{1d}(k-2) \\
x_{4d}(k) &= x_{1d}(k-3)
\end{aligned}
\tag{11}
$$

Clearly, the input will excite the resonance frequency which results in an undesirable trajectory for the states. Therefore, the first equation of (11) is reduced to the steady state relationship between x_1 and u_{ff}, but taking into account the filter dynamics. The steady state relationship is calculated by taking all $x_{1d}(k-i) = x_{1d}(k)$. In more general terms this relationship is defined by:

$$\frac{x_{1d}}{u_{ff}} = \lim_{z \to 1} [1\,0\,0\,0\,0\,0](zI - A)^{-1} B = 526 \tag{12}$$

To take into account the filter dynamics, $H_q(z^{-1})u_{ff}$ is taken instead of u_{ff} in equation (11):

$$x_{1d} = 526\, H_q(z^{-1})\, u_{ff} \tag{13}$$

The desired state variables x_{5d} and x_{6d} can be calculated using the output equation of the model (7).

$$
\begin{aligned}
x_{5d}(k) &= \alpha_{df}(k) \\
x_{6d}(k) &= x_{5d}(k-1) = \alpha_{df}(k-1)
\end{aligned}
\tag{14}
$$

with α_{df} the filtered motor angle. The trajectory of the motor angle, α_d, is determined by using the experimentally found relationship (6) between the end point position, the strain gauge signal and the motor angle:

$$
\alpha_d = \frac{y_d - a_{p1}\varepsilon_d}{a_{p2}}
\tag{15}
$$

where for the calculation of ε_d the steady state relationship between input and strain gauge signal is used:

$$
\frac{\varepsilon_d}{u_{ff}} = \lim_{z \to 1} H_1(z^{-1}) = 170.96
\tag{16}
$$

The filtered motor angle, α_{df} is given by:

$$
\alpha_{df} = H_q(z^{-1})\, \alpha_d
\tag{17}
$$

4.3 Reduction of the steady state error by integral action

Using this control law, the steady state position error measured at the end point amounts to $1.5\,mm$, which is higher than the required error tolerance of $0.5\,mm$. This error is due to the high static friction of the direct drive motor which makes the motor stop before the desired position is reached.

To eliminate this steady state error an integral term, based on a trapezoid integration rule, may be added to the controller output:

$$
u_i(k) = k_i \frac{T}{2}\left(y_d(k) + y_d(k-1) - y(k) - y(k-1)\right) + u_i(k-1)
\tag{18}
$$

where k_i is the integral feedback gain

$\qquad T$ is the sampling period

An alternative approach consists of adding a seventh state variable x_7 to the state space model. This state variable corresponds to the integration of the position. The integral feedback gain is then calculated together with the state feedback gains using the pole placement algorithm. Comparable results are obtained by using both approaches.

In order to reduce the overshoot caused by the integral action, the integral action is activated only when the position is in the vicinity of the desired final position.

Another problem caused by the nonlinear friction are the limit cycles which occur when the end point is in the vicinity of the final position. This effect occurs when the motor comes to stop due to the fact that the control torque becomes smaller than the friction torque. The residual energy in the system is transferred to a limit cycle at a frequency different from the system frequencies. The small oscillations are still amplified by adding the integral action to the controller. To eliminate these oscillations a local modification of the control in the vicinity of the desired final position is performed. The modification consists of phasing out the energy in the system and reinstating the asymptotic stability properties of the open loop system. This is performed by modifying the magnitude of the control input $u(k)$ according to the ratio $V(x(k))/\delta_m$ when the state $x(k)$ is in the vicinity of the desired final position [14].

$$u^*(k) = \left\{ \begin{array}{ll} u(k) & V(x) \geq \delta_m \\ u(k)\frac{V(x)}{\delta_m} & V(x) < \delta_m \end{array} \right. \tag{19}$$

$V(x)$ is a continuously differentiable locally positive definite function of the controlled states. δ_m is a small positive constant defining the vicinity of the final state $x = x_d$.
The intention to phase out the energy in the system suggests to use the total energy in the controlled modes as function $V(x)$. However, a simpler positive definite function of the controlled states is used, here;

$$V(x) = (\alpha - \alpha_d)^2 \tag{20}$$

This choice gives satisfactory results.

4.4   Control results

A tracking test has been executed to validate the controller performance. The reference trajectory is based on a parabolic acceleration profile. It gives intermediate positions between the initial and final position along a smooth path, and therefore avoids unnecessary excitation of the eigenfrequencies. Prevention is better than cure. The maximum end point acceleration and end point velocity during the tracking are $24.2\frac{m}{s^2}$ and $8.0\frac{m}{s}$. These values are near to the limitations of the motor. The total displacement of the end point is $5.28\,m$, which corresponds to one motor revolution. The final position should be reached after 1.34 seconds.

A healthy approach to pole placement is to replace only the poles of the open loop system that result in a bad dynamic behaviour under tracking conditions. This avoids unnecessary large feedback gains. The two complex conjugated poles of the resonance frequency and the two poles at $z = 1$ are on or too close to the unit circle in the z-plane. They are not suited for a tracking control action and must be replaced. The system poles that originate from the Butterworth filter $H_q(z^{-1})$ result in an acceptable dynamic behaviour [8], and can be copied to the closed loop system. The choice of the remaining four pole positions is motivated by robustness considerations. Two poles are placed at $\omega_0 = 2\pi f \frac{rad}{sec}$ and $\zeta = 1.0$ for good tracking dynamics, and two poles are placed at $\omega_0 = 2\pi f \frac{rad}{sec}$ and $\zeta = 0.707$ for good regulator dynamics [5]. f is the frequency of the poles. These poles are dominant if f is smaller than the cut-off frequency of the Butterworth filter ($10Hz$). A frequency f of $7Hz$ for the dominant poles of the controller resulted in the best performance for the tracking test. The feedback gains for this controller are:

$$K = \left[\begin{array}{cccccc} 0.641 & -1.85 & 1.46 & -0.248 & 1.12\,10^4 & -1.06\,10^4 \end{array} \right]$$

The closed loop bandwidth is approximately $7\,Hz$. The controller becomes unstable if the dominant poles are larger then $8\,Hz$.

The values of the integral feedback gain k_i, the value δ_i defining the vicinity of the final position with respect to the integral action, and the value δ_m defining the vicinity of the final position with respect to the phasing out strategy, are determined experimentally. Good results are obtained with:

$$k_i = 7.73\,10^{-5}$$
$$\delta_i = 2.5\,mm$$
$$\delta_m = 2.35\,rad^2$$

Figure 5 shows the control results. There are no oscillations at a frequency equal to the resonance or anti-resonance frequency. The steady state error is equal to zero. The maximum tracking error is $11\,mm$.

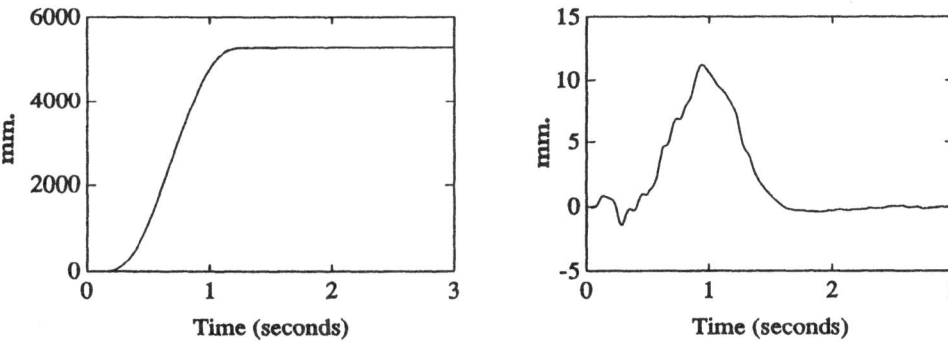

Figure 5: Control results. Left figure: measured (solid line) and desired (dashed line) end point position. Right figure: tracking error.

Figure 6 shows the measured end point position, when the end point is in the vicinity of the final position. This position is directly measured using a digitizer board [17].

Let Δt be $t - t_{fin}$ where

t_{fin} = time at which the desired final position should be reached $(1.34\,s)$
t = time

The following results are obtained:

- The maximum overshoot is equal to $0.42\,mm$ and occurs at $\Delta t = 0.56\,s$.

- The position error remains within the set error tolerance of $0.5\,mm$ after $\Delta t = 0.19\,s$.

- The amplitude of the remaining oscillation is less than $0.1\,mm$ after $\Delta t = 0.86\,s$. This oscillation is due to the resolution of the strain gauge signal (see equation 6). (The influence of encoder resolution is much smaller.)

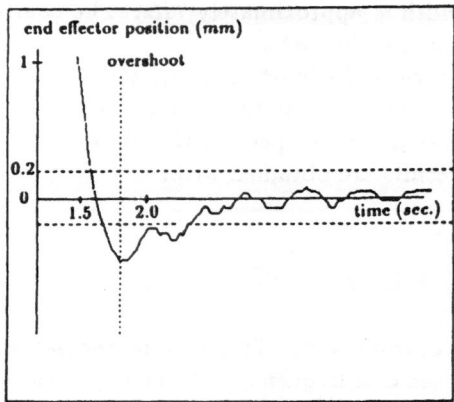

Figure 6: End point position in the vicinity of the final position, measured using a digitizer board.

5 Conclusion

In this paper a controller is designed for a flexible one-link robot based on an experimentally identified model. The control law consists of state feedback, integral feedback, and acceleration feedforward. Nonlinear joint friction has a large effect on the design of a linear controller for a flexible robot.

On the one hand it influences the accuracy of the identified linear model on which the controller is based. Nonlinear friction causes identified transfer functions to be dependent on the level of excitation. This effect is most outspoken in the neighbourhood of the resonance and anti-resonance frequencies.

On the other hand nonlinear friction causes limit cycles and steady state errors, which can not be corrected with plain state feedback. Limit cycles inhibit accurate positioning and appear when the feedback gains are too high. These limit cycles are avoided by a restricted closed loop bandwidth and by a phasing out of the feedback once the final position is reached. Steady state errors are eliminated with integral feedback.

Large tracking errors are another defect of plain state feedback controllers. Acceleration feedforward is the right remedy for this defect. It reduces the tracking error to negligible values.

Acknowledgement
This paper presents work carried out in the framework of ESPRIT Project 1561 ("SACODY") with partners: Bertin & Cie (F), Katholieke Universiteit Leuven (B), AEG AG (D), Leuven Measurements and Systems (B), KUKA Schweissanlagen und Robotor GmbH (D), and University College Dublin (IR). The contractors would like to acknowledge the support of the Commission of the European Communities, as represented by DG XIII, on Telecommunications, Informations Industries and Innovation, and more especially its CIM Group.

References

[1] Adams, M. and Swevers, J. 1988. Identification and control of a single degree of freedom positioning system. 13th International seminar on modal analysis, K.U.Leuven, Leuven, Belgium.

[2] Åström, K. J. and Wittenmark, B. 1984. Computer controlled systems. Englewood Cliffs, N.J.: Prentice Hall.

[3] Clary, J. 1984. Robust algorithms in adaptive control. Ph.D. Thesis, Stanford University, Department of Electrical Engineering.

[4] Demeester, F. 1987. Selection of applicable off-line computation algorithms for trajectory control. Technical report, task 3154, Esprit project 1561:Sacody.

[5] Duque, M. and Samaan, M. 1988. Partial state LQ and GPC adaptive control: An experimental evaluation. Eighth International Conference on Analysis and Optimization of Systems. Springer-Verlag.

[6] Eppinger, S. D. and Seering, W. P. 1988. Modeling robot flexibility for end point force control. IEEE International Conference on Robotics and Automation.

[7] Faillot, JL. 1987. Examine state of the art on control of flexible structures. Technical report, task 3121, Esprit project 1561:Sacody.

[8] Franklin G. F. and Powell J. D. 1980. Digital control of dynamic systems. Addison-Wesley.

[9] Goodwin, G. C. and Sin, K. S. 1984. Adaptive filtering predicion and control. Englewood Cliffs, N.J.: Prentice Hall.

[10] Lembregts, F. Swevers, J. 1988. Development and evaluation of off-line identification techniques. Technical report, task 612, Esprit project 1561:Sacody.

[11] Ljung, L. and Soderstrom, T. 1983. Theory and practice of recursive identification. Cambridge: MIT Press.

[12] Ljung, L. 1987. System identification: theory for the user. Englewood Cliffs, N.J.: Prentice Hall.

[13] Mertens, M. and Van der Auweraer, H. 1989. The complex stiffness method to detect and identify nonlinear dynamic behaviour of SDOF systems. Mechanical Systems and Signal Processing, 3(1):37-54.

[14] Preumont, A. 1987. Spillover alleviation for nonlinear active control of vibration. J.Guidance, 11(2):124-130.

[15] Ramakrishnan, S. 1985. Experimental identification and control of the tip position of a flexible, single link manipulator. Ph.D. Thesis, University of California, Berkeley, Department of Mechanical Engineering.

[16] Rovner M. D. 1987. Experiments toward on-line identification and control of a very flexible one-link manipulator. Int. J. Robotics Res. 6(3):3-19.

[17] Van Den Bossche, J. 1989. RODYM: a new approach to robot metrology. Technical report, K.U.Leuven.

[18] Schmitz, E. 1984. Initial experiments on the end-point control of a flexible one-link robot. Int. J. Robotics Res. 3(3):62-75.

[19] Sidman, M. 1986. Adaptive control of a flexible structure. Ph.D. Thesis, Stanford University, Department of Electrical Engineering.

[20] Swevers, J. Moonen, M. 1988. A black box identification algorithm for analysis of structural dynamics. 13th International seminar on modal analysis, K.U.Leuven, Leuven, Belgium.

[21] Van Der Auweraer, H. 1987. Development and evaluation of advanced measurement methods for experimental modal analysis. Ph.D. Thesis, Dept. of Mechanical Engineering, K.U. Leuven, Leuven, Belgium.

A Simple Juggling Robot:
Theory and Experimentation

M. Bühler, D. E. Koditschek, and P. J. Kindlmann *
Center for Systems Science
Yale University, Department of Electrical Engineering

Abstract

We have developed a formalism for describing and analyzing a very simple representative of a class of robotic tasks which involve repeated robot-environment interactions, among them the task of juggling. We review our empirical success to date with a new class of control algorithms for this task domain that we call "mirror algorithms." These new nonlinear feedback algorithms were motivated strongly by experimental insights after the failure of local controllers based upon a linearized analysis. We offer here a proof that a suitable mirror algorithm is correct with respect to the local version of a specified task — the "vertical one-juggle" — but observe that the resulting ability to place poles of the local linearized system does not achieve noticeably superior transient performance in experiments. We discuss the further analysis and experimentation that should provide a theoretical basis for improving performance.

*This work has been supported in part by PMI Motion Technologies, INMOS Corporation and the the National Science Foundation under a Presidential Young Investigator Award held by the second author.

1 Introduction

Progress in robotics, as in any other science, requires an interplay of theory and experimentation. The inevitable tension between these opposites pulls analysis away from sterile abstractions and pragmatism from blind tinkering. This paper, presented in a symposium devoted to that interplay, offers an account of research partaking strongly of both modes of inquiry in the effort to achieve a level of dexterity relatively new to robotics: the ability to manipulate intermittent dynamical environments. We have built a one degree of freedom robot that can "juggle" two degree of freedom bodies falling otherwise freely in the earth's gravitational field and bring them to a desired periodic orbit from a wide range of initial conditions and in-flight perturbations by a controlled schedule of impacts. Our efforts might be seen as a case study in this productive tension at work.

We began with an abstract formalism [4] for describing and analyzing the particular task under investigation, the "vertical one juggle." This representation suggested a simple control strategy arising from straightforward application of standard tools of linear systems theory. The strategy did not work when implemented on our physical apparatus [6]. A very different strategy born of entirely intuitive reasoning and some physical tinkering achieved notable experimental success in the vertical one juggle [5] and generalized equally successfully to a variety of interesting extensions [7]. We have been able to furnish a local correctness proof of the empirically validated strategy with respect to the original formal model, and will present this proof in the present paper. We will show as well how the same theory entitles us (after further analytical tinkering with the successful strategy) to claim complete control over the local transients. However, the data we present reveals little empirical distinction between the theoretically slower and faster local transient gain settings. A paper presently in preparation shows how a return to the formalism affords a global correctness proof of the successful scheme assuming a radically simplified model of the robot in its dynamical environment. Further experimentation will reveal whether or not the control over global transient behavior resulting from this newest analysis is physically discernible.

The direct comparisons we display in the paper between simulation studies and experimental data show that our original formal model is empirically relevant despite its many simplifications and blindness to known physical effects. Thus, we have reached a critical juncture in our program of research. If our recent global analysis of the radically simplified model does not result in experimentally observable improvements in the intuitively conceived control strategy then we will be forced to pursue a global analysis of the physically validated model that may well prove intractable. In contrast, empirically successful predictions of this simplified model would strongly suggest that a generalizable body of physically relevant theory for robotics in intermittent dynamical environments is at hand. For this "simplistic" global theory both contains the linearized analysis and has a formal claim to a certain "universality" within discrete dynamical systems theory [8]. Thus, we would be well along the road to a geometric formalism that translates abstract human goals into automatic robot controllers.

We have no doubt that such a body of theory is both essential and possible. The first systematic work in this domain of robotic tasks has been the pioneering research of Raibert whose careful experimental studies verify the correctness of his elegant control strategies for legged locomotion [14]. Our analysis of simplified models of Raibert's hopper [9] uses the same global tools as in the simplified model of our juggler to make strong assertions about the transient and limit behavior of his machines that also remain to be empirically validated. On the other hand, McGeer has successfully used the kind of local linearized analysis that failed in our experiments to build passive (unpowered) walking robots [13, 12], and feels that similarly tractable analysis should suffice for controlling running machines as well [11]. Wang [18] has proposed to use the same local techniques for studying *open loop* robot control strategies in intermittent dynamical environments although his ideas remain to be tested as well. Research by Atkeson et al. on juggling [1] suggests that task level learning methods may relieve dynamics based (or any other parametric) controller synthesis methods of the need to achieve precise performance requirements as long as a basically functioning system has been assured. Thus, increasing numbers of researchers have begun to explore the problems of robotics in intermittent dynamical environments with increasingly successful results.

This paper is organized as follows. The experimental setup and original mathematical model are reviewed in Section 2. Analysis of the "contact geometry" between robot and environment gives rise to an impact model — the "environmental control system" — with respect to which the "vertical one-juggle" task is formally defined and proven to be achievable in Section 3 via local linear analysis. Section 4 accounts for the failure of linear algorithm founded on this contact geometry: resulting impact schedules, while provably correct, are not robust and produce basins of attraction around the desired task point that are insufficiently large to be physically observable. Section 4.3 then motivates the intuitively developed successful class of nonlinear control algorithms. Briefly, trajectories in the puck phase space, are projected down to the robot phase space, yielding explicit robot reference trajectories. The projected trajectories have an intuitively appealing character which lead us to name this procedure the *mirror algorithm*. When our puck-robot system is forced toward this surface, that is, when we force the robot to track the trajectory specified by a mirror algorithm applied to a free falling puck, the result is a successful vertical one-juggle. In Section 5 we provide the local linear analysis of why our robot control strategy works. Specifically, we are able to prove that the family of environmental control laws arising from the robot "mirror algorithm" is correct with respect to the vertical one-juggle task set as defined in the original paper [4]. We first demonstrate that this family has the effect of inducing a lower dimensional invariant submanifold of the nonlinear discrete puck dynamics — the impact dynamics originally studied in [4]. We next show that the mirror algorithm family ensures that the local linearized dynamics around any valid task point, when restricted to this invariant submanifold, defines a completely controllable linear time invariant system. We finally observe that the new construction includes enough design freedom to afford pole placement with respect to that local linearized reduced dimensional control system. Thus, for any valid vertical one-juggle, we are able to synthesize a mirror law which attains the desired task while theoretically achieving arbitrarily specified local transient behavior. The conclusion offers some speculations upon the larger implications of this work for robotic tasks in more general intermittent dynamical environments.

2 The Empirical and Analytical Setting

This section describes our experimental setup and presents a simplified mathematical model which enables us to pose and solve robot juggling tasks as formal problems in control theory. The physical apparatus we use to obtain the data reported in this paper is depicted in Figure 1, described in Section 2.1. The mathematical models are presented in Section 2.2.

2.1 Experimental Apparatus

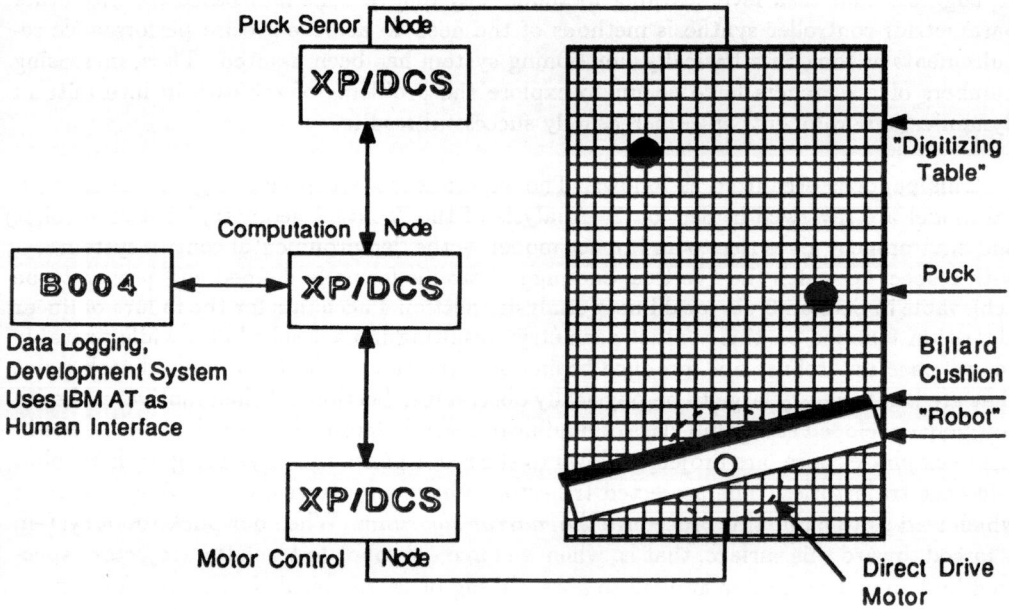

Figure 1: The Yale Juggler

The physical apparatus consists of a puck, which slides on an inclined plane and is batted successively by a simple "robot:" a bar with billiard cushion rotating in the juggling plane as depicted in Figure 1.

2.1.1 A Distributed Real Time Controller

All intelligent sensor and controller functions are performed by a four node distributed computational network formed from the INMOS transputer based Yale XP/DCS control node [10]. The INMOS T800 is a 32 bit 10 MIPS RISC chip with an integral floating point unit capable of a sustained 1.5 Mflop computational rate. Each transputer provides four independent 5, 10 or 20 MBits/sec serial DMA channels, so any node can communicate with up to four neighbors while simultaneously executing its own program with no effect upon either computational or communications rate (after initial start up overhead). Our XP/DCS CPU board complements the transputer's modular and flexible character by providing fast external memory, support for the four serial communication links, two fiber optic links, and an I/O expansion connector. The board's backplane connector is pin compatible with the INMOS ITEM Development System. The plug-in XP/DCS I/O board enhances the transputer's computational and communication power with a bidirectional latched 32 bit I/O bus with full handshaking support. Half of this board is allotted to a prototyping area allowing for easy customization to specific I/O needs. The cost of each mother/daughter board set at the time of writing is slightly over \$2000.

2.1.2   Distribution of Computational Resources to Mechanical Hardware

In order to move the bar according to some puck dependent control algorithm, the puck's position and velocity in both directions on the plane must be measured. Presently, this is accomplished by placing an oscillator inside the puck and burying a grid in the juggling plane, thus imitating a big digitizing tablet. On the back of the plane, a simplified XP/DCS system, the Puck Sensor Node, is used as a smart sensor. It measures the voltages induced in the sensing grid by the puck. The puck position in the plane is computed from the zero and first order moments. This information is used to estimate the puck's state: we use a standard linear observer to reduce measurement noise in position and velocity data. Each puck state measurement is communicated asynchronously via fiber optics to the Computation Node. This sampling and communication process is performed at a rate of 1kHz (when tracking one puck). We constructed an XP/DCS based real time (60Hz) stereo vision system that will enable us to move off the plane into three space soon.

The main node is the Computation Node which receives puck state information from the sensor node, reports logging data to the logging node, implements the control algorithm and issues the resulting desired robot states to the Motor Control Node. Various additional tasks like detecting the puck motion status (up, top, down, impact), predicting puck states (both used for extracting logging data, not for control) as well as extensive error checking and housekeeping tasks have to be performed on this node as well. The sampling time can vary between 500 and 1000 μs. The Motor Control Node is dedicated to commanding a high torque dc servo actuator (donated to the robotics laboratory by PMI Motion Technologies) at a rate of 2kHz.

The experiences with the XP/DCS, the transputer and the development environment derived from this application are very encouraging. No single number can capture the ease of use and the little time spent with system overhead. Given the T800's intrinsic floating point capability, and the mathematical function library, formulas were programmed (in OCCAM, the "C-like" native compiler) almost directly from the blackboard with no attempt at code optimization. In spite of substantial calculations, and a great deal of data logging and error handling overhead, very high sampling rates were achieved. The system operates capably in a high EMI environment in consequence of the low cost 5MBits/sec fiber optic units from Hewlett Packard built into the Yale XP/DCS boards.

2.2 Mathematical Models

It is clear that this simple apparatus entails a multitude of complex physical effects: collisions between two partially elastic bodies; stiction, coulomb friction, and spin effects of the puck at impact; bouncing of the puck perpendicular to the plane; and so on. Yet since we desire a theory of robotics in intermittent dynamical environments, it is critical that our mathematical models of any particular physical setup be analytically tractable: simple enough to be amenable to formal proof; abstract enough to permit insights into a broader range of robotic devices and environments. In this section we present a simplified model which abstracts away as much of the extraneous physics as is possible while retaining the essential aspects of a robot and environment whose dynamics are intermittently coupled. The vindication of our claim to have retained the essential aspects, of course, is only possible by recourse to physical experiment, and will be provided in Section 4.

The central model of the robot and the environment is developed in Section 2.2.1: it preserves the disparity in degrees of freedom between them but abstracts away most of the dynamical complications introduced by spin, collision, and friction. Section 2.2.2 presents a greatly simplified account of the manner in which collisions between the robot and environment affect their behavior.

2.2.1 A Revolute Robot in a Two Degree of Freedom Environment

Locate a frame of reference, $\mathcal{F}_0$, at the center of the robot shaft, with x-axis perpendicular to the plane, and z-axis defined by the projection of a vertical ray pointed directly into the earth's gravitational field onto the plane, as depicted in Figure 2. Define r so that it measures the angle of the right hand portion of the robot's bar (with the hitting surface — the billiard cushion — facing up) away from the x-axis on the juggling plane.

The configuration space of the entire problem is the cross product, $\mathcal{C} \triangleq \mathcal{B} \times \mathcal{R}$, of the body and the robot configurations. We will represent the location of the falling body on the plane $\mathcal{B} = \mathbb{R}^2$ with the coordinates (b_1, b_2) denoting, respectively, the position of its centroid relative to the "horizontal" (y) and "vertical" (z) axes of the reference frame, $\mathcal{F}_0$. For present purposes, without restriction of generality, we will only consider the right half of the juggling plane, $b_1 > 0$. We will model the robot's configuration space as $\mathcal{R} \triangleq [-\pi/2, \pi/2] \subset \mathbb{R}$, which we restrict to a half revolution where the hitting billiard cushion is facing up on the right side of the juggline plane.

In isolation, the robot's dynamics occur in its phase space, $\mathcal{V} \triangleq T\mathcal{R} \approx \mathcal{R} \times \mathbb{R}$, of angular positions and velocities, and may be modeled simply by the equations

$$\begin{bmatrix} \dot{v}_1 \\ \dot{v}_2 \end{bmatrix} \triangleq \begin{bmatrix} \dot{r} \\ \ddot{r} \end{bmatrix} = \begin{bmatrix} v_2 \\ \frac{\tau}{\rho} \end{bmatrix}, \tag{1}$$

(where τ denotes the commanded torque from the motor control node, ρ, denotes the moment of inertia of the bar) since the PMI motor, with its high bandwidth and power, low shaft friction and inertia deployed in the absence of a transmission comes close to providing a source of "pure torque." [1]

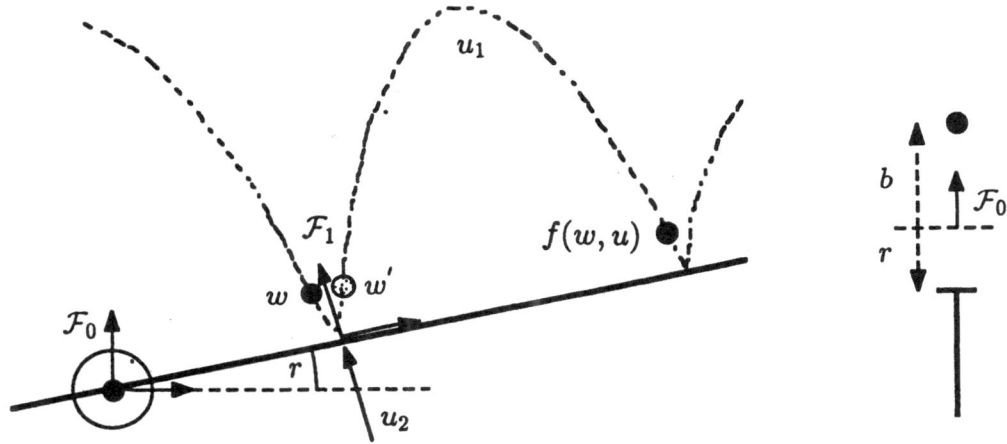

Figure 2: The Impact Event

The absence of gravitational torques on the robot's bar reflects its symmetric distribution of mass about the joint.

In isolation, the puck's dynamics occur in its phase space, $\mathcal{W} \triangleq T\mathcal{B} \approx \mathcal{B} \times \mathbb{R}^2$, and may be modeled simply by the equations

$$\begin{bmatrix} \dot{w}_1 \\ \dot{w}_2 \end{bmatrix} \triangleq \begin{bmatrix} \dot{b} \\ \ddot{b} \end{bmatrix} = \begin{bmatrix} w_2 \\ a \end{bmatrix} \triangleq n(w), \tag{2}$$

(where $a = [0, -\gamma]^{\mathrm{T}}$, and $\gamma = \gamma_{grav} \sin\beta$ denotes the projection of the gravitational constant of acceleration, γ_{grav}, onto the sliding plane inclined away from vertical by the angle β) since we assume that the puck is a point of unit mass sliding on a frictionless surface.

---

[1]Unfortunately, the large mass of this motor mitigates against its role in a multi-jointed direct drive robot.

In fact, this idealized model is overly simplistic since there is noticeable coulomb friction on the sliding plane, and a closer approximation to reality would replace n in (2) by

$$n'(w) \triangleq \begin{bmatrix} w_2 \\ c(w_2) + a \end{bmatrix}; \qquad c(w_2) \triangleq \gamma_{coul} \begin{bmatrix} step(\dot{b}_1) - step(-\dot{b}_1) \\ step(\dot{b}_2) - step(-\dot{b}_2) \end{bmatrix},$$

where *step* denotes the unit step function. One of the objectives of our study is to develop a control procedure which is robust enough to succeed even in the face of such unmodeled dynamics, and we will use only n from (2) in the formal analysis. However, in the sequel, we will find it interesting to compare numerical simulations of the robot control laws in the idealized environment, n, with the same strategies run in the more realistic simulation model, n', as against empirical data.

The set of all possible impact configurations may be described by a smooth surface in the puck-robot configuration space, C, as formalized by the following statement which gives an implicit and explicit representation.

Lemma 2.1 *The set $\mathcal{I} \subset C$ of configurations where the robot is in contact with the puck is a smooth submanifold specified as the zero set, $\iota^{-1}[0]$, of a scalar valued map, $\iota : C \to \mathbb{R}$, which may be globally parametrized by a function of the body configurations, $\tilde{\theta} : \mathcal{B} \to C$.*

Proof: Using the reference frame, $\mathcal{F}_0$, the puck, $(b_1, b_2) \in \mathcal{B}$, contacts the bar, $r \in \mathcal{R}$ if and only if $\tan r = b_2/b_1$. Thus, denoting the "body angle" by,

$$\theta(b_1, b_2) \triangleq atan(b_2/b_1), \tag{3}$$

it is clear that $\mathcal{I}$ is the graph of θ in C. That is to say, $\mathcal{I}$ is simply the image of the map

$$\tilde{\theta}(b) \triangleq \begin{bmatrix} b \\ \theta(b) \end{bmatrix},$$

or, alternatively, the zero level set of the map

$$\iota(b, r) \triangleq r - \theta(b).$$

□

An impact configuration, $(r, b) \in \mathcal{I}$, implicitly defines the robot's "virtual gripper" — the point of contact on the billiard cushion — and it is useful to define a new "virtual gripper frame," $\mathcal{F}_1$ whose origin is in the body's center, b, whose x-axis is parallel to that of $\mathcal{F}_0$, but whose y-axis is aligned with the robot bar, all depicted in Figure 2. The new frame has a representation with respect to the "base frame" given (in two dimensional homogeneous coordinates) by

$$^0\mathcal{F}_1 = \begin{bmatrix} R & b \\ 0^\mathsf{T} & 1 \end{bmatrix}; \qquad R \triangleq \begin{bmatrix} \cos r & -\sin r \\ \sin r & \cos r \end{bmatrix}. \tag{4}$$

2.2.2 The Impact and Flight Models

We now develop a simplified model of the dynamics of repeated puck-robot impacts based upon the following assumptions. First, we assume here that all interactions between ball and robot during impact can be modeled as an instantaneous event: à posteriori velocities are related to à priori velocities via a simple "coefficient of restitution," $\alpha \in (0,1)$ [17]. We have previously considered dynamical interactions in a one degree of freedom problem [3] for which this assumption is invalid: we believe that a similar treatment will be possible in the present more complicated situation, but have not yet addressed the matter carefully. Second, we assume that the robot mass is sufficiently large as to make the puck's mass negligible. We are obviously more interested in the general situation wherein this assumption is invalid, and, moreover, wherein the actual mass of the body is poorly known. Such problems will be considered in future experiments and analysis.

More formally, within the puck-robot phase space, [2]

$$\mathcal{X} \triangleq T\mathcal{C} = \mathcal{W} \times \mathcal{V},$$

consider the set of all possible velocities at each impact configuration,

$$T_{\mathcal{I}}\mathcal{C} \triangleq \bigcup_{(b,r) \in \mathcal{I}} T_{(b,r)}\mathcal{C} = \left\{ \left((b,r), (\dot{b}, \dot{r}) \right) \in \mathcal{X} : (b,r) \in \mathcal{I} \right\}. \qquad (5)$$

The first assumption amounts to the hypothesis of a "collision map," $c : T_{\mathcal{I}}\mathcal{C} \to T_{\mathcal{I}}\mathcal{C}$ which takes an à priori phase at contact, $x \in T_{\mathcal{I}}\mathcal{C}$, into a new phase, $x' = c(x) \in T_{\mathcal{I}}\mathcal{C}$, in the same contact configuration — in coordinates, $(b', r') = (b, r)$. The second assumption may be expressed in coordinates as the hypothesis that $\dot{r}' = \dot{r}$. It is left to describe a map

$$\dot{b}' = c\left((b,r), (\dot{b}, \dot{r}) \right).$$

Consider, first, the simplified case shown in Figure 2. The à posteriori velocity of the body after impact, $\dot{b}'$ is related to the à priori velocity of the body $\dot{b}$, and that of the robot's gripper, $\dot{r}$ as

$$\dot{b}' = -\alpha\dot{b} + (1 + \alpha)\dot{r}. \qquad (6)$$

In the full model the velocity of the revolute robot's virtual gripper is $u_2 \triangleq \|b\| \cdot \dot{r}$ and will be considered in the subsequent section a robot control input. Let us now further assume that the puck's velocity component parallel to the robot bar is unchanged by the impact. Then the à posteriori velocity of the two degree of freedom body after impact, $\dot{b}'$, is related to the à priori velocity of the body, $\dot{b}$, and that of the robot's virtual gripper, in the $\mathcal{F}_1$ coordinates as

$$^1\dot{b}' = \begin{bmatrix} 1 & 0 \\ 0 & -\alpha \end{bmatrix} {}^1\dot{b} + \begin{bmatrix} 0 \\ (1+\alpha) \end{bmatrix} \|b\|\dot{r} = \bar{C}\, {}^1\dot{b} + \bar{c}u_2, \qquad (7)$$

---

[2]Throughout the paper, given a smooth manifold $\mathcal{M}$, we will denote its "tangent bundle" — the union over $\mathcal{M}$ of the tangent vectors over each of its elements — by $T\mathcal{M}$.

which, expressed with respect to $\mathcal{F}_0$, the world frame, is

$$\dot{b}' = C\dot{b} + cu_2 \triangleq \mathbf{c}\left((b,r),(\dot{b},\dot{r})\right). \tag{8}$$

where

$$C \triangleq R\bar{C}R^{\mathbf{T}}; \qquad c \triangleq R\bar{c}.$$

Note that for each fixed configuration, $(b,r) \in \mathcal{I}$, $\mathbf{c}$ is linear in the velocities. [3] Note that R, and, hence, C, c, are all functions on the configuration space. According to Lemma 2.1, $\mathcal{I}$ is an immersion of $\mathcal{B}$: in other words, we may substitute $\theta(b)$ for all instances of r in (8), so that $\mathbf{c}$ is independent of the robot configuration value. It depends only upon b (in a nonlinear fashion) $\dot{b}$ and $\dot{r}$ (in a linear fashion as has been already noted). The independence of $\mathbf{c}$ from r enables us to consider the past and future history of the body at time of collision with no regard to the robot's behavior other than its velocity at the moment of impact.

The forward trajectory of the body is now obtained by integrating its motion in $\mathcal{W}$ starting from the initial conditions, $w = \left(b,\dot{b}\right), \dot{r}$, according to the isolated dynamics, $n(w)$, given in (2),

$$w(t) = \begin{bmatrix} b + \mathbf{c}(b,\dot{b},\dot{r})t + \frac{1}{2}at^2 \\ \mathbf{c}(b,\dot{b},\dot{r}) + at \end{bmatrix}. \tag{9}$$

---

[3]In formal parlance, $\mathbf{c}$ is a *vector bundle morphism* of $T_{\mathcal{I}}\mathcal{C}$.

3 The Environmental Control Problem

We may now investigate the response of the puck to all logically possible impact events by examining the *environmental control system* . This results from considering the effect of repeated puck-robot impacts on the future puck trajectory and is described completely by the discrete impact map (9), assuming arbitrarily assigned values for the robot inputs, $(r, \dot{r}) \in \mathcal{V}$, at each impact event. In other words, we treat the robot as an independent external "agent of control" and consider the various puck behaviors resulting from the robot's "perfect" actions.

The formal control problem is developed in Section 3.1. The particular control task considered in the sequel is stately formally and shown to be logically achievable in Section 3.2. This mode of investigation exhibits some parallels to the seminal work of Schaefer and Cannon [15]. In both cases the problem involves an inherently unstable mechanical system — the stabilization of inverted pendulum setpoints and the stabilization of periodic juggling motions. In both cases, the linearized controllability analysis provides formal verification of the unituitive knowledge that the problems are solvable. There exist basic differences as well. In Schaefer and Cannon's worl, linear control theory is also employed successfully to solve the task. In contrast, in this paper we will show that linear theory does not provide a satisfactory design tool.

3.1 The Impact Schedule

From (9) it is clear that the first time, t_{j+1}, after the jth impact at time t_j, at which the robot and body again make contact, $(b(t_{j+1}), r(t_{j+1})) \in \mathcal{I}$, is a function of the robot's future position trajectory, $r(t_{j+1})$. Moreover, (8) shows that the velocity of the virtual gripper at impact is determined by choice of the robot's velocity at impact, $\dot{r}(t_j)$. In the sequel we will use the term *impact schedule* to denote a sequence of pairs,

$$\{u(t_j)\}_{j=0}^{\infty}; \qquad u(t_j) = \left[\begin{array}{c} u_1(t_j) \\ u_2(t_j) \end{array} \right] \in \mathcal{U} \triangleq \mathbb{R}^2, \tag{10}$$

in the robot's *action set*, $\mathcal{U}$, where $u_2(t_j) = \|b\| \cdot \dot{r}$ denotes the velocity of the virtual gripper at the moment of the jth impact which occurs at time t_j, and $u_1(t_j) \triangleq t_{j+1} - t_j$ denotes the interval of time which elapsed between that impact event and its successor. An impact schedule gives rise to a sequence of puck states measured just before impact,

$$\{w(t_j)\}_{j=0}^{\infty}; \qquad w(t_{j+1}) = f(w(t_j), u(t_j)),$$

where $f : \mathcal{W} \times \mathcal{U} \to \mathcal{W}$, is derived by substituting (8) into (9) to obtain

$$f(w, u) = \left[\begin{array}{c} w_1 + \dot{b}'(w)u_1 + \frac{1}{2}au_1^2 \\ \dot{b}'(w) + au_1 \end{array} \right]. \tag{11}$$

where, recalling from (8),

$$\dot{b}'(w) = C(w_1)w_2 + c(w_1)u_2.$$

This nonlinear discrete dynamical control system constitutes the *environmental control system.*

An *environmental control problem* results from prescribing some desired sequence of puck states, $\{w^*(t_j)\}_{j=0}^{\infty}$, and asking for an impact sequence, $\{u^*(t_j)\}_{j=0}^{\infty}$, which results in asymptotic convergence of $w(t_j)$ to $w^*(t_j)$.

Clearly, any control problem may be solved by a great variety of controller structures. In this paper we shall be solely concerned with solutions via pure feedback compensation. In this particular section, we shall abstract away all physical properties of the robot and presume it to be an "ideal" feedback agent which measures puck states, $w(t_j)$, and delivers control inputs, $u(t_j)$, accordingly. This point of view affords a precise definition of the juggling task as well as the demonstration that the task is at least logically achievable in Proposition 3.1.

3.2   The Vertical One-Juggle

Before proceeding with the theoretical discussion, it seems worthwhile to provide a more intuitive explanation of what it means to stabilize a fixed point of the environmental control system (11). Probably the simplest systematic behavior of this environment imaginable (after the rest position), is a periodic vertical motion of the puck in its plane. Specifically, we want to be able to specify an arbitrary "apex" point, and from arbitrary initial puck conditions, force the puck to attain a periodic trajectory which impacts at zero position and passes through that apex point. This corresponds exactly to the choice of an appropriate fixed point, w^*, of (11).

To see this, consider first the one degree of freedom environment, $\mathcal{B} = \mathbb{R}$ as depicted in Figure 2. Selecting $w^* = \left(b^*, \dot{b}^*\right)$ as the desired constant set point indicates that we want the impact to occur at the position b^* and with the velocity just before impact given by $\dot{b}^*$. If w^* is truly a fixed point of the closed loop dynamics, then the velocity just after impact must be $-\dot{b}^*$, and this "escape velocity" leads to a free flight puck trajectory whose apex occurs at the height $b_{apex} = b^* + \frac{\dot{b}^{*2}}{2\gamma}$, assuming the simple ballistic model of free flight with no friction (2). Thus, a constant w^* "encodes" a periodic puck trajectory which passes forever through a specified apex point, b_{apex}.

In the two degree of freedom environment, $\mathcal{B} = \mathbb{R}^2$, the same notion applies, only in this case, a purely vertical steady state trajectory requires zero horizontal velocity, $\dot{b_1}^* = 0$. Again, a fixed vertical impact velocity, $\dot{b_2}^*$, from a specified impact height, b_2^*, implies a specified apex position. However, since the robot has only one degree of freedom, impacts at a position from which the virtual gripper frame is not oriented identically to the base frame will impart some non-zero horizontal velocity, violating the condition that the steady state trajectory be purely vertical. Thus, in contrast to the one degree of freedom case, it is clear that not every constant set point, w^* is attainable in steady state for the two degree of freedom environment.

We may now restate these observations more formally. A robot feedback strategy is a map, $g : \mathcal{W} \to \mathcal{U}$, from the body's state to the robot's action set, $\mathcal{U}$ (10), resulting in the impact strategy $u(t_j) = g(w(t_j))$. The robot-environment closed loop dynamical system is formed from the composition of f (11) with g,

$$w(t_{j+1}) = f_g(w(t_j)); \qquad f_g(w) \triangleq f(w, g(w)). \tag{12}$$

According to the definition in the previous section, a robot feedback strategy, g, accomplishes the constant set point environmental control problem, $w(t_{j+1}) = w(t_j) = w^*$, if and only if w^* is an asymptotically stable fixed point of f_g. We are now led to ask first which set points, w^*, can be made fixed points of (12) and second, of these, which can be made asymptotically stable by an appropriate choice of g. Given the specific juggling task at hand, we are only interested in a subset of the full puck impact phase space. As mentioned before we are limiting the setpoints to the right hand side $b_1 > 0$ of the juggling plane. Also, it only makes sense to admit negative vertical puck velocities $\dot{b}_2 < 0$ just before impact — i. e. those which which point towards the robot's hitting bar. Thus we define the *working* puck phase space $\tilde{\mathcal{W}} \subset \mathcal{W}$ as

$$\tilde{\mathcal{W}} \triangleq \left\{ w \in \mathcal{W} : b_1 > 0, \ \dot{b}_2 < 0 \right\}.$$

In answer to the first question, our previous analysis reveals that only a subset of points in the task set, $\mathcal{T} \subseteq \tilde{\mathcal{W}}$, may be made fixed points of the closed loop system.

Proposition 3.1 *Given the discrete dynamical control system, (11), and a point, $w^* \in \tilde{\mathcal{W}}$, there exists a robot feedback strategy, $g : \mathcal{W} \to \mathcal{U}$ such that w^* is a fixed point of the closed loop map, f_g (12) if and only if*

(i) $w^* \in \mathcal{T}$, the vertical one-juggle task set, where

$$\mathcal{T} \triangleq \left\{ w \in \tilde{\mathcal{W}} : b_2 = 0, \ \dot{b}_1 = 0 \right\}.$$

(ii) $g(w^*) = u^* \triangleq \begin{bmatrix} -2/\gamma \\ -\frac{1-\alpha}{1+\alpha} \end{bmatrix} \dot{b}_2^*.$

Proof: Consider the fixed point condition

$$w = f(w, g(w)) = \begin{bmatrix} b + \dot{b}' u_1 + \frac{1}{2} a u_1^2 \\ \dot{b}' + a u_1 \end{bmatrix}$$

Elimination of $\dot{b}' = -\frac{1}{2}au_1^*$ directly results in

$$(1) \quad \dot{b_1}^* = 0$$
$$(2) \quad u_1^*(w^*) = -\frac{2}{\gamma}\dot{b_2}^*$$

With these conditions and

$$\dot{b}' = C\dot{b} + cu_2 = -\frac{1}{2}au_1$$

we obtain

$$(3) \quad \dot{b_2}^* = 0$$
$$(4) \quad u_2^*(w^*) = -\frac{1-\alpha}{1+\alpha}\dot{b_2}^*$$

where conditions (1),(3)[(2),(4)] are equivalent to (i)[(ii)].

$\square$

Notice that within the allowable puck phase space, T corresponds exactly to those constant set points which our previous intuitive thinking led us to understand would cause a purely vertical periodic puck trajectory which returns to the same apex point again and again. Thus we are led to call any set point $w^* \in T$ a *vertical one-juggle task*, and say that a robot feedback strategy, g, constitutes a *vertical one-juggle* if such a point is an asymptotically stable fixed point of (12). In answer to the second question, our previous analysis showed that all points in the vertical one-juggle task set may be achieved.

Proposition 3.2 *If*

$$w^* \in T$$

and g fixes w^*, $f_g(w^*) = w^*$, (12) then system (11) is locally controllable at $(w^*, g(w^*))$.

Proof: According to Proposition 3.1, g fixes w^* if and only if $g(w^*) = u^*$. Thus the system is locally controllable if and only if

$$\begin{aligned} A &= [D_w f](w^*, g(w^*)) = [D_w f](w^*, u^*) \\ B &= [D_u f](w^*, g(w^*)) = [D_u f](w^*, u^*) \end{aligned} \tag{13}$$

comprise a completely controllable pair.

Taking partial derivatives of (11) gives

$$A = \begin{bmatrix} 1 & -\frac{4}{\gamma}\frac{\dot{b_2}^{*2}}{\dot{b_1}^*} & -\frac{2}{\gamma}\dot{b_2}^* & 0 \\ 0 & 1 & 0 & \frac{2}{\gamma}\alpha\dot{b_2}^* \\ 0 & 2\frac{\dot{b_2}^*}{\dot{b_1}^*} & 1 & 0 \\ 0 & 0 & 0 & -\alpha \end{bmatrix} ; \quad B = \begin{bmatrix} 0 & 0 \\ \dot{b_2}^* & -\frac{2}{\gamma}(1+\alpha)\dot{b_2}^* \\ 0 & 0 \\ -\gamma & 1+\alpha \end{bmatrix}.$$

It suffices to show that four of the eight columns of the matrix (B, AB, A^2B, A^3B) are linearly independent. The four columns (B, A^2B) we consider are

$$
\begin{bmatrix}
0 & 0 & \frac{4\dot{b}_2^{*3}}{\gamma b_1^{*}}(2\alpha - 3) & \frac{8\dot{b}_2^{*3}}{\gamma^2 b_1^{*}}(1+\alpha)(3-\alpha) \\
\dot{b}_2^{*} & -\frac{2}{\gamma}\dot{b}_2^{*}(1+\alpha) & \dot{b}_2^{*}(2\alpha^2 - 2\alpha + 1) & -\frac{2}{\gamma}\dot{b}_2^{*}(1-\alpha+\alpha^2) \\
0 & 0 & 4\frac{\dot{b}_2^2}{b_1^2}(1-\alpha) & -\frac{4\dot{b}_2^{*2}}{\gamma b_1^{*}}(1+\alpha)(2-\alpha) \\
-\gamma & 1+\alpha & -\alpha^2\gamma & \alpha^2(1+\alpha)
\end{bmatrix}.
$$

The determinant of this matrix,

$$
\frac{16}{\gamma^2}\frac{\dot{b}_2^{*6}}{b_1^{*2}}\alpha(1+\alpha)^2
$$

is nonzero for any $w^* \in \mathcal{T}$.

$\square$

According to linear control theory, if (A, B) is a completely controllable pair then for any desired set of poles whose complex elements appear in conjugate pairs,

$$
\Lambda = \{\lambda_i\}_{i=1}^n \subset \mathbb{C},
$$

there exists a matrix, $K_\Lambda \in \mathbb{R}^{2 \times n}$ such that the closed loop spectrum achieves that set,

$$
spectrum\,(A + BK_\Lambda) = \Lambda,
$$

where n is the dimension of the dynamical system. Now suppose that the feedback algorithm, g, is chosen to be

$$
g(w) \triangleq u^* + K_\Lambda(w - w^*). \tag{14}
$$

Since
$$
\begin{aligned}
[D_w f_g](w^*) &= [D_w f](w^*, g(w^*)) + [D_u f](w^*, g(w^*))\,[D_w g](w^*) \\
&= A + BK_\Lambda,
\end{aligned}
$$

it follows that any K_Λ for which $\Lambda \subset \mathcal{D}^1 \subset \mathbb{C}$ (the open unit disk in the complex plain) yields a feedback law, g, which achieves the vertical one-juggle as defined in Section 3.2.

We have now shown that the vertical one-juggle is logically achievable. This discrete analysis formally confirms the intuition that only state information at impact should be required for a successful juggling algorithm — that full trajectory information is redundant.

However, we have said nothing yet concerning the ability of the robot to realize any particular feedback strategy, g, much less one which stabilizes a desired set point, w^*: it is completely up to the designer to solve the robot control problem and achieve an approximation to the required impact strategy.

4 Robot Implementation: Controller Synthesis in Task Geometry

The preceding analysis employed a geometric representation of the task domain in terms of a discrete dynamical control system on puck velocities over the contact set, $\mathcal{I}$. That analysis permitted a rigorous definition of the task at hand and the logical assurance of its possibility. We now turn our attention to the *robot control problem* — the synthesis of robot control laws that result in impact schedules which accomplish a specified task.

4.1 The Robot Control Problem

The robot control problem can be formulated as follows. Given a desired impact schedule — a solution to the environmental control problem, $\{u^*(t_j)\}_{j=0}^{\infty}$, which results in asymptotic convergence of $w(t_j)$ to $w^*(t_j)$, find an appropriate continuous robot control input $\tau(t)$ that forces the robot with the dynamics

$$\begin{bmatrix} \dot{v}_1 \\ \dot{v}_2 \end{bmatrix} \triangleq \begin{bmatrix} \dot{r} \\ \ddot{r} \end{bmatrix} = \begin{bmatrix} v_2 \\ \frac{\tau}{\rho} \end{bmatrix}, \tag{15}$$

to implement the desired discrete impact sequence.

4.2 Direct Implementation of a Discrete Time Environmental Controller

The controllability analysis in Section 3.2 not only demonstrates that our task is possible: it also furnishes a robot control synthesis procedure based upon standard linear time invariant control theory. It seemed entirely plausible to us that a physical implementation of this straightforward procedure would result in empirical success. Indeed, our appeal to the mathematical formalism of Section 3 was partially motivated by the expectation that controller design could be based upon such well understood theoretical principles. To our surprise, these controllers failed when implemented on the experimental apparatus.

For a variety of task points, $w^* \in \mathcal{T}$, we chose a variety of stable spectra, Λ, in the open unit disk, $\mathcal{D}^1$, determined K_Λ using a numerical procedure, and determined an ideal impact schedule according to the affine feedback law, g, described by (14). We then induced our robot to deliver a close approximation of this impact schedule via a reasonable ad hoc procedure. In no case of this series of experiments did we observe a successful vertical one-juggle. In Section 4.2.1 we describe in detail the implementation procedure. In Section 4.2.2 and 4.2.3 we offer a critique of this procedure and suggest several reasons for this unexpected outcome.

4.2.1 Implementation Procedure

Of course, the key to success or failure of even a practicable feedback scheme stemming from the discrete analysis would depend upon the details of how the robot is commanded to implement the impact schedule required by (14). For, as we have previously described, the analysis of the environmental control problem in Section 3, provides an abstract characterization of what is logically possible assuming that the robot is a perfect "feedback agent." It remains entirely silent concerning the manner in which a particular impact schedule is achieved. Several different implementation procedures were attempted; the best performance seemed to result from the following procedure:

1. Since the feedback algorithm (14) is based upon the states of the puck just before impact, a "start-up" procedure at time $j = 1$ is required. For simplicity, we start from the desired apex point, and assume that the first impact occurs at $\hat{w}_1 = w^*$: the robot is commanded to hit the puck with a velocity obtained from applying the feedback law (14) to the estimated impact state: $u_{j=1} = g(\hat{w}_1)$. Of course, in this case, $g(w^*)$ yields $u_{2,j=1} = u_2^*$ given in Proposition 3.1.

2. Just before the actual first impact, we measure and estimate the true state of the puck before impact, w_1, and evaluate the feedback law, $g(w_1)$ to get a desired time inverval to next hit, $\hat{u}_{1,j=1}$. Now we use the dynamical model (11) to predict $\hat{w}_2$ with u_2 set to the actual measured impact velocity, and u_1 set at the nominal value, $\hat{u}_{1,j=1}$.

3. The desired velocity of the next impact, $u_{2,j=2}$, is determined by applying the feedback law to the predicted next impact point, $g(\hat{w}_2)$. Now $\hat{w}_2$ prescribes the robot's angle, r, at second impact, and, together with $u_{2,j=2}$, prescribes the robot's angular velocity, $\dot{r}$ at second impact.

4. To implement the desired robot states for the second impact, we use a simple open loop fixed torque control strategy which works as follows. Acceleration from a rest position to $\dot{r}$ takes time t_{acc}. Together with r, this yields the robot's rest position. During the puck flight, its states are predicted t_{acc} in advance. When the predicted position crosses r, the robot starts accelerating until the second impact occurs. Measurements made after the fact show that this strategy yields satisfactory accuracy.

5. The procedure continues using the measured estimate of the forthcoming impact, w_j, to obtain a value for $u_{1,j}$ from $g(w_j)$, (14), and a predicted next impact state, $\hat{w}_{j+1}$, from the model (11), to generate a value for $u_{2,j+1}$ from $g(\hat{w}_{j+1})$.

4.2.2 Limitations of Linear Control Theory

The local nature of our task definition in Section 3.2 — asymptotic stability of a fixed point — afforded an easy proof that it may actually be attained, based upon standard arguments from linear control systems theory. However, (11) is a highly nonlinear control system. The determination that a particular equilibrium state is asymptotically stable provides very little help in estimating the domain of attraction and the domain of "containment" around that state. The former is comprised of those initial conditions which tend asymptotically toward the equilibrium state, while the latter is comprised of those initial conditions which are guaranteed to remain in some specified neighborhood of the equilibrium state. Since our juggling plane, in contrast to its analytical model developed in Section 2.2, is not truly infinite, achieving a particular containment region will be crucial to the success of any real juggling strategy, regardless of its domain of attraction.

In fact, numerical simulation showed that the domain of containment of the idealized closed loop discrete dynamical system resulting from (14) was in the order of our sensor resolution and thus unacceptably small. For certain arbitrarily chosen values of w^*, pole assignments, Λ, could be found that resulted in a large domain of attraction. However, we could find no pole assignments for physically realizable settings of w^* which achieved a domain of attraction whose diameter along the $\dot{b}_2$ axis of $\mathcal{W}$ was greater than 6 % of the value $\dot{b}_2{}^*$. In all of these simulations, the trajectories within the domain of attraction leave the physical boundaries of actual juggling plane. For typical settings, the domain of containment within the juggling plane was no larger than 2 % of the desired fixed point magnitude. This is smaller than the error tolerance of the puck position sensing sensing system. We conclude that the linearized analysis of even this simplified discrete nonlinear system is inadequate to the desired task. In formal terms, the local definition of the task in 3.2 may be too weak.

Moreover, since the parameters of the ultimate closed loop system with spectrum Λ are chosen according to a numerical procedure that finds a matrix K_Λ, as a function of the pair (A, B), a controllable but *ill conditioned* pair may result in large departures of the closed loop parameters from their desired values either because the numerical procedure is very sensitive or because of small departures in implementation from the numerically determined value of K_Λ. Thus, pole placement in the face of ill conditioned data is not robust. Indeed, for all values of w^* examined, numerical tests revealed that the pair (A, B), while completely controllable as guaranteed by Proposition 3.2, was poorly conditioned. We conclude that our experimental gain settings were not sufficiently close approximations of K_Λ even to guarantee stability, much less containment.

4.2.3 Limitations of the Discrete Impact Model

As mentioned at the end of Section 3, an initially attractive feature of the discrete analysis is that it confirms the intuition that only state information at impact should be required for a successful juggling algorithm — that full trajectory information is redundant. Conceptual appeal notwithstanding, in reality state information at or very near the impact event is exceedingly difficult to measure. Moreover, just as they necessarily rely upon the accuracy of the impact state measurement, algorithms resulting from the discrete analysis critically depend upon the accuracy of the impact parameters and idealizations of the model — the coefficient of restitution; the validity of the zero friction assumption; a zero puck diameter — because this information is explicitly used to compute the next desired impact states and the next robot inputs. It is necessary, for example, to correct for the error between the next predicted and the actual impact position. How to do that, is left to the designer. These features do not contribute to a practicable scheme.

More fundamentally, as pointed out in the very beginning of the paper, this algorithm only specifies desired robot control inputs to the environment at the moment of impact. It is completely up to the designer to solve the robot control problem, and this leads to the kind of ad hoc implementations to which we resorted as described in Section 4.2.1.

Thus, even had the linearized analysis resulted in a robust feedback law with a physically realizable region of containment we feel that it would fail to provide a satisfactory synthetic framework in the present task domain. We were led to seek, instead, some means of generating successful and provably correct juggling behavior in the continuous time framework of the robot itself.

4.3 A New Algorithm

We now introduce a procedure which meets the desiderata of the previous paragraph. We introduce a new synthesis procedure defined on the entire cross product puck-robot phase space, $\mathcal{W} \times \mathcal{V}$, which represents the continuous physical trajectories of both rather than the discrete time evolution of their mutual impacts. This synthesis procedure gives rise to a family of robot control algorithms which we demonstrate empirically here and prove mathematically (in Section 5) to be correct. For reasons that will soon become apparent, we call this synthesis procedure the "mirror algorithm."

4.3.1 Intuitive Motivation

Intuitively, two different ideas are combined to produce an algorithm that is implemented on the robot by recourse to standard trajectory tracking techniques. First, we "reflect" the desired periodic puck trajectory in $\mathcal{W}$ into a "distorted mirror image" in $\mathcal{V}$: the "distortion" is so designed that the "cross product" trajectory of the puck and robot reflection in $\mathcal{X} = \mathcal{W} \times \mathcal{V}$ intersects the impact set, $T_{\mathcal{I}}\mathcal{C}$, with characteristic tangent vectors, $(\dot{b}, \dot{r})$, whose image under the collision map, c (8), is "favorable" to the task at hand. Second, we borrow from Raibert [14, 3] the idea of modifying the robot's trajectory by "servoing" on the discrepancy between the (constant) total mechanical energy of the puck in its desired steady state, and the currently measured value.

To better convey the nature of the new algorithm, we will first discuss the one degree of freedom case (Figure 2). Let the puck drop from the desired steady state height corresponding to a steady state value of $w^* = (0, \dot{b}^*) \in \mathcal{T} \subseteq \mathcal{W}$. Suppose the robot tracks exactly the "distorted mirror" trajectory of the puck,

$$r = -\kappa_{10}b,$$

where κ_{10} is a constant. Contact between the two can occur only when $(r, b) = (0, 0)$. Therefore, in this case, the contact configurations are limited to a single configuration $\mathcal{I} = (r = 0, b = 0)$. Since $\dot{r} = -\kappa_{10}\dot{b}$, an impact at this configuration occurs at the phase point $s = \big((0,0), (\dot{b}, -\kappa_{10}\dot{b})\big) \in T_{\mathcal{I}}\mathcal{C} \subseteq \mathcal{X}$. Now solving the equation $c(s) = -\dot{b}^{\bullet}$ for κ_{10}, yields a choice of that constant,

$$\kappa_{10} = \frac{1 - \alpha}{1 + \alpha},$$

which ensures a return of the puck to the original height. Thus a properly tuned "distortion constant," κ_{10} will maintain a correct puck trajectory in its proper periodic course — formally, as we will soon show, it achieves the fixed point condition of Proposition 3.1.

In the absence of friction, the desired steady state periodic puck trajectory is completely determined by its total vertical energy,

$$\eta(w) = \frac{1}{2}\dot{b}^2 + \gamma b,$$

in this case,

$$\eta^{\bullet} \triangleq \eta(w^{\bullet}) = \frac{1}{2}\dot{b}^{\bullet 2}.$$

This suggests the addition to the the original mirror trajectory,

$$r = -\kappa(w)b; \qquad \kappa(w) \triangleq \kappa_{10} + \kappa_{11}[\eta^{\bullet} - \eta(w)], \tag{16}$$

of a term which "servos" around the desired steady state energy level. Since we neglect friction during the puck's flight, we may assume that $\dot{\eta} \equiv 0$, hence,

$$\dot{r} = -\kappa(w)\dot{b}.$$

At steady state, $\eta(w) = \eta^{\bullet}$, the fixed point condition is still preserved. However, deviations of η away from $\eta^{\bullet}$ cause proportionately harder or softer robot impacts than the steady state condition requires. It is plausible that these proportionately adjusted deviations will cause convergence toward $\eta^{\bullet}$: we will shortly prove that this is indeed the case.

We now describe the manner in which this idea "scales" to the particular case at hand — the two degree of freedom cartesian environment presented in Section 3.

The basic idea carries over into this environment by just adding linear PD feedback compensation terms for the horizontal component. Define the "puck angle" as

$$\theta(b) \triangleq atan\frac{b_2}{b_1}.$$

Now, as opposed to controlling the robot height as a function of puck height, we control the robot angle r as a function of puck angle θ as shown in (17) where κ_{ij} are fixed scalar gains, and M is a symmetric matrix in $\mathbb{R}^{4 \times 4}$.

The first two terms in κ_2 are borrowed from standard linear feedback control theory, implementing proportional derivative feedback. Analyzing the linearized system at a fixed point with just these two first terms in κ_2 results in an ill conditioned system. The last two terms in this expression were introduced to ensure complete controllability locally without interfering with the global nature of the algorithm.

Thus, implementing a mirror algorithm is an exercise in robot trajectory tracking wherein the reference trajectory is a function of the puck's state.

$$
\begin{aligned}
r_d(t) &= -\kappa_1(w)\theta + \kappa_2(w). \\[6pt]
\kappa_1(w, w^*) &\triangleq \kappa_{10} + \kappa_{11}[\eta(w^*) - \eta(w)] \\
\kappa_2(w, w^*) &\triangleq \kappa_{21}(b_1 - b_1^*) + \kappa_{22}(\dot{b}_1 - \dot{b}_1^{\,*}) + \\[6pt]
&\quad \frac{1}{2}\frac{(w-w^*)^{\mathrm{T}} M(w-w^*)}{[\kappa_{31}+((w-w^*)^{\mathrm{T}} M(w-w^*))^2]^2} + \frac{\kappa_{41}(\dot{b}_2-\dot{b}_2^{\,*})}{[\kappa_{42}+(b_2-b_2^{\,*})^2]^2},
\end{aligned}
\tag{17}
$$

4.3.2 Data Quality

Before discussing the plots, some remarks about the quality of the experimental data are called for. The inductive puck position sensor as described in Section 2.1, is accurate to about 0.5 inch, due to a systematic measurment error resulting from the nonlinear distribution of the puck's generated flux across the measurment loops as well as measurment noise in the electromagnetically sensitive inductive grid. The latter is exacerbated during operation due to noise emitted by the high power motor, most of which fortunately can be filtered out using a standard observer. This explains the quality of the data in Figure 3. For the impact data, two further detrimental effects occur. First, the impact states are somewhat difficult to obtain accurately: during the short but not truly infinitesimal impact event, the puck-robot system is moving, making it difficult to decide, when to take the measurment *during* the time of contact; in contrast, the velocity data must be measured *just before* the impact. Second, the solid metal·robot distorts the magnetic field of the puck at time of impact. Of course all of these effects can be compensated for in a very time consuming approach. A simple satisfactory solution to compensate for the errors introduced at impact is to use a very slow filter for estimating the puck states and arguing that the dynamics of these effects is much faster, resulting in an improved measurment. Of course it has to be checked that the slow filter has converged to the true puck states well before the impact. Concluding this paragraph we can say that first, the actual puck trajectory during flight (Figure 3) is obviously smooth, and second, that the steady state error band around the desired impact states is actually smaller than depicted on the Figures 4 to 7.

We now present plots of simulated and experimental data in order to validate our simplified model used for analysis and to illustrate the utility of our analytical results for both the idealized model as well as the real system. In Section 5.2 we describe how some gain settings in (17) are derived readily from our analysis.

4.3.3 Validation of the Simplified Model

Figure 3, a continuous "recording" of a vertical one-juggle nicely depicts the rapid convergence typical for initial conditions (in drop-off position) from any region within the puck's workspace not too close to the origin — a kinematic singularity. Due to the nonzero puckradius and robot juggling bar dimensions, the vertical impact position b_2^*, the sum of the two, is nonzero as well, as can be seen in this plot. Despite departures from the idealized model and the relatively large sensor noise discussed in the previous section, it may be observed from this and the subsequent plots that our algorithm produces steady reliable juggling performance. We have recorded vertical one-juggle runs with hundreds of impacts without encountering any failures.

Next, we present data plots of the puck states just before impact. Each experimental data curve displays statistical information (mean plus/minus one standard deviation) obtained from 20 successive runs (without handpicking). We feel that this presentation offers a closer rendering of true performance than one based upon a handpicked best run. Such presentation methods are standard practice in other natural sciences, but seem only slowly to be entering the field of robotics [13].

Figures 4 to 7 compare the responses of the analytical model with and without friction to the responses of our experimental setup (with friction) for two different initial conditions. These two initial conditions for the puck impact states result from dropping the puck from two extremes of the right juggling half plane, the upper left (initial condition one) and the lower right corner (initial condition two). The steady state values in the horizontal position, b_1, are very close around the desired value for all curves. The plots of the vertical impact velocity, $\dot{b}_2$, demonstrate, first, as we expected, that the effect of the unmodeled friction is a steady state deviation, which second, is rather accurately predicted by the model. Examining the transients, notice that the experimental transient responses for $\dot{b}_2$ (lower plots) consistently match the responses of the model with friction, as expected. However, for b_1 (upper plots) the experimental transient responses are closer to the much faster transient model responses *without* friction than to those of the model with friction. This favorable discrepancy is not completely understood at present. We suspect that the unmodeled effect of spin at impact might be responsible for this benign discrepancy.

4.3.4 Confirming the Limited Utility of Linear Analysis

Recall that the last two terms in (17) were introduced to arbitrarily specify the local behavior — that is, to place the poles of the linearized system around the fixed point.

For the experiments reported here, they were placed at $[0, \pm 0.1j]$ for fast convergence and very close to the unit circle at $[0.9. \; 0.9 \pm 0.1j]$ for slow convergence. Figures 4 to 7 show a side by side comparison of the "fast" system with the "slow" system for the two initial conditions. The experimental data confirms our limited ability to affect the local behavior. There is no clear improvement in the steady state error of the fast system. Our ability to place the local poles is demonstrated more drastically by the simulated data in Figure 8: We place the poles outside the unit circle and verify the unstable local dynamics. The same plot also reveals the strong nonlinear (global) properties of the mirror algorithm: the injected instable behavior is bounded to a small neighborhood around the fixed point. However, while we can verify control over local transients via simulations, the experimentally observed changes are rather small when taking into account the large difference in pole settings. For this system the interesting dynamic behavior seems to result from its global properties and therefore the local linear behavior is only of limited usefulness. Further remarks about the utility of the linear analysis can be found at the beginning of Section 5.

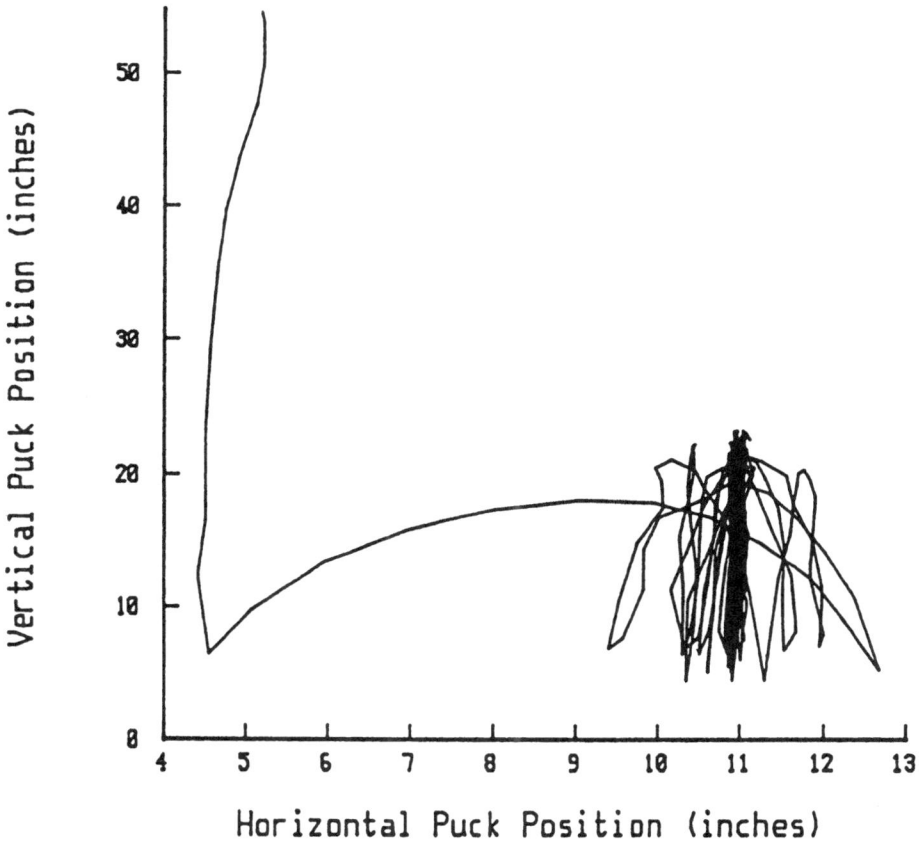

Figure 3: Sample continuous data

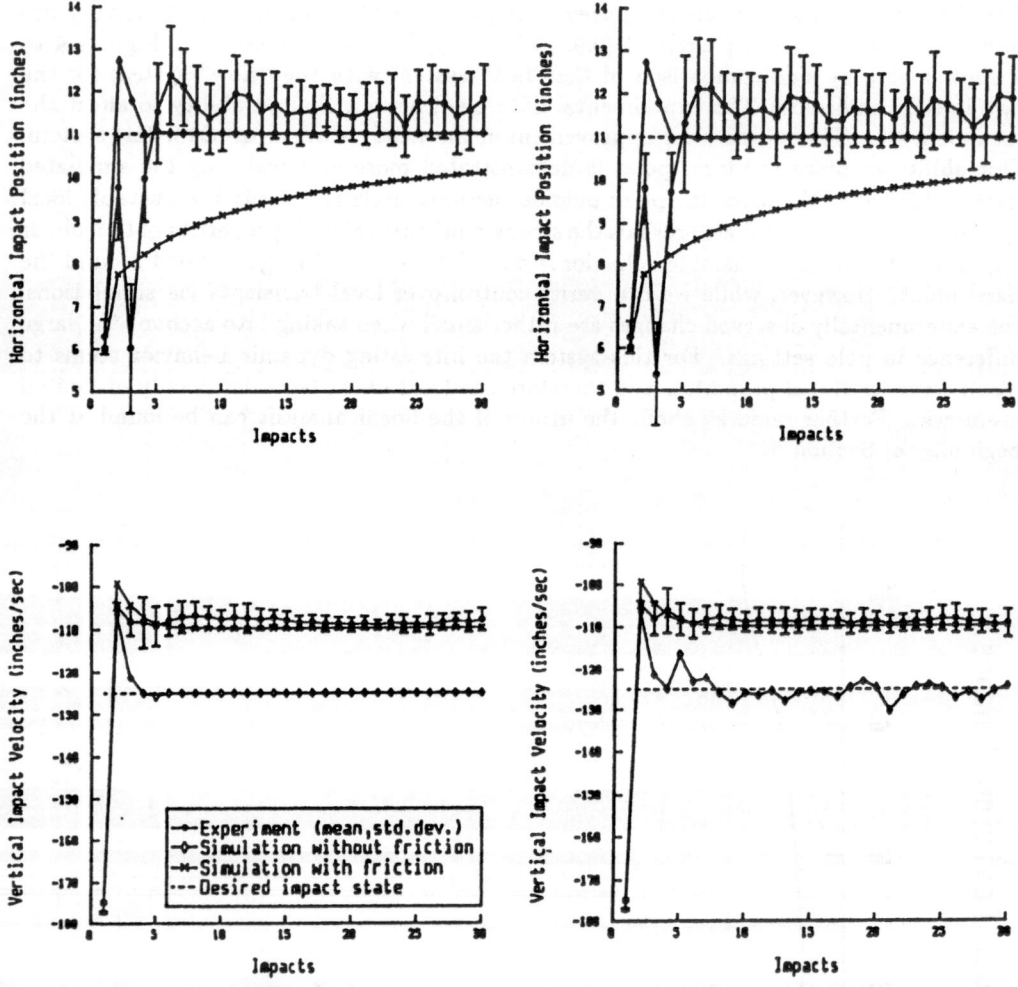

Figure 4: Fast local transient setting, Initial condition one.

Figure 5: Slow local transient setting, Initial condition one.

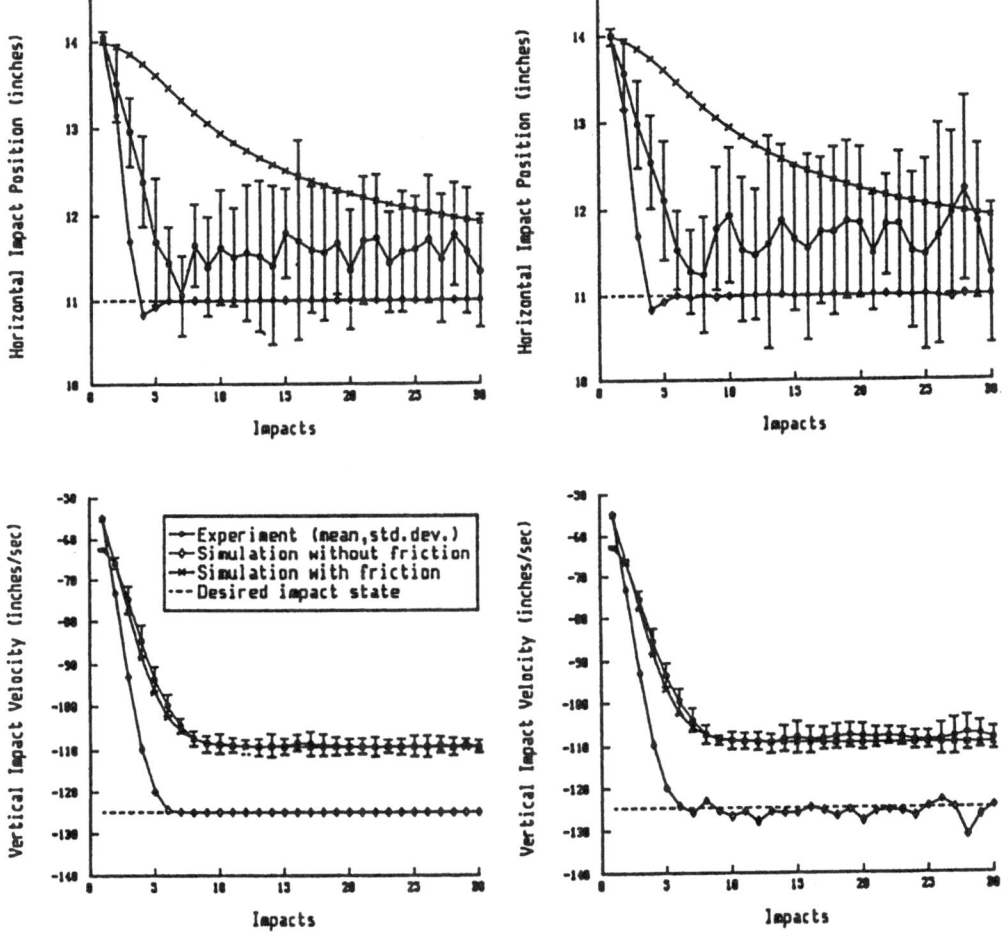

Figure 6: Fast local transient setting, Initial condition two.

Figure 7: Slow local transient setting, Initial condition two.

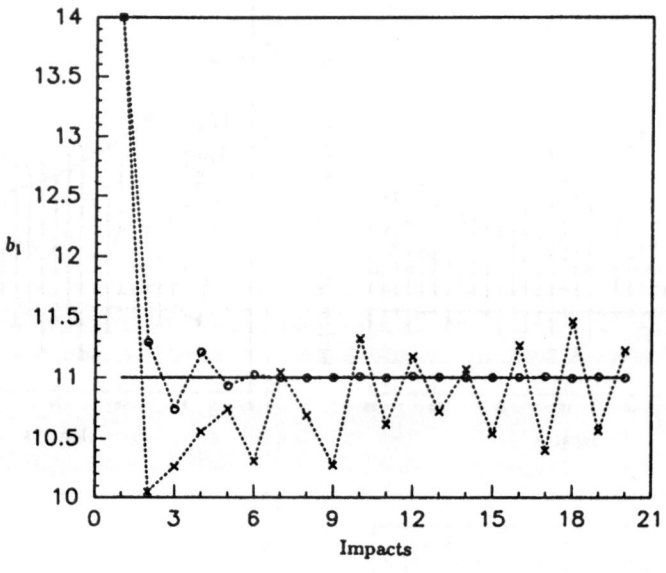

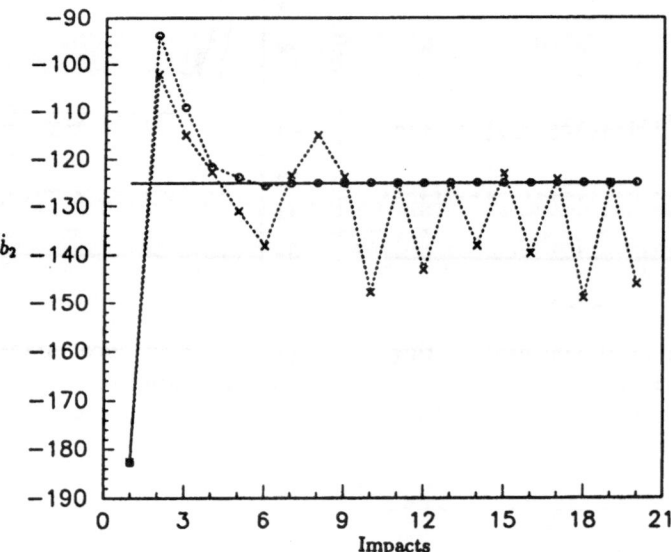

o   Local Stability: Poles set to $[-0.1, \ 0, \ 0.1]$
x   Local Instability: Poles set to $[2.0, \ 0, \ 0.1]$

Figure 8: Simulation of a locally unstable, but globally stable system.

5 Analysis of the Mirror Algorithm: Controller Synthesis in Phase Space

In this section we provide a formal demonstration that the mirror algorithm succeeds in producing a correct vertical one-juggle: a locally asymptotically stable fixed point for any desired position on the task plane, $\mathcal{T}$. In Section 5.1 we demonstrate that the mirror law achieves a correct vertical one-juggle. In Section 5.2 we provide an example of how to "place poles" of the invariant submanifold restriction.

Before proceeding, some remarks about the usefulness of our analysis are called for. For the one degree of freedom system we can readily examine the nonlinear discrete impact map and derive the region of attraction [7]. However, for the planar system, the nonlinear analysis of the impact map is intractable. When confronted with such nonlinear systems, we first have to investigate the behavior of the linearized system around the fixed point in Section 5.1. This is an important first step and often provides a sufficiently accurate description of the system around a large enough region of the fixed point to be useful. In our case, even though we can command the poles of the linearized system using the results of the linear analysis, we cannot significantly affect the physically observable behavior of the system. This is because the influence of the algorithmic terms which affect the local behavior is so small that any changes are dwarfed by the error band of steady state performance in our specific implementation. Nonetheless, the linear analysis is valid and useful for investigating global behavior. For example, we started with a globally and locally stable one juggle algorithm. After detuning several of the gains k_{ij} to destabilize the system, investigating the poles of the local linear system exhibited instable poles which confirms our expectation that the global behavior is reflected in the local linear characteristics.

5.1 Local Stabilization of an Invariant Submanifold

The local proof is divided in two parts; Section 5.1.1 establishes that the algorithm induces a three dimensional invariant submanifold of the effective closed loop environmental dynamics in Proposition 5.2. In Section 5.1.2 we provide relatively unrestrictive sufficient conditions under which the poles of the lower dimensional discrete control system — the restriction of (11) to this submanifold — may be placed by adjusting the mirror algorithm gains in Proposition 5.6. Finally, in Section 5.2 we choose gain settings of the mirror algorithm which meet the requirements of Proposition 5.6, and show how the pole locations of the data in Section 4.3 are obtained.

In Section 4.3 we introduced the mirror algorithm as an exercise in robot trajectory tracking where the reference trajectory is a function of the puck's state. We now assume that the robot has achieved exact tracking via standard linear PD theory applied to (1), and interpret the consequences in terms of invariant surfaces in the puck-robot phase space, $\mathcal{X} \triangleq \mathcal{W} \times \mathcal{R}$, introduced in Section 2.2.2. Of course, in practice, one obtains merely asymptotically exact tracking: we will suppose that the robot has "enough time" in between impacts to achieve a sufficiently close approximation to this asymptotic state at each successive impact.

5.1.1 The Induced Invariant Submanifold

We define a new surface, the "mirror surface," $\mathcal{M} \subset \mathcal{X}$ to be the zero set, $\mathcal{M} \triangleq \tilde{\mu}^{-1}[0]$, of a new scalar valued function,

$$\tilde{\mu} : \mathcal{X} \to \mathbb{R} : (r, \dot{r}, w) \mapsto r - \mu(w) \tag{18}$$

where

$$\mu : \mathcal{W} \to \mathbb{R} : w \mapsto \kappa_1(w) \cdot \theta - \kappa_2(w)$$

θ is the "puck angle," and κ_i are the gain functions detailed in (17). The robot implementation strategy is to control the robot's state, $v = (r, \dot{r})$ so that the coupled robot-environment state of $\mathcal{X}$ remains within the mirror set, $\mathcal{M}$, for all time. As described above, we now assume that this objective is attained. By studying the intersection of $\mathcal{M} \subset \mathcal{X}$ with the impact velocities $T_{\mathcal{I}}\mathcal{C} \subset \mathcal{X}$, we may now determine the effective environmental control law that the robot realizes in so doing.

Lemma 5.1 *Suppose the puck-robot phase space trajectory, $x(t_j)$, remains in the mirror surface, $\mathcal{M} \subset \mathcal{X}$, defined in (18). Then the resulting impact schedule defines an environmental feedback law, $g : \mathcal{W} \to \mathcal{U}$ which realizes the fixed point condition of Proposition 3.1 for any $w^* \in \mathcal{T}$ if and only if the gain settings (17) satisfy*

$$\kappa_{10} = \frac{1 - \alpha}{1 + \alpha} - \gamma \frac{\kappa_{41}}{\kappa_{42}^2} \frac{b_1^*}{\dot{b}_2^*}$$

Proof: Assuming that the robot has achieved the mirror surface we have

$$(r, \dot{r}) = (\mu, \dot{\mu})$$

when $(b, r) \in \mathcal{I}$. Therefore the robot bar velocity at a point of impact is

$$u_2 = g_2(w) = \|b\|\dot{r} = \|b\|\dot{\mu} = \|b\| \cdot d\mu(w) \cdot n(w).$$

Evaluating this expression at $w^* = [b_1^*, \quad 0, \quad 0, \quad \dot{b}_2^*]'$ with the above κ_{10} yields the desired robot control velocity at impact,

$$u_2^* = -\frac{1 - \alpha}{1 + \alpha} \dot{b}_2^*.$$

Using (8), the resulting velocity vector after impact is

$$\dot{b}^{*'} = \bar{C}\dot{b}^* + \bar{c}u_2^* = \begin{bmatrix} 0 \\ -\dot{b}_2^* \end{bmatrix}.$$

If the robot remains on $\mathcal{M}$, the next impact will occur again at w^*. Integrate the motion of the puck to compute the time of flight, as desired as,

$$u_1^* = -\frac{2}{\gamma}\dot{b}_2^*.$$

□

Proposition 5.2 *If the puck-robot phase trajectory, $x(t_j)$ remains in $\mathcal{M}$ then the resulting impact sequence realizes an environmental feedback control law, $g : \mathcal{W} \to \mathcal{U}$ whose closed loop dynamics, f_g, in $\mathcal{W}$ possesses an invariant co-dimension one set, $\mathcal{S} \subset \mathcal{W}$, that is*

$$f_g(\mathcal{S}) \subset \mathcal{S},$$

which is smooth in some open neighborhood of the fixed point, w^*.

Proof: Observe that the intersection of the mirror surface, $\mathcal{M}$, with the tangents over the impact surface, $T_{\mathcal{I}}\mathcal{C}$, is given by the cross product,

$$\mathcal{M} \cap T_{\mathcal{I}}\mathcal{C} = \sigma^{-1}[0] \times \mathcal{R},$$

of the robot phase space, $\mathcal{R}$, with the zero set of the smooth map

$$\sigma : \mathcal{W} \to \mathbb{R} : w \mapsto \theta - \mu.$$

For if $x \in T_{\mathcal{I}}\mathcal{C}$ then

$$x = \big((b, r), (\dot{b}, \dot{r})\big) \text{ and } r = \theta(b),$$

while if $x \in \mathcal{M}$ then

$$x = \big((b, r), (\dot{b}, \dot{r})\big) \text{ and } r = \mu(b, \dot{b}),$$

and it follows that $v = (r, \dot{r})$ is unconstrained, while $w = (b, \dot{b})$ satisfies

$$\theta(b) = \bar{\mu}(b, \dot{b}).$$

The result now follows when we define $\mathcal{S} \triangleq \sigma^{-1}[0]$. For the differential,

$$\begin{aligned} d\sigma(w) &= d\theta - d\mu \\ &= \left[-\frac{b_2}{b_1^2 + b_2^2}, \frac{b_1}{b_1^2 + b_2^2}, 0, 0 \right] \end{aligned}$$

$$-\left[\frac{\kappa_1(w)b_2}{b_1^2 + b_2^2} + \kappa_{21}, \quad \kappa_{11}\gamma atan\frac{b_2}{b_1} - \frac{\kappa_1(w)b_1}{b_1^2 + b_2^2}, \quad \kappa_{22}, \quad \kappa_{11}\dot{b}_2 atan\frac{b_2}{b_1}\right]$$

$$-\frac{\kappa_{31} - 3\cdot[(w - w^*)^{\mathsf{T}}M(w - w^*)]^2}{[\kappa_{31} + [(w - w^*)^{\mathsf{T}}M(w - w^*)]^2]^3}\cdot(w - w^*)^{\mathsf{T}}M$$

$$-\kappa_{41}\cdot\frac{\kappa_{42} - 3(\dot{b}_2 - \dot{b_2}^*)^2}{[\kappa_{42} + (\dot{b}_2 - \dot{b_2}^*)^2]^3}\cdot[0, \quad 0, \quad 0, \quad 1]$$

is non-zero at any fixed point, $w^* \in \mathcal{T}$,

$$d\sigma(w^*) \;=\; d\theta(w^*) - d\mu(w^*) = \left[0, \quad -\frac{1}{b_1^*}, \quad 0, \quad 0\right] - \left[\kappa_{21}, \quad -\frac{\kappa_{10}}{b_1^*}, \quad \kappa_{22}, \quad \frac{\kappa_{41}}{\kappa_{42}^2}\right]$$

$$= \left[-\kappa_{21}, \quad \frac{2}{(1 + \alpha)b_1^*}, \quad -\kappa_{22}, \quad -\frac{\kappa_{41}}{\kappa_{42}^2}\right] \tag{19}$$

and by continuity, also non-zero in some open neighborhood, $\mathcal{O} \subset \mathcal{W}$. Thus $\mathcal{S} \cap \mathcal{O}$ is a smooth co-dimension one submanifold of $\mathcal{W}$ according to the Implicit Function Theorem [16]. Moreover, under the assumption that the puck-robot phase remains in $\mathcal{M}$, all impact events must occur with puck phase $w \in \mathcal{S}$. Thus, whatever environmental control law, g is induced, it has the property that $w \in \mathcal{S}$ implies $f_g(w) \in \mathcal{S}$.

□

An immediate consequence of this result is that the puck dynamics may be studied in three dimensional space rather than $\mathcal{W} = \mathbb{R}^4$. For present purposes, since we are merely interested in asymptotic stability of the fixed point, it suffices to consider the linearized system on an open subset in $\mathbb{R}^3$. Since w^* is a regular point of σ, it follows from the implicit function theorem that an open neighborhood of that point (in the induced topology of $\mathcal{S} \subset \mathcal{W}$) is parametrized by some smooth function, $p : \widehat{\mathcal{W}} \to \mathcal{S}$, where $\widehat{\mathcal{W}}$ is some open neighborhood around the origin of $\mathbb{R}^3$. Let $p^{-1} : p(\widehat{\mathcal{W}}) \to \widehat{\mathcal{W}}$ denote the local coordinate chart corresponding to p, and $\hat{w}^* \triangleq p^{-1}(w^*)$ denote the local coordinate representation of the fixed point, w^*.

Corollary 5.3 *Assume that the puck-robot trajectory remains on the mirror surface. Define the arrays,*

$$P \triangleq [Dp](\hat{w}^*); \qquad P^\dagger \triangleq \left[Dp^{-1}\right](w^*),$$

and recall from Proposition 3.2 that

$$A \triangleq [D_w f](w^*, u^*); \qquad B \triangleq [D_u f](w^*, u^*); \qquad K \triangleq [D_w g](w^*).$$

The induced discrete closed loop puck dynamics, f_g, on $\mathcal{W}$, achieves the vertical-one juggle if and only if the 3×3 array,

$$P^\dagger [A + BK] P,$$

has eigenvalues inside the unit circle.

Proof: According to the assumption, the puck trajectory w (discrete sequence of puck states at impact) remains in S. Thus, a successful vertical one-juggle obtains if and only if the restriction of f_g to S is locally asymptotically stable at w^*. The latter may be expressed in $\hat{\mathcal{W}}$ coordinates as

$$\hat{f}_g \triangleq p^{-1} \circ f_g \circ p,$$

hence, trajectories, w_j near w^* in S are attracted to w^* if and only if

$$\left[D_{\hat{w}} \hat{f}_g \right] (\hat{w}^*)$$

has eigenvalues inside the unit circle. The desired result follows by applying the chain rule.

$\square$

5.1.2 Pole Placement on the Invariant Submanifold

The linearized feedback gains at the fixed point, w^*, may now be obtained as follows. The second row of

$$\hat{K} = KP = \left[\begin{array}{c} \kappa_1^{\mathsf{T}} \\ \kappa_2^{\mathsf{T}} \end{array} \right] P,$$

is immediately available.

Lemma 5.4 *For $w^* \in T$ the mirror algorithm induces the local linearized feedback gains*

$$\hat{\kappa}_2 \triangleq P^{\mathsf{T}}(w^*) \left[\left[\begin{array}{cccc} 0 & \dot{b}_2^* & 0 & -\gamma \\ 0 & 0 & 0 & 0 \\ b_1^* & 0 & 0 & 0 \\ 0 & b_1^* & 0 & 0 \end{array} \right] d\mu(w^*)^{\mathsf{T}} + \|b_1^*\| \cdot \left[D_w^2 \mu \right] (w^*) \left[\begin{array}{c} 0 \\ \dot{b}_2^* \\ 0 \\ -\gamma \end{array} \right] \right].$$

Proof: Recall that the velocity of the robot bar at point of impact with the puck is given by

$$u_2 = \|w_1\| \dot{r}$$

Since the impact must occur on S, we have, $\dot{r} = \dot{\mu}$, hence,

$$u_2 = g_2(w) = \|w_1\| \cdot d\mu(w) \cdot n(w),$$

where n is the puck dynamical law (2). The result obtains from elementary calculus, since

$$k_2 \triangleq dg_2(w^*)^{\mathsf{T}}$$

and thus

$$
\begin{aligned}
\hat{k}_2 &= P^\mathsf{T} k_2 \\
&= P^\mathsf{T} \left(d\mu \cdot n \frac{1}{\|w_1^*\|} \begin{bmatrix} w_1^* \\ 0 \end{bmatrix} + \|w_1^*\| \cdot \begin{bmatrix} 0 & 0 \\ I & 0 \end{bmatrix} d\mu^\mathsf{T} \right. \\
&\qquad \left. + \|w_1^*\| \cdot \left[D_w^2 \mu \right] (w^*) n \right) \\
&= P^\mathsf{T} \left(\frac{1}{\|w_1^*\|} \begin{bmatrix} w_1^* w_2^{*\mathsf{T}} & w_1^* a^\mathsf{T} \\ w_1^{*\mathsf{T}} w_1^* \cdot I & 0 \end{bmatrix} d\mu^\mathsf{T} + \|w_1^*\| \cdot \left[D_w^2 \mu \right] (w^*) \begin{bmatrix} w_2^* \\ a \end{bmatrix} \right).
\end{aligned}
$$

Substituting

$$
w^* = \begin{bmatrix} w_1^* \\ w_2^* \end{bmatrix} = \begin{bmatrix} b_1^* \\ 0 \\ 0 \\ \dot{b}_2^* \end{bmatrix}
$$

yields the desired result.

□

Since $g_1(w)$ may be impossible to obtain in analytical closed form (depending upon the nature of μ), we must compute its jacobian implicitly.

Lemma 5.5 *Denote the canonical basis of* $\mathbb{R}^2$ *by*

$$
\{e_1, e_2\} = \left\{ \begin{bmatrix} 1 \\ 0 \end{bmatrix}, \begin{bmatrix} 0 \\ 1 \end{bmatrix} \right\},
$$

and

$$
s^\mathsf{T} \triangleq d\sigma(w^*).
$$

If

$$
s^\mathsf{T} B e_1 \neq 0,
$$

then the mirror algorithm induces the local closed loop feedback gains in the restricted discrete dynamical system, $\hat{f}_g$, *given by*

$$
\hat{K} = \begin{bmatrix} \hat{\kappa}_1^\mathsf{T} \\ \hat{\kappa}_2^\mathsf{T} \end{bmatrix} = -\frac{1}{s^\mathsf{T} B e_1} \left(\begin{bmatrix} s^\mathsf{T} A P \\ 0 \end{bmatrix} + \begin{bmatrix} s^\mathsf{T} B e_2 \\ -s^\mathsf{T} B e_1 \end{bmatrix} \hat{k}_2^\mathsf{T} \right),
$$

where $\hat{k}_2$ *is given in Lemma 5.4.*

Proof: All impact states w under the mirror law (17) will occur on S which is parametrized by $w = p(\hat{w})$ and thus $\sigma \circ p \equiv 0$. As S is an invariant submanifold of the closed loop environmental dynamics,

$$
\sigma \circ f_g \circ p \equiv 0,
$$

we have

$$
\begin{aligned}
0 &= [D_w\,(\sigma \circ f_g \circ p)]\,(\hat{w}^*) \\
&= [D_w\sigma]\,(w^*)\,(A + BK)\,[D_{\hat{w}}p]\,(\hat{w}^*) \\
&= s^{\mathsf{T}}\left(A + Be_2 k_2^{\mathsf{T}} + Be_1 k_1^{\mathsf{T}}\right)P,
\end{aligned}
$$

Under the hypothesis, it follows that

$$
\hat{\kappa}_1^{\mathsf{T}} = \kappa_1^{\mathsf{T}} P = -\frac{1}{s^{\mathsf{T}} Be_1}\left(s^{\mathsf{T}} AP + s^{\mathsf{T}} Be_2 \hat{\kappa}_2^{\mathsf{T}}\right),
$$

and the result follows after substitution for $\hat{\kappa}_1$ and rearrangement of terms in $\hat{K}$.

□

Proposition 5.6 *Let $\|w_1^*\| \neq 0$ and $s^{\mathsf{T}} Be_1 \neq 0$. Define the arrays*

$$
\bar{A} \triangleq \left(I - \frac{Be_1 s^{\mathsf{T}}}{s^{\mathsf{T}} Be_1}\right)A; \qquad l \triangleq \left[\begin{array}{c} -\frac{s^{\mathsf{T}} Be_2}{s^{\mathsf{T}} Be_1} \\ 1 \end{array}\right].
$$

If $(P^\dagger \bar{A} P, P^\dagger Bl)$ is a completely controllable pair then the local transient behavior of $\hat{f}_g$ may be arbitrarily determined by proper choice of the mirror algorithm gains, (17).

Proof: Using the results of Lemma 5.4 and Lemma 5.5 it suffices to demonstrate that $\hat{k}_2$ may be arbitrarily specified without changing $s, P, P^\dagger$, all of which also depend upon μ. But notice that the latter are determined by $\mu^* \triangleq \mu(w^*)$ and $d\mu^* \triangleq d\mu(w^*)$, while the former depends upon its hessian, $\bar{M} \triangleq [D_w^2 \mu]\,(w^*)$, as well:

$$
\hat{k}_2 = P^\dagger\left(h(\mu^*, d\mu^*) + \|w_1^*\| \cdot \bar{M} n(w^*)\right). \tag{20}
$$

Taking the hessian of (17) we obtain

$$
\bar{M} = M_1 + \frac{1}{k_{31}^2} M, \tag{21}
$$

where $M_1 \triangleq -[D_w^2 k_1]\,(w^*)$, $M_1^{\mathsf{T}} = M_1$. Notice that M is an arbitrary symmetric matrix introduced in (17) — which does not appear in μ^* or $d\mu^*$. To achieve any desired value, $\hat{k}_2^*$, we need merely solve the linear equations

$$
\frac{1}{\|w_1^*\|}\left(\hat{k}_2^* - P^\dagger h(\mu^*, d\mu^*)\right) = P^\dagger \bar{M} n(w^*), \tag{22}
$$

for the symmetric matrix, $M \in \mathrm{sym}(4)$. Denote the left hand side of this equation by the symbol $\bar{h}$, an arbitrary $\mathbb{R}^{3\times 1}$ array. Now, using the kronecker product and stack notation [2], this system of equations is equivalent to

$$
\bar{h} = \left[n(w^*)^{\mathsf{T}} \otimes P^\dagger\right] \bar{M}^{\mathsf{S}},
$$

where $\bar{M}^{\text{s}}$ — an array in $\mathbb{R}^{16 \times 1}$ — is in the image of $\Pi_{\text{sym}(4)}$ — the linear isomorphism from $\mathbb{R}^{10}$ to sym(4) — the symmetric matrices of $\mathbb{R}^{4 \times 4}$. Thus, the equation may be rewritten as

$$\bar{h} = \left[n(w^*)^{\text{T}} \otimes P^\dagger \right] \Pi_{\text{sym}(4)} \tilde{m},$$

$\tilde{m} \in \mathbb{R}^{10}$. With $n(w^*) = [0 \ \ b_2^* \ \ 0 \ \ -\gamma]^{\text{T}}$ the expression $(n^{\text{T}} \otimes P^\dagger)$ has no rows in skew(4). Thus the array

$$\left[n(w^*)^{\text{T}} \otimes P^\dagger \right] \Pi_{\text{sym}(4)}$$

has rank 3. For, skew(4) — the set skew symmetric matrices on $\mathbb{R}^4$ — is the orthogonal complement of sym(4) (both considered as linear subspaces of $\mathbb{R}^{4 \times 4}$). Thus, any $\hat{k}_2$ may be achieved by proper choice of M without changing $s, P, P^\dagger$.

According to Corollary 5.3, the local behavior of $\hat{f}_g$ is determined by $P^\dagger A P + P^\dagger B \hat{K}$. Substituting for $\hat{K}$ as given in Lemma 5.5 yields

$$P^\dagger A P + P^\dagger B \hat{K} = P^\dagger \bar{A} P + P^\dagger B l \hat{k}_2^{\text{T}}.$$

Thus, since $\hat{k}_2$ may be freely assigned, if $(P^\dagger \bar{A} P, P^\dagger B l)$ is completely controllable, it follows that the poles of the jacobian of $\hat{f}_g$ may be placed arbitrarily.

□

5.2 Example: Choosing the Pole Locations for a Particular Run

In Section 4.3 we presented data for two different runs which we distinguished by the labels "slow" and "fast" local transient settings. In this section we use the theory developed above to show how to place poles of the local reduced dimensional discrete control system, and, in particular report the gain settings required to obtain the pole locations corresponding to those data.

5.2.1 Selection of Gains for Reduced Dimensional Controllability

In order to examine the controllability of our system according to Proposition 5.6 we must ensure that $(P^{\text{T}} \bar{A} P, P^{\text{T}} B l)$ is a completely controllable pair.

While the local analysis of section 5.1 gives us detailed instructions on how to choose the symmetric matrix M for arbitrary poleplacement, the rest of the gains — except κ_{10} — have to be determined experimentally. They were selected as follows:

$$\kappa_{11} = 3 \cdot 10^{-5}; \ \ \kappa_{21} = 7 \cdot 10^{-3}; \ \ \kappa_{22} = 5 \cdot 10^{-3}; \ \ \kappa_{31} = 100; \ \ \kappa_{41} = 0.01; \ \ \kappa_{42} = 10. \quad (23)$$

With an inclination of the juggling plane of 30 degrees off the vertical, the effective gravitational constant is $\gamma = 334.36 \frac{in}{sec^2}$ and the dry friction coefficient is $\mu_{fric} = 0.16$. The desired one juggle was chosen to be

$$w^* = [11in, \quad 0, \quad 0, \quad -125\frac{in}{sec^2}]^{\mathsf{T}}$$

The coefficient of restitution was experimentally found to be $\alpha = 0.7$. Now, from Lemma 5.1 we determine $\kappa_{10} = 0.1765$.

A and B are as given in Section 3.2,

$$A = \begin{bmatrix} 1 & -\frac{4}{\gamma}\frac{\dot{b}_2^{*2}}{\dot{b}_1^*} & -\frac{2}{\gamma}\dot{b}_2^* & 0 \\ 0 & 1 & 0 & \frac{2}{\gamma}\alpha\dot{b}_2^* \\ 0 & 2\frac{\dot{b}_2^*}{\dot{b}_1^*} & 1 & 0 \\ 0 & 0 & 0 & -\alpha \end{bmatrix}; \quad B = \begin{bmatrix} 0 & 0 \\ \dot{b}_2^* & -\frac{2}{\gamma}(1+\alpha)\dot{b}_2^* \\ 0 & 0 \\ -\gamma & 1+\alpha \end{bmatrix}.$$

With w^* as given as above the first and with $s^{\mathsf{T}} \equiv d\sigma(w^*)$ as given in (19), the second condition in Proposition 5.6 is satisfied.

In order to construct a matrix P notice that the condition

$$\sigma \circ p \equiv 0,$$

holds for all $\hat{w} = p^{-1}(w)$ by definition of the impact surface $\mathcal{S}$. Therefore, on $\mathcal{S}$, it must also be that

$$d\sigma Dp = s^{\mathsf{T}}P = 0.$$

We now construct a matrix P, whose transpose has the kernel s^{T}, satisfying the previous equation:

$$P \triangleq [(J \otimes I)s; \quad (L \otimes J)s; \quad (K \otimes J)s] \cdot \frac{1}{\|s\|},$$

where

$$I \triangleq \begin{bmatrix} 1 & 0 \\ 0 & 1 \end{bmatrix}; \quad J \triangleq \begin{bmatrix} 0 & -1 \\ 1 & 0 \end{bmatrix}; \quad K \triangleq \begin{bmatrix} 0 & 1 \\ 1 & 0 \end{bmatrix}; \quad L \triangleq \begin{bmatrix} 1 & 0 \\ 0 & -1 \end{bmatrix}.$$

This particular choice of P has the additional property that $P^{\mathsf{T}}P = I_{3\times 3}$, thus we have

$$P^{\dagger} = P^{\mathsf{T}}.$$

Finally we can compute the controllability matrix of the pair $(P^{\dagger}\bar{A}P, P^{\dagger}Bl)$. For the above settings of all parameters, variables and system constants its rank is 3 and its condition number is 0.001.

5.2.2 Selection of Gains for Reduced Dimensional Pole Placement

For the experimental and simulation data presented in Section 4.3 we selected two pole locations: a setting for slow local transients,

$$\hat{\Lambda}_{slow} \triangleq \{0.9, \quad 0.9 \pm 0.1j\}$$

and a setting for fast local transients,

$$\hat{\Lambda}_{fast} \triangleq \{0, \quad \pm 0.1j\}.$$

Given any desired pole locations for the linearized system, a standard numerical pole placement routine will now give us the desired $\hat{\kappa}_2^*$. Now we use (20) to compute the necessary hessian $\bar{M}$ and (21) to obtain the gain matrix M for (17).

By chosing

$$\bar{M} = \begin{bmatrix} 0 & \bar{m}_1 & 0 & 0 \\ \bar{m}_1 & \bar{m}_2 & 0 & 0 \\ 0 & 0 & 0 & \bar{m}_3 \\ 0 & 0 & \bar{m}_3 & \bar{m}_4 \end{bmatrix},$$

we can rewrite the right side of (22)

$$P^\dagger \bar{M} n(w^*) = P^\dagger \bar{M} \begin{bmatrix} w_2^* \\ a \end{bmatrix} = P^\dagger \begin{bmatrix} \dot{b}_2^* & 0 & 0 & 0 \\ 0 & \dot{b}_2^* & 0 & 0 \\ 0 & 0 & -\gamma & 0 \\ 0 & 0 & 0 & -\gamma \end{bmatrix} \begin{bmatrix} \bar{m}_1 \\ \bar{m}_2 \\ \bar{m}_3 \\ \bar{m}_4 \end{bmatrix} \triangleq Q\bar{m}.$$

As $rank(Q) = 3$, we can find a solution for $\bar{m}$:

$$\bar{m} = Q^{\mathrm{T}}(QQ^{\mathrm{T}})^{-1} \frac{1}{\|w_1^*\|} (\hat{k}_2^* - P^\dagger h).$$

With $\bar{M}$ available, the yet unassigned gain matrix M is determined by (21):

$$M_{slow} = \begin{bmatrix} 0 & 559.39 & 0 & 0 \\ 559.39 & 20.55 & 0 & 3.4 \\ 0 & 0 & 0 & 68.65 \\ 0 & 3.4 & 68.65 & 2.98 \end{bmatrix}, \quad M_{fast} = \begin{bmatrix} 0 & 25.95 & 0 & 0 \\ 25.95 & -15.55 & 0 & 3.4 \\ 0 & 0 & 0 & 1.2 \\ 0 & 3.4 & 1.2 & -1.32 \end{bmatrix}.$$

Now we have completely specified a " mirror algorithm " which is garanteed to implement a vertical one-juggle according to our definition.

Conclusion

We have shown how the geometry of the contact configurations, $\mathcal{I}$, leads to a discrete dynamical model of the effect of robot impact strategies upon the behavior of an otherwise free falling puck. This model provides a framework for rigorously defining dexterous robotic tasks — for example, the "vertical one juggle" — and determining their feasibility. Although prescriptions for explicit impact strategies may be extracted from this model as well, it does not seem to offer an empirically viable framework for synthesis of robot control laws.

After many failed attempts to implement a logically correct but physically under-constrained and non-robust algorithm extracted from the discrete dynamics arising out of this "contact geometry," we were led to a new type of control algorithm based on a completely different "mirror geometry" inhabiting the continuous phase space of the robot-environment pair. Experiments attest to the effectiveness of this control design. Moreover, analysis of the intersection between the mirror surface and the impact surface results in a correctness proof with respect to the discrete dynamical "environmental control system" that formally defines the task.

The central notion of robot controller synthesis via a "mirror geometry" in phase space appears to generalize to other interesting robotic tasks in this domain. For example, we have extended it to the task of catching falling objects and have applied it successfully to the task of juggling two pucks simultaneously as well. Retrospective correctness proofs notwithstanding, the generation of algorithm geometry is completely heuristic at present: each synthesis is empirically hand-tailored to fit the given task. Nevertheless, the ana-lytical tractability of the resulting robot-environment closed loop as demonstrated here raises the hope that sufficient understanding may soon be realized to afford automatic translation of suitably expressed task definitions into provably correct and empirically valid robot controller designs.

In the longer term, we believe these ideas will have still wider application. For ex-ample, analytical techniques similar to those employed here result in correctness proofs for (simplified versions of) Raibert's empirically verified legged locomotion algorithms [3]. Our juggler and Raibert's hopper "settle down" to a characteristic steady state pattern because that pattern is an attracting periodic orbit of the closed loop robot-environment dynamics. Very likely, similar "natural" control mechanisms would make good candi-dates for gait regulation and other more complex tasks requiring controlled intermittent collisions with a dynamical environment.

References

[1] E. Aboaf, S. Drucker, and C. Atkeson. Task-level robot learning: juggling a tennis ball more accurately. In *Proc. IEEE International Conference on Robotics and Automation*, pages 1290–1295, Scottsdale, AZ, May 1989.

[2] R. Bellman. *Introduction to Matrix Analysis*. McGraw Hill, New York, 1965.

[3] M. Bühler and D. E. Koditschek. Analysis of a simplified hopping robot. In *IEEE International Conference on Robotics and Automation*, pages 817–819, Philadelphia, PA, Apr 1988.

[4] M. Bühler and D. E. Koditschek. A prelude to juggling. In 26th *IEEE Conference on Decision and Control*, pages (paper available from authors — not in proceedings), Los Angeles, CA, Dec 1987.

[5] M. Bühler, D. E. Koditschek, and P. J. Kindlmann. A family of robot control strategies for intermittent dynamical environments. In *IEEE International Conference on Robotics and Automation*, pages 1296–1301, Scottsdale, AZ, May 1989.

[6] M. Bühler, D. E. Koditschek, and P. J. Kindlmann. A one degree of freedom juggler in a two degree of freedom environment. In *Proc. IEEE Conference on Intelligent Systems and Robots*, pages 91–97, Tokyo, Japan, Oct 1988.

[7] M. Bühler, D. E. Koditschek, and P. J. Kindlmann. Planning and Control of Robotic Juggling Tasks. In H. Miura, editor, *Fifth International Symposium on Robotics Research*, page (to appear), MIT Press, Tokyo, Japan, 1989.

[8] P. Collet and J. P. Eckmann. *Iterated Maps on the Interval as Dynamical Systems*. Birkhäuser, Boston, 1980.

[9] D. E. Koditschek and M. Bühler. Analysis of a simplified hopping robot. *The International Journal of Robotics Research*, (to appear).

[10] F. Levin, M. Bühler, and D. E. Koditschek. The Yale Real-Time Distributed Control Node. In *Second Annual Workshop on Parallel Computing*, Oregon State University, Portland, OR, Apr 1988.

[11] T. McGeer. *Passive Bipedal Running*. Technical Report IS-TR-89-02, Simon Fraser University, Centre for Systems Science, Apr 1989.

[12] T. McGeer. Powered flight, child's play, silly wheels and walking machines. In *Proc. IEEE International Conference on Robotics and Automation*, pages 1592–1597, May 1989.

[13] T. McGeer. *Stability and Control of Two-Dimensional Biped Walking*. Technical Report IS-TR-88-01, Simon Fraser University, Centre for Systems Science, Sep 1988.

[14] Marc H. Raibert. *Legged Robots That Balance*. MIT Press, Cambridge, MA, 1986.

[15] J. F. Schaefer and R. H. Cannon Jr. On the control of unstable mechanical systems. In *IFAC*, pages 6c.1–6c.13, London, Jun 1966.

[16] I. M. Singer and J. A. Thorpe. *Lecture Notes on Elementary Topology and Geometry*. Springer-Verlag, NY, 1967.

[17] J. L. Synge and B. A. Griffith. *Principles of Mechanics*. McGraw Hill, London, 1959.

[18] Yu Wang. *Dynamic Analysis and Simulation of Mechanical Systems with Intermittent Constraints*. PhD thesis, Carnegie-Mellon University, 1989.

Passive Dynamic Running

Clay M. Thompson

Marc H. Raibert

Artificial Intelligence Laboratory
Massachusetts Institute of Technology
Cambridge, MA

Abstract

Previous work has considered how springy legs can improve the efficiency of the vertical motions of running, making them into resonant spring-mass oscillations that recycle energy from one step to the next. This paper considers how springy hips can be used to improve the efficiency of the legs' fore and aft swinging motions in running. We have studied a passive hopping machine model, composed of links, masses, and springs, but with no actuators. By tuning the mechanical parameters of the system and choosing appropriate initial conditions, we find reentrant trajectories for the system that coordinate the vertical body motions with the leg sweeping motions, and that accommodate ground interaction constraints. Data are presented from a computer simulation of the model.

1 Introduction

Running is a motion that combines a vertical oscillation of the body with a fore-aft oscillation of the legs. Previous work by us and by others has considered how elastic energy storage can be used to generate vertical motion of the body, without requiring a large expenditure of energy on each step. The body can bounce on springy legs during the stance phase, storing a portion of the kinetic energy as strain in the leg springs, and releasing it later to help power the next step. This approach is appealing because it offers energetically efficient vertical motions of the body, and contributes to simplified control. This approach is used by some legged robots (Raibert 1986) and by many animals (Alexander 1988).

In this paper we consider how elastic energy storage might also be used to generate the fore-aft oscillations of the legs, without large energy expenditure. The goal is to avoid losing the kinetic energy of the legs each time they reverse their fore-aft sweeping motion. The leg's kinetic energy increases with the square of running speed, so these losses are particularly severe at high running speeds. Our approach is to turn the legs into harmonic oscillators which approximate the motions needed for running. The legs can be made into harmonic oscillators by introducing torsional springs at the hip joints. The resulting leg oscillations move the foot backward with respect to the hip during the stance phase, and forward in preparation for the next step during the flight phase.

The objective of this study was to see if we could design a simple passive system that moved its legs with suitable trajectories for running. We implemented computer simulations of a planar one-legged model composed entirely of springs, masses, and linkages. The model is shown in Figure 1. We manipulated the running trajectories by tuning the natural frequency of the vertical bouncing motion to be a specific fraction of the natural frequency of the leg swinging oscillation, and by choosing initial conditions according to the running speed. We manipulated the parameters until phase plots of the variables indicated behavior that repeated on itself, one

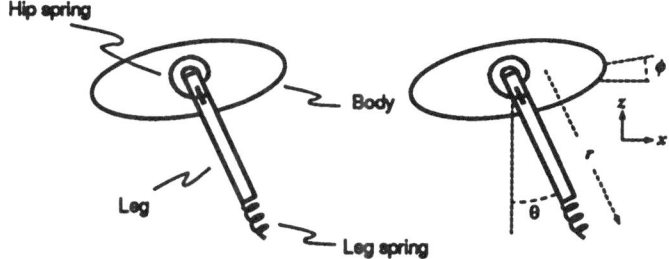

Figure 1: Diagram of planar one-legged model used in simulations.

step after another. The observed running trajectories had a high degree of reentrance, with nearly no energy losses.

The systems we consider are passive in that they are made up of springs, links, and masses, with no actuators or other sources of external energy. In doing these studies we do not suggest that a physical legged system can operate passively for sustained periods of time. A source of energy is needed to make up for mechanical losses, some of which are unavoidable, and a source of control is needed to maintain the reentrant running trajectory. Once the passive part of the system is understood, it should be possible to introduce actuators and algorithms that provide energy and control. We expect physical legged systems that use this approach to have a *tuned gait*, for which the energy efficiency will be highest. At other gaits the system will perform with reduced efficiency, depending on how far the gait deviates from the tuned gait.

2 Background

Means for providing efficient fore-aft motion of the legs have been considered in previous work. Mochon and McMahon (1980, 1981) modeled the human leg as a compound pendulum. They showed that the behavior of the leg during the swing phase of human walking could be accomplished as a passive ballistic motion requiring no energy other than that delivered through forward motion of the hip. McGeer (1989a) built a nearly passive walking machine that used ballistic swing motions not too different from those modeled by Machon and McMahon. His machine had two single-link pendulum legs and it used gravity to sustain the walking motion. More recently, McGeer (1989b) analyzed a two-legged passive dynamic running model and gave conditions for reentrant behavior and stability.

Ivan Sutherland discussed a *tuning fork* model of locomotion in 1983 (Sutherland 1983). He noticed that the motions of the tines of a tuning fork were somewhat like the leg motions used by animals during walking and running.

Alexander (1988) has studied the broad question of how springs are used in animal locomotion. He has proposed that the aponeurosis, a sheet of tendon found in the backs of some quadrupeds, might act like a tension spring when the back is bent, with the vertebral column acting as a compression spring. These springs could reverse the direction of the legs during the gathered phase of galloping. He estimates that about half of the internal energy could be stored in the aponeurosis and vertebral column of a fast galloping deer (Alexander *et al.* 1985).

3 Models

To study passive dynamic running we used a computer simulation of a planar one-legged model. The model is shown in figure 1. The model has a body of mass m_b and moment of inertia J_b, measured about the hip, and a leg of mass m_ℓ and moment of inertia J_ℓ, also measured about the hip. The model has two springs. The hip spring acts between the leg and the body, exerting torque about the hip axis. The hip spring has stiffness k_h. A leg spring acts along the leg axis, between the lower part of the leg and the support surface. The leg spring is massless, it exerts force only during the stance phase, and has stiffness k_ℓ.

There is a third spring that acts tangent to the leg axis. This spring, stiffness k_t, represents the combined lateral compliance of the foot and the ground. The constitutive relations for the leg and tangent springs determine the forces applied to the foot during contact. The equations of motion that describe the system are:

$$\ddot{x}(m_b + m_l) = F_x \tag{1}$$

$$\ddot{z}(m_b + m_l) = F_z - (m_b + m_l)g \tag{2}$$

$$\ddot{\phi}J_b = k_h(\theta - \phi) \tag{3}$$

$$\ddot{\theta}J_\ell = r(F_x \cos\theta + F_z \sin\theta) - k_h(\theta - \phi) \tag{4}$$

where $[x\ z\ \phi]$ are the position and orientation of the body in the plane, and $[F_x\ F_z]$ is the ground force acting on the foot.

A variable-step Runge-Kutta routine was used to integrate the equations of motion to obtain behavior as a function of time. For each simulation, we chose initial conditions and adjusted parameters to get the desired reentrant behavior. In the following paragraphs we describe the behavior of the model when it runs, and the methods we used for choosing the parameters and initial conditions needed to obtain reentrant passive dynamic running.

Symbol	Description	Nominal Value
g	acceleration of gravity	$9.81\,\mathrm{m/sec^2}$
m_b	body mass	10.0 kg
J_b	body moment of inertia	$2.5\,\mathrm{kg \cdot m^2}$
r	leg length	
r_0	leg rest length	0.7 m
m_ℓ	leg mass	1.0 kg
J_ℓ	leg moment of inertia	$0.25\,\mathrm{kg \cdot m^2}$
k_h	hip spring constant	
k_ℓ	leg spring constant	
k_t	leg tangent spring constant	$k_\ell/10$
x	forward position of hip	
z	vertical position of hip	
θ_b	body pitch angle w.r.t horizontal	
θ_ℓ	leg angle w.r.t. vertical	
$\dot{z}_{l_0}$	vertical liftoff velocity	
F_x	horizontal ground force	
F_z	vertical ground force	

Table 1: Parameters of One-Legged Passive Model

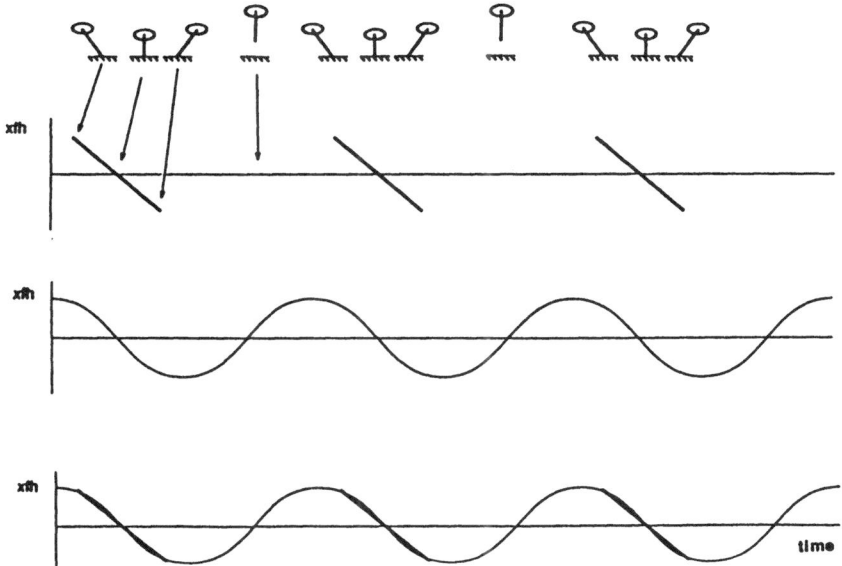

Figure 2: Harmonic oscillation of the hip moves foot approximately as desired for constant speed forward running. A) Plot of horizontal position of the foot with respect to the hip during the stance phase, for constant speed forward travel. B) Plot of horizontal position of foot with respect to the hip for harmonic hip oscillation, assuming a fixed leg length. C) Comparison of the two curves shows that harmonic hip motion would provide a good approximation to constant speed forward travel.

Vertical Bouncing

During flight, the center of mass of the system travels along a parabolic trajectory determined by the vertical position and velocity at liftoff

$$z(t) = z_{lo} + \dot{z}_{lo}t - \frac{gt^2}{2} \tag{5}$$

where z_{lo}, $\dot{z}_{lo}$ are the vertical position and velocity of the body at liftoff, and g is the acceleration of gravity. The peak altitude is

$$z_{max} = \frac{\dot{z}_{lo}^2}{2g}. \tag{6}$$

The duration of the flight phase is

$$T_f = \sqrt{\frac{8z_{max}}{g}} = \frac{2\dot{z}_{lo}}{g}. \tag{7}$$

During stance, the vertical motion is a harmonic rebound determined by the system mass bouncing on the leg spring. The natural frequency of this rebound is

$$\omega_\ell = \sqrt{\frac{k_\ell}{m_b + m_\ell}}. \tag{8}$$

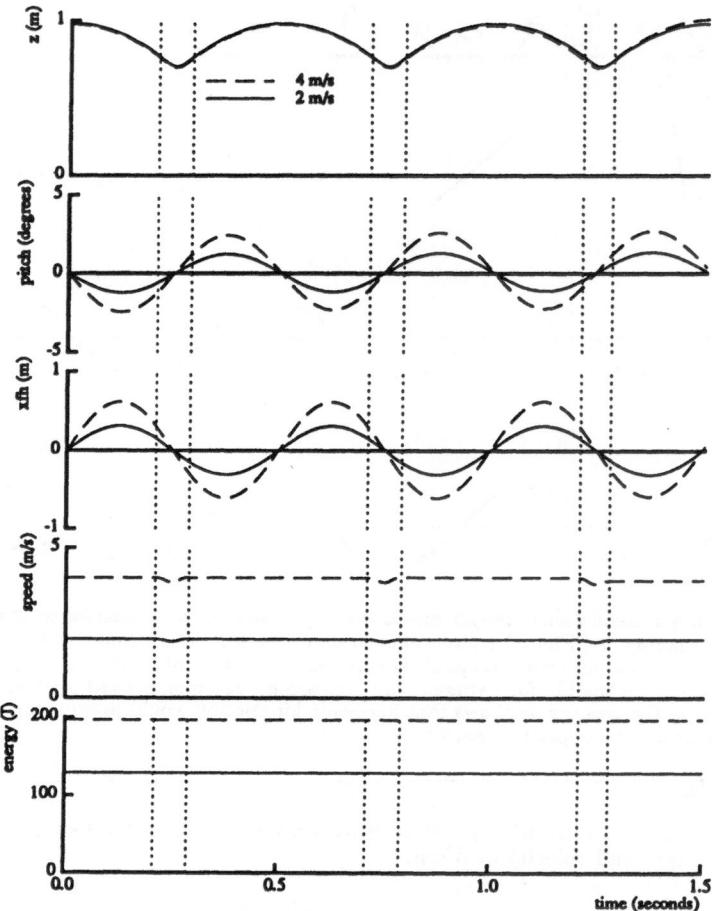

Figure 3: Passive dynamic running of planar one-legged hopper. Data are from computer simulations of model shown in figure 1a and described in table 1. Data are shown for running speeds of 2 and 4 m/s. Vertical dotted lines bracket the stance phase. Solid) $\dot{x} = 2\,\text{m/s}$, $k_h = 35.9\,\text{kg}\cdot\text{m}$, $k_\ell = 29,750\,\text{kg/m}$, $z_{max} = 0.98\,\text{m}$, $\dot{\theta}_0 = 2.67\,\text{rad/sec}$. Dashed) $\dot{x} = 4\,\text{m/s}$, $k_h = 35.9\,\text{kg}\cdot\text{m}$, $k_\ell = 25,500\,\text{kg/m}$, $z_{max} = 0.97\,\text{m}$, $\dot{\theta}_0 = 5.33\,\text{rad/sec}$.

From McMahon and Cheng (1989) we know that the duration of the stance phase in vertical hopping is

$$T_s = \frac{2(\pi - \arctan(|\dot{z}_{to}|\omega_\ell/g))}{\omega_\ell} \tag{9}$$

In this paper we approximate the stance phase as one half cycle of the natural oscillation:

$$T_s \approx \frac{\pi}{\omega_\ell}. \tag{10}$$

This approximation is valid when the vertical velocity at touch down is large compared to g/ω_ℓ, or when the ratio of flight duration to stance duration is greater than 1. This analysis of the vertical motion is strictly valid only for hopping in place, without forward travel and sweeping motions of the legs. Closed form solutions for the stance duration of non-vertical hopping are not known (McMahon and Cheng, 1989).

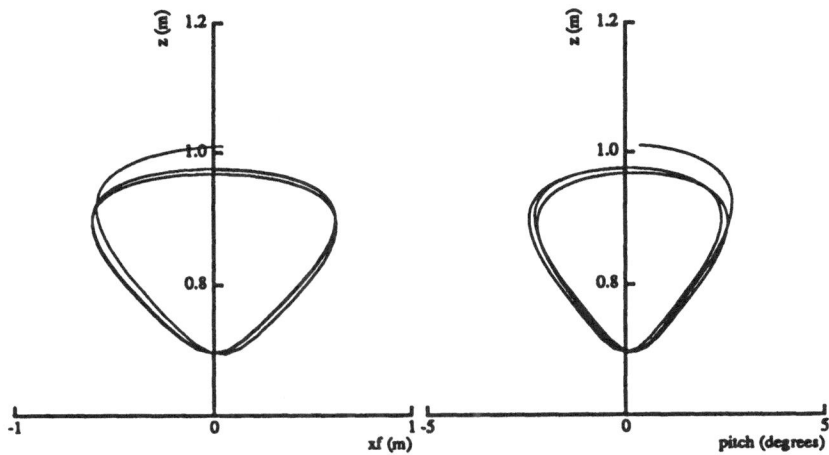

Figure 4: Data from figure 3 (4 m/s) replotted as one variable against another. If the behavior were perfectly reentrant, the plotted trajectories would perfectly superimpose. The outlying branches at the top of each plot indicate that after about three running cycles the behavior began to diverge from the reentrant trajectory. Both plots are for the same run of the one-legged model. $\dot{x} = 4\,\text{m/s}$, $k_h = 35.9\,\text{kg}\cdot\text{m}$, $k_\ell = 25,500\,\text{kg/m}$, $z_{max} = 0.97\,\text{m}$, $\dot{\theta}_0 = 5.33\,\text{rad/sec}$.

Hip Oscillation

The natural frequency of the hip oscillation is given by

$$\omega_h = \sqrt{\frac{k_h}{J_{eff}}}, \qquad (11)$$

where k_h is the hip spring constant and $J_{eff} = J_b J_\ell/(J_b + J_\ell)$ is the effective moment of inertia of the combined leg and body about the hip. The characteristic period of the hip oscillation is

$$T_h = \frac{2\pi}{\omega_h}. \qquad (12)$$

The horizontal displacement of the foot from the hip is

$$x_f = r\sin(\theta_{max}\sin(\omega_h t)) \qquad (13)$$

where r is leg length and θ_{max} is the amplitude of the hip oscillation. We have assumed that the hip spring is at rest when $\theta = \phi = 0$. This function, plotted in figure 2, is approximately linear in time for small values of time and θ_{max}.

4 Choosing Parameters for Passive Dynamic Running

Figure 2 plots foot motion for an ideal legged system traveling forward at constant speed. During the stance phase, the foot does not move with respect to the ground, so the velocity of the foot with respect to the body is the negative of the body's forward velocity. Therefore, for forward travel at constant speed, foot position with respect to the body is a linear function of time. During the flight phase, the primary constraint on foot motion is that the foot be moved forward in time for the next stance phase.

Our basic approach to finding passive reentrant running trajectories is based on the fact that harmonic hip motion can generate foot motion that closely approximate the foot motion found in constant speed forward travel. Figure 2 shows a foot motion produced by a harmonic hip oscillation that was tuned to approximate the foot motion used in ideal constant speed travel. The approximation is based on the linearity of the *sin* function for small values of its argument.

The remainder of this section describes how we choose system parameters and initial conditions to find reentrant trajectories for the model. We assumed fixed values of body mass m_b, body moment of inertia J_b, leg mass m_ℓ, leg moment of inertia J_ℓ, and nominal leg length r_0. Given a desired running speed $\dot{z}_d$ and step period T_{step}, we choose the spring constant for the hip k_h, spring constant for the leg k_ℓ, initial leg pitch rate $\dot{\theta}_0$, initial body pitch rate $\dot{\phi}_0$, and initial body altitude z_{max}.

Hip spring constant k_h

The stiffness of the hip spring is chosen so the hip undergoes one complete oscillation during one complete step. The natural frequency of the hip oscillation is

$$\omega_h = \frac{2\pi}{T_{step}} = \sqrt{\frac{k_h}{J_{eff}}}. \tag{14}$$

From (14) we see that

$$k_h = (2\pi/T_{step})^2 J_{eff} \tag{15}$$

where T_{step} must be specified.

Leg spring constant k_ℓ

The stiffness of the leg spring is chosen to establish the duration of the stance phase as a fraction of the stride period. We define a duty factor ρ, which expresses the duration of the stance phase as a fraction of the step period

$$\rho = T_s/T_{step} = \frac{\omega_h}{2\omega_\ell}, \tag{16}$$

assuming the stance phase is one half cycle of leg spring oscillation. We choose a value for the duty factor to ensure that the stance phase occurs during the roughly linear portion of the foot's fore-aft travel. Small values of ρ give the best linearity, but result in larger peak leg forces and longer flight durations. We experimented with values of ρ between 0.125 and 0.35. A good value for ρ is 0.125, but the smaller the better.

Given the natural frequency of the hip and a value of ρ, the leg spring constant k_ℓ can be found

$$k_\ell = m(\frac{\pi}{\rho T_{step}})^2. \tag{17}$$

Because we do not have an exact expression for T_s, (17) gives a value for k_ℓ that only approximates the desired value for ρ. The desired value of ρ is obtained by adjusting k_ℓ iteratively on a series of trials.

Initial values for leg angular rate $\dot{\theta}_0$ and body angular rate $\dot{\phi}_0$

The amplitude of the hip oscillation and the speed of the foot as it moves back and forth are both determined by the initial value of the leg angular rotation rate, $\dot{\theta}_0$. This parameter is selected so the speed of the foot moving backward during the stance phase matches the desired forward speed of the body.

The simulation is begun by dropping the model from a specified height. Therefore, the initial state of the system is equivalent to the state at mid flight. The angular leg rate at mid flight equals the angular rate at mid stance, with a sign reversal. The angular leg rate is chosen so the backward foot velocity at mid stance is matched to the desired forward speed

$$\dot{\theta}_0 = \dot{z}_d/r. \tag{18}$$

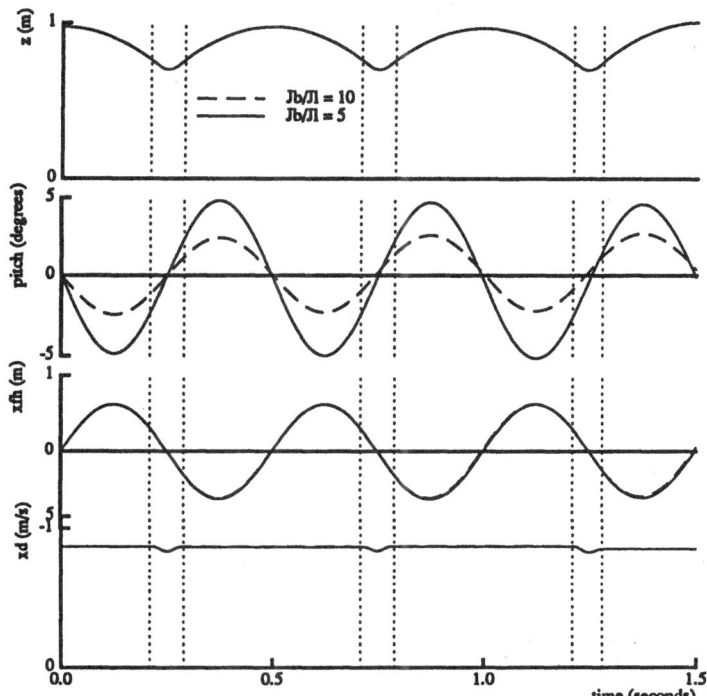

Figure 5: The ratio of body moment of inertia to leg moment of inertia determines the amount of body pitching. A larger body moment of inertia results in smaller body pitch amplitude. In all other respects, the two running motions are indistinguishable. $\dot{x} = 4\,\text{m/s}$, $k_h = 65.8\,\text{kg}\cdot\text{m}$, $k_\ell = 25,500\,\text{kg/m}$, $z_{max} = 0.97\,\text{m}$, $\dot{\theta}_0 = 5.33\,\text{rad/sec}$.

From (18) and (13) we find the maximum leg angle $\theta_{max} = \dot{\theta}_0/\omega_h$.

To maintain zero angular momentum during flight, the body and leg must counteroscillate, with rates and amplitudes inversely related to their moment of inertia

$$\frac{\dot{\phi}}{J_\ell} = -\frac{\dot{\theta}}{J_b}. \tag{19}$$

The initial value of body pitch rate is therefore $\dot{\phi}_0 = (\dot{\theta}_0 J_\ell)/J_b$.

Initial value for vertical position of body, z_{max}

The duration of the flight phase is manipulated by choosing the initial altitude of the body. The duration of a step is $T_{step} = T_s + T_f$. We choose the parameters of the system so that $T_s = \rho T_{step}$, which gives a flight phase duration

$$T_f = (1 - \rho)T_{step}. \tag{20}$$

The initial altitude that provides the correct flight duration is

$$z_{max} = \frac{g}{8}(1 - \rho)^2 T_{step}^2 + \frac{r_0 \dot{\theta}_0}{\omega_h} \cos \frac{\omega_h T_s}{2}, \tag{21}$$

where the second term is the altitude of the body at touchdown. Examining (14) through (21), we see that three independent parameters are required to specify the passive dynamic running

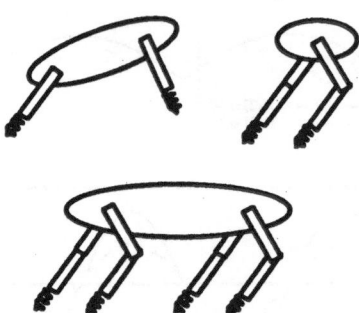

Figure 6: Models with compound pendulum legs and springy hips. The leg would fold during the swing phase, and pogo during the stance phase.

motion: T_{step}, ρ, and $\dot{x}$. Other parameters needed to specify the motion can be calculate from these three.

5 Results

Figures 3 and 4 show the results of a typical simulation of the model. The initial conditions used to generate the figures were determined from equations (5) – (21) as described above, and through a hand optimization process. During optimization, the leg spring constant was changed until we obtained the required stance duration and a reentrant running cycle.

Once the hip and leg spring stiffnesses are chosen, it is possible to manipulate the initial conditions to run at different speeds. Figure 3 includes data for two running speeds. The physical parameters of the model were the same for both running speeds, with adjustments made only in the initial conditions. This suggests that a single machine could run at a range of speeds, without requiring mechanical tuning.

Figure 5 shows data for simulations with two different ratios of leg moment of inertia to body moment of inertia. The magnitude and rate of body pitching are the only variables affected by this change. The behavior of all other variables remains unchanged.

The trajectories explored in this paper represent unstable equilibria. Although the behavior is reentrant if the system is undisturbed, there is no mechanism to return the system to the passive trajectory if there is a disturbance that causes it to deviate. This effect is seen more clearly in figure 4, where the position of the foot with respect to the hip and pitch angle are plotted against altitude. These data show a gradual drift in the phase trajectory. If the trajectory were stable, it would return to an equilibrium limit cycle. A complete implementation of passive dynamic running would include a control mechanism to stabilize the oscillation and eliminate this sort of drift.

So far we have considered passive dynamic running in the context of a planar one-legged hopping machine with a telescoping leg. We have also considered a planar two-legged system with telescoping legs, as shown at the top left in figure 6. So far, we have found reentrant trajectories for this planar two-legged system when the legs are used in synchrony. We would like to find a passive bounding trajectory that uses the legs out of phase, but have not yet done so. We intend to study systems with compound pendulum legs, like those studied by Mochon and McMahon (1980) and shown in figure 6.

6 References

Alexander, R. McN. 1988. *Elastic Mechanisms in Animal Movement.* Cambridge: Cambridge University Press.

Alexander, R. McN., Dimery, N. J., Ker, R. F. 1985. Elastic Structures in the back and their role in galloping in some mammals. *J. Zoology (London)* 207:466–482.

McGeer, T., 1989a. Passive Dynamic Walking, *International J. Robotics Research.*

McGeer, T., 1989b, Personal communications.

McMahon, T. A., Cheng, G. C., 1989, The mechanics of running: how does stiffness couple with speed?, Submitted to *Journal of Biomechanics.*

Mochon, S., McMahon, T. A. 1980. Ballistic walking. *J. Biomechanics* 13:49–57.

Mochon, S., McMahon, T. A. 1981. Ballistic walking: An improved model. *Mathematical Biosciences* 52:241–260.

Raibert, M. H. 1986. *Legged Robots That Balance* Cambridge: MIT Press.

Sutherland, I. E. 1983. *A Walking Robot.* Pittsburgh: The Marcian Chronicles, Inc.

Cooperative and Learning Control for Complex Robot Systems

Fumio Miyazaki*, Osamu Sonoyama*,Toshiro Manabe** and Tetsuya Manabe*

* Faculty of Engineering Science, Osaka University

Toyonaka 560,Osaka Japan

** West Japan Railway Corp.

Abstract: The control problem of robot systems with unmodeled factors is discussed by using two examples. One example of such complex systems is a robot manipulator with Rubbertuators which have complicated dynamical characteristics such as the hysterisis and the compressibility of air. Another one is a multi-fingered robot hand manipulating an object under the influence of the task environment. The applicability of a learning control scheme to the former and a cooperative control scheme to the latter is examined through several experiments.

1 Introduction

The control problem of robot systems is easy to deal with if their mathematical models are available. For example, the Computed Torque Method for the trajectory tracking control exercises its power if a robot system can be regarded as a rigid mechanical system. Even if physical parameters in the model are unknown, we can find several solutions of that problem, namely, parameter identification methods or adaptive control[1] [2] [3]. Whether or not such decisive techniques can be found depends on the a priori information of the system structure. If the actual system includes unknown factors difficult to model, what we can do is to derive a candidate for the effective control strategy only by using a partial knowledge of the system and to check its applicability through experiments.

In this paper we deal with two robot systems as the examples of such complex systems difficult to model. The first example is a robot manipulator with Rubbertuators[4] which have complicated dynamical characteristics such as the hysterisis and the compressibility of air. Referring to the dynamic properties of Rubbertuators, we discuss the applicability of a learning control scheme. The second example is a multi-fingered robot hand manipulating an object. In general, it is difficult to represent the contact relationship between the object and fingertips or between the object and the environment exactly. We will give a practical cooperative control scheme for such a complicated system and verify its effectiveness through several manipulation tasks.

2 A Robot System with Rubbertuators

First we consider a robot manipulator driven by Rubbertuators (developed by Bridgestone) as an example of complex robot systems. **Fig.1** shows the fundamental system structure. This one link manipulator is driven by means of pulling and pushing motions(antagonism)generated by a pair of the Rubbertuators. This can be also regarded as a cooperative control of two Rubbertuators. In **Fig.1**, θ is the rotational angle from an equilibrium point with initial pressure p_0 to $p_0 \pm \Delta p$ (one is $p_0 + \Delta p$ and another is $p_0 - \Delta p$) while control signals to proportional type servo valves are from u_0 to $u_0 \pm \Delta u$,where u_0 and p_0 are the initial control current and pressure respectively and Δu and Δp are the variations. Rubbertuator has a lot of different features from conventional actuators,which are light weight with a high power-to-weight ratio and have compliant properties controllable by the internal pressure. So this actuator is expected to be suitable for high speed motion control,force control and compliance control. However it is also considered to be unsuitable for high precision control due to the complicated dynamical

characteristics such as the hysterisis and the compressibility of air. Although the static properties of the Rubbertuator have been discussed so far, the dynamical analysis of it has not been done so much. This is one of the reasons that effective control schemes have not been proposed for manipulators with the Rubbertuators.

In this section, first we investigate dynamic properties of a robot system driven by Rubbertuators experimentally. The frequency-response approach is one of the advantageous way to this end. As a result, it is shown that the dynamical model needs to be described as a higher order system than it is considered . This acquired model clarifies the difficulty of changing the system properties with the conventional high gain position feedback from the viewpoint of stability,which means feedforward compensation is necessary for high-speed and high-precision trajectory tracking control. To obtain the feedforward term without using the explicit system model, we apply a learning control scheme which uses practice to improve movement by altering the stored data on the basis of previous errors at the execution. This scheme makes it possible for pneumatic robot system with the Rubbertuators to perform high speed and high precision control,which could not have been done by feedback control.

We also apply this learning control scheme to hybrid position/force control for which the Rubbertuator is supposed to be suitable because of its inherent compliance. Compliant properties adapt the end-tip of the robot arm to unknown surfaces of environment. The robot system with Rubbertuators originally has passive compliance and can perform hybrid control successfully with no active compliance control. Finally, we will show the experimental results of trajectory and hybrid control based on the learning scheme and verify the usefulness of the robot systems with Rubbertuators.

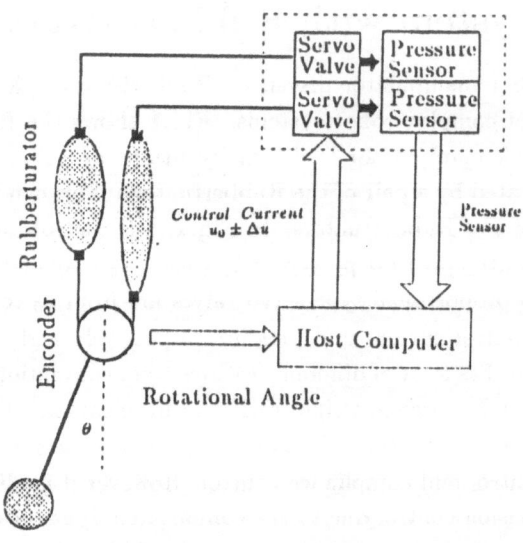

Fig.1: System Configuration

2.1 System Identification

The exact model of Rubbertuator which is complicated because of its non-linearities such as compressibility and saturation of the air, and so on, is difficult to obtain. Even if we could utilize it, such a complicated model won't be useful because reliable control schemes for such a system have not been proposed. On the other hand, the frequency-response approach,which provides the dynamical characteristics of a control system in the frequency domain, is a readily available technique for analysis and design of such a complicated system. This technique is applied to obtain a dominant system properties of a manipulator with Rubbertuators.

Servo Valve

Although it is said that high order and time lag characteristics must be considered for the servo valve[5],good performance of that used in this system makes it possible to approximate as a first-order system from the result of frequency response test. Eq.(1) is a transfer function between the control current and the output pressure of the servo valve.

$$\Delta P(s) = \frac{K_v}{1 + T_v s} \Delta U(s) \tag{1}$$

where K_v is Current-Pressure constant, T_v is Time constant of servo valve, $\Delta P(s)$ is Laplace transform of $\Delta p(t)$, $\Delta U(s)$ is Laplace transform of $\Delta u(t)$.

Rubbertuator

Generally, it is said that the dynamic relationship between the control pressure and the output rotational angle θ in **Fig.1** is described as a second-order system. The frequency response test, however, shows that the phase lag of this system is more than 180 degrees. So an acceptable model needs to be described as a higher-order system than it is considered. Then the transfer function between the control pressure and the output rotational angle is described as follows.

$$\Theta(s) = G_a(s)G(s)\Delta P(s) \qquad (2)$$

$$G_a(s) = \frac{K_m}{s^2 + 2\zeta\omega_n s + \omega_n^2} \qquad (3)$$

where K_m is Pressure-Rotation constant, ω_n is Resonant frequency, ζ is Viscous coefficient, $\Theta(s)$ is Laplace transform of $\theta(t)$, $G_a(s)$ is Transfer function between the joint torque and the rotational angle, $G(s)$ is Transfer function between the control pressure and the joint torque.

It is difficult to identify $G(s)$ exactly. So we approximate this transfer function $G(s)$ as a first-order system given in Eq.(4), which is supposed to be an enough approximation.

$$G(s) = \frac{1}{1 + T_m s} \qquad (4)$$

where T_m is Time constant.

Fig.2 and **Fig.3** show the comparison between the frequency response characteristics of the actual system and that of the model, where the bias pressure p_0 is $2.4(\mathrm{kgf/cm^2})$ and the parameters of Eq.(3), Eq.(4) are as follows.

$$
\begin{aligned}
K_m &= 57(\mathrm{rad/sec^2}) \\
\omega_n &= 9.7(\mathrm{rad/sec}) \\
\zeta &= 0.125 \\
T_m &= 25(\mathrm{msec})
\end{aligned}
$$

Note: It should be noticed that the linearized model identified by the frequency response technique represents only the local properties. In fact, the compliance of Rubbertuator changes dependently on the bias pressure. This feature resembles that of human muscles.

2.2 Learning Control

In designing learning controllers for high-order systems such as Rubbertuators, we should give care to the fact that unmodeled dynamics might influence the convergence. In order to design a learning controller without considering the mathematical model of a system, we employ the way Uchiyama[6] proposed , that is to design a controller by using Bode diagrams of a system. A sufficient condition of the convergence for SISO system is

$$M < 2\cos\phi \qquad (5)$$

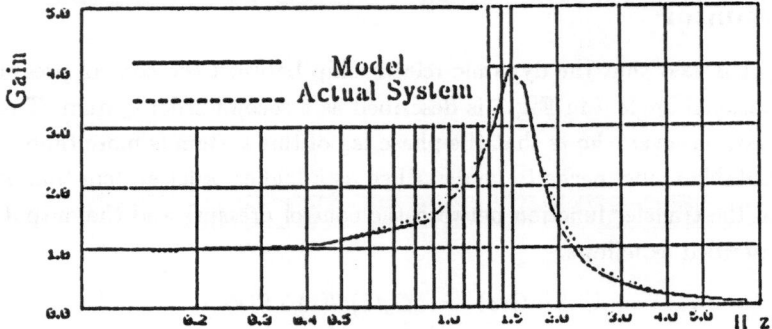

Fig.2: Comparison of Gain

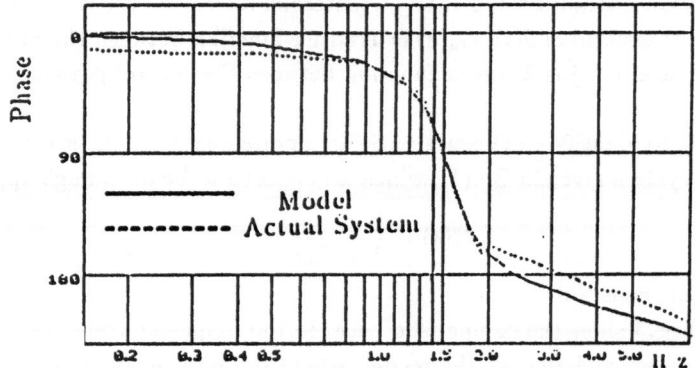

Fig.3: Comparison of Phase

where M and ϕ are defined by G and H as follows,

$$M = |GH| \tag{6}$$

$$\phi = \angle(GH) \tag{7}$$

$$\Delta U_{k+1}(s) = \Delta U_k(s) + H(s)(\Theta_d(s) - \Theta_k(s)) \tag{8}$$

where G is transfer function of the system and H is a learning operator formulated in Eq.(8). Now we employ the translation of a time function and Eq.(8) is rewritten as

$$\Delta u_{k+1}(t) = \Delta u_k(t) + \beta T(h)(\theta_d(t) - \theta_k(t)) \tag{9}$$

$$\theta(t + h) = T(h)\theta(t) \tag{10}$$

where $\Delta u_k(t)$ is the k_{th} input, β is the learning gain, T is the translation operator defined as Eq.(10) and h is the time to translate.

Fig.4 shows the convergence condition of the designed controller for one-link manipulator system with Rubbertuators to achieve a desired motion. In case that the characteristics of servo valves are different, the hierarchical learning scheme[7] is available. The

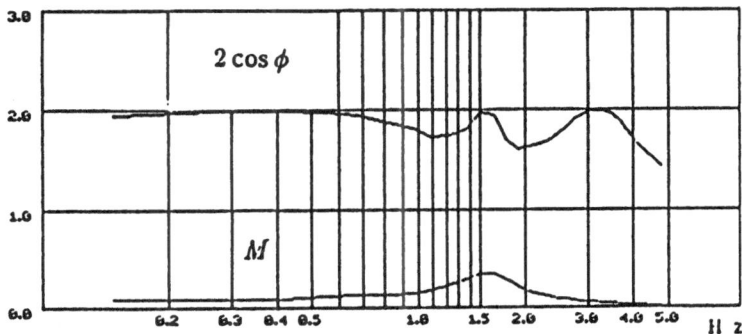

Fig.4: Convergence Condition

learning control scheme like Eq.(10) was applied to trajectory and hybrid control of two-link manipulator system with Rubbertuators. The learning control scheme for trajectory control is given by

$$\Delta u_{k+1}(t) = \Delta u_k(t) + BT(\theta_d(t) - \theta_k(t)) \tag{11}$$

where $u_k = [u_{k1}, u_{k2}]^T$ is the k_{th} input, $B = \text{diag}[\beta_1, \beta_2]$ is the learning gain, $\theta_d = [\theta_{d1}, \theta_{d2}]^T$ is the desired trajectory, $\theta_k = [\theta_{k1}, \theta_{k2}]^T$ is the k_{th} joint rotational angle, $T = \text{diag}[T(h_1), T(h_2)]$ is the time translation operator. The learning control scheme for hybrid control is

$$\begin{aligned} \Delta u_{k+1}(t) = \\ \Delta u_k(t) + K_p T_p J^{-1} S(P_d(t) - P_k(t)) \\ + K_f T_f J^T (I - S)(F_d(t) - F_k(t)) \end{aligned} \tag{12}$$

where J is the Jacobian matrix, $K_p = \text{diag}[K_{p1}, K_{p2}]$ is the learning gain of position, $K_f = \text{diag}[K_{f1}, K_{f2}]$ is that of force, $S = \text{diag}[1, 0]$ is the selection matrix, $T_p = \text{diag}[T(h_{p1}), T(h_{p2})]$ and $T_f = \text{diag}[T(h_{f1}), T(h_{f2})]$ are the time translation operators, $P_d = [P_d, 0]^T$ and $F_d = [0, F_d]^T$ are the desired position and force respectively, $P_k = [P_{kx}, P_{ky}]^T$ and $F_k = [F_{x,k}, F_{y,k}]^T$ are the actual responses, which are all expressed in terms of a task oriented coordinates. Next we show the experimental results of trajectory and hybrid control. Though the convergence for multi-link robot system has not been proved, we obtained successful results.

Results of Trajectory Control

Fig.5 shows the fundamental structure of the experimental system. The desired trajectory (5_{th} order time function) of the end tip of manipulator is given so as to draw a circle,

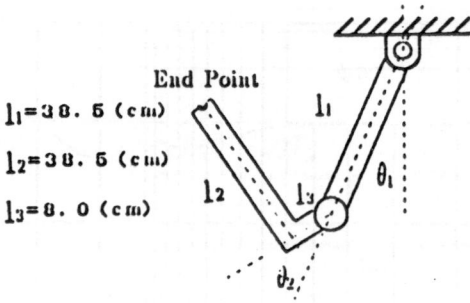

Fig.5: System Configuration (Two-Link Manipulator)

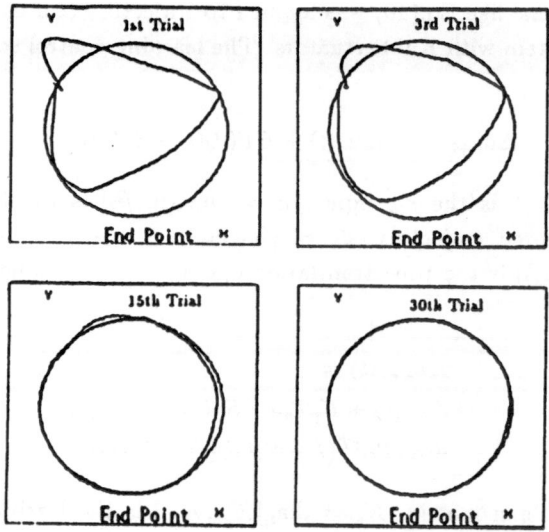

Fig.6: Trace of End Point

$$B = \mathrm{diag}[0.5, 0.5], h_1 = 165(ms), h_2 = 50(ms)$$

and is transformed to the joint trajectory through inverse kinematics. **Fig.6** shows the response at the $1_{st}, 3_{rd}, 15_{th}$ and 30_{th} trial stage, and **Fig.7** does a tendency of the error to decrease at each trial stage.

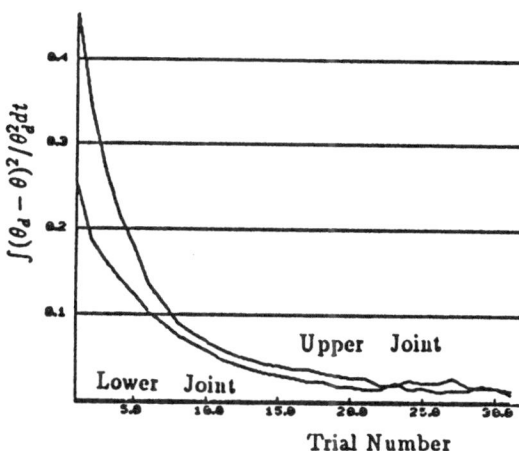

$\int(\theta_d - \theta)^2/\theta_d^2 dt$

Upper Joint

Lower Joint

Trial Number

Fig.7: Tendency of the error to decrease at each trial stage

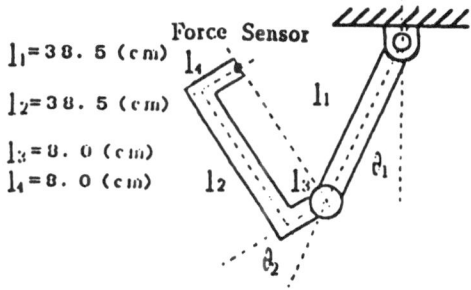

$l_1 = 3\,8.\ 5\ (c\,m)$
$l_2 = 3\,8.\ 5\ (c\,m)$
$l_3 = 8.\ 0\ (c\,m)$
$l_4 = 8.\ 0\ (c\,m)$

Force Sensor

Fig.8: System Configuration (Hybrid Control)

Results of Hybrid Control

In this experiment, the force is controlled in Y direction and the position is in X direction. Force signal is taken by a three-axis force sensor through a 50_{th}-order FIR digital filter (Cut off frequency is 15 Hz). Desired trajectories are the 5_{th} order time functions. **Fig.8** shows the experimental system schematically. Responses at the $1_{st}, 3_{rd}, 15_{th}$ and 30_{th} trial stage are shown in **Fig.9** and a tendency of the error to decrease at each trial stage is confirmed in **Fig.10**.

92

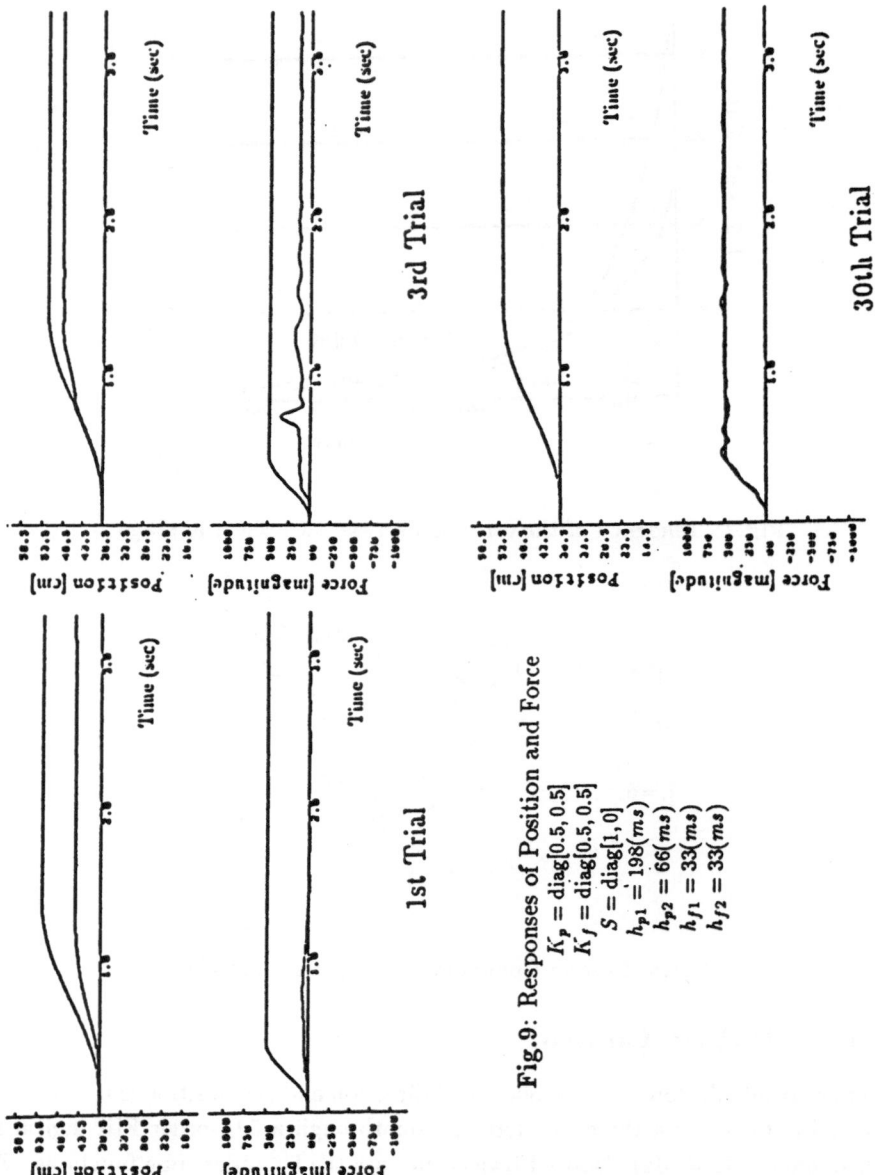

Fig.9: Responses of Position and Force

$K_p = \text{diag}[0.5, 0.5]$
$K_f = \text{diag}[0.5, 0.5]$
$S = \text{diag}[1, 0]$
$h_{p1} = 198(ms)$
$h_{p2} = 66(ms)$
$h_{f1} = 33(ms)$
$h_{f2} = 33(ms)$

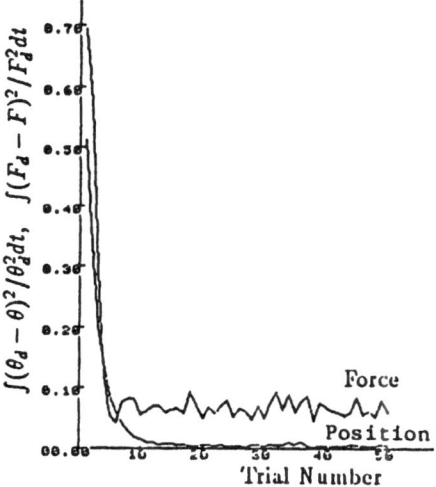

Fig.10: Tendency of the error to decrease at each trial stage

3 Multiple Robot Fingers

In this section we consider another complex robot system, that is, multiple robot fingers manipulating an object. One of the interesting problem concerning such a system is how to generate cooperative motion of each fingers. A great deal of analysis has been done so far on such a problem. Most of them suppose that the ideal contact relation between the object and finger-tips and all physical parameters are known beforehand. Although this ideal model is easy to deal with, the resultant control strategies can be unreliable due to unmodeled factors. Practical contol strategies should work well even if the actual model includes unmodeled factors. As a practical cooperative control scheme for multiple robot fingers, we propose a hierarchical structured control strategy shown in **Fig.11** which consists of four levels. In the lowest level A, a velocity-based inner servo loop is closed at each joint of fingers. The command to the inner loop is generated in the level B so as to achieve a desired fingertip position. This level also involves feedback loops of grasping forces and active compliance control loops based on the 3-axis force sensors attached to each fingertip. In the level C, the desired motion pattern of each fingertip is determined by taking account of not only the target position and orientation of the grasped object but also the grasping force. In the highest level, a control strategy to perform a certain task is updated according to the change of environment. In the following we explain the details of level B, referring to the actual robot hand (**Fig.12**).

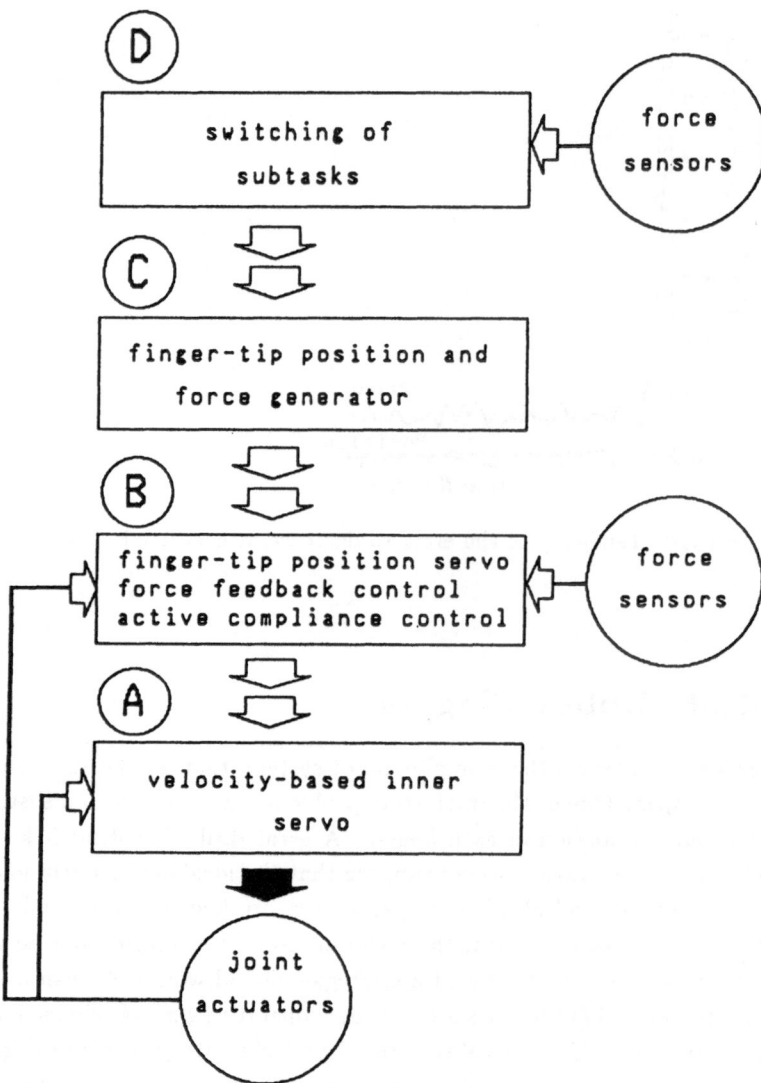

Fig.11: Hierarchical Structured Cooperative Control Scheme

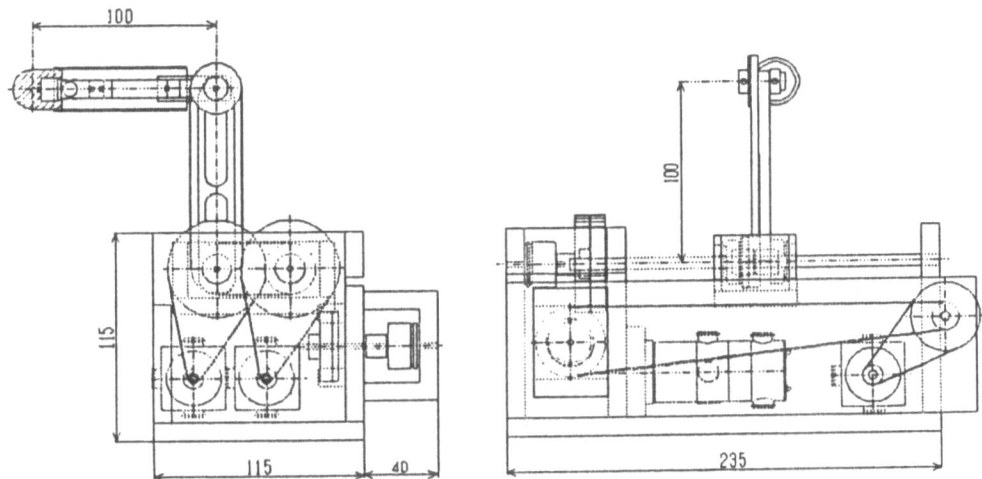

Fig.12: Mechanism of Robot Fingers

3.1 Control Strategies in Level B

This level includes three control functions.

<< 1 >> Positioning of the Object

To achieve the desired position and orientation of the object by using three fingers each of which has 3 d.o.f., we construct a position servo with respect to the mass center of the object X_g and two relative displacement vectors X_{01} & X_{02} among three fingertips (see **Fig.13**). This is a kind of Jacobian Transpose Controller and its stability can be proved under a restricted condition [9].

<< 2 >> Force Feedback

Force error at each fingertip is fedback to joint actuators. This can meet unknown factors associated with the contact relationship between the object and fingertip.

<< 3 >> Active Compliance Control

To accommodate each finger to the interaction between the object and unknown environment, we control the compliance of object. The outline of this method is as follows:

(step 1) Calculate the net force and moment $[F_c^T, M_c^T]^T$ about an arbitrary chosen compliance center X_c such that

$$[F_c^T, M_c^T]^T = P[F_0^T, F_1^T, F_2^T]^T \tag{13}$$

where P is the transformation matrix.

(step 2) Determine the desired variation of the position and orientation of the object due to the external force $[F_c^T, M_c^T]^T$ by using the relation

$$[dX_c^T, dZ_c^T]^T = K_c[F_c^T, M_c^T]^T \tag{14}$$

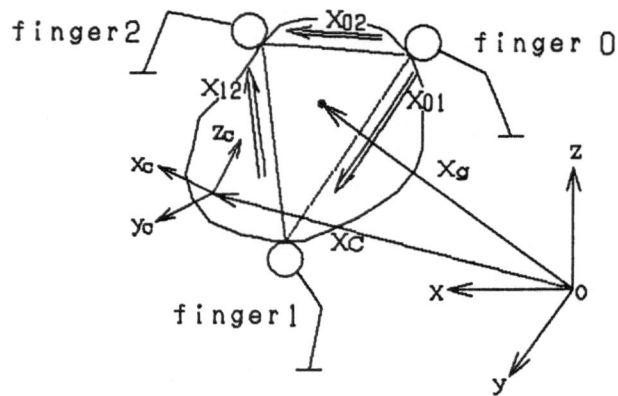

Fig.13: Generalized Coordinate in Workspace

where K_c means a desired compliance which is predetermined according to the objective of manipulation.

(step 3) Modify each fingertip position so as to achieve the desired variation of the object and perform the positioning servo scheme of $<< 1 >>$. This compliance control is used only when the object is in contact with the environment.

3.2   Execution of Several Tasks

The above-mentioned cooperatice control system was employed to carry out several tasks, rotating a crank, drawing with a felt pen, inserting a peg into a hole, fastening a bolt with a nut, and so on. Each of these tasks consists of a sequence of subtasks prepared beforehand as a result of task analysis. A subtask is automatically switched to another one based on the force signal.

Drawing

Observing how we draw with a pen, we can see that the drawing consists of three subtasks; (1)pushing a pen against a surface, (2)actually drawing, (3)pulling up a pen. A period of transition from (1) to (2) is when the net force in the normal direction of the surface exceeds a certain threshold. If the tip position of the pen is within a neighborhood of the final target point, the subtask (2) is switched to (3). Of course the compliance control is used only when the subtask (2) is performed. The flow chart of the task is shown in Fig.14.

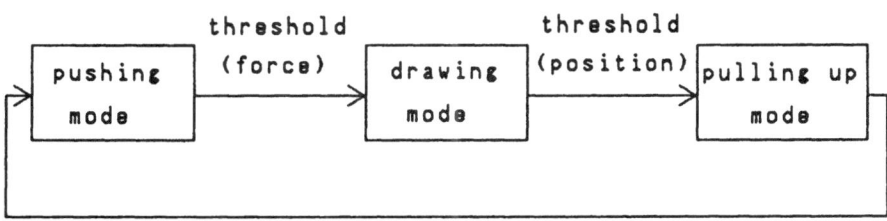

Fig.14: Flow Chart of Drawing Task

Fastening

The fastening task is to rotate a grasped object intermittently. In this task, a fundamental motion pattern is used repeatedly. The total task can be separated into three subtasks; (1)grasping a nut (2)rotating (3)releasing and initializing the fingertip position. Transition from the subtask (1) to (2) happens when the internal force exceeds a threshold. The compliance control is employed in the subtask (2). In this subtask, the desired fingertip position is updated so that the nut moves forward by a pitch. Release motion begins just after the fingertip reaches the final target point. The total fastening task ends if the movement along the axis of rotation exceeds a threshold. (see **Fig.15**)

Rotating

Rotating a crank is one of the simple task for the multi- fingered robot hand, provided that the compliance can be controlled. **Fig.16** shows an experimental result of this task.

Owing to the compliance control, the robot hand can accommodate the trajectory error of the crank due to the insufficient estimation of the fixed point of the crank. If the grasped object can move freely, the robot hand achieves almost perfect trajectory tracking (see **Fig.17**).

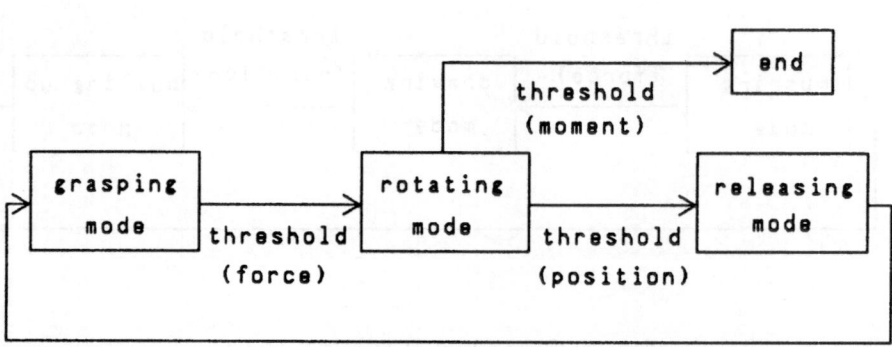

Fig.15: Flow Chart of Fastening Task

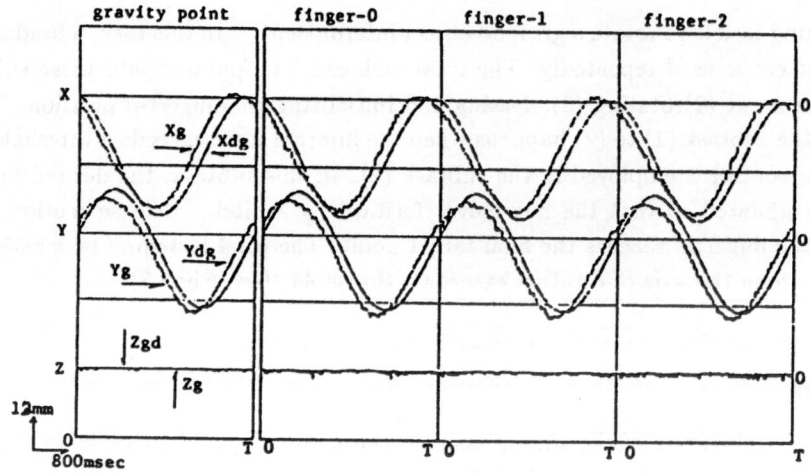

Fig.16: Result of Rotating a Crank

4 Conclusion

We have tried to find effective control schemes for complex robot systems. Candidates for them were derived from a partial knowledge of the system. Although we could not give the theoretical bases of their applicability, the most important thing is the fact that these schemes actually work well. The resultant successful results, while restricted to specific robot systems, can help us to derive the more general approach to the control problem of such complex systems.

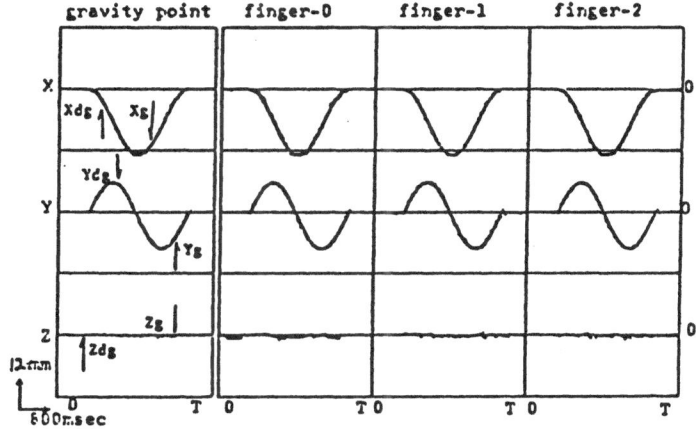

Fig.17: Result of the Same Motion as Fig.16 in Free Enviromment

References

[1] J.J.Craig: Robotics-Mechanics and Control, Addison-Welscy Publishing Co.,1986

[2] H.Asada and J.J.E.Slotine: Robot Analysis and Control, Wiley Interscience,1986

[3] J.J.E.Slotine, Putting Physics in Control - The Example of Robotics, IEEE Control System Magazine,12/18,December 1988

[4] K.Inoue: Rubbertuators and Applications for Robots, Proc. of the 4th Int. Symp. of Robotics Research,1987

[5] Sheare,et al.: Fluid Power Control, MIT Press,1960

[6] M.Uchiyama: Formation of High-Speed Motion Pattern of a Mechanical Arm by Trial, Trans. of SICE,Vol.21,No.6 ,706/712,1978 (in Japanese)

[7] O.Sonoyama et al.: Learning Control of Pneumatic Rubber Artificial Musles, Proc. of the 6th Conf. of Robotics Society of Japan,277/278,1988 (in Japanese)

[8] M.T.Mason and J.K.Salisbury: Robot Hands and the Mechanics of Manipulation, MIT Press,1985

[9] S.Arimoto,F.Miyazaki and S.Kawamura: Cooperative Motion Control of Multiple Robot Arms on Fingers, Proc. of the 1987 IEEE Int. Conf. on Robotics and Automation,Raleigh, 1407/1412,1987

[10] D.E.Whitney, Historical Perspective and State of the Art in Robot Force Control, Int. Jnl. of Robotics Research,Vol.6 No.1,Spring,1987

ASYMMETRIC HYBRID CONTROL OF POSITIONS AND FORCES OF A DUAL ARM ROBOT TO SHARE LOADS

Masaru Uchiyama and Toshiaki Yamashita
Department of Precision Engineering
Tohoku University
Aramaki Aza-Aoba, Aoba-ku, Sendai, 980, Japan

Abstract

Experiments have been presented to verify the asymmetric hybrid control scheme presented by Uchiyama for the coordination of a dual arm robot to hold a single object. The control scheme implements both hybrid control of forces and positions and load sharing between the two arms. An industrial dual arm robot with four degrees-of-freedom for each arm is used in the experiment. An adaptive load sharing algorithm is obtained experimentally and is integrated with the asymmetric control scheme. The experiment shows that, using the adaptive load sharing function, the dual arm robot can hold the object robustly to the disturbance on the object even when relatively soft holding forces are commanded. Tasks such as to slide an object held by the arms, pressing on another object, are achieved successfully.

1. Introduction

There are strong needs for dual arm manipulators to be applied to tasks in unstructured environments such as space and the ocean since, unlike in the factory, no auxiliary equipments like belt conveyers or parts feeders to help them are expected in those environments. Therefore, aiming those applications, research activities with respect to the dual arm manipulator have increased rapidly [1]–[18]. The research area have extended over task scheduling, trajectory planning, kinematics [14], dynamics [5], and force control [1]–[4], [6], [7], [9]–[13], [15]–[18]. Those research issues are discussed in [8]. Among those issues, what this paper is particularly concerned with is the force control for the coordination of two manipulators to carry a single object. This problem includes issues on hybrid control of positions and forces [1]–[4], [6], [7], [10], [12], [13], [15]–[18], control of load sharing between two manipulators [9], [11], [12], [17], and kinematic analysis to give a basis to those coordinated control schemes [7], [10], [17], [18].

The hybrid control of multiple arm robots was studied by Hayati [4]. He proposed a hybrid scheme for the control of compliant motion of an object constrained by other objects in the environment. What he considered in his scheme, however, is only external positions and forces on the object. He did not consider the control of forces and moments applied to the object internally by the multiple arms. Hence, in his scheme, the robot may crush the object if it is brittle. To solve this problem, Uchiyama and Dauchez defined task vectors systematically to include both the internal and the external forces, velocities and positions, and, using those task vectors, formulated a hybrid control scheme, called symmetric scheme, for the coordination of two manipulators [10]. The word "symmetric" comes from that the task vectors are symmetric functions of the joint vectors of each manipulator.

The effectiveness of this scheme was verified experimentally by Uchiyama and Naka-mura [13]. The experiments showed, as had been expected theoretically, the effective-ness of the scheme; simultaneous control of holding forces and positions of the object was achieved successfully by the scheme. At the same time, however, the experiments showed an unexpected problem. The problem is that the holding of the object is very much vulnerable to disturbances (forces) on the object; namely, the holding collapsed very easily and the robot dropped the object when the forces are applied to the object. This exposed a defect of the scheme; the holding of the object is not adaptable to the changes of the loads on the object. Thus, modification of the scheme to provide that adaptability was required.

Recently, Uchiyama proposed to achieve it by adding load sharing capability to the scheme [17]. This was done by replacing the Moore-Penrose inverse, used in the Uchiyama and Dauchez's scheme [10] to decompose the forces on the object into the internal and external forces, with a generalized inverse including a load sharing matrix. The new scheme is asymmetric for the two manipulators in the sense that the task vectors are defined as asymmetric functions of the joint vectors of each manipulator. Although load sharing is discussed in [17], only theoretical results have been presented; no experimental verification of the scheme has been presented. The adaptability may be achieved by tuning the load sharing matrix but how to tune it has not been clarified yet.

The purpose of this paper is to present experimental verification of the asymmetric scheme after finding an algorithm to tune the load sharing matrix. An industrial dual arm robot with four degrees-of-freedom for each arm is used in the experiment. The robot has 6-axis force/torque sensors at each wrist of the arms. The load sharing algorithm is obtained through the experiment and is integrated with the asymmetric control scheme. The experiment shows that, using the adaptive load sharing function, the dual arm robot can hold an object robustly to disturbing forces on the object even when relatively soft holding forces are commanded. Tasks such as to slide an object held by the robot, pressing on another object, are achieved successfully.

This paper is organized as follows: The asymmetric control scheme presented in [17] to give the framework for the adaptive load sharing is summarized in Section 3, after historical perspective of how the asymmetric control scheme was derived, presented in Section 2. In Section 4, adaptive load sharing algorithms are presented for a few examples of tasks. Experimental verification of those algorithms are presented in Section 5. The conclusions of this paper and future problems regarding force control of dual arm robots are presented in Section 6.

2. From Symmetric Control Scheme to Asymmetric Control Scheme

One of well-known control schemes for the coordination of dual manipulators to hold a common object is a master/slave scheme [1], [3], so called, in which a master arm is position-controlled to determine the position of the object and a slave arm is force-controlled, for the forces on the object to be controlled, to follow the master arm. This scheme was criticized because of the following reasons:

- The force controlled slave arm is always under position disturbances due to the motion of the master arm. Therefore it is difficult to achieve accurate control of forces on the object.

- Basis for how master or slave modes should be assigned to each arm is not clear. Nor is found any effective automatic mode assignment algorithm.

To solve the above problems, non-master/slave schemes in which neither a master arm nor a slave arm is distinguished have been proposed by many researchers [2], [4], [6], [7], [10], [13], [16]–[18]. The symmetric hybrid control scheme presented by Uchiyama and Dauchez [10] is one of those non-master/slave schemes.

The scheme employs a set of generalized task vectors consisting of force, velocity and position vectors. A key idea to derive those task vectors is a concept of a virtual stick which allows us to note a representative point on the object and dramatically simplifies the mathematics for the derivation. Thanks to it, the decomposition of the motions of the two arms is achieved theoretically to yield internal and external task vectors; the internal task vectors describe the internal forces on the object and the relative motions of the two arms, while the external task vectors the external forces on the object and the absolute motions of the object held by the two arms. Using those task vectors, a symmetric hybrid control scheme is formulated. The theory is summarized as follows:

The situation depicted in Figure 1 is considered; the two arms hold a common object. Each arm can apply both forces and moments to the object. The virtual sticks are imaginary rigid sticks fixed on each hand to reach the representative point of the object. Using those virtual sticks, the positions, velocities and forces of each hand are represented at the tips of the virtual sticks. Let us denote those forces and moments at the tips of the virtual sticks as F_{bi} and N_{bi}, translational and rotational velocities as v_{bi} and ω_{bi}, and position and orientation as p_{bi} and ϕ_{bi}, respectively, where $i = 1$ and 2 to represent the arm number. Exactly speaking, ϕ_{bi} is not a vector because it does

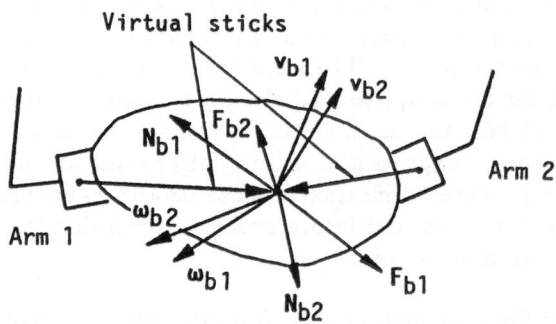

Figure 1: Two arms holding a common object.

not satisfy the axiom of the vector. Therefore we call it a pseudo-vector. Using the above vectors (or pseudo-vectors), we define the following vectors (or pseudo-vectors):

$$f_{bi} = [\, F_{bi}^T \;\; N_{bi}^T \,]^T \tag{1}$$

$$s_{bi} = [\, v_{bi}^T \;\; \omega_{bi}^T \,]^T \tag{2}$$

$$x_{bi} = [\, p_{bi}^T \;\; \phi_{bi}^T \,]^T. \tag{3}$$

The resultant forces and moments on the object caused by f_{bi} ($i = 1$ and 2) are given by

$$\begin{aligned}
f_a &= f_{b1} + f_{b2} \\
&= [I_6 \; I_6] \, [f_{b1}^T \; f_{b2}^T]^T \\
&= W [f_{b1}^T \; f_{b2}^T]^T \\
&= W q_b
\end{aligned} \tag{4}$$

where I_n (here $n = 6$) represents the $n \times n$ unit matrix. Solving (4) for a given f_a yields

$$\begin{aligned}
q_b &= W^+ f_a + (I_{12} - W^+ W) \zeta \\
&= W^+ f_a + [I_6 \; -I_6]^T f_r
\end{aligned} \tag{5}$$

where W^+ is the Moore-Penrose inverse of W and ζ is an arbitrary 12-dimensional vector. Since W is a 6×12 matrix with rank 6, it has a 6-dimensional range and a 6-dimensional null space. From (5), defining the range as the external force/moment vector space, and the null space as the internal force/moment vector space, we have the external force/moment vector f_a and the internal force/moment vector f_r as follows:

$$f_a = f_{b1} + f_{b2} \tag{6}$$

$$f_r = \frac{f_{b1} - f_{b2}}{2}. \tag{7}$$

Using the force-velocity duality derived by the principle of virtual work, we have the external velocity vector s_a and the internal velocity vector Δs_r as follows:

$$s_a = \frac{s_{b1} + s_{b2}}{2} \tag{8}$$

$$\Delta s_r = s_{b1} - s_{b2}. \tag{9}$$

Finally, integrating (8) and (9), respectively, we have the external position vector x_a and the internal position vector Δx_r as follows:

$$x_a = \frac{x_{b1} + x_{b2}}{2} \tag{10}$$

$$\Delta x_r = x_{b1} - x_{b2}. \tag{11}$$

It is noted that careful attention should be paid to the integration of the angular velocities because the direct integration of the angular velocities does not give meaningful orientation angles. See [10] for the details of the integration process. Using the external

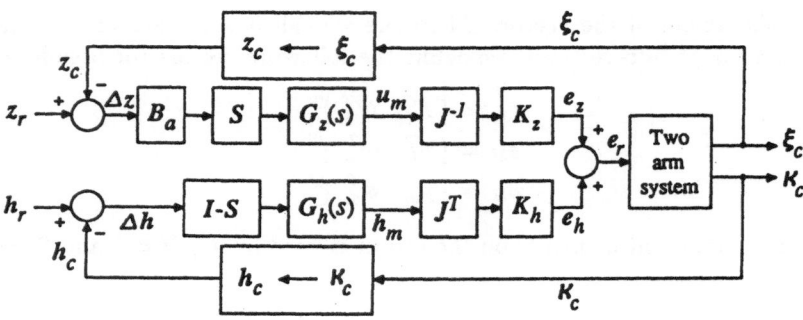

Figure 2: A hybrid position/force control scheme.

and internal forces, velocities and positions given by (6) through (11), task vectors to describe the coordination of the two arms are defined as follows:

$$h = [\, f_a^T \quad f_r^T \,]^T \tag{12}$$

$$u = [\, s_a^T \quad \Delta s_r^T \,]^T \tag{13}$$

$$z = [\, x_a^T \quad \Delta x_r^T \,]^T. \tag{14}$$

It is convenient that f_a, s_a and x_a be represented with respect to a base frame and f_r, Δs_r and Δx_r with respect to an object frame. See [10] for the details.

Using the above task vectors, a hybrid control scheme of positions and forces for the single arm robot is easily extended to implement the coordinated control of the dual arm robot. Figure 2 shows a typical control scheme diagramatically. Since the task vectors are symmetric functions of the force, velocity and position vectors at the tips of the virtual sticks, the scheme is called a symmetric scheme. The vector ξ represents the joint angles (or displacements for slide joints) of the two arms. The vector κ represents the forces and moments measured by the force sensors at the wrists of each arm. The suffixes r and c represent the reference values and the current values, respectively. B_a is the matrix to transform errors in orientation angles into the corresponding rotation vector. J is the Jacobian matrix of the transformation from ξ to z. The details of B_a and J are found in [10]. $G_z(s)$ and $G_h(s)$ are diagonal operator matrices to represent the position control law and the force control law, respectively. K_z and K_h are diagonal constant coefficient matrices to transform the velocity command and the torque command into the motor commands, respectively. S is a diagonal matrix of which diagonal elements take values 1 or 0. When 1 is taken, the corresponding component is position-controlled; when 0 is taken, force-controlled. I is the unit matrix of which size is the same as S.

The symmetric hybrid control scheme was implemented by an industrial dual arm robot to verify its effectiveness. Experimental results show that simultaneous control of holding forces and carrying positions of the object can by achieved successfully [13]. Comparison of those results with results obtained for a master/slave scheme shows the superiority of the symmetric scheme over the master/slave scheme; large errors in the control of holding forces were observed in the master/slave scheme with the change of the object positions, while both the holding forces and the object positions were controlled accurately in the symmetric scheme even when the object was moved [18].

A problem, however, was also found in the experiment; the robot dropped the object when a relatively strong force was applied to the object. This shows that the holding of an object by this scheme is vulnerable to the disturbance. It means that it is impossible to implement those tasks such as filing by two arms in which the object held by the arms is exposed to disturbing forces. Through investigation we found that the problem comes from the fact that the load to the robot is always shared equally by the two arms, as can be easily shown by (6) and (7); therefore, the loads may break the contact condition of the arms on the object to lead collapse of the holding. To incorporate load sharing capability into the symmetric scheme is essential for the realization of robust holding and, hence, for the enhancement of the task performability of the robot.

3. Framework of Asymmetric Control Scheme

The incorporation of load sharing capability is achieved simply by replacing the Moore-Penrose inverse W^+ in (5) with a generalized inverse W^-. As a result the symmetric control scheme becomes non-symmetric, or asymmetric, for the two arms, but still within the framework of the non-master/slave scheme. In this sense the incorporation bring the symmetric scheme into a unified version in which non-master/slave and load sharing features coexist. The details of the asymmetric scheme is presented in [17]. In what follows of this section we present a summary of the asymmetric scheme.

The highlight in the derivation of the asymmetric scheme is to find a solution of q_b in (4) for f_a to share loads, which is obtained by

$$
\begin{aligned}
q_b &= W^- f_a + (I_{12} - W^- W)\zeta \\
&= W^- f_a + [I_6 \ -I_6]^T f_r
\end{aligned}
\tag{15}
$$

where W^- is a generalized inverse of W. The solution is much the same as (5) except that W^+ is replaced with W^-. The difference is important since W^- is defined by

$$
W^- = [\, K^T \ (I_6 - K)^T \,]^T
\tag{16}
$$

where K is a 6×6 load sharing matrix to define how loads are shared between the two arms. $W^- = W^+$ when $K = \frac{1}{2}I_6$. From (15), f_a and f_r are calculated as follows:

$$
f_a = f_{b1} + f_{b2}
\tag{17}
$$

$$
f_r = (I_6 - K)f_{b1} - K f_{b2}.
\tag{18}
$$

From the force-velocity duality, the external velocity s_a and the internal velocity Δs_r are calculated as follows:

$$
s_a = K^T s_{b1} + (I_6 - K^T)s_{b2}
\tag{19}
$$

$$
\Delta s_r = s_{b1} - s_{b2}.
\tag{20}
$$

Integrating those velocities with the assumption that the relative motion is small and that

$$
K = \mathrm{diag}\,[\,K_1, \ K_2\,]
\tag{21}
$$

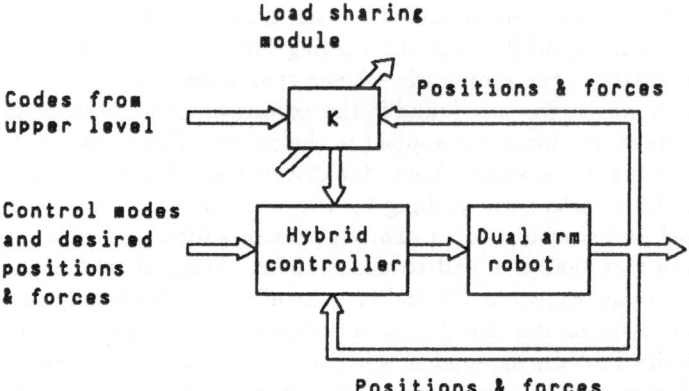

Figure 3: A hybrid controller with a load sharing module.

where K_1 and K_2 are 3×3 matrices yields

$$p_a = K_1^T p_{b1} + (I_3 - K_1^T) p_{b2} \tag{22}$$

$$n_a = K_2^T n_{b1} + (I_3 - K_2^T) n_{b2} \tag{23}$$

$$o_a = K_2^T o_{b1} + (I_3 - K_2^T) o_{b2} \tag{24}$$

$$a_a = K_2^T a_{b1} + (I_3 - K_2^T) a_{b2} \tag{25}$$

and

$$\Delta p_r = p_{b1} - p_{b2} \tag{26}$$

$$\Delta \Omega_r = \frac{(n_{b2} \times n_{b1} + o_{b2} \times o_{b1} + a_{b2} \times a_{b1})}{2} \tag{27}$$

where

$$H_a = \begin{bmatrix} n_a & o_a & a_a & p_a \\ 0 & 0 & 0 & 1 \end{bmatrix} \tag{28}$$

$$H_{bi} = \begin{bmatrix} n_{bi} & o_{bi} & a_{bi} & p_{bi} \\ 0 & 0 & 0 & 1 \end{bmatrix} \tag{29}$$

to represent the positions and orientations of the frames to accompany the velocity vectors s_a and s_{bi}, respectively. Δp_r and $\Delta \Omega_r$ represent differential translations and rotations corresponding to f_r (or Δs_r).

Using the task vectors defined by (17) through (20) and (22) through (27), a hybrid control scheme can be organized. The organization of the scheme is essentially the same as the one presented in Figure 2 except that the load sharing matrix K is included and should be determined somehow. Figure 3 shows a load sharing module to tune K annexed to the hybrid controller. The load sharing module gets information from the plant (the dual arm robot) and the upper level motion planner. How to tune K, however, has not been clarified yet, and is an important research issue. Generally, K has to be changed case by case according to the loading conditions. The next section presents an adaptive load sharing algorithm for robust holding.

4. Adaptive Load Sharing

We consider those situations shown in Figure 4 where the two arms hold an object by pressing it in order to do tasks. The internal task vectors should be force-controlled because the internal positions are all constrained. Force references to the internal forces are zero except the force in the direction of pressing. The reference to the pressing force is non-zero but not very large for the object to be held softly. The external task vectors are position-controlled for the case (a) because the external positions (the positions of the object) is not constrained although they are disturbed by the forces due to the collision with an obstacle in the environment, while the external task vectors are hybrid-controlled for the case (b) because the external positions are constrained by the triangular block. External forces disturbing the object held by the arms is relatively large in this case due to the pressing forces of the object on the triangular block and also the friction forces between the object and the triangular block when the object is slided on the block. In those situations adaptive load sharing is needed to support the external forces without loosing the holding conditions: the contacts of the arms on the object. Otherwise the holding of the object will fail as the experiment presented in [13] suggested.

Before deriving an algorithm to share loads for those cases, we assume that, for the sake of simplicity, the load sharing matrix K is diagonal:

$$K = \text{diag} [\alpha_j] \tag{30}$$

where α_j is the load sharing coefficient for the jth component of f_a. Substituting (30) into (17) through (20) yields decoupled equations for each component of the task vectors as follows:

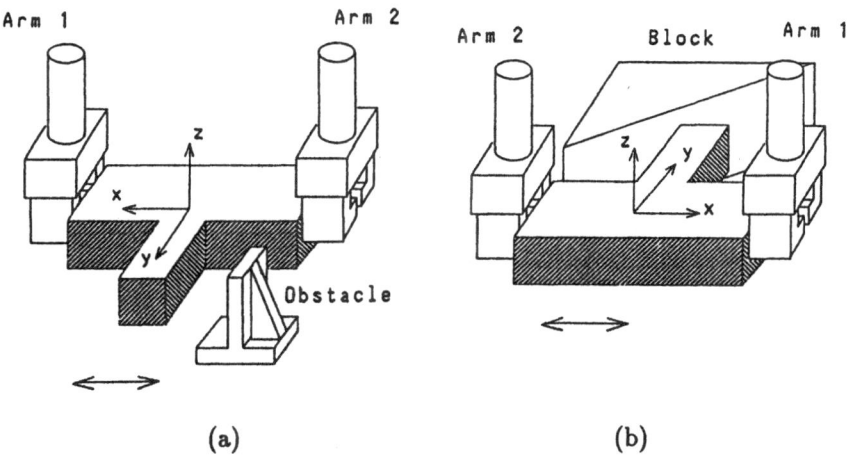

(a)   (b)

Figure 4: Tasks to be considered. (a) Carrying an object where there is an obstacle. (b) Pressing and sliding an object on another object.

$$f_{aj} = f_{b1j} + f_{b2j} \tag{31}$$

$$f_{rj} = (1 - \alpha_j)f_{b1j} - \alpha_j f_{b2j} \tag{32}$$

$$s_{aj} = \alpha_j s_{b1j} + (1 - \alpha_j)s_{b2j} \tag{33}$$

$$\Delta s_{rj} = s_{b1j} - s_{b2j}. \tag{34}$$

The inverse transformation of (31) through (34) are obtained as follows:

$$f_{b1j} = \alpha_j f_{aj} + f_{rj} \tag{35}$$

$$f_{b2j} = (1 - \alpha_j)f_{aj} - f_{rj} \tag{36}$$

$$s_{b1j} = s_{aj} + (1 - \alpha_j)\Delta s_{rj} \tag{37}$$

$$s_{b2j} = s_{aj} - \alpha_j \Delta s_{rj}. \tag{38}$$

Based on the above condition, we propose a load sharing algorithm for the tasks presented in Figure 4 as follows: First, as to the external forces, since the forces in the y and z directions are supported by the friction forces on the contact surfaces, supposing that the contact conditions for the two arms are the same, we let

$$\alpha_j = \frac{1}{2}. \tag{39}$$

The forces in the x direction, however, is supported by the end-effector pressed on the object. The pulling forces on the object cannot be yielded on the contact surface. Therefore, in principle, we should let

$$\alpha_j = \left\{ \begin{array}{ll} 1 & f_{aj} \geq 0 \\ 0 & f_{aj} < 0 \end{array} \right. \tag{40}$$

where f_{aj} is calculated by (31) from f_{b1j} and f_{b2j} measured by the force sensors. From the practical point of view, however, such abrupt switching of the load sharing coefficient α_j causes hazardous effects to the system. So we propose to calculate α_j by the following equation:

$$\alpha_j = \left\{ \begin{array}{ll} 1 & f_{aj} \geq f_o \\ \frac{1}{2}\left(1 + \frac{f_{aj}}{f_o}\right) & f_o > f_{aj} > -f_o \\ 0 & -f_o \geq f_{aj} \end{array} \right. \tag{41}$$

where f_o is a threshold to be specified properly. As to the external moments, it is obvious that the moment around the x axis be equally shared by the two arms; α_j is calculated by (39). However, it is not clear how the moments around the y and z axes be supported by the two arms. Load sharing algorithms for the moment around the z axis are investigated in the next section.

5. Experiments

The dual arm robot used in the experiment is the same as the one used in the previous experiments [6], [13], [18]. It is a dual arm industrial robot with 4 degrees-of-freedom controlled by a 16 MHz, 32-bit, and 80386 CPU personal computer. Each arm is a

Figure 5: Carrying an object where there is an obstacle.

Figure 6: Pressing and sliding an object on another object.

gantry-type Cartesian arm having x, y and z motors and a motor to rotate the wrist around the z axis. At each wrist of the arms is installed a 6-axis force/torque sensor.

Figures 5 and 6 are photographs to show the robot doing the tasks presented in Figure 4 (a) and (b), respectively. The force and position task vectors are defined for the coordinate systems shown in Figure 4 (a) and (b), respectively, as follows:

$$h = [\, F_{xa}, \; F_{ya}, \; F_{za}, \; N_{za}, \; F_{xr}, \; F_{yr}, \; F_{zr}, \; N_{zr} \,]^T \tag{42}$$

$$z = [\, x_a, \ y_a, \ z_a, \ \theta_a, \ \Delta x_r, \ \Delta y_r, \ \Delta z_r, \ \Delta \theta_r \,]^T \qquad (43)$$

where N_{za} and θ_a represent the external moment and the rotation angle of the object around the z axis, respectively, while N_{zr} and $\Delta \theta_r$ represent the internal moment and the relative rotation angle between the two arms, also around the z axis, respectively. The rest of the notations for the components of h and z represent the x, y and z components of the external and internal forces and the external and internal positions, respectively. The selection matrix is set as follows:

$$S = \left\{ \begin{array}{ll} \mathrm{diag} \, [\, 1, \ 1, \ 1, \ 1, \ 0, \ 0, \ 0, \ 0 \,] & \text{for Figure 5} \\ \mathrm{diag} \, [\, 1, \ 0, \ 1, \ 1, \ 0, \ 0, \ 0, \ 0 \,] & \text{for Figure 6.} \end{array} \right. \qquad (44)$$

Experimental data for Figure 5 are presented in Figure 7. The hybrid control schemes with and without the adaptive load sharing function are compared in the figure. Figure (a) shows data with adaptive load sharing and (b) without adaptive load sharing. The reference to the holding force is 2 N for both cases. The adaptation was done only for α_1 for F_{xa}. The change of α_1 is shown in the figure. α_4 for N_{za} was set constantly $\frac{1}{2}$; namely, no adaptation for α_4 in both cases. It is observed that the internal position was changed by the external forces for the case (b). This means that the holding of the object failed. From this experiment the robustness of holding in the asymmetric control scheme with adaptive load sharing has been proved.

Experimental data for Figure 6 are presented in Figures 8 and 9. The reference to the holding force is again 2 N for both cases. Figure 8 shows data with adaptive load sharing. But in this case α_4 for N_{za} was set constantly $\frac{1}{2}$; no adaptation was made for α_4. The data shows that the internal positions are disturbed by the strong external forces and a moment. This means that the robot may drop the object and, actually, the task failed. The moment N_{za} was supposed to be the main cause of the change of the internal positions. Adaptation in load sharing of N_{za} was thought of to improve the task and, experimentally, α_4 was set as

$$\alpha_4 = \alpha_1 \qquad (45)$$

to follow the result of the adaptation in α_1. Results are presented in Figure 9. It is observed that the change of the internal positions in the previous experiment has been significantly improved. Intuitively this suggests that, since the pressing force by the arm not supporting F_{xa} is weak, N_{za} should be supported by the other arm supporting F_{xa}. No theory for that, however, has been found yet.

6. Conclusions

Control schemes that have been presented mainly by the authors' group for the coordination of a dual arm robot to hold a single object has been reviewed to clarify the problems regarding the coordination of the dual arm robot. And, it has been shown that the robust holding of the object is the problem to be studied in this paper. A solution to the robust holding is to achieve it by sharing loads adaptively. In this sense, the asymmetric control scheme presented in [17] is the most comprehensive scheme since it solves both the problem of hybrid control of positions and forces and the problem of adaptive load sharing. Then, experimental verification of the scheme has been presented, after load sharing algorithms for a few special cases were considered. An effective load sharing algorithm for the robust holding of the object was found experimentally. Although the algorithm seems to be intuitively reasonable, no theory for it has been found yet.

Future research will be directed to theoretical investigation of the problem of load sharing to find an algorithm for it more theoretically. Dynamic problems in the coordination should also be investigated in future because a couple of malfunctions were observed that seemed to come mainly from dynamic effects of the system.

Acknowledgment

This research is supported by the Grant-in-Aid for Scientific Research from the Ministry of Education under the Project Number 63550178. The authors express their thanks for the support.

References

[1] E. Nakano, S. Ozaki, T. Ishida, and I. Kato, "Cooperational Control of the Anthropomorphous Manipulator 'MELARM'," in *Proc. 4th Int. Symp. on Industrial Robots*, Tokyo, pp. 251–260, November 1974.

[2] S. Fujii and S. Kurono, "Coordinated Computer Control of a Pair of Manipulators," in *Proc. 4th IFToMM World Congress*, Newcastle upon Tyne, pp. 411–417, September 1975.

[3] H. Kikuchi, T. Niinomi, M. Sato, and Y. Matsumoto, "Heavy Parts Assembly by Coordinative Control of Robot and Balancing Manipulator," in *Proc. IFAC 9th World Congress*, Budapest, vol. VI, pp. 175–180, July 1984.

[4] S. Hayati, "Hybrid Position/Force Control of Multi-Arm Cooperating Robots," in *Proc. 1986 IEEE Int. Conf. on Robotics and Automation*, San Francisco, pp. 82–89, April 1986.

[5] Y. F. Zheng and J. Y. S. Luh, "Joint Torques for Control of Two Coordinated Moving Robots," in *Proc. 1986 IEEE Int. Conf. on Robotics and Automation*, San Francisco, pp. 1375–1380, April 1986.

[6] M. Uchiyama, N. Iwasawa, and K. Hakomori, "Hybrid Position/Force Control for Coordination of a Two-Arm Robot," in *Proc. of 1987 IEEE Int. Conf. on Robotics and Automation*, Raleigh, pp. 1242–1247, April 1987.

[7] P. Dauchez and M. Uchiyama, "Kinematic Formulation for Two Force Controlled Cooperating Robots," in *Proc. '87 Int. Conf. Advanced Robotics*, Versailles, pp. 457–467, October 1987.

[8] A. J. Koivo and G. A. Bekey, "Report of Workshop on Coordinated Multiple Robot Manipulators: Planning, Control, and Applications," *IEEE J. of Robotics and Automation*, vol. 4, no. 1, pp.91–93, 1988.

[9] Y. F. Zheng and J. Y. S. Luh, "Optimal Load Distribution for Two Industrial Robots Handling a Single Object," in *Proc. 1988 IEEE Int. Conf. on Robotics and Automation*, Philadelphia, pp. 344–349, April 1988.

[10] M. Uchiyama and P. Dauchez, "A Symmetric Hybrid Position/Force Control Scheme for the Coordination of Two Robots," in *Proc. 1988 IEEE Int. Conf. on Robotics and Automation*, Philadelphia, pp. 350–356, April 1988.

[11] M. E. Pittelkau, "Adaptive Load-Sharing Force Control for Two-Arm Manipulators," in *Proc. 1988 IEEE Int. Conf. on Robotics and Automation*, Philadelphia, pp. 498–503, April 1988.

[12] T. E. Alberts and D. I. Soloway, "Force Control of a Multi-Arm Robot System," in *Proc. 1988 IEEE Int. Conf. on Robotics and Automation*, Philadelphia, pp. 1490–1496, April 1988.

[13] M. Uchiyama and Y. Nakamura, "Symmetric Hybrid Position/Force Control of Two Cooperating Robot Manipulators," in *Proc. 1988 IEEE Int. Workshop on Intelligent Robots and Systems*, pp. 515–520, November 1988.

[14] S. Lee, "Dual Redundant Arm Configuration Optimization with Task-Oriented Dual Arm Manipulability," *IEEE Trans. on Robotics and Automation*, vol. 5, no. 1, pp. 78–97, 1989.

[15] S. A. Schneider and R. H. Cannon, Jr., "Object Impedance Control for Cooperative Manipulation: Theory and Experimental Results," in *Proc. 1989 IEEE Int. Conf. on Robotics and Automation*, Scottsdale, pp. 1076–1083, May 1989.

[16] K. Kosuge, M. Koga, K. Furuta, and K. Nosaki, "Coordinated Motion Control of Robot Arm Based on Virtual Internal Model," in *Proc. 1989 IEEE Int. Conf. on Robotics and Automation*, Scottsdale, pp. 1097–1102, May 1989.

[17] M. Uchiyama, "A Unified Approach to Load Sharing, Motion Decomposing, and Force Sensing of Dual Arm Robots," *5th Int. Symp. of Robotics Research*, Tokyo, August 1989 (to be presented).

[18] K. Nakamura, *A Study on Cooperational Control of a Dual Arm Manipulator for Assembly Tasks*, M.Sc. Thesis, Department of Precision Engineering, Tohoku University, February 1989.

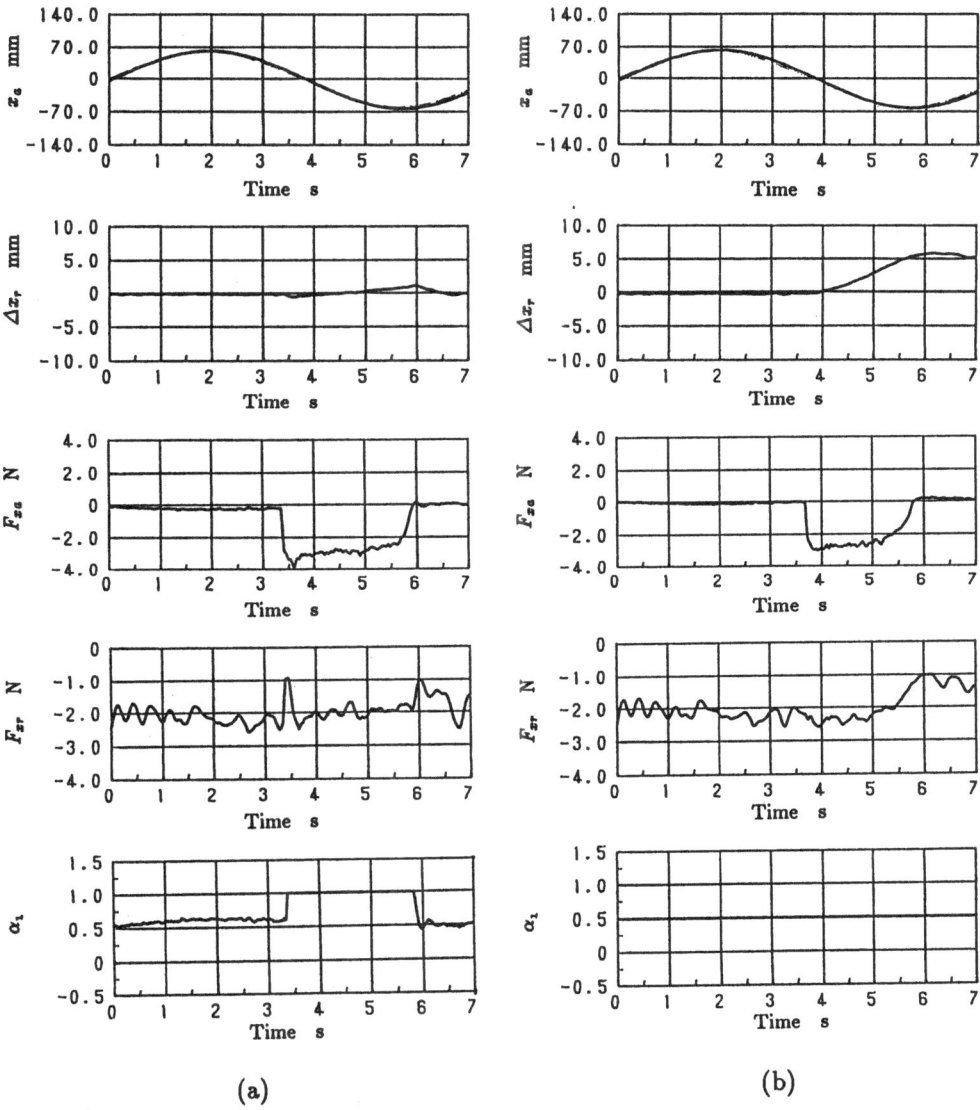

Figure 7: Experimental data for the task presented in Figure 5. (a) With adaptive load sharing. (b) Without adaptive load sharing.

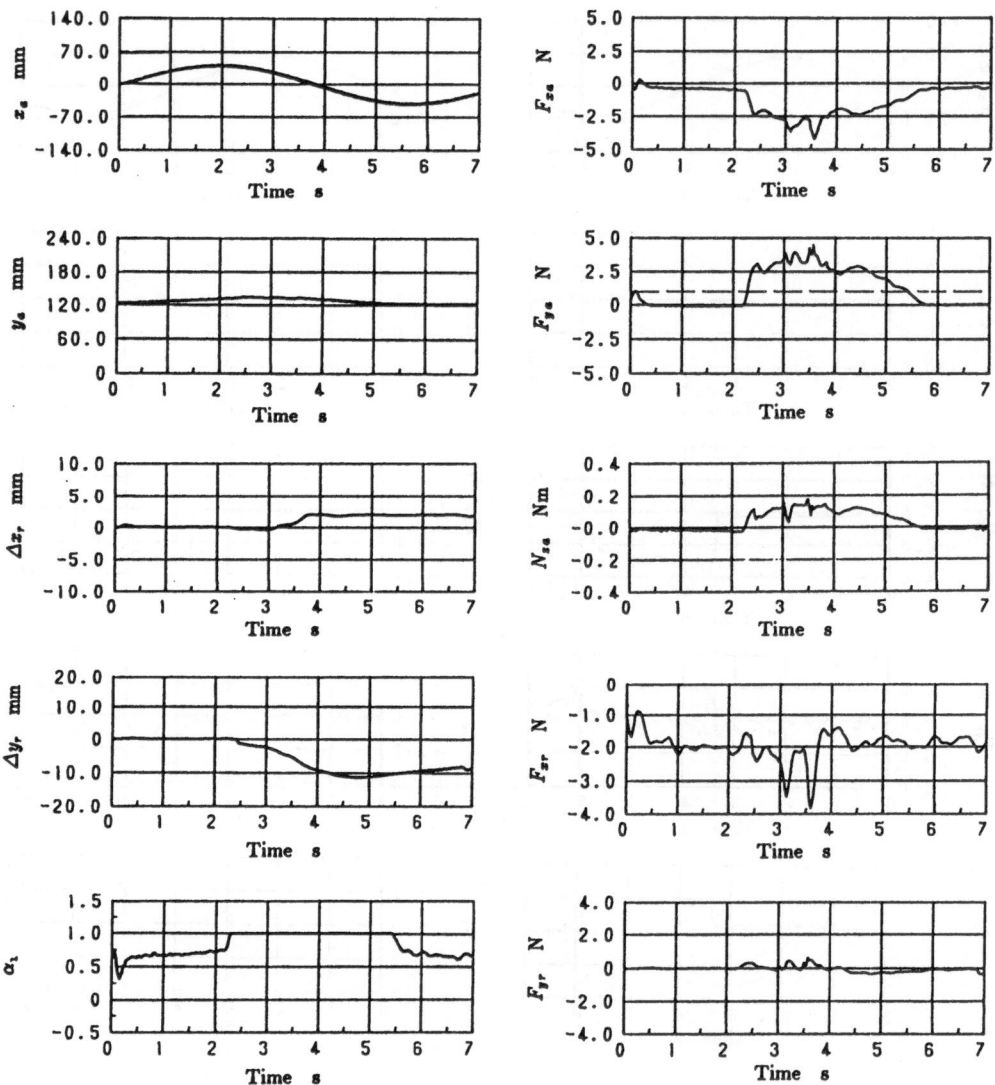

Figure 8: Experimental data (1) for the task presented in Figure 6 where only F_{xa} is adaptively shared by the two arms.

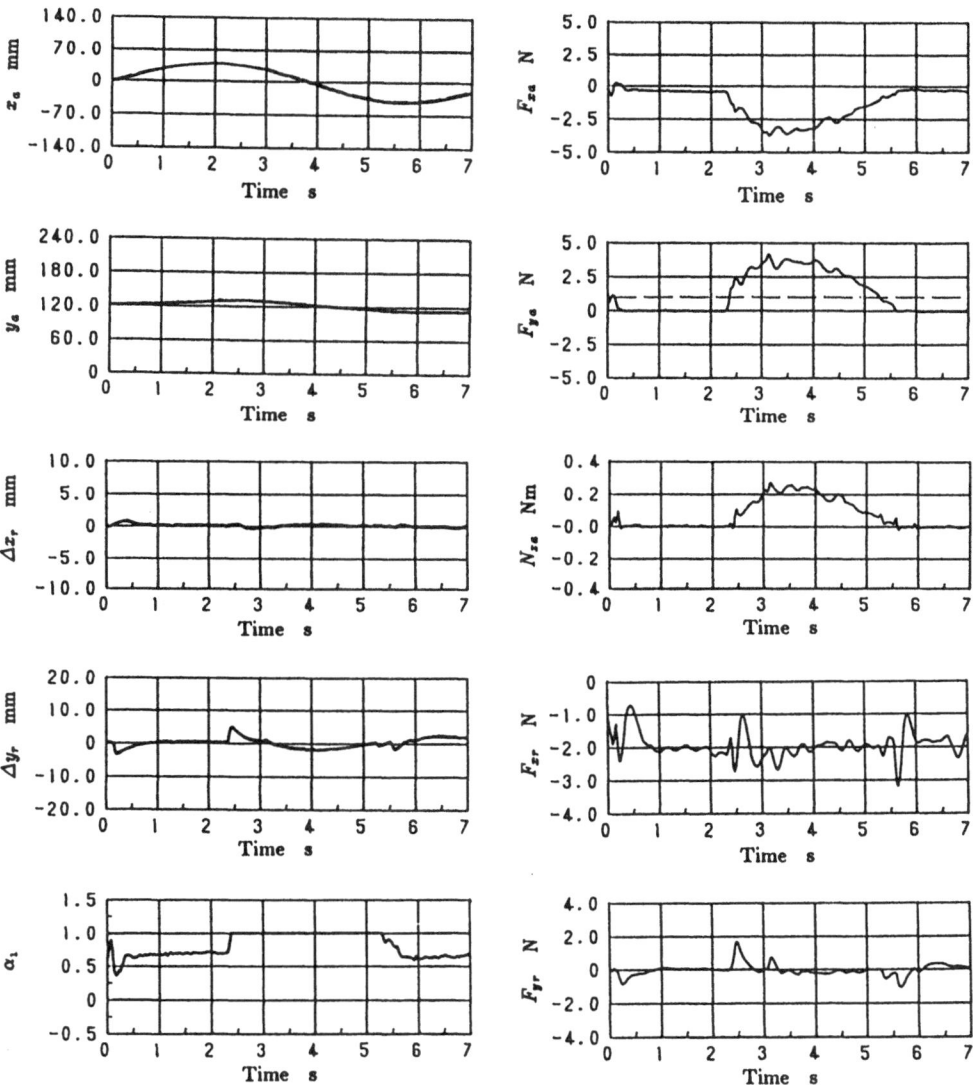

Figure 9: Experimental data (2) for the task presented in Figure 6 where N_{za} as well as F_{xa} is adaptively shared by the two arms.

Dynamic Hybrid Position/Force Control of Robot Manipulators: On-Line Estimation of Unknown Constraint

Tsuneo Yoshikawa* and Akio Sudou**

*Division of Applied Systems Science
Faculty of Engineering, Kyoto University
Uji, Kyoto 611, Japan

Abstract

For the application of robot manipulators to complex tasks, it is often necessary to control not only the position of a manipulator but also the force exerted by the end effector on an object. For this purpose, Raibert and Craig proposed the hybrid position/force control method. Extending their method, we proposed the dynamic hybrid control method which takes into consideration the manipulator dynamics and the constraints on the end effector specified by the given task. One difficulty in implementing our method is that we usually do not have precise information on the size and position of the object with which the end effector contacts. To cope with this difficulty a problem of dynamic hybrid control with unknown constraint is studied in this paper.

After introducing the dynamic hybrid control approach, we first develop an on-line estimation algorithm which estimates the local shape of the constraint surface from measured data. Then we show by experiments using a SCARA type robot that the combination of this algorithm with the dynamic hybrid control method works fairly well. This approach decreases the burden on the operator of giving precise data on the constraint and makes the dynamic hybrid control approach more practical.

1 Introduction

In order to apply robot manipulators to a wider class of tasks, it will be necessary to control not only the position of a manipulator but also the force exerted by the end effector on an object.

---

*Currently with Mitsubishi Heavy Industries, Ltd., Hyogo-ku, Kobe 652, Japan

Force control of manipulators has been studied by many researchers [2],[4], [6],[8],[9],[14],[15]. The hybrid control method proposed by Raibert and Craig[8] deals with the situation where the position of end effector must be controlled in certain directions and the force in other directions. In this approach, however, the manipulator dynamics has not been taken into account rigorously. A dynamic hybrid control method which is based on the dynamic equation of robot manipulators and a description of the end effector constraint by constraint hypersurfaces has been proposed in a previous paper[15]. It has been shown that, if the manipulator is not in a singular configuration, the desired position and force trajectories can be simultaneously realized.

For position control of manipulators, several controller designs have been proposed which are based on a two-step procedure: first, the system dynamics is linearized by an appropriate nonlinear state feedback, and then, a servo compensator is designed for this linearized model to cope with modelling errors and disturbances [10]-[13]. However, this procedure has not been established for constrained motion.

Based on the basic equations given in [15] and the two-step procedure mentioned above, an approach to designing controllers for dynamic hybrid control has been presented in [16]. Firstly, we have given a nonlinear state feedback law which linearizes the manipulator dynamics. Formulation of the constraint by the constraint hypersurfaces plays an essential role in establishing the linearizing law. Secondly, the position and force servo controllers for the linearized model have been designed using the concept of two-degree-of-freedom servo controllers [13]. The merit of this servo controller is that it can take into account both the command response and the robustness of the controller to modelling error and disturbance.

In spite of these efforts, we still have several difficulties in implementing our approach. One major difficulty is that we usually do not have precise information on the size and position of the objects which the end effector contacts. This prompted us to study the problem of dynamic hybrid control with unknown constraint, which is the main theme of this paper.

After discussing the theoretical background of our approach, we first develop an estimation algorithm which estimates the local shape of the constraint surface from data obtained on-line. Then we show by experiments using a SCARA robot that the combination of this algorithm with the dynamic hybrid control method works fairly well. This approach decreases the burden on the operator of giving precise data on the constraint surfaces and makes the dynamic hybrid control approach more practical.

There have been several research efforts related to estimation of the constraint surface for force control [1],[3],[5],[7]. Merlet [7] proposed using force measurement to determine the surface normal. Blaner and Belanger [1] reduced the problem to that of estimating some unknown parameters representing the constraint surface and proposed using the extended Kalman filter. Kazanzides et al.[5] proposed a method of determining the surface normal and tangent from the measured force and end effector velocity. The estimation algorithm in this paper is in the same line as [7] and [5]. However our method combines the position and force measurements to treat the three-dimensional situation directly and provides a means for cancellation of frictional force. Ish-Shalom [3] discussed the problem in a wider framework of robot languages.

2 Dynamic Hybrid Control Approach

2.1 Linearization by State Feedback

In this section, based on [15] and [16], we describe the dynamic hybrid control approach which consists of linearization of robot dynamics by a state feedback control law and a servo compensation of the resulted linearized system.

We consider a manipulator with n_q degrees of freedom, whose joint vector is denoted by an n_q-dimensional vector q and whose end effector position with respect to a fixed reference frame is denoted by an n-dimensional vector r $(n \leq n_q)$.

We make the following three assumptions:

(i) The constraint on the end effector position can be expressed by a set of m $(m \leq n)$ hypersurfaces

$$s_i(r) = 0, \quad i = 1,2,\ldots,m \tag{1}$$

which are differentiable twice and mutually independent in a subset S of n-dimensional end effector position space. These constraint surfaces are independent of the force applied by the end effector, in other words, the surfaces and the end effector (including the force sensor) have infinitely large stiffness. No friction acts between the constraint surfaces and the end effector.

(ii) The relation between the end effector position r and joint vector q is given by

$$r = c(q) \tag{2}$$

(iii) The arm dynamics is given by

$$M(q)\ddot{q} + h(q,\dot{q}) + V\dot{q} + g(q) = \tau \tag{3}$$

where " · " denotes the time derivative, $M(q)$ is the nonsingular inertia matrix ($M(q)$ will be written as M hereafter), $h(q, \dot{q})$ represents the centrifugal and Coriolis forces, V is the viscous friction matrix, $g(q)$ represents the effect of gravity, and τ is the joint driving force. The actuator dynamics can be neglected and τ can be treated as the input to the system.

To express the end effector position on the constraint surfaces we further assume that there exists a set of $(n-m)$ twice-differentiable scalar functions $\{p_1(r),\ p_2(r),\ \cdots,\ p_{n-m}(r)\}$ such that $\{s_i(r),\ i=1,\ 2,\ \cdots,\ m;\ p_j(r),\ j=1,\ 2,\ \cdots,\ (n-m)\}$ are mutually independent in S. We define a vectors r_p and r_p by $r_p=[p_1(r),\ p_2(r),\ \cdots,\ p_{n-m}(r)]^T$ and $r_p=[s_1(r),\ s_2(r),\ \cdots,\ s_m(r)]^T$. We also define $r_c=[r_p{}^T,\ r_p{}^T]^T$. The vector r_p specifies the end effector position on the constraint hypersurfaces (1), and that $s_i(r)=\text{const.}$ and $p_j(r)=\text{const.}$ provide a curved coordinate system fit for the given constraint.

The assumption of no frictional force does not hold in reality. This force will be taken care of by the servo-controller discussed later.

A very simple example will now be given to illustrate the concept of constraint hypersurfaces and the vector r_p. Consider a task of tracking the contour of a circular plate of radius $\alpha\ (>0)$ by a two-degree-of-freedom planar maipulator as shown in Fig.1(a). The end effector position r is denoted by $r=[x,\ y]^T$ and the constraint surface is specified by

$$s_1(r)=x^2+y^2-\alpha^2=0. \tag{4}$$

An example of scalar function $p_1(r)$ is given by

$$p_1(r)=\phi=\text{atan2}(y,x) \tag{5}$$

where ϕ denotes the angle between the X axis and the vector r as shown in Fig. 1(b). By taking, for example, $S=\{(x,\ y)^T \mid x^2+y^2 \geq \alpha^2/2\}$ it is obvious that $s_1(r)$ and $p_1(r)$ are mutually independent and $r_p=p_1(r)$ specifies the end effector position on the constraint surface. Fig.1(c) shows the curved coordinates for this case.

In preparation for giving a nonlinear state feedback control law to linearize the robot dynamics (3), we obtain the relations among velocities and accelerations of the joint vector q, the end effector position r, and the constraint frame vector r_c. First of all, differentiating r_c with respect to time, we obtain

$$\dot{r}_c = \begin{bmatrix} \dot{r}_p \\ 0 \end{bmatrix} = E\dot{r} \tag{6}$$

where

$$E = \begin{bmatrix} E_p \\ E_p \end{bmatrix} = [e_1,\ e_2,\ \cdots,\ e_n]^T \tag{7}$$

$$E_p = [e_1, e_2, \cdots, e_{n-m}]^T \tag{8a}$$

$$e_j = \frac{\partial p_j(r)}{\partial r} \ , \ j=1, 2, \cdots, n-m \tag{8b}$$

$$E_F = [e_{n-m+1}, e_{n-m+2}, \cdots, e_n]^T \tag{9a}$$

$$e_{n-m+i} = \frac{\partial s_i(r)}{\partial r} \ , \ i=1, 2, \cdots, m. \tag{9b}$$

Note that E specifies a local coordinate frame corresponding to the constraint frame introduced by Mason [6]. Matrices E_p and E_F represent the position and force conrol directions, respectively. Differentiating (6) once more we obtain

$$\ddot{r}_c = \begin{bmatrix} \ddot{r}_p \\ 0 \end{bmatrix} = E \ddot{r} + \dot{E} \dot{r}. \tag{10}$$

On the other hand, by differentiating (2) we obtain

$$\dot{r} = J \dot{q} \tag{11}$$

$$\ddot{r} = J \ddot{q} + \dot{J} \dot{q} \tag{12}$$

where

$$J \triangleq \partial c(q) / \partial q^T. \tag{13}$$

Let f, and $f_c = [f_p{}^T, f_F{}^T]^T$ be the generalized force representations of the force exerted by the end effector, corresponding to the generalized coordinate vectors r and $r_c = [r_p{}^T, r_F{}^T]^T$, respectively. Note that the constraint force is given by $-f$, or $-f_c$. Then from (6) we obtain

$$f_, = E^T f_c. \tag{14}$$

Since $f_p = 0$ due to the assumption of no friction force on the constraint surface, f_F represents the force applied to the constraint surfaces by the end effector. From (7), (14), and $f_p = 0$, we obtain

$$f_, = E_F{}^T f_F. \tag{15}$$

On the other hand from (11), joint driving force τ, which is equivalent to the force f, is given by

$$\tau_, = J^T f_,. \tag{16}$$

Therefore, when an arbitrary driving force command τ_c is applied to the manipulator under constraint (1), the manipulator dynamics is given by

$$M(q)\ddot{q} + h(q,\dot{q}) + V\dot{q} + g(q) = \tau - J^TE_r{}^Tf_r \qquad (17)$$

With these preparations, we can show that the joint driving force τ_c given by the following state feedback control law makes the closed loop system characteristics linear [16]:

$$\tau_c = \tau_p + \tau_r \qquad (18a)$$

$$\tau_p = M\ddot{q}_d + h(q,\dot{q}) + V\dot{q} + g(q) \qquad (18b)$$

$$\ddot{q}_d = J^+\{E^{-1}[\begin{smallmatrix}u_1\\0\end{smallmatrix}] - \dot{E}\dot{r}) - \dot{J}\dot{q}\} + (I-J^+J)k \qquad (18c)$$

$$\tau_r = J^TE_r{}^Tu_2 \qquad (18d)$$

where u_1 and u_2 are new input vectors whose physical meanings will be made clear later. In (18c) J^+ denotes the pseudoinverse of J, I is the identity matrix of appropriate dimension, and the vector k is an arbitrary time function. Then assuming that the manipulator is not in a singular configuration, using (10), (12), and (17), we can show that the characteristics of the closed loop system is given by a set of linear, decoupled equations:

$$\ddot{r}_p = u_1 \qquad (19)$$
$$f_r = u_2 \qquad (20)$$

From the above result it is now clear that the desired position trajectory can be specified through u_1 and the desired force trajectory can be specified through u_2. Note that in the linearized system given by (19) and (20), although the characteristics of the positional part described by (19) is dynamic, that of the force part described by (20) is static.

Assume that the desired position trajectory for r_p is given by $r_{pd}(t)$ and the desired force trajectory for f_r is given by $f_{rd}(t)$ where t is the time variable. Then if Eqs.(1)~(3) are exactly correct, the inputs $u_1 = \ddot{r}_{pd}$ and $u_2 = f_{rd}$ will achieve both the desired position and force. Due to modelling errors and unpredictable disturbances, however, the real response of this system may deviate from the desired one. In order to cope with this deviation, we have to design a servo controller for the linearized system (19) and (20) [10]-[13]. This will be done in the next subsection.

Remark 1: Khatib [4] has also noticed the importance of considering the arm dynamics in position and force control of manipulators. He proposed to specify given tasks by a generalized task specification matrix and to unify the active force control term into an operational space command vector. The main difference between the present approach and his is that the end effector constraint is described explicitly by the constraint hypersurfaces. This makes it possible to rigorously linearize and decouple the system dynamics with respect to position and force components by a nonlinear state feedback law.

2.2 Servo-Controller for Linearized Model

For the servo-controller design problem for linear systems there are a variety of approaches, such as optimal regulator, pole-assignment, PID controller, two-degree-of-freedom servo system, and so on. Any of these approaches can be applied to the linearized model (19) and (20). Here, we adopt the following control algorithm:

$$u_1 = \ddot{r}_{rd} + K_{rd}(\dot{r}_{rd} - \dot{r}_r) + K_{rp}(r_{rd} - r_r) \tag{21}$$

$$u_2 = f_{rd} + K_{fi} \int_0^t [f_{rd}(t') - f_r(t')]dt' \tag{22}$$

where $K_{rd} = \text{diag}(k_{rd})$, $K_{rp} = \text{diag}(k_{rp})$, and $K_{fi} = \text{diag}(k_{fi})$. Note that these algorithms consist of feedfoward terms ($\ddot{r}_{rd}$ and f_{rd}) and error feedback terms (PD feedback for position loop and I feedback for force loop). Also note that when we define the position and force errors by

$$e_r = r_{rd} - r_r \tag{23}$$

$$e_f = f_{rd} - f_r \tag{24}$$

the closed loop dynamics of the control system is obtained as

$$\ddot{e}_r + K_{rd}\dot{e}_r + K_{rp}e_r = 0 \tag{25}$$

$$\dot{e}_f + K_{fi}e_f = 0. \tag{26}$$

Hence by a proper choice of the gain matrices K_{rd}, K_{rp}, and K_{fi}, the position and force trajectories converge to the desired ones as time goes to infinity.

Remark 2: In [16] we have discussed a servo-controller design based on the theory of two-degree-of-freedom servo-systems. In that context, (21) and (22) corresponds to specifying the compensators C_1 and C_2 as follows in the basic structure of the two-degree-of-freedom servo-controller shown in Fig.2. For the position control loop ($G_r = (1/s^2)I$; s is the Laplace operator),

$$C_{1r} = I \tag{27}$$

$$C_{2r} = (k_{rd}s + k_{rp})I \tag{28}$$

and for the force control loop ($G_F = I$)

$$C_{1F} = I \tag{29}$$
$$C_{2F} = (k_{Fi}/s)I. \tag{30}$$

where subscripts P and F denote the position and force control loop, respectively. By properly selecting the parameters k_{Pd}, k_{Pp}, and k_{Fi}, we can obtain satisfactory compensators from the viewpoint of sensitivity and stability margin.

With these considerations, the total position/force hybrid control system is given by Fig.3.

3 Unknown Constraint Surfaces

One major difficulty with the dynamic hybrid control approach described in the previous section is that precise equation (1) of the constraint hypersurfaces is difficult to obtain, either due to lack of information on the shape and size of the object or due to inaccurate positioning of the otherwise precisely known object. Furthermore, even when it is possible to give the constraint hypersurfaces, it would be much better to be able to perform the hybrid conrol task without very precise description of the constraint.

Taking these points into account we consider in this section the problem of dynamic hybrid control for unknown constraint hypersurfaces. We first note that for calculating τ_c of the linearizing control law (18) at any instant of time, we do not need to know the whole function $s_i(r)$ but only the current values of E and $\dot{E}$ which represent a local property of the function $s_i(r)$. Hence if we can estimate the current values of E and $\dot{E}$ from available measurements, we would be able to perform dynamic hybrid control for unknown constraint hypersurfaces. In the following, such an estimation algorithm will be given.

We assume:

(i) The end effector position and the translational forces in three-dimensional space are of concern. The orientation need not be considered because either it is arbitrary or it is fixed. So the position vector r and force vector f_t are three-dimensional vectors (n=3).

(ii) The real trajectory of the end effector position r and force f_t is measurable.

(iii) The number of constraint hypersurfaces is known to be one (m=1) but its equation is not known. This means that the end effector position is constrained on an unknown smooth two-dimensional curved surface.

(iv) The end effector is required to track the intersection of an appropriately specified plane Q and the constrained surface in a specified direction while applying a specified force in the normal direction of the constrained surface. This plane is represented by

$$W(r)=0 \tag{31}$$

and will be called the virtual constraint plane hereafter. (Extension of Q to a curved surface could be done easiy.)

(v) An estimate of matrix E at the initial time, $\hat{E}_0$, is given.

Suppose that the end effector is moving on the constraint surface near the virtual constraint plane Q. Let $\triangle r$ be the vector expressing the current end effector position $r(t)$ from a position $r(t-\triangle t)$ on its trajectory which is a small distance back from the current position; $\triangle r = r(t) - r(t-\triangle t)$. We adopt the normalized vector of $\triangle r$

$$\tilde{e}_1 = \frac{\triangle r}{\| \triangle r \|} \tag{32}$$

as an estimate of the current tangent vector of the constraint surface. Based on the measured force $f_1(t)$ which is exerted on the surface by the end effector, we wish to determine the outward normal vector e_3 of the constraint surface. If the constraint surface is frictionless, we can obtain e_3 by $e_3 = -f_1 / \| f_1 \|$. However, the force f_1 usually contains a frictional force component. Therefore, by assuming that the frictional force is in the direction of end effector motion, we calculate the estimated normal vector $\hat{e}_3$ by

$$\hat{e}_3 = -\hat{f}_1 / \| \hat{f}_1 \| \tag{33}$$

where

$$\hat{f}_1 = f_1 - (f_1^T \tilde{e}_1) \tilde{e}_1 = f_1 - (f_1^T \triangle r) \triangle r. \tag{34}$$

Since we wish to have the end effector move on the virtual constraint plane Q, it is desirable to select a position control axis of the constraint frame in such a way that the axis is tangent to the constraint surface and parallel to the plane Q. A good estimate of this axis is given by

$$\hat{e}_1 = \frac{w \times \hat{e}_3}{\| w \times \hat{e}_3 \|} \tag{35}$$

where $\times$ denotes the vector product and w is a normal vector of the virtual plane Q:

$$w = \frac{\partial W(r)}{\partial r} . \tag{36}$$

Note that $\tilde{e}_1$ of (32) is not appropriate for e_1 because, although $\tilde{e}_1$ is normal to $\hat{e}_3$, it is not necessarily parallel to Q due to possible trajectory error. Also note that the direction of $\hat{e}_1$ is uniquely determined by the shape of the constraint surface and the normal direction of the virtual plane (see Fig 4). More specifically, when we look at the object from the negative side $(W(r)<0)$ of the virtual constraint plane, that is, in the direction of w, $\hat{e}_1$ is always surrounding the object in the clockwise direction. We determine the last direction by

$$\hat{e}_2 = \hat{e}_3 \times \hat{e}_1 \tag{37}$$

so that $\hat{e}_1, \hat{e}_2,$ and $\hat{e}_3$ form a right-hand coordinate system. Vectors $\hat{e}_1$ and $\hat{e}_2$ represents the position control direction and $\hat{e}_3$ is the force control direction. Let

$$\hat{E} = [\hat{e}_1, \hat{e}_2, \hat{e}_3] . \tag{38}$$

and

$$\widehat{\dot{E}} = \frac{\hat{E}(t) - \hat{E}(t - \triangle t)}{\triangle t} . \tag{39}$$

Using these values $\hat{E}$ and $\widehat{\dot{E}}$ for E and $\dot{E}$ in (18), we can calculate the linearizing control input τ_c.

Next we consider the servo-control algorithm. Let the velocity and acceleration of the end effector in the $\hat{e}_1$ and $\hat{e}_2$ directions be $\dot{r}_p = [\dot{r}_{p1}, \dot{r}_{p2}]^\tau$, $\ddot{r}_p = [\ddot{r}_{p1}, \ddot{r}_{p2}]^\tau$. Also let $u_1 = [u_{11}, u_{12}]^\tau$. We give the position control algorithm in the $\hat{e}_1$ direction by

$$u_{11} = \ddot{r}_{pd1} + K_{pd1}(\dot{r}_{pd1} - \dot{r}_{p1}) + K_{pp1} \triangle r_{p1} \tag{40a}$$

$$\triangle r_{p1} = \int_0^t (\dot{r}_{pd1} - \dot{r}_{p1}) dt' \tag{40b}$$

where $\dot{r}_{pd1}$ and $\ddot{r}_{pd1}$ are desired values for $\dot{r}_{p1}$ and $\ddot{r}_{p1}$, and matrics K_{pd1} and K_{pp1} are constant feedback gains. Since the desired velocity and acceleration in the $\hat{e}_2$ direction is zero, we give the control algorithm in that direction by

$$u_{12} = -K_{pd1} \dot{r}_{p2} + K_{pp2} \triangle r_{p2} + K_{pi2} \int_0^t \triangle r_{p2} dt' \tag{41a}$$

$$\triangle r_{p2} = -\frac{W(r)}{\hat{e}_2^\tau w} \tag{41b}$$

where $\triangle r_{r,i}$ is an approximated deviation of the end effector position from the virtual plane measured in the direction of e_i, and matrices $K_{P,i,i}$, $K_{P,i,i}$, and $K_{P,i,i}$ are constant feedback gains.

As for the force control algorithm in the $\hat{e}_i$ direction, we construct the control input u_i by

$$u_i = f_{r,i} + K_{P,i} \int_0^t (f_{r,i} - f_r) dt \qquad (42)$$

where matrix $K_{P,i}$ is a constant gain matrix. These servo-controllers are the same as (21) and (22), except for the integral term in (41a) which has been added to decrease the steady-state deviation of the end effector position from the virtual plane.

4 Experiments

4.1 Outline of the Robot System

In order to examine the effectiveness of the proposed approach, we conducted some simple experiments using a SCARA type robot. The overview of the robot system is shown in Fig. 5. The controller is a 16-bit personal computer (CPU 80286 and FPU 80287). Only the first three joints, which will be called joints 1,2 and 3, hereafter, of the robot were used to control the motion in three-dimensional space (see Fig.6). The joints are driven by D.C. servo motors with rotary encoders. Gear reductions are used with gear ratios 1/36.232 for the first joint and 1/24.068 for the third joint. The second joint is prismatic and a ball screw mechanism with 10mm/rotation is used. The lengths of links are $\ell_1 = 0.3$m and $\ell_3 = 0.25$m. As shown in fig.7, attached to the tip is a force sensor (manufactured by Omron Tateishi Electronics Co., force capacity 5kgf, torque capacity 20kgf·cm).

When joint variables q_i (i=1.2.3) are taken as in Fig. 6, and r is defined by $r = [x, y, z]^\tau$, then the Jacobian matrix J is given by

$$J = \begin{bmatrix} -\ell_1 \sin q_1 & 0 & -\ell_1 \sin q_1 - \ell_3 \sin(q_1 + q_3) \\ \ell_1 \cos q_1 & 0 & \ell_1 \cos q_1 + \ell_3 \cos(q_1 + q_3) \\ 0 & 1 & 0 \end{bmatrix} . \qquad (43)$$

For this arm, the identified dynamic equation is given by

$$\tau_1 = (0.740 + 0.096 \cos q_3) \ddot{q}_1 + (0.021 + 0.048 \cos q_3) \ddot{q}_3$$
$$- 0.048 \sin q_3 (2\dot{q}_1 + \dot{q}_3) \dot{q}_3 + 1.589 \dot{q}_1$$

$$\tau_2 = 22.152\,\ddot{q}_2 + 76.225\,\dot{q}_2 + 91.430$$
$$\tau_3 = (0.021 + 0.048\cos q_3)\,\ddot{q}_1 + 0.391\,\ddot{q}_3 \tag{44}$$
$$+ 0.048\sin q_3\,\dot{q}_1^2 + 0.210\,\dot{q}_3$$

in MKS units.

Parameters of the servo-controller have been selected by trial and error as

$$k_{P,1} = 625, \quad k_{P,4,1} = 30$$
$$k_{P,2} = 847, \quad k_{P,4,2} = 44, \quad k_{P,t,2} = 1331 \tag{45}$$
$$k_{F,t} = 10.$$

The C language is used and the sampling time is 9.8 ms.

4.2 Experimental Results

As is shown in Fig. 8, the constraint surface is the bottom of a stainless steel kitchen bowl which is set upside-down on a table. Its exact shape and position is unknown to the robot.

(A) Experiment 1

The initial position of the end effector is $[0.526, -0.118, -0.062]^T$. The virtual constraint plane is specified as a horizontal plane, i.e. a plane including the above initial position and having $[0,0,-1]^T$ as the normal vector. The desired trajectory is to remain at the initial position for 1.0 second, then to move in the e_1 direction with a trapezoidal velocity curve (with constant acceleration of $0.2 m/s^2$ for the first 0.02m, with zero acceleration for the middle 0.16m, and with constant deceleration of $-0.2 m/s^2$ for the last 0.02m). The desired force trajectory is 10N. An experimental result is shown in Fig.9. Fig.9(a) is the velocity response in the $\hat{e}_1$ direction, Fig.9(b) is the position error in the $\hat{e}_2$ direction, and Fig.9(c) is the force response in $\hat{e}_3$ direction. Except for some error in force response at the begining of acceleration and of deceleration, the robot tracks the desired trajectory fairly well. The $\hat{e}_3$ directional position error is invisible in the figure probably because the desired trajectory does not need a motion in Z direction and the Coulomb friction is dominant.

Figs.10 and 11 show the estimated directions $\hat{e}_1$ and $\hat{e}_3$ of the constraint frame. Although there is a little fluctuation at the beginning of motion, these estimates are otherwise quite smooth and seem to be good ones.

(B) Experiment 2

This experiment is the same as Experiment 1 except that the virtual constraint plane is now specified as a vertical plane including the initial position of the end effector and having $[-1, -1, 0]^{\tau}$ as the normal vector. The result is shown in Fig.11, 12, and 13. As is seen from Fig.12(b) the position error in the $\hat{e}_z$ direction appears in this case, because the desired trajectory needs a motion in Z direction.

The results of the above two experiments show the validity of the proposed method.

5 Conclusion

The problem of dynamic hybrid control with unknown end effector constraint has been studied. Assuming that the force exerted on the constraint surface by the end effector is measurable, an algorithm has been proposed for on-line estimation of the local shape of the constraint surface from the data of position and force. It has been shown by experiments using a SCARA robot that the combination of the estimation algorithm with the dynamic hybrid control method works well. This approach saves the operator from the tedious task of giving precise constraint hypersurfaces to the robot controller. It is hoped that this makes the dynamic hybrid control approach more practical.

References

[1] M. Blauer, and P.R. Belanger, "State and Parameter Estimation for Robotic Manipultors Using Force Measurements", IEEE Trans. on Automatic Control, Vol.AC-32, No.12, December 1987.

[2] H. Hogan, "Impedance Control; An Approach to Manipulation: Part I-Theory; Part II-Implementation; Part III-Applications", ASME Journal of Dynamic Systems, Measurement and Control, vol. 107, pp.1-24, March 1985.

[3] J. Ish-Shalom, "The CS Language Concept: A New Approach to Robot Motion Design", The International Journal of Robotics Research, Vol.4, No.1, Spring 1985.

[4] O. Khatib "A Unified Approach for Motion and Force Control of Robot Manipulators; The Operational Space Formulation", IEEE Journal of Robotics and Automation, vol. RA-3, no.1, pp.43-53, February 1987.

[5] P. Kazanzides, N.S. Bradley, and W.A. Wolovich, "Dual-Drive Force/Velocity Control; Implementation and Experimental Results", IEEE Int. Conf. on

Robotics and Automation, pp.92-97, May, 1989.

[6] M. T. Mason, "Compliance and Force Control for Computer Controlled Manipulators", IEEE Trans. on Systems, Man, and Cybernetics, vol.SMC-11, pp.418-432, June 1981.

[7] J.P. Merlet, "C-Surface Applied to the Design of a Hybrid Force-Position Robot Controller", Proc. 1987 IEEE Int. Conf. on Robotics and Automation, Vol.2, pp.1055-1059, Raleigh, NC, April 1987.

[8] M. H. Raibert and J. J. Craig, "Hybrid Position/Force Control of Manipulators", Trans. ASME, J. DSMC, vol.103, pp.126-133, June 1981.

[9] K. J. Salisbury, "Active Stiffness Control of A Manipulator in Cartesian Coordinates", 19th IEEE Conference on Decision and Control, pp.95-100, Nov. 1980.

[10] C. Samson, "Robust Nonlinear Control of Robotic Manipultors", Proc. 22nd IEEE Conf. on Decision and Control, San Antonio, pp.1211-1216, December 1983.

[11] J.J. Slotine, "Robust Control of Robot Manipulators", Int. J. Robotics Res., vol.4, no.2, pp.49-64, Summer 1985.

[12] M.W. Spong and M. Vidyasagar, "Robust Linear Compensator Design for Nonlinear Robotic Control", IEEE Journal of Robotics and Automation, vol.RA-3, no.4, pp.345-351, August 1987.

[13] T. Sugie, T. Yoshikawa and T. Ono, "Robust Controller Design for Robot Manipultors", ASME Journal of Dynamic Systems, Mesurement and Control, vol.110, no.1, pp.94-96, March 1988.

[14] D.E. Whitney, "Historical Perspective and State od the Art in Robot Force Control", Proc. 1985 IEEE Int. Conf. on Robotics and Automation, pp.262-268, 1985.

[15] T. Yoshikawa, "Dynamic Hybrid Position/Force Control of Robot Manipulators — Description of Hand Constraints and Calculation of Joint Driving Force ", IEEE Journal of Robotics and Automation, vol.RA-3, no.5, pp.386-392, October 1987.

[16] T. Yoshikawa, T. Sugie, and M. Tanaka, "Dynamic Hybrid Position/Force Control of Robot Manipulators -Controller Design and Experiment", IEEE Journal of Robotics and Automation RA-4, 6, pp.699-705, December, 1988.

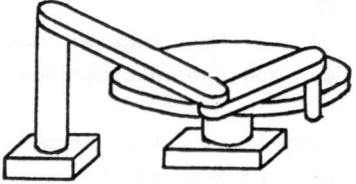

(a) Overview of the task.

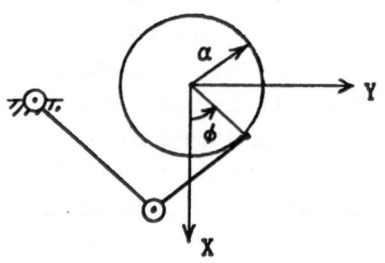

(b) Schematic diagram in XY plane.

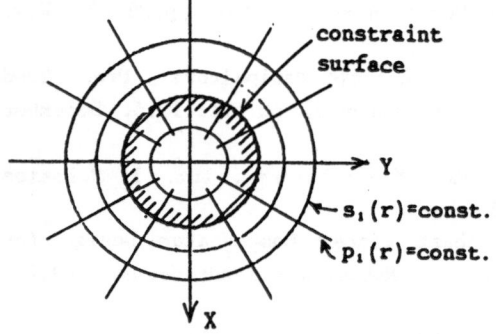

(c) Curved coordinates.

Figure 1   Constraint hypersurface and vector $r_,$.

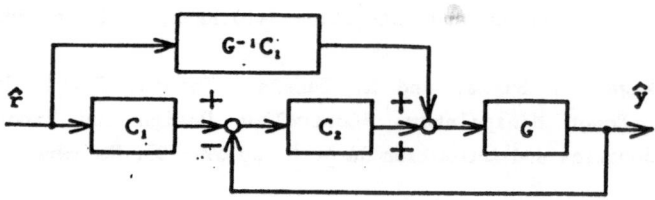

Figure 2   Two-degree-of-freedom servo system.

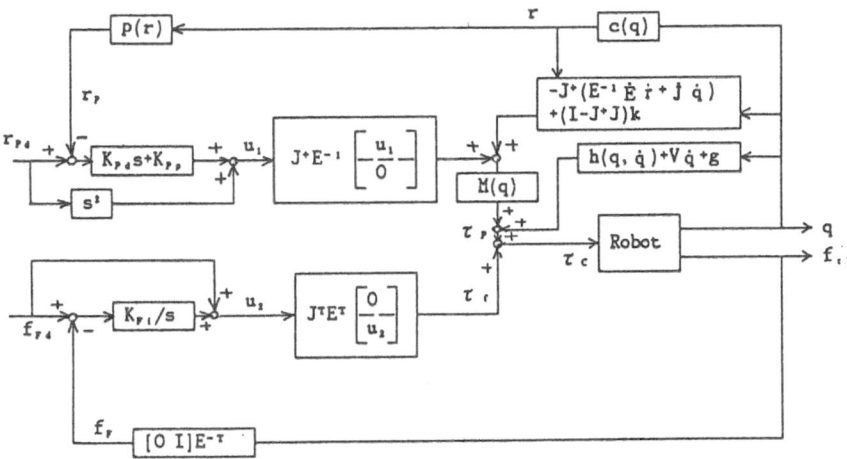

Figure 3 Block diagram of dynamic hybrid control system.

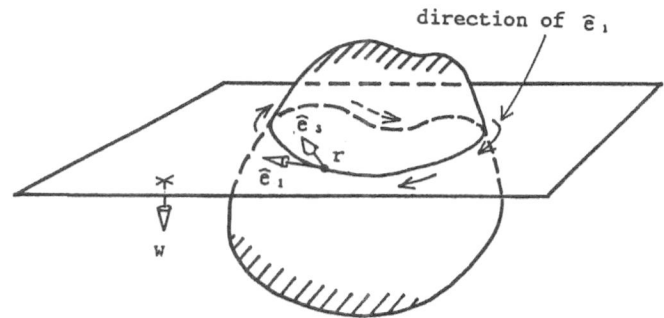

Figure 4 Vectors $\hat{e}_1$ and $\hat{e}_3$.

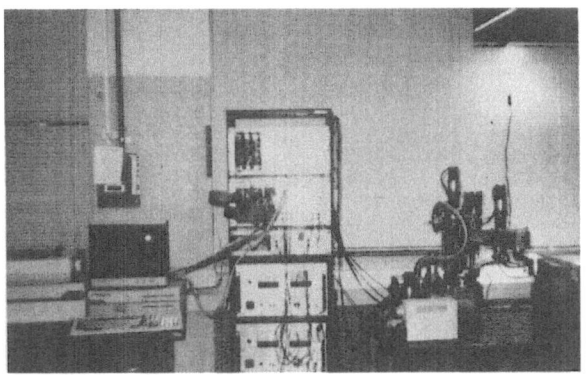

Figure 5 Overview of the robot system.

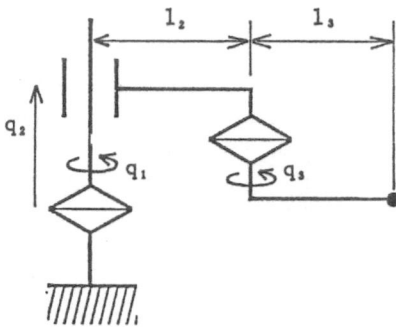

Figure 6 Mechanism of the manipulator.

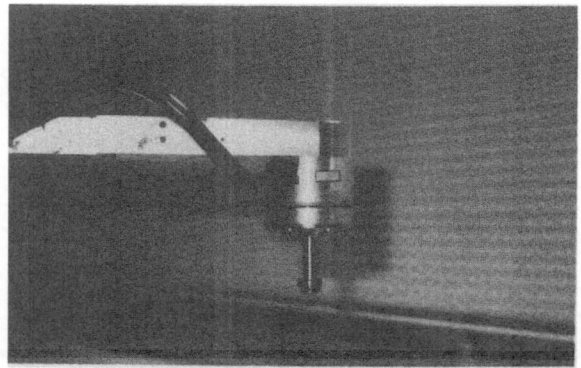

Figure 7 Six-axis force sensor.

Figure 8 Constraint surface used in experiment.

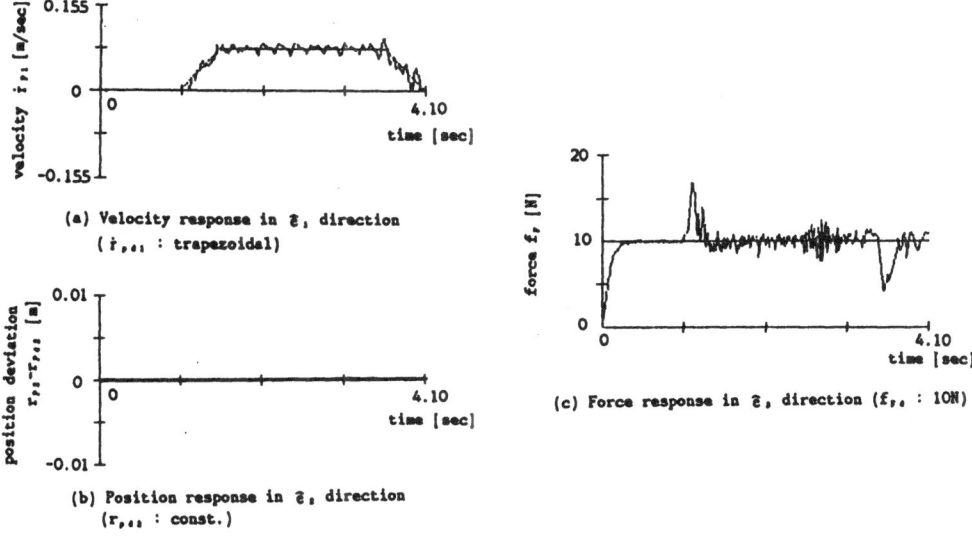

(a) Velocity response in $\bar{z}_1$ direction
($\dot{r}_{,a_1}$: trapezoidal)

(b) Position response in $\bar{z}_1$ direction
($r_{,a_1}$: const.)

(c) Force response in $\bar{z}_1$ direction ($f_{,a}$: 10N)

Figure 9 Result of Experiment 1.

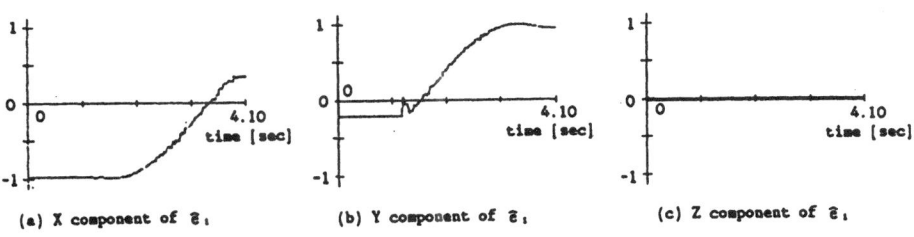

(a) X component of $\hat{c}_1$

(b) Y component of $\hat{c}_1$

(c) Z component of $\hat{c}_1$

Figure 10 Estimate $\hat{c}_1$ for Experiment 1

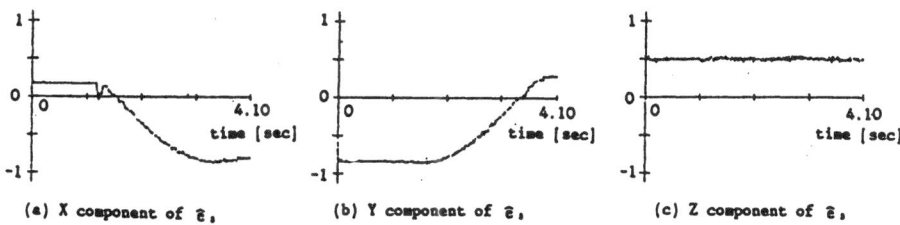

(a) X component of $\hat{c}_2$

(b) Y component of $\hat{c}_2$

(c) Z component of $\hat{c}_2$

Figure 11 Estimate $\hat{c}_2$ for Experiment 1

134

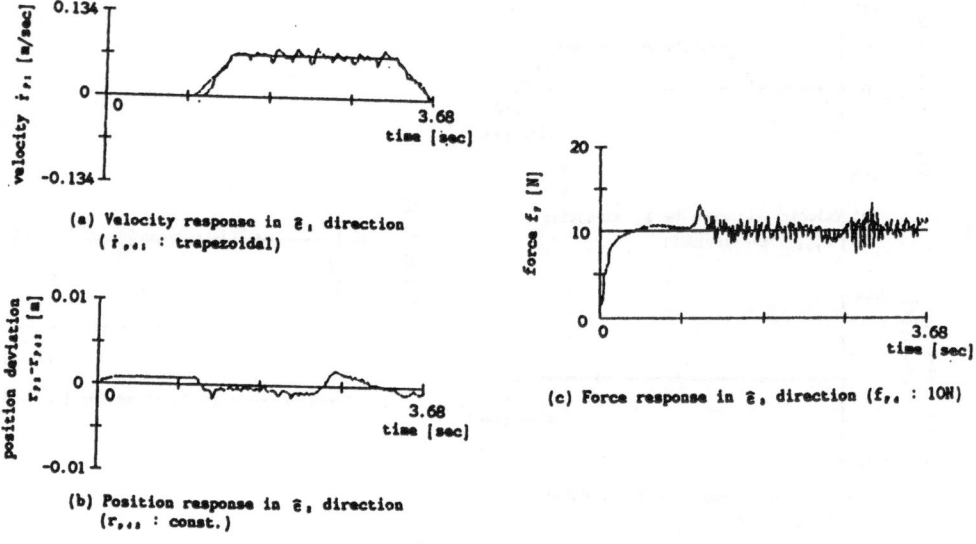

Figure 12 Result of Experiment 2

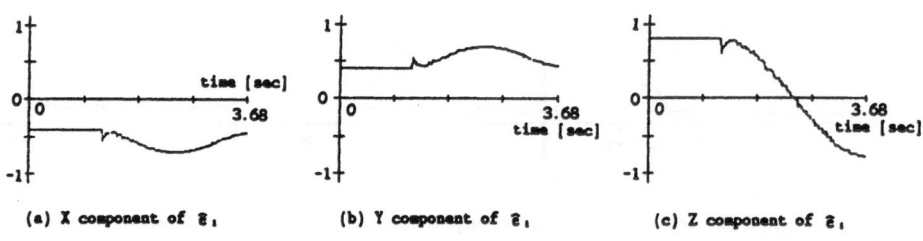

Figure 13 Estimate $\hat{e}_1$ for Experiment 2

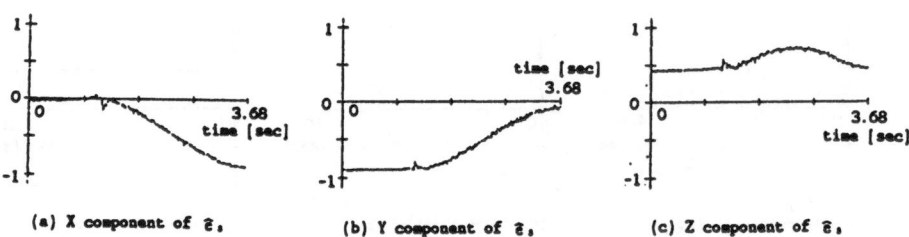

Figure 14 Estimate $\hat{e}_2$ for Experiment 2

Hidden Markov Model Analysis of Force/Torque Information in Telemanipulation

Blake Hannaford and Paul Lee

Jet Propulsion Laboratory,
California Institute of Technology, Pasadena, CA 91109

Abstract

A new model is developed for prediction and analysis of sensor information recorded during robotic performance of tasks by telemanipulation. The model uses the Hidden Markov Model (Stochastic functions of Markov Nets) to describe the task structure, the operator or intelligent controller's goal structure, and the sensor signals such as forces and torques arising from interaction with the environment. The Markov process portion encodes the task sequence / sub-goal structure, and the observation densities associated with each sub-goal state encode the expected sensor signals associated with carrying out that sub-goal. Methodology is described for construction of the model parameters based on engineering knowledge of the task. The Viterbi algorithm is used for model based analysis of force signals measured during experimental teleoperation and achieves excellent segmentation of the data into sub-goal phases. The Hidden Markov Model achieves a structured, knowledge based model with explicit uncertainties and mature optimal identification algorithms.

1   Introduction

Telemanipulation is the ability to control a remote robotic manipulator to perform tasks. Interest has recently increased in the measurement of telemanipulation performance and the modeling of remote task performance. Of course performance critically depends on the quality of sensory information fed back to the operator from the robotic workcell. Force and torque information is best sensed with a 6 axis load cell at the robot wrist. An important channel for this information to reach the operator is kinesthetic force feedback (KFF). In this modality, the operator commands manipulator motion through a single multi-dof joystick and the joystick exerts forces and torques on the operator proportional to those sensed at the robot wrist.

Recent experimental work in telemanipulation (Hannaford, 1987, Draper, 1988, Hannaford & Wood, 1989) has used force-torque information as a source of task performance information by recording force/torque time functions during task performance and computing performance measures on the information.

The operator performs different sub-tasks during a task. For example, she moves the manipulator through free space, makes contact with a surface, applies a force, etc. Each sub-task corresponds to a particular mental state of the operator during which she has a specific sub-goal in mind. Each sub-task contains a mixture of task-defined and exploratory movement. By the standards of production line robotics and automation, typical tasks performed in telemanipulation are poorly characterized. Usually, little or no task related information such as dimensions, location, or orientation of the manipulated object is known to the control system. All of this information must be determined by the operator through an interactive sensory/motor process.

The typical task considered here emphasizes contact between manipulator and environment. Because the tools and tasks are typically made of metal, their stiffness is high and contact is discontinuous. The operator's inputs to the system are typically increments of motion corresponding to sampled control outputs (Bekey, 1962, Navas & Stark, 1968).

The resulting force/torque information is a fast changing signal which although arising from deterministic processes, has much of the appearance of randomness because it is a high-gain function of the operator's mental state. Task knowledge is of some use in predicting these forces/torques. For example, if the task is to insert a connector into a socket, it is reasonable to expect a certain sequence of force and torque vectors based on the design of the connector and socket. These forces are the minimum necessary in an engineering sense to complete the task with perfect knowledge. However, to these minimum forces are added the forces generated by operator actions carried out for the purpose of exploration and identification of the task.

This paper presents a model of the forces generated by an operator performing telemanipulation tasks. It is based on the Hidden Markov Model (HMM) formulation of stochastic functions of a Markov Process. HMM's have recently scored successes applied to the identification of speech (Baker, 1975, Rabiner, et.al., 1983, Reddy, 1988, Rabiner, 1989). Speech is in many ways a similar signal to the force/torque signals considered here because it is a noise-like signal modulated by actions corresponding to mental states. In the introduction of this paper we will use the term "force" to refer to the complete 6 vector of forces and moments. The computations performed below on actual data use only the x axis force signal. We make the case below that the usefulness of HMMs is not limited to force information but can be used for multiple modalities and potentially for the fusion of their data.

A HMM is a Markov random process which cannot be observed directly. Instead, each state of the Markov model specifies a probability density from which observations are generated. In the case in which these densities are non-overlapping, the observation problem is trivial. In the more interesting cases, the densities overlap and the problem of observing the state sequence of the underlying Markov model depends on optimally combining the information on state transition probabilities with the state dependent observation densities and the observation at a particular time.

In an excellent overview of the HMM work in speech recognition, Rabiner & Juang (1986) discussed three problems: to compute the probability of an observation sequence given a particular HMM, to compute the most likely state sequence from an observation sequence and HMM, and to identify the HMM from an observation sequence or set of sequences.

In speech recognition with HMM's the hidden states may or may not correspond to phonemes. The observations are the spectral parameters or linear prediction coefficients computed using a data window centered at a given time. Training data can be used to generate HMMs for each word to be recognized (problem 3) (Levinson, et.al., 1983). The probability of each utterance can be computed for the stored HMMs (problem 1) and the most likely model can be selected. Alternatively, the models may be ranked according to how likely each one is to have generated the utterance, and further probabilistic reasoning can be performed with a higher level HMMs (Baker, 1975).

Our model is to associate the Markov states with operator's mental states or sub-goals. The transition probabilities encode the difficulty with which the operator completes each sub-goal of the task. Our knowledge of the task is sufficient to come up with approximate probability density functions for the forces and torques expected in each state of manipulation.

Our goal is not recognition of the complete force/torque sequence but tracking of the transitions between sub-goals in the operator's mind en route to task completion (problem 2). An algorithm presented in Rabiner and Juang (1986) which is ideally suited to this task is the Viterbi algorithm (Viterbi, 1967). This Dynamic Programming method finds the most likely sequence of state transitions given a HMM and a data set. It also yields a computation of the probability of this state sequence.

A previous approach to segmentation of the force record is a syntactic method (Hannaford & Wood, 1989) which encodes task knowledge as a strict sequence of states specified by the task descriptions. Each state defines an appropriate boolean combination of force thresholds which permit transition to the next state. The main problem with this method is that in practice the task-related forces are highly variable and thus no fixed set of thresholds is adequate to describe the full range of forces encountered in many repetitions of a task. The HMM formalism increases the robustness with which state sequences can be identified because it explicitly allows for uncertainty in the force signals.

This modeling study draws on a body of experimental experience arising from the performance evaluation of experimental manipulation systems (Hannaford, 1987, Hannaford & Wood 1989). In these experiments, forces and torques were recorded from multiple operators performing a mix of tasks in several control modes. The primary tools of analysis in these studies were two functionals computed from the force data, Completion Time (CT), and Sum of Squared Force (SOSF). The lower the values of these functionals, the better the manipulation performance. These quantities could be averaged among different populations of experiments to produce reduced statistical results of the effects of different control modes on performance.

In this study, we use the raw force information and attempt to find models which generate realistic force time functions for a given task, and which allow segmentation of the experimental task data into phases in a robust and knowledge driven way.

2   Methods

We collected experimental data during remote manipulation experiments conducted at the JPL Man-machine Systems Laboratory (Hannaford and Wood, 1989). The basic data consist of 6 channels of force/torque information in addition to jaw opening and finger clamping force information acquired at 100 samples per second.

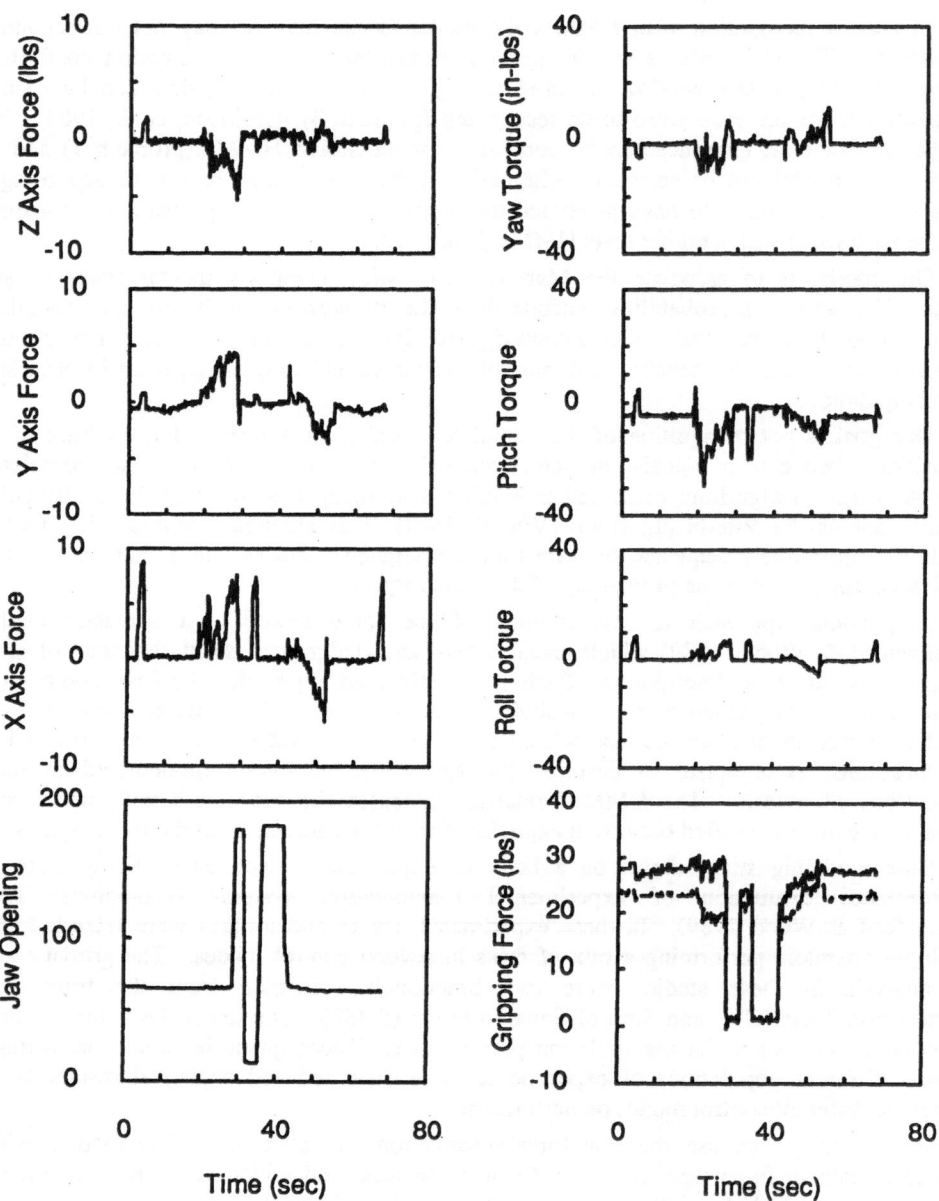

Figure 1. Complete data set recorded from a single repetition of the peg-in-hole experiment (Hannaford & Wood, 1989). Shown are *x*, *y*, and *z* values of force, and the moments about those axes, roll, pitch, and yaw, expressed in the tool frame. Task definition included taps — momentary contact with the task board — to provide reference points in the task. Peg insertion corresponded to predominantly positive forces in the *x* direction, extraction predominantly negative forces.

An example of a data record obtained from a single repetition of a high precision peg-in-hole task (Figure 1) shows that although much of the signal is rapidly changing or noise-like, there is also a structure apparent in the data.

This particular task was performed according to a specified sequence. The sequence began with the peg grasped in the hand. The steps were as follows:

Tap the peg on a marked location on the task board. Translate to the hole opening. Insert and release the peg. Translate to the back of the peg. Tap the peg end. Regrasp and extract the peg. Translate back to the mark. Tap the mark.

The noisy force signal can be interpreted with knowledge of the task. For example, the x axis of the force sensing coordinate system is aligned with the direction of successfully inserting or extracting the peg. Positive forces therefore correspond to a force exerted into the task board. The task related forces alternate between quiet periods corresponding to translation of the end effector in free motion and periods of large sustained forces during contact. The task was designed to include the taps which provide easily recognizable benchmarks in the force record. At each tap point, the operator made momentary contact between the manipulator (or the grasped object) and the task board. Taps are evident in the force record (especially in the x axis of force) at t=5 sec., t=34 sec., and t=68 seconds. The contact forces (t=18 through 28 seconds and t=42 through 54 seconds) correspond to insertion and extraction of the peg. During each phase, forces are predominantly in the direction required to perform the task, i.e. positive during insertion and negative during extraction.

With the hidden Markov model, we seek a way to encode this knowledge of the task and to model this force signal. The essential analogy is that different phases of the task correspond to different sub-goals of the operator, and that during each such mental state we expect to see a certain pattern of forces. The transitions between mental states or sub-goals are modeled by the Markov state transition probability matrix,

$$
A = \begin{bmatrix} a_{11} & a_{12} & \cdots \\ a_{21} & a_{22} & \\ & & a_{NN} \end{bmatrix}
$$

where a_{ij} = the probability of transition from state i to state j at a given time step. In each state we know to expect certain force signals, F(t), to be generated. The expected forces are specified by the state dependent observation densities, $B_i(f)$. Where

$$B_i(f) = Prob(F(t) = f \mid State = i)$$

For discrete valued observations, f_d, $B_i(f_d)$ is a probability distribution. For continuous valued observations, f_c, $B_i(f_c)$ is a probability density function. For practical purposes, these can be treated equivalently (Rabiner et al., 1985). Thus, we statistically define the force observations we expect from each sub-goal, as well as the transition dynamics between sub-goal states.

Classes of HMMs: As in the case of speech recognition (Rabiner, 1989), in modeling manipulation it is useful to constrain the state transition matrix to enforce the sequential nature of the tasks. One of the simplest forms of the constrained HMMs is the "Simple Left to Right" (SLR) model (Figure 2-a). In this case,

$$a_{ij} = 0 \qquad i < j \text{ or } i > j+1$$

This constraint means that each state can transition only to itself or the next highest numbered state. An additional constraint on all HMMs is of course that

$$\sum_j a_{ij} = 1$$

A slightly relaxed form of this model is the "Augmented Left to Right" (ALR) model (Figure 2-b) in which

$$a_{ij} = 0 \qquad i < j \text{ or } i > j+3$$

and thus transitions are permitted to the same state, to the next state, or which skip ahead one or two states.

Computer Implementation

Our software implemented these algorithms in a flexible C language package in which the HMM could be prepared in a human readable text file (Figure 3). This file contains the number of states in the HMM, N, a naming scheme in which the states are identified by their specific instatiation of a generic sub-task (such as translation) which specifies their observation densities, the transition probability matrix: A, and the observation densities parameterized as Gaussian distributions, $B_j(f)$.

(a) Simple Left-to-Right (SLR) model

(a) Augmented Left-to-Right (ALR) model

Figure 2. Two types of constrained Hidden Markov Models useful in modeling task progression.

```
# THIS MODEL DESCRIBES A PEG-IN-HOLE TASK
# WITH ALL TAPS, INSERTS, & EXTRACTS.
#
no_states:11
#
#   states:   1    2    3    4    5    6    7    8    9    10   11
state_no:    m1   t1   m2   i    m3   t2   m4   x    m5   t3   m6
#
init_dis:    .9  .09  .01   0    0    0    0    0    0    0    0
#
#   A[i,j]:   1    2    3    4    5    6    7    8    9    10   11
#
a[1,j]:      .75  .25   0    0    0    0    0    0    0    0    0
a[2,j]:       0   .75  .25   0    0    0    0    0    0    0    0
a[3,j]:       0   .0   .85  .15   0    0    0    0    0    0    0
a[4,j]:       0    0    0   .95  .05   0    0    0    0    0    0
a[5,j]:       0    0    0    0   .85  .15   0    0    0    0    0
a[6,j]:       0    0    0    0    0   .75  .25   0    0    0    0
a[7,j]:       0    0    0    0    0    0   .85  .15   0    0    0
a[8,j]:       0    0    0    0    0    0    0   .95  .05   0    0
a[9,j]:       0    0    0    0    0    0    0    0   .85  .15   0
a[10,j]:      0    0    0    0    0    0    0    0    0   .75  .25
a[11,j]:      0    0    0    0    0    0    0    0    0    0   1.
#
no_state_types:4
#            code    mu      sigma
move:        m       0.0     0.2
tap:         t       7.0     1.5
insert:      i       2.0     1.2
extract:     x      -2.      1.7
#
no_data:             200
delt(data/sec):2
seed:                379
```

Figure 3. Model Description File. ASCII text file describing the SLR HMM used in this study. File defines state names and types, state transition matrix, A, and parameters for Gaussian observation densities for the four state types.

This file was an input to two modules. One simulated the output of a given HMM, and the other applied the Viterbi algorithm to a data stream. We could thus verify the Viterbi algorithm software by generating HMM outputs and tracking their state transitions with the Viterbi algorithm.

Heuristic HMM Construction

To develop a HMM for a given task, the first step is to define a set of sub-tasks corresponding to distinct observable patterns of sensor information (force in this case). For example, for the peg in hole task, the sub-tasks are free space motion (negligible contact forces), tapping (brief positive force), insertion (sustained predominantly positive forces), extraction (sustained predominantly negative forces).

The observation probability density function (PDF) for each sub-task is selected from the task knowledge of the experimenter. In our implementation, these PDFs are approximated as Gaussians although this is not an assumption of the mathematical method, and any appropriate PDF could be used. The means and variances in force based on the above descriptions are:

Sub-task	Mu	Sigma
move:	0.0	0.3
tap:	8.0	1.7
insert:	6.0	2.6
extract:	-4.0	2.5

The next step is to decide on the HMM topology. A strict state sequence can be enforced by selecting a simple left-to-right model. A less constrained but still forward moving configuration is the augmented left-to-right model. In this example, we will apply the simple left-to-right model using the sequence:

m,t,m,i,m,t,m,x,m,t,m

where the sequence is encoded by:

m	Move
t	Tap
i	Insert
x	Extract

This leads to an 11 state HMM where the states are named:

1	2	3	4	5	6	7	8	9	10	11
m1	t1	m2	i	m3	t2	m4	x	m5	t3	m6

to keep them distinct.

Finally, we must determine the non-zero entries of the state-transition matrix, A. For this we use a simple heuristic: choose the probabilities so that the mean time in each state is equal to the mean observed time in the state.

Consider a state in a HMM which at each time sample has probability p of remaining in the same state and 1 - p of exiting. The mean number of time samples spent in the state, $\bar{n}$ is

$$\bar{n} = \frac{p}{1 - p}$$

and it can easily be derived that

$$p = \frac{\bar{t}}{dt + \bar{t}}$$

where dt is sampling period, and $\bar{t}$ is the mean time in the state.

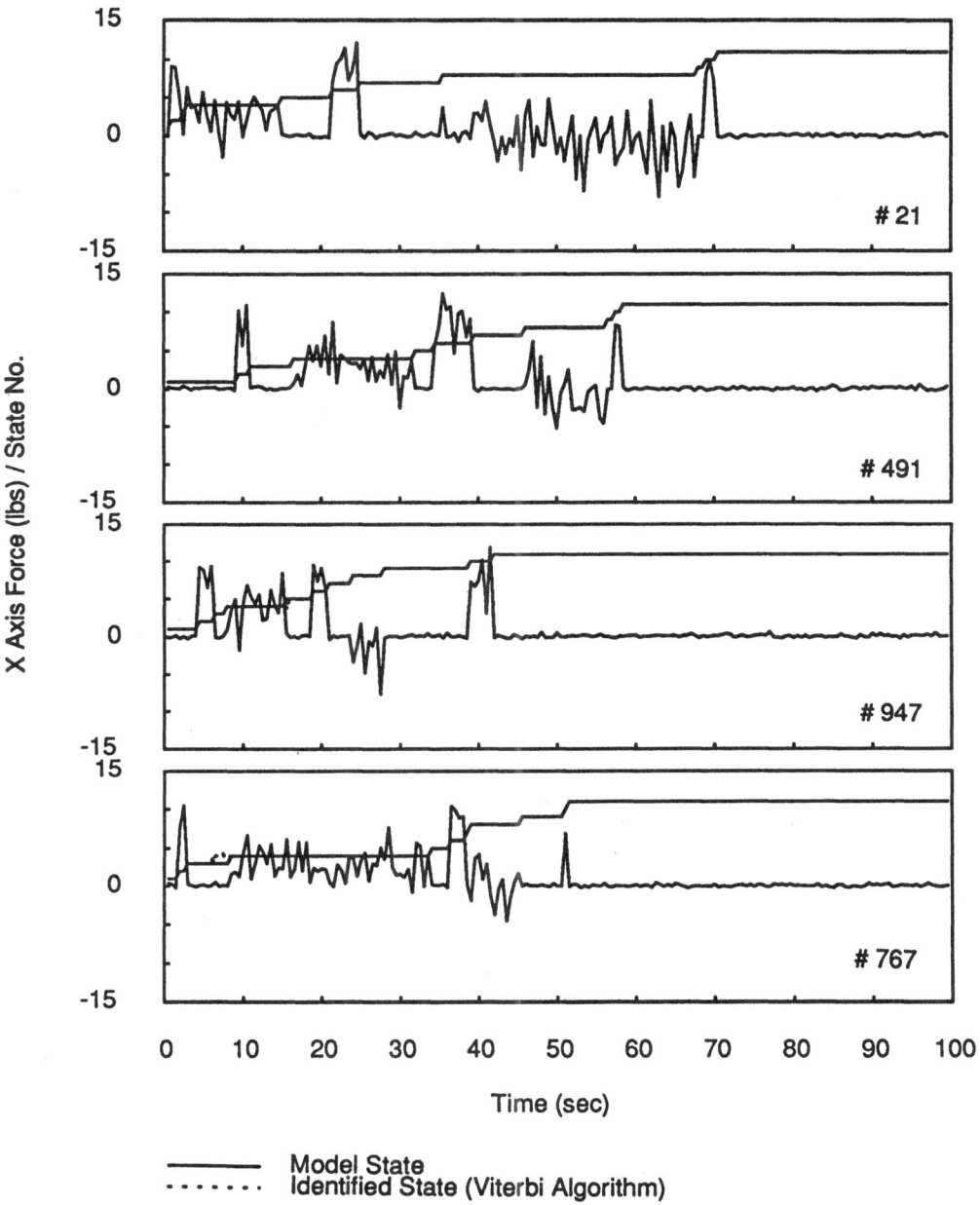

Figure 4. Four simulations of x axis of force in the peg-in-hole task by the HMM specified in Figure 3. In addition to force, the state number is plotted; as actually occurred in the model (solid) and as estimated by the Viterbi algorithm (dashed). The two are in perfect registration except at one point.

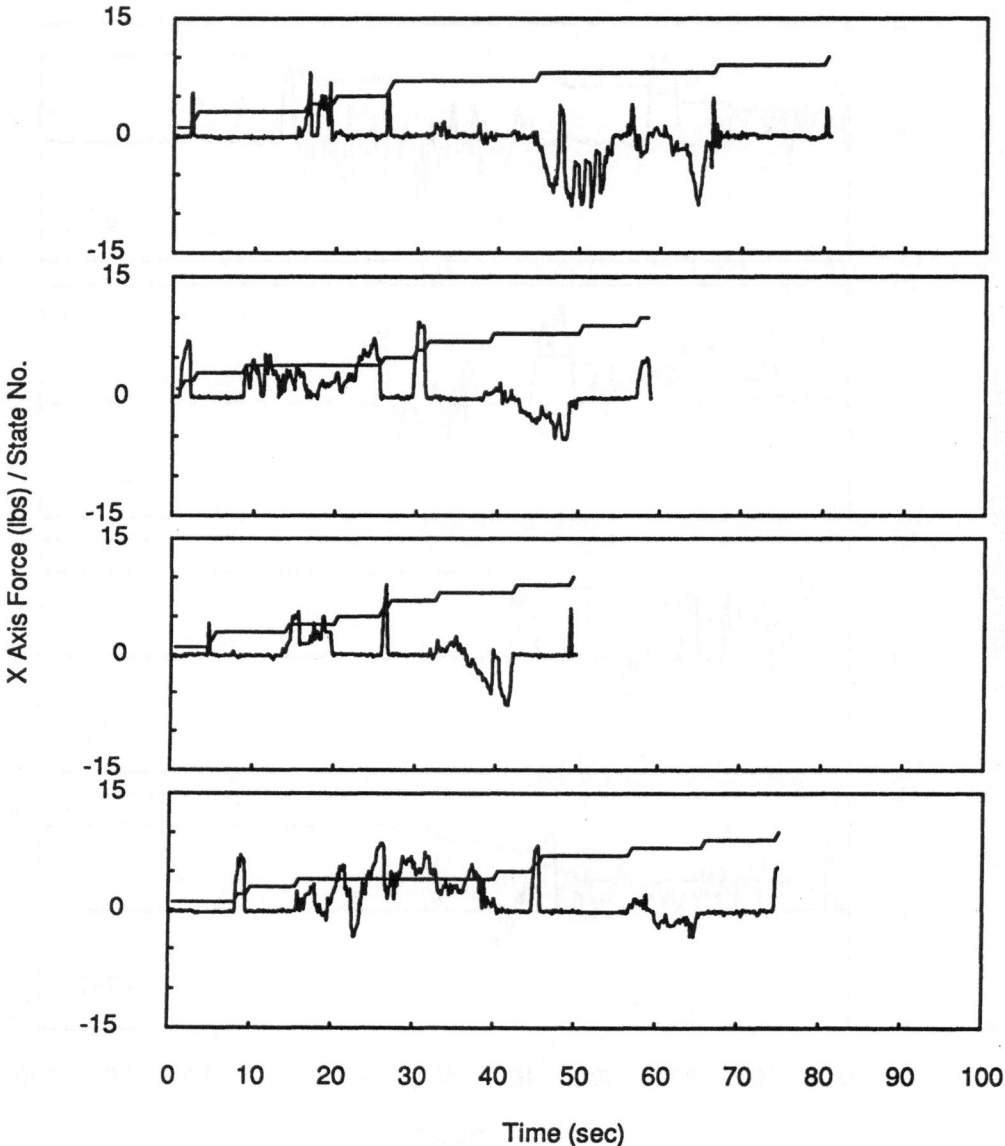

Figure 5. Four experimental data records segmented into task states by the Viterbi algorithm and the HMM in Figure 3. The records are selected to be qualitatively similar to the HMM simulations of Figure 4.

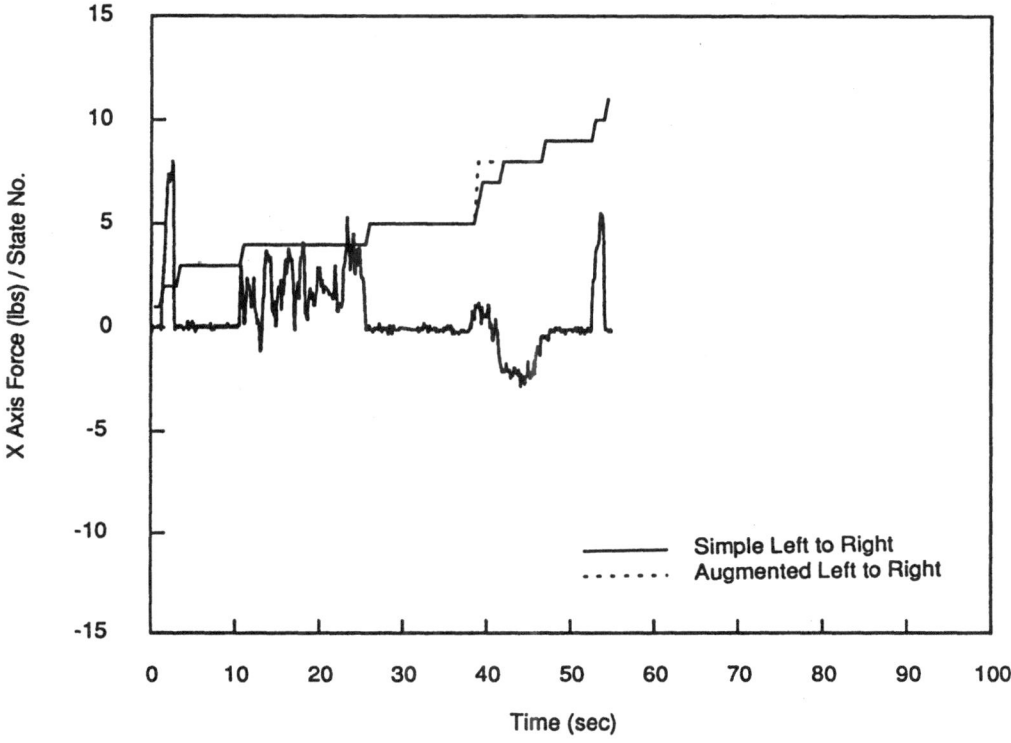

Figure 6. Response of Viterbi algorithm to experimental data modified by removing the central tap. The ALR model (dashed line) explicitly detected the missed event. The SLR model strained to find a segmentation which did not skip a state.

3 Results

A heuristically constructed HMM for the peg-in-hole insertion/extraction task gave realistic results in simulating the force signals observed from teleoperation of the task and guided the Viterbi algorithm to perform accurate segmentation of the experimental force records. Four simulations generated from the HMM (Figure 4) produce force signals which are quite similar to those observed in experimental data. For comparison, four experimental records (Figure 5) have been selected which are as close as possible to the simulations. The model output is qualitatively very similar to the experimental data except that the taps vary more in width and in number of peaks.

Also plotted with the data is a series of states which show the transitions of the underlying Markov model. The line is actually two superimposed ones indicating the "actual" state of the simulation (solid line), and the identified state from the Viterbi algorithm (dashed line). The two lines are exactly superimposed except for one place (t=8 sec, #767), indicating near perfect tracking. Observing the identified state number (read on the pounds scale) and referring to the task numbers and names given in the example, it can be verified that the correct sub-tasks are identified consistently and robustly by the Viterbi algorithm.

Effects of Topology and Incomplete Data

The robustness of the HMM can be evaluated by observing its behavior in the case of incomplete data. Two HMMs were developed, the simple left-to-right model described above, and an augmented left-to-right model. The augmented left-to-right model was obtained by adding a small probability (0.01) for the transition from state n to n + 3, and subtracting a corresponding amount from the self state transition. The Viterbi algorithm was used with each HMM to identify the state transitions in force data records from six repetitions of the peg-in-hole task and in a modified version of the same records in which the middle tap was removed. The removal of the central tap corresponds to the removal of two sub-task states, the tap itself, and the post-tap movement which is no longer distinct from the pre-tap movement. The ALR model explicitly allows for this transition but does not require it. The SLR model does not allow for transitions skipping states, but the Viterbi algorithm can track the state sequence in the altered data by finding a plausible place in which the skipped state can be accommodated for a very brief time interval (1 sample). Since we have used the Gaussian distribution to model the output signal of each state, this can always be accomplished without the total sequence probability going to zero.

In an example using one of the data records (Figure 6) the two identified state sequences are overplotted and show differences only at the positive transient at the beginning of the extraction phase (t = 40 seconds). At the initial positive transient, the SLR model (solid line) identifies a tap and a move skipping almost directly from state 5 (free motion) to state 7 (also free motion) and then about 2 seconds later to state 8, (extraction). The ALR model is not constrained to make single jumps but instead makes a double jump and one single jump to transition from state 5 to state 8 in less than 1 second.

Task Segmentation Performance

This computation was repeated for the six raw data files, and the six modified versions in which the central tap had been removed by editing the data. The correctness of the segmentation was assessed by visual inspection. The results (Table 1) showed that both the SLR and ALR models achieved perfect segmentation of the force data for the six intact data files, and that the more general topology of the ALR model enabled segmentation in two thirds of the data files in which the tap was removed.

TABLE 1

Percent correct state segmentation.

Data	Model Type SLR	ALR
Complete	100%	100%
Partial	0%	67%

4 Discussion

This paper has applied Hidden Markov Models to force/torque data measured during telemanipulation experiments. The model and it's associated optimal state estimation algorithm, the Viterbi algorithm, have been very successful at the task of segmenting the data record into phases corresponding to sub-goals of the task. This technique has proved robust in spite of the noise-like appearance of the force signal.

We believe that this modeling technique has much broader applicability in robotics. The HMM provides rich modeling structure within a statistical framework. This allows algorithms incorporating the model to represent complex systems but at the same time be robust to "real world" sensory signals.

In autonomous manipulation, accomplishment of a goal will lead to a sequence of operations and a resulting sensory trace. When the task involves interaction with the environment, detection of exception conditions becomes difficult because of the quickly changing and noise-like nature of sensor signals. The HMM allows statistical treatment of these signals while at the same time incorporating task knowledge. The statistical nature of the state observation densities improves robustness by avoiding hard syntactic constraints on the sensor data.

Note however, that absolute constraints can also be encoded into an HMM through the structure of the state observation densities. In our examples, these densities were Gaussians and thus had a non-zero probability for all observation values. Another model could include densities which are non-zero on only a sub-set of the observation range. This would apply a hard constraint on the Viterbi algorithm to avoid certain states for pre-specified sensor readings.

Two applications that will be briefly considered here are execution monitoring for fault detection, and sensor fusion.

During autonomous execution of a task, a software agent could track incoming sensory information (in parallel with any other control or sensor processing algorithms) by using the Viterbi algorithm against a previously stored HMM of the task. An exception can detected when the probability value of the identified state sequence falls below a threshold.

This type of execution monitoring is also necessary during teleoperation phases of a task carried out by an advanced telerobot system. In a task containing one phase which requires direct teleoperation, the HMM and Viterbi algorithm could be used by the autonomous portion of the system to track the human's progress in accomplishing the task. Finally, under conditions of substantial time delays between control station and slave robot, a HMM will be useful to monitor the progress of human generated command sequences. HMM detection of task exceptions could prevent destructive conditions arising before the communication link would allow supervisory response.

Task level HMM monitoring can make use of multiple sensors. The HMM and the Viterbi algorithm makes probability a common language in which all sensor readings can be expressed. Each sensor value such as a force reading, a proximity measurement, an image feature, can have a probability of occurrence (or a probability density) defined for each state of task completion. For the multi-dimensional case, a multi-dimensional Probability such as the jointly Gaussian can be used. Although more complex to specify having $j + j^2$ parameters (where j is the number of dimensions), this formulation gives tremendous potential power to perform sensor fusion. Probability becomes the common language for signals from multiple sensor systems. The Viterbi algorithm uses the HMM to "reason" about the "meaning" of each sensor input through combining their a priori probabilities.

Construction of HMMs

This paper has presented HMMs which were constructed heuristically based on our knowledge of the task. This can be improved upon in several ways. First, the Baum-Welch algorithm (Levinson, et.al., 1983) is an iterative procedure for improving an HMM fit to a given set of data. Given a HMM and data set, the algorithm incrementally changes the model in such a way that the probability that the data was generated by the HMM is increased. This method is susceptible to sub-optimal model fits at local maxima in the parameter space. The heuristically generated model is thus valuable to provide a starting point for the Baum-Welch algorithm. Thus, given a set of sensor traces from "normal" executions of a task, the HMM can be estimated for later use by an application.

The heuristic method of HMM construction is also valuable because it provides a structure for approximate human knowledge about the task. Viewed this way, heuristic construction of a HMM is a form of knowledge engineering. This process could be facilitated with a software tool which would enforce constraints on the HMM or statistically estimate probability density functions from manually identified segments of data.

The basic Markov model structure and the heuristic given above for choosing state transition probabilities has a fundamental limitation on the accuracy with which it can model sequential tasks. That limitation is that the Markov property specifies an exponential probability density function on the time in a given state. This is a poor model of processes which have a symmetrical distribution of state times around a non-zero mean. For example tasks whose component sub-tasks have minimum valid durations. The exponential duration density of the HMM manifests itself in large variances in state duration and significant probabilities for very long durations. This can be seen for example in the abnormally long duration taps generated by the HMM simulation (Figure 4).

Rabiner (1989) has developed extensions to the HMM which allow explicit state duration densities. This allows the HMM designer to characterize any distribution of state durations for more flexible modeling of tasks such as manipulation which have for example a minimum duration. Exploration of explicit state duration models is one of the future directions for this research.

Acknowledgements

This research was performed at the Jet Propulsion Laboratory, California Institute of Technology, under contract with the National Aeronautics and Space Administration. The authors would like to thank Dr. Yves Goussard of CNRS, France for introducing us to Hidden Markov Models, and Dr. Antal Bejczy, and Laurie Wood of JPL for help and encouragement.

References

Baker, J.K., "The DRAGON System - An Overview," IEEE Trans. Acoustics, Speech & Signal Processing, vol. ASSP-23, no. 1, 1975.

Bekey, G., IRE Trans. Human Factors in Electronics, vol. HFE-3, p. 43, 1962.

Draper, J.V., C.A. Wrisberg, and L.M. Blair, "Measuring Operator Skills and Teleoperator Performance," Proc. Intl. Symposium Teleoperation and Control, pp. 341-349, Bristol, England, July, 1988.

Hannaford, B., "Task Level Testing of the JPL-OMV Smart End Effector," Proceedings of the JPL - NASA Workshop on Space Telerobotics, vol. 2, pp. 371-380, JPL Publication 87-13, Pasadena, CA, July 1, 1987.

Hannaford, B. and L. Wood, "Performance Evaluation of a 6 Axis High Fidelity Generalized Force Reflecting Teleoperator," Proceedings JPL/NASA Conference on Space Telerobotics, JPL Publication 89-7, Pasadena, CA, January 1989.

Juang, B.H. and L.R. Rabiner, "Mixture Autoregressive Hidden Markov Models for Speech Signals," IEEE Trans. Acoustics, Speech & Signal Processing, vol. ASSP-33, no. 6, 1985.

Levinson, S.E., L.R. Rabiner, and M.M. Sondhi, "An Introduction to the Application of the Theory of Probabalistic Functions of a Markov Process to Automatic Speech Recognition," Bell Sys. Tech. Journal, vol. 62, no. 4, pp. 1035-1074, 1983.

Navas, F. and L. Stark, "Sampling or Intermittency in Hand Control System Dynamics," Biophysical Journal, vol. 8, no. 2, pp. 252-302, 1968.

Rabiner, L.R., S.E. Levinson, and M.M. Sondhi, "On the Application of Vector Quantization and Hidden Markov Models to Speaker-Independent Isolated Word Recognition," Bell Sys. Tech. Journal, vol. 62, no. 4, pp. 1074-1105, 1983.

Rabiner, L.R., B.H. Juang, S.E. Levinson, and M.M. Sondhi, "Recognition of Isolated Digits Using Hidden Markov Models with Continuous Mixture Densities," Bell Sys. Tech. Journal, vol. 64, no. 6, pp. 1211-1234, 1985.

Rabiner, L.R. and B.H. Juang, "An Introduction to Hidden Markov Models," IEEE ASSP Magazine, pp. 4-16, January 1986.

Rabiner, L.R., "A Tutorial on Hidden Markov Models and Selected Applications in Speech Recognition," Proceedings of the IEEE, vol. 77, no. 2, pp. 257-286, Feb. 1989.

Reddy, R., "Foundations and Grand Challenges of Artificial Intelligence: 1988 AAAI Presidential Address," AI Magazine, pp. 9-21, Winter 1988.

Viterbi, A.J., "Error Bounds for Convolutional Codes and an Asymptotically Optimal Decoding Algorithm," IEEE Transactions on Information Theory, vol. IT-13, pp. 260-269, 1967.

Adaptation to Environment Stiffness in the Control of Manipulators

Laeeque Daneshmend, Vincent Hayward,
and Michel Pelletier

McGill Research Centre for Intelligent Machines
&
Department of Electrical Engineering, McGill University
3480 University Street, Montréal, Québec, Canada H3A 2A7

ABSTRACT

The dynamic performance of fixed-gain force or impedance manipulator control systems in constrained situations is very dependent on the environment parameters: e.g. the force response at low stiffnesses may be sluggish, while high stiffnesses give rise to bouncing and instability. An adaptive controller, based on the model reference approach, for multi-axis damping control is presented. Simulation results show that the adaptive scheme can significantly improve the performance in force tracking and enhance stability at high stiffnesses, by rendering the behavior independent of the environment stiffness. Experimentation on a PUMA manipulator shows the limitations of the approach when applied to industrial manipulators with significant nonlinearities and flexibilty in the transmission.

1. Introduction

Sensitivity to environment parameters is a well known problem in the force control of manipulators in constrained situations [1, 2, 3, 4, 5]. Results show that stability and performance are very dependent on the environment and force sensor stiffnesses. As stiffness increases, the controllers tend to become less and less damped and with very high natural frequencies, eventually becoming unstable. The bandwidth of such systems has to be limited for the sake of stability by adding active or passive compliance.

This paper focuses on Whitney's damping control scheme [1], with the aim of achieving uniform behaviour for variable environment stiffness. Damping control is a particular case of impedance control where the desired impedance is chosen to make the constrained manipulator behave like a generalized damper:

$$F = B(\dot{x}_d - \dot{x}) \qquad (1)$$

The input is velocity and the force applied is proportional to the velocity error. In a constrained situation, the velocity is zero in the direction of the surface normal and the force applied is directly proportional to the demanded input velocity, which may be pre-programmed or input via a kinesthetic interface.

Because of the compliance to external forces without explicit selection of compliant frames, damping control can simplify the planning of assembly and insertion tasks [6]. Hence it is a good choice for compliant teleoperation, as well as autonomous compliance.

In our approach, robustness to variations in system stiffness is achieved using a discrete-time, single-input single-output model-reference adaptive control (MRAC) scheme within the damping control loop. Formulation and simulations of this adaptive scheme for operational space hybrid control have been presented earlier [7]. A single-axis version of the adaptive damping scheme has also been investigated in an earlier paper [8]. The complete formulation and simulation results for the multi-axis adaptive damping controller can be found in [9].

The modelling of the force and damping control loops is first reviewed. This is followed by design of the adaptive controller, and presentation of simulation results. Next, implementation issues are addressed. Finally, experimental results are presented and analyzed.

2. System Modelling

The block diagram of the multi-axis damping controller is shown in Figure 1. In the following discussions and tests, point contact is assumed between the robot end-effector and environment surface, i.e. the environment cannot generate torques. The robot wrist is also considered frozen so that orientation terms are ignored. The environment is modeled as a smooth, low curvature surface which can be considered to be locally plane.

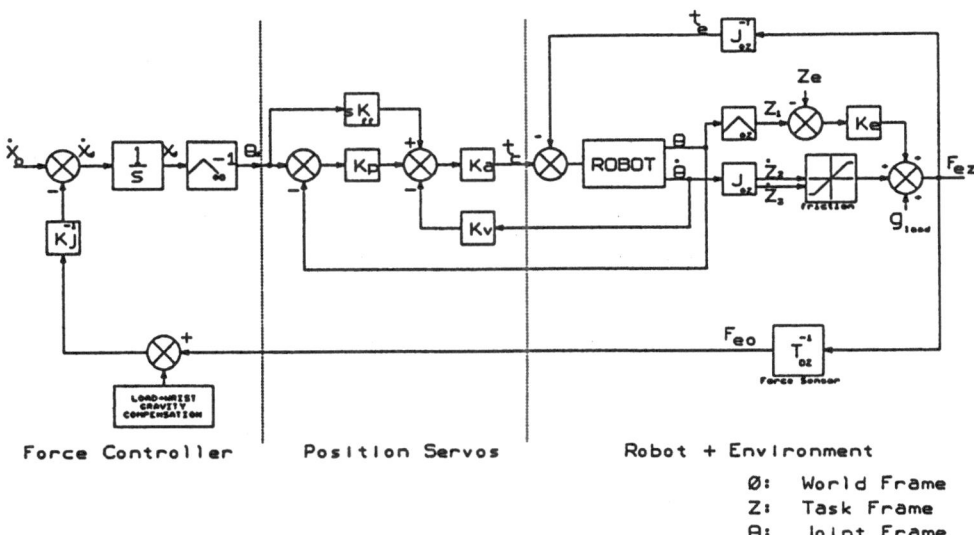

Figure 1: The Multi-Axis Damping Controller

The major differences between the single-axis [8] and multi-axis cases are introduced by the non-linear kinematic and dynamic transformations. Also, the force feedback signal from the wrist contains more than the external contact force. The wrist sensor picks up the gravity forces on the gripper and payload: this effect has to be compensated to avoid droop. In practice, it can be handled by compensating the force feedback with the actual or estimated value of the wrist and payload weight. Small errors in the weight value can be overcome with the addition of a deadband in the feedback loop.

The inertial forces sensed at the wrist caused by the wrist and payload weight are also neglected in the following formulation. These terms are related to acceleration and should have no significant effect on the behavior, when performing typical tasks requiring compliant motion.

The dynamical response of the system in joint coordinates is given by:

$$M_\theta(\theta)\ddot{\theta} + C_\theta(\theta, \dot{\theta}) + g_\theta = \tau_c + \tau_u + \tau_e \qquad (2)$$

where

$\theta, \dot{\theta}, \ddot{\theta} \equiv$ Joint positions, velocities and accelerations (rad, rad/s, rad/s^2);
$M_\theta \quad \equiv$ Manipulator Inertia Matrix;
$C_\theta \quad \equiv$ Coriolis and Centrifugal Terms;
$g_\theta \quad \equiv$ Gravity Terms;
$\tau_c \quad \equiv$ Control Torques at Output Shaft;
$\tau_u \quad \equiv$ Disturbance and Friction Torques;
$\tau_e \quad \equiv$ Environment Contact Torques.

A better understanding of the system dynamics is achieved by using the task frame formulation. The task coordinates are defined as $x = (x_1, x_2, x_3)$, where the x_1 axis coincides with the environment surface normal and points towards the inside of the surface. The x_2 and x_3 axes are tangent to the environment surface. The dynamics of the manipulator can be expressed directly in task coordinates:

$$M_x \ddot{x} + C_x + g_x = F_{cx} + F_{ux} + F_{ex} \qquad (3)$$

with

$$M_x = J_{\theta x}^{-T} M_\theta J_{\theta x}^{-1} \qquad (4)$$

$$C_x = J_{\theta x}^{-T} \left\{ C_\theta(\theta, \dot{\theta}) - M_\theta J_{\theta x}^{-1} \dot{J}_{\theta x} \dot{\theta} \right\} \qquad (5)$$

$$g_x = J_{\theta x}^{-T} g(\theta) \qquad (6)$$

where

$J_{\theta x} \quad \equiv$ Manipulator Jacobian Matrix in Task Frame.

The controller force is given by:

$$
\begin{aligned}
F_{cx} &= J_{\theta x}^{-T} K_a \left\{ K_p(\theta_d - \theta) + K_{ff}\dot{\theta}_d - K_v\dot{\theta} \right\} \\
&= J_{\theta x}^{-T} K_a \left\{ (K_p + sK_{ff})\Lambda_{\theta 0}^{-1} \left(\frac{\dot{x}_d}{s} \right) - K_p\theta - K_v\dot{\theta} \right\}
\end{aligned}
\qquad (7)
$$

where

$K_a \quad \equiv$ Actuator Torque Constant;

153

K_p ≡Proportional Gain of Joint Servo;
K_{ff} ≡Velocity Feedforward Gain of Joint Servo;
K_v ≡Velocity Feedback Gain of Joint Servo;
x_d ≡Desired Cartesian Position;
θ_d ≡Desired Joint Position;
$\Lambda_{\theta 0}$ ≡Manipulator Forward Kinematics Operator.

The environment surface is modeled as a pure stiffness of value K_e with friction in the tangential directions. i.e.:

$$F_{ex} = -\overline{K_e}(x - \overline{x_\epsilon}) - f_{fx} \tag{8}$$

and

$$\overline{K_e} = \begin{pmatrix} K_e & 0 & 0 \\ 0 & 0 & 0 \\ 0 & 0 & 0 \end{pmatrix} \tag{9}$$

$$\overline{x_e} = \begin{pmatrix} x_e \\ 0 \\ 0 \end{pmatrix} \tag{10}$$

$$f_{fx} = \begin{pmatrix} 0 \\ f_{fx2} \\ f_{fx3} \end{pmatrix} \tag{11}$$

where
x_e ≡Environment Equilibrium Position in Task Frame;
f_{fxi} ≡Environment Friction Forces (Tangential to Surface).

The basic behavior of the non-adaptive damping controller will similar to the single-axis case analyzed in [8], i.e. the response will still be very dependent on the environment stiffness.

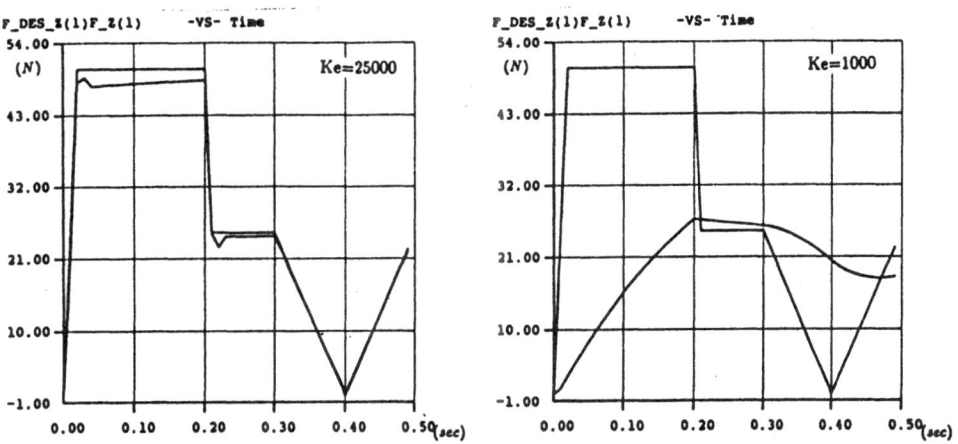

Figure 2: Force Response in Normal Axis, Non-Adaptive Multi-Axis Damping Controller, for Medium (25 kN/m) and Low (1 kN/m) Stiffness Environments

This sensitivity is seen in the simulation results of Figure 2, which are for a PUMA 560 manipulator performing damping control at 100 Hz sample rate, where the controller is well tuned for a medium stiffness environment, but performs extremely sluggishly for low environment stiffness. Similarly, at high stiffnesses the response becomes highly oscillatory [9].

However, because of the varying kinematic and dynamic terms, the closed-loop response will depend on the robot configuration and velocity as well. This requires a *worst case* tuning to assure stability for all robot states. This will require excessive damping to ensure global stability even for known, constant, environment stiffness, and the robot will respond sluggishly for most configurations.

In order to develop the adaptive controller equations and to determine the MRAC plant model, it is assumed that the wall orientation, and thus the world to task transformation matrix, is known. The task frame dynamical equation (3) can be linearized around an equilibrium point defined by $x = x_0$, $\dot{x} = 0$, $x_d = x_{d0}$ and $\dot{x}_d = 0$, corresponding to the joint positions θ_0, θ_{d0}, and velocities $\dot{\theta}_0 = \dot{\theta}_{d0} = 0$. This is done using the Taylor series expansion, where the higher order terms are dropped. In order to allow easy reading, the following linear operators are defined under the assumption that the Jacobian matrices are slowly varying. (The terms that reduce to zero have been omitted.)

$$g_{x_O} \equiv g_x \Big|_{x=x_0} \tag{12}$$

$$G_x^* \equiv \frac{\partial g_x}{\partial x} \Big|_{x=x_0} \tag{13}$$

$$C_x^* \equiv \frac{\partial C_x}{\partial \dot{x}} \Big|_{\substack{x=x_0 \\ \dot{x}=0}} \tag{14}$$

$$K_v^* \equiv J_{\theta x}^{-T} K_a K_v J_{\theta x}^{-1} \Big|_{x=x_O} \tag{15}$$

$$K_{ff}^* \equiv J_{\theta x}^{-T} K_a K_{ff} J_{\theta_d x}^{-1} \Big|_{\substack{x=x_0 \\ x_d=x_{d0}}} \tag{16}$$

$$K_{p0}^* \equiv J_{\theta x}^{-T} K_a K_p \Lambda_{\theta x}^{-1}(x) \Big|_{x=x_0} \tag{17}$$

$$K_p^* \equiv J_{\theta x}^{-T} K_a K_p J_{\theta x}^{-1} \Big|_{x=x_0} \tag{18}$$

$$K_{p1}^* \equiv J_{\theta x}^{-T} K_a K_p \Lambda_{\theta x}^{-1}(x_d) \Big|_{\substack{x=x_0 \\ x_d=x_{d0}}} \tag{19}$$

$$K_{pd}^* \equiv J_{\theta x}^{-T} K_a K_p J_{\theta_d x}^{-1} \Big|_{\substack{x=x_0 \\ x_d=x_{d0}}} \tag{20}$$

$$K_{pc}^* \equiv K_{p1}^* - K_{pd}^* x_{d0} - K_{p0}^* + K_p^* x_0 \tag{21}$$

where
$J_{\theta_d x} \equiv$ Manipulator Jacobian of Desired Position in Task Frame.
The linearized task frame dynamical equation of the constrained manipulator under damping

control is thus given by:

$$D_x(s)x = (K_{pd}^* + K_{ff}^* s)x_d + K_{pc}^* + F_{ux} + \overline{K_e x_e} + G_x^* x_0 - g x_0 - f_{fx} \tag{22}$$

where

$$D_x(s) \equiv M_x s^2 + (K_v^* + C_x^*)s + (\overline{K_e} + K_p^* + G_x^*). \tag{23}$$

As seen from this equation, the system is coupled in general. Motions and inputs in a particular direction influence motion in all directions, because matrices D_x, K_{pd}^* and K_{ff}^* are not diagonal. In order to be able to apply the MRAC controller in the normal contact force direction, it is necessary to assume that the system is decoupled, i.e. that tangential motion does not affect the normal contact force. This is physically justifiable: intuitively, the motion in a particular direction should depend mostly on inputs in that direction. In particular, this approximation should be valid for manipulators with simple configurations, where the axes are relatively well decoupled kinematically, especially if the position servos are well tuned and the velocities and accelerations are low. This assumption of decoupling has been verified in simulation for the PUMA 560 manipulator (see Chapter 5 of [9]).

With this assumption, the off-diagonal terms of matrices mentioned above can be ignored diagonal. From the above linearized equation, the first line is extracted to obtain a relation between the input and the contact force in the surface normal direction. This yields:

$$D_{x(1,1)}(s)x_1 = \left(K_{pd}^* + K_{ff}^* s\right)_{(1,1)} x_{d1} + K_{pc(1)}^*$$
$$+ F_{ux(1)} + K_e x_e + G_{x(1,1)}^* x_{01} - g x_0(1). \tag{24}$$

The model now depends only on the normal direction terms. The contact force is found by applying equation (8) and, as for the single axis case [8], the constant bias terms are eliminated by multiplying both sides by s, i.e. differentiating.

$$F_{x1} = K_e \left[\frac{\left(K_{pd}^* + K_{ff}^* s\right)_{(1,1)}}{D_{x(1,1)}(s)} x_{d1} + \frac{F_{ux(1)}}{D_{x(1,1)}(s)} \right.$$
$$\left. + \frac{K_e x_e}{D_{x(1,1)}(s)} + \frac{K_{pc(1)}^* + G_{x(1,1)}^* x_{01} - g x_0(1)}{D_{x(1,1)}(s)} - x_e \right] \tag{25}$$

$$sF_{x1} = \frac{K_e \left(K_{pd}^* + K_{ff}^* s\right)_{(1,1)}}{D_{x(1,1)}(s)} \dot{x}_{d1} + \frac{K_e \dot{F}_{ux(1)}}{D_{x(1,1)}(s)}. \tag{26}$$

If it is assumed that the disturbance term F_{ux} can be compensated, or that its effect is negligible because it is slowly varying, the SISO plant model used for the MRAC is given by

$$sF_{x1} = \frac{K_e \left(K_{pd}^* + K_{ff}^* s\right)_{(1,1)}}{D_{x(1,1)}(s)} \dot{x}_{d1}. \tag{27}$$

This gives the expression relating the velocity input and the contact force in the normal direction.

3. Adaptive Control

This section develops a model reference adaptive control (MRAC) strategy to maintain constant, stable, desired dynamics of the closed-loop damping controller. The MRAC scheme is intended to work in conjunction with the conventional damping loop, as shown in Figure 3, to achieve the desired compliant motion response regardless of variations in environment stiffness.

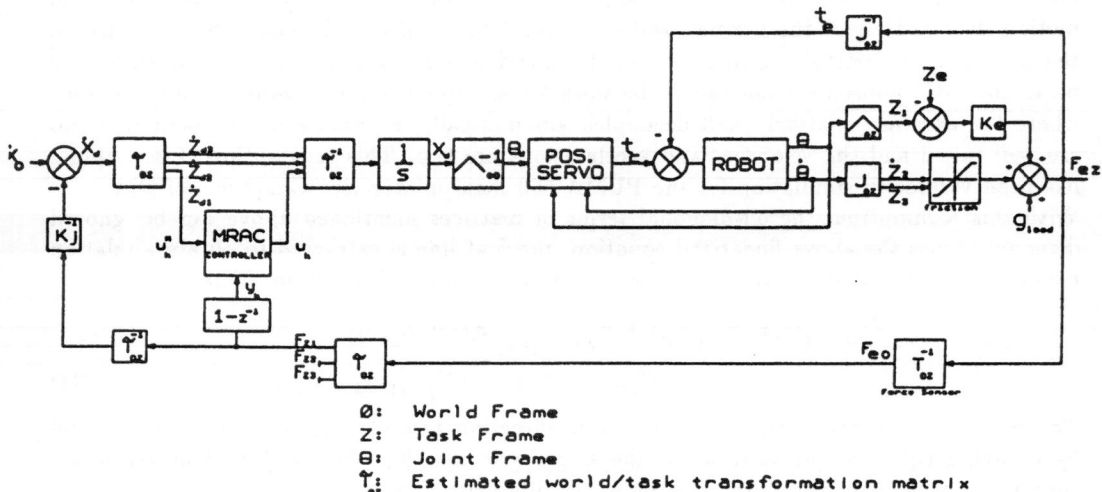

Ø: World Frame
Z: Task Frame
θ: Joint Frame
$\hat{T}_{\theta z}$: Estimated world/task transformation matrix

Figure 3: Adaptive Multi-Axis Damping Controller

3.1 The MRAC Design Technique

It is desirable to maintain constant, damped (non-oscillatory), dynamics for each compliant axis. We propose to apply the explicit MRAC scheme of Landau and Lozano [10] for discrete-time linear SISO plants. This design is only applicable to minimum-phase plants, since it relies on cancellation of the plant zeroes to achieve perfect model-following. For purposes of proving stability, it assumes that the time-delay and upperbounds of the plant polynomials are known. The linear SISO plant $G_p(z)$ is defined by

$$G_p(z) = \frac{y(z)}{u(z)} = \frac{z^{-d}B(z^{-1})}{A(z^{-1})} \tag{28}$$

where $z^{-d} \equiv$ a pure time delay of d sample periods, and

$$A(z^{-1}) = 1 + \sum_{i=1}^{n_A} a_i z^{-i} \tag{29}$$

$$B(z^{-1}) = \sum_{i=0}^{n_B} b_i z^{-i}. \tag{30}$$

The adaptive feedback loops are defined as

$$B_S(z^{-1}) = B(z^{-1})S(z^{-1}) - b_0 \tag{31}$$

where

$$S(z^{-1}) = 1 + \sum_{i=1}^{n_S} s_i z^{-i} \tag{32}$$

and

$$R(z^{-1}) = \sum_{i=0}^{n_R} r_i z^{-i}. \tag{33}$$

For the plant defined by (28) the design objectives of this scheme are:

(i) The control should be such that, in tracking a command input, the plant output satisfies

$$\frac{y(z)}{u^M(z)} = \frac{z^{-d} D(z^{-1})}{C_1(z^{-1})} \tag{34}$$

where

$y(z)$ $\equiv$ plant output;
$u^M(z)$ $\equiv$ command input;

$$C_1(z^{-1}) = 1 + \sum_{i=1}^{n_{C_1}} c_i^1 z^{-i} \tag{35}$$

$$D(z^{-1}) = \sum_{i=0}^{n_D} d_i z^{-i}; \tag{36}$$

(ii) The control should be such that, in regulation, an initial output disturbance, $\omega(k)$, defined by

$$\omega(k) = y(0) \neq 0 \quad for \ k = 0, \qquad \omega(k) = 0 \quad \forall k \neq 0 \tag{37}$$

(i.e. an impulse, present for the first sample only) is eliminated with the dynamics defined by

$$C_2(z^{-1})z^{-d}y(z) = 0 \quad k \geq 0 \tag{38}$$

where

$$C_2(z^{-1}) = 1 + \sum_{i=1}^{n_{C_2}} c_i^2 z^{-i} \tag{39}$$

is an asymptotically stable polynomial.

To satisfy the design objectives the identity

$$A(z^{-1})S(z^{-1}) + z^{-d}R(z^{-1}) = C_2(z^{-1}) \tag{40}$$

must hold. It can be shown that $S(z^{-1})$ and $R(z^{-1})$ are uniquely defined by (40) (see Appendix A of [10]), with the degree of the polynomials given by

$$n_S = d - 1 \tag{41}$$

$$n_R = \max\{(n_A - 1), (n_{C_2} - d)\}. \tag{42}$$

Ideally $C_2(z^{-1})$ should equal unity, since this would ensure the best possible performance in regulation for perfect model-following. However, as shown by simulation results [10], this polynomial is crucial to the performance of the MRAC scheme. It smooths out the response of the adaptation mechanism to both parameter changes and to disturbances.
The adaptive control law is given by

$$\tilde{p}^T(k).\phi(k) = C_2(q^{-1}).y^M(k+d) \tag{43}$$

where "q^{-1}" denotes the backward shift operator. The regression (or measurement) vector is given by

$$\phi^T(k) = \{u(k), u(k-1), \dots, u(k-d-n_B+1), y(k), \dots, y(k-n_R),\} \tag{44}$$

and the estimated parameter vector is given by

$$\tilde{p}^T = \{\tilde{b}_0, \tilde{b}_0.s_1 + \tilde{b}_1, \tilde{b}_0.s_2 + \tilde{b}_1.s_1 + \tilde{b}_2, \dots, \tilde{b}_{n_B}.s_{n_S}, \tilde{r}_0, \dots, \tilde{r}_{n_R},\}. \tag{45}$$

The adaptive controller adjustment (or parameter estimation) algorithm is

$$\tilde{p}(k) = \tilde{p}(k-1) + F_k.\phi(k-1).\epsilon^*(k) \tag{46}$$

where the adaptation gain matrix is given by

$$F_{k+1} = \frac{1}{\lambda_1(k)}.\left[F_k - \frac{F_k.\phi(k-d).\phi^T(k-d).F_k}{\frac{\lambda_1(k)}{\lambda_2(k)} + \phi^T(k-d).F_k.\phi(k-d)} \right] \tag{47}$$

$$0 < \lambda_1(k) \le 1, \qquad 0 \le \lambda_2(k) < 2 \quad \forall k, \qquad F_0 > 0$$

with the *a posteriori* filtered plant-model error given by

$$\epsilon^* = \frac{\tilde{e}(k)}{1 + \phi^T(k-d).F_k.\phi(k-d)} \tag{48}$$

where

$$\tilde{e}(k) = C_2(q^{-1})y(k) - \tilde{p}^T(k-1)\phi(k-d). \tag{49}$$

3.2 Adaptive Damping Controller Design

For the damping control problem, we have

$$G_p(z) = \mathcal{Z}\{G_{zoh}G(s)\} = (1 - z^{-1})\mathcal{Z}\{\frac{1}{s}G(s)\} \qquad (50)$$

where $G(s)$ is the linearized transfer function between rate of change of force and demand velocity, from (27). After some algebraic manipulation [9], we obtain:

$$n_A = 2, \qquad n_B = 1, \qquad d = 1. \qquad (51)$$

Similarly we can represent the desired performance of the system using an explicit second-order reference model with

$$n_{C_1} = 2, \qquad n_D = 1, \qquad d = 1. \qquad (52)$$

Assuming we want second-order regulation dynamics defined by

$$C_2(z^{-1}) = 1 + c_1^2 z^{-1} + c_2^2 z^{-2}. \qquad (53)$$

Using (41) and (42) $\Rightarrow$

$$n_S = 0, \qquad n_R = 1. \qquad (54)$$

Hence, the general MRAC scheme for damping control in the presence of variable enviromemnt stiffness, using second-order process and reference models and second-order regulation dynamics, is defined by:

(i) Regression vector

$$\phi^T(k) = \{u(k), u(k-1), y(k), y(k-1)\}; \qquad (55)$$

(ii) Estimated parameter vector

$$\tilde{p}^T(k) = \{\tilde{b}_0(k), \tilde{b}_1(k), \tilde{r}_0(k), \tilde{r}_1(k)\}; \qquad (56)$$

(iii) Adaptive control law

$$\tilde{p}(k)^T.\phi(k) = C_2(q^{-1})y^M(k+1); \qquad (57)$$

(iv) Adaptive controller adjustment (parameter estimation) algorithm given by (46), with the adaptation gain, F, being a 4×4 matrix;

(v) Adaptation gain calculation algorithm given by (47) and (48) with $d = 1$.

This adaptive multi-axis damping controller was simulated extensively [9]. As can be seen from Figure 4, it works remarkably well, considering that it has no *a priori* knowledge of environment stiffness. These simulations also used a 100 Hz sample rate for both the adaptation and the damping loops. As can be seen, good tracking is achieved over a wide range of environment stiffnesses. (The initial lag in the low stiffness response is due to the finite time taken for the adaptation mechanism to converge, and the limiting of the maximum velocity for the manipulator.)

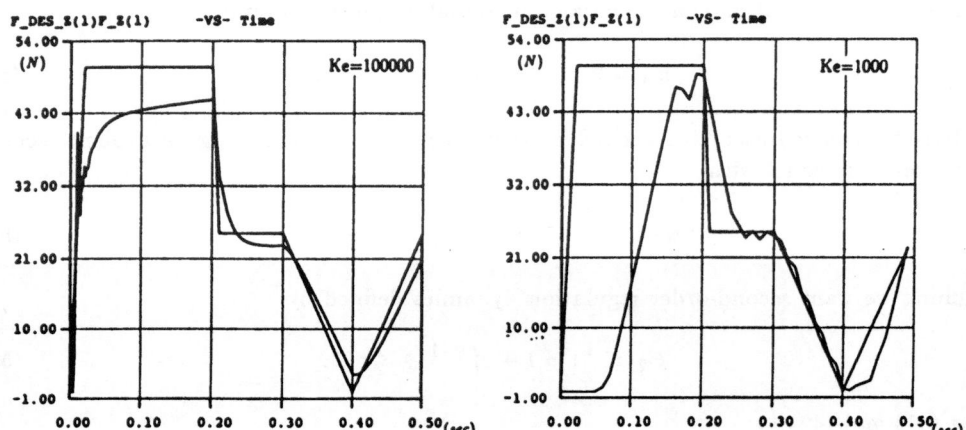

Figure 4: Force Response in Normal Axis, Adaptive Multi-Axis Damping Controller, for Medium (100 kN/m) and Low (1 kN/m) Stiffness Environments

4.   Implementation

The adaptive multi-axis damping controller was implemented on the *KALI* multiprocessor robot programming and control system, developed at McRCIM [11, 12]. *KALI* performs trajectory generation, model-based dynamic compensation, and servocontrol functions, for one or more manipulators. For this experimentation, we used one PUMA 560 manipulator, connected directly to the *KALI* system.

Two of *KALI's* CPUs were used to implement the joint feedback loops for the PUMA: proportional-plus-velocity (PV) feedback was used. The PV loops ran at at a 500 Hz sampling rate. Set-points from the trajectory generator CPU were generated at 100Hz, and interpolated by the servo level at every servocontrol sample period.

Force feedback was provided by a LORD 6 degree-of-freedom force sensor, mounted on the PUMA's wrist flange. This communicated force/torque readings in tool frame coordinates to the controller over a parallel port. The sampling rate of the force sensor in this configuration was limited to 100 Hz. Gravity offset compensation was achieved using the force sensor's zero-bias command, which was adequate since the kinematic configuration of the manipulator did not change significantly during the course of each experimental run.

The damping controller itself was implemented on the trajectory generator CPU, and ran at 100 Hz. The adaptive controller was implemented on the same CPU, and ran at the same rate. Model-based dynamic compensation was not implemented for the experimental results in this paper.

"Jacketing" software for the adaptive controller was used to make it more robust in practice. This ensured that the adaptation mechanism was only active when in contact with the environment, by detecting contact using thresholding of the force signal. It also placed limits on the velocity deviation that could be commanded by the adaptation mechanism.

5.   Experimentation

The first step in experimentation was to verify the sensistivity of the non-adaptive damping controller to variations in environment stiffness. Tests were conducted with various materials: soft and firm foam, stiff cardboard, and steel. The results confirmed the expectations of the modelling: response was highly dependant on surface stiffness. For example, for a damping controller gain of 100 N/m/s in the surface normal axis, contact with a steel surface resulted in a large ampitude oscillation, with a period of roughly 1 Hz. However, with the same damping gain, contact with a firm foam surface resulted in well-damped behaviour. Increasing the damping gain to 500 N/m/s resulted in well-damped behaviour for the steel surface, but sluggish response for the foam surface.

The next step was to test the estimation mechanism. The non-adaptive controller was run, and the demand velocity and differentiated force signals fed to the estimator. The results for the second-order model using recursive least-squares (RLS – $\lambda_1 = \lambda_2 = 1$) are shown in Figure 5. The estimates do not converge, due to amplification of high-frequency noise by the differentiation of the force signal.

The differentiation of the force signal was abandoned, and replaced by subtraction of d.c. values of demanded position and force when in equilibrium contact with the surface. The estimates using the second-order model and RLS are shown in Figure 6. Although convergence is achieved, analysis of the estimated values reveals that they correspond to an unstable zero in the plant, which would lead to instablity if these values were used in closed-loop adaptation. Indeed, tests using this estimator in adaptive control did go unstable.

A reduced, first-order, model was tried next for the estimator. Results for exponentially weighted recursive least squares (EWLS – $\lambda_1 = 0.99, \lambda_2 = 1$) are shown in Figure 7. While the estimates are slightly noisier than for the second-order case of Figure 6, they correspond to reasonable magnitudes of physical values.

The EWLS first-order MRAC of Figure 7 was then tried in closed-loop control. As seen from Figure 8, the estimates diverged, and the system became unstable. It is thus evident that closing the loop with the adaptation mechanism in the forward path introduces some factor that results in unstable behaviour, though the estimator works well in open-loop.

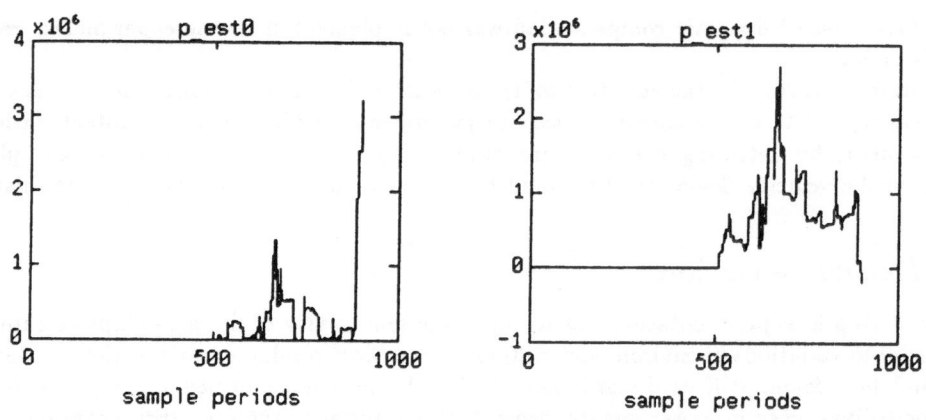

Figure 5: Parameter Estimates using RLS, 2nd-Order Model, differentiated force signal (Only the first Two elements of the parameter estimate vector are shown)

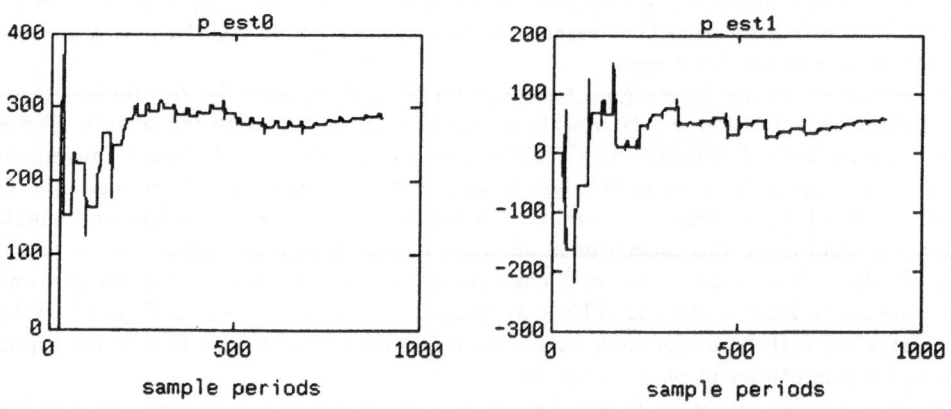

Figure 6: Parameter Estimates using RLS, 2nd-Order Model, offsets removed using averaging (Only the first Two elements of the parameter estimate vector are shown)

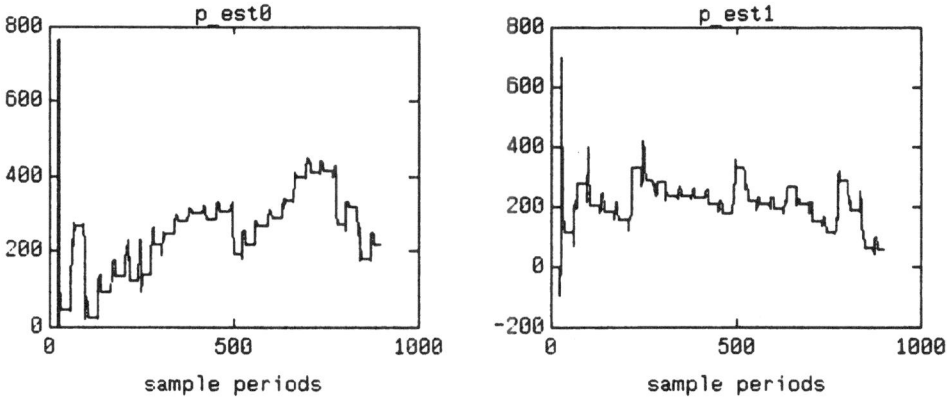

Figure 7: Parameter Estimates using EWLS, 1st-Order Model, offsets removed using averaging (Only the first Two elements of the parameter estimate vector are shown)

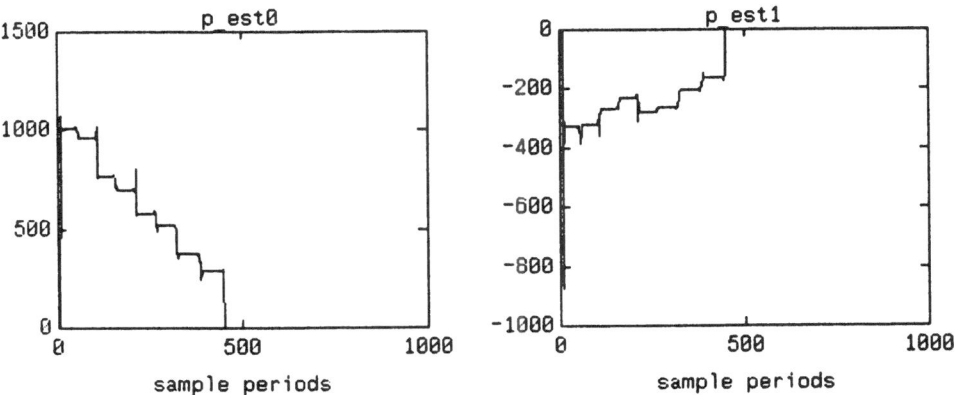

Figure 8: Closed-loop MRAC using EWLS, 1st-Order Model, offsets removed using averaging (Only the first Two elements of the parameter estimate vector are shown)

This behaviour is believed to be due to the non-ideal nature of the actual PUMA manipulator, in comparison with the model used for the simulation results. The PUMA exhibits significant stiction at the joint actuators, and does not actually possess perfectly rigid dynamics, particularly in the transmission mechanism from motors to joints. Hence, the MRAC estimator is trying to impose a linear model on a system with significant non-linearities. It may succeed in converging in open-loop, but its convergence is not theoretically guaranteed in closed-loop in cases where there is such a severe plant-model mismatch.

6. Conclusion

This experimental study has been useful in highlighting the limitations of verification of robot control algorithms by simulation. Concepts that appear feasible in simulation, such as differentiation of the force signal, have been shown to be of little practical use by experiment results. Assumptions that appear valid in simulation, are often not so in practice.

On the other hand, experimentation has provided insights into overcoming many of the shortcomings of the adaptive multi-axis damping control design. For example, averaging to remove d.c. bias, instead of differentaiating, appears feasible in practice, as does use of a first-order model.

The further development of the adaptive damping control concept has also benefited from the experimental results. Two options appear open to achieving practical implementation of the scheme. One is the application of this scheme to better manipulators with structural rigidity and low friction actuators, and even local force feedback at the joints to linearize the system still further. Several such advanced manipulators have been developed or are under construction (as mentioned in other papers in these proceedings), including one under development here at McGill by Vincent Hayward. The second option, which we are also pursuing, is to explicitly compensate for the non-linear effects in the formulation of the adaptive damping control scheme.

Acknowledgements

This work was partially supported by the Natural Sciences and Engineering Research Council (NSERC) of Canada, and by the Province of Québec's MESST Actions Structurantes Program. The *KALI* experimental apparatus was partially funded by the Jet Propulsion Laboratory, NASA. The force sensor used in the experimentation was generously donated by LORD Corp. Michel Pelletier was supported by the National Science and Engineering Research Council of Canada, through a graduate studies scholarship, and by CAE Electronics Ltd. of Ville St-Laurent. The authors would like to thank Tony Topper for his efforts in making the *KALI* system a usable research tool.

References

1. Whitney, D. E., *Force Feedback Control of Manipulator Fine Motions*, ASME Journal of Dynamic Systems, Measurement, and Control, June 1977, pp. 91-97.

2. An, C.H., Hollerbach, J.M., *Dynamic Stability Issues in Force Control ¡of Manipulators*, Proceedings of the IEEE International Conference on Robotics and Automation, 1987, pp. 890-896.

3. Eppinger, S.D., Seering, W.P., *On Dynamic Models of Robot Force Control*, Proceedings IEEE International Conference on Robotics and Automation, 1986, pp. 29-34.

4. Maples, J.A., Becker, J.J, *Experiments in Force Control of Robotic Manipulators*, Proceedings of the IEEE International Conference on Robotics and Automation, 1986, pp. 695-702.

5. Eppinger, S.D., Seering, W.P., *Understanding Bandwidth Limitations in Robot Force Control*, Proceedings of the IEEE International Conference on Robotics and Automation, 1987, pp. 904-909.

6. Lozano-Perez, T., Mason, M.T., Taylor, R.H., *Automatic Synthesis of Fine Motion Strategies for Robots*, International Journal of Robotic Research, Vol. 13, No. 1, Spring 1984, pp. 3-24.

7. Daneshmend, L.K., Hayward, V., *Adaptation in the Control of Multiple Coordinated Manipulators*, Proceedings of the 2nd ISRAMM, Albuquerque, New Mexico, Nov. 1988, pp. 219-226.

8. Pelletier,M., and Daneshmend, L.K., *Dynamic Performance of Robot Contact Tasks Using Damping Control*, Proceedings of the 12th Biennial ASME Conference on Mechanical Vibration and Noise, September 17-20, 1989, Montréal.

9. Pelletier,M., *Adaptive Damping Control for Robotic Teleoperation*, M.Eng. Thesis, Department of Electrical Engineering, McGill University, Montréal, July 1989.

10. Landau,I.D., and Lozano,R., *Unification of Discrete Time Model Reference Adaptive Control Designs*, Automatica, vol. 17, no. 4, 1981, pp. 593-611.

11. Hayward, V., and Hayati, S., *KALI: AN environment for the Programming and Control of Cooperating Manipulators*, Proceedings of the American Control Conference, June 15-17 1988, Atlanta, GA, pp. 473-478

12. Hayward, V., Daneshmend, L.K., and Hayati, S., *An Overview of KALI: A System to Program and Control Cooperative Manipulators*, Proceedings of the Fourth International Conference on Advanced Robotics, Columbus, Ohio, June 1989.

EXPERIMENTAL STUDIES OF
ADAPTIVE MANIPULATOR CONTROL

Günter Niemeyer and Jean-Jacques E. Slotine
Nonlinear Systems Laboratory
Massachusetts Institute of Technology
Cambridge, MA 02139, USA.

ABSTRACT

Effective adaptive controller designs potentially combine high-speed and high-precision in robot manipulation, and furthermore can considerably simplify high-level programming by providing consistent performance in the face of large variations in loads or tasks. Previously, a simple, globally tracking-convergent, direct adaptive manipulator controller was derived and demonstrated experimentally. It was then further refined into a "composite" version, whose adaptation law is driven by both tracking error in joint motion and prediction error in joint torques, and therefore represents a combination of a direct and an indirect approach. An effective implementation for multi degrees-of-freedom manipulators was later achieved by developing appropriate recursive algorithms for both joint-space and direct cartesian space control. This paper summarizes the implementation and gives experimental results, performed on a 4-degrees-of-freedom articulated robot arm, verifying the performance of the controller.

1. INTRODUCTION

Adaptive control of linear time-invariant single-input single-output systems has been extensively studied, and a number of globally convergent controllers have been derived. Extensions of the results to nonlinear or multi-input systems have rarely been achieved. Yet, research in adaptive robot control has been very active in recent years (e.g., [Slotine and Li, 1986], [Craig, *et al.*, 1986], [Middleton and Goodwin, 1986]; [Hsu, *et al*, 1987], [Sadegh and Horowitz, 1987], [Bayard and Wen, 1987], [Koditschek, 1987], [Slotine and Li, 1987a-e]; [Li and Slotine, 1988], [Walker, 1988]), and led to the remarkable result that global convergence properties similar to those of single-input linear systems can indeed be obtained for robot manipulators, which represent an important class of nonlinear, time-varying, multi-input multi-output dynamic systems. In [Slotine and Li, 1986], we derived a simple, globally tracking-convergent, direct adaptive manipulator controller, which was demonstrated experimentally in subsequent work [Slotine and Li, 1987b]. Based on the observation that parameter information is extracted from tracking error in direct adaptive control and from prediction error in indirect adaptive control, the algorithm was further refined into a "composite" version, whose adaptation law is driven by both tracking error in joint motion and prediction error in joint torques, and therefore represents a combination of a direct and an indirect approach [Slotine and Li, 1987e]. To allow the effective implementation for multi-d.o.f. systems, appropriate recursive algorithms were then developed and studied in [Niemeyer and Slotine, 1988, 1989]. Extensions were also presented for direct cartesian control of possibly redundant manipulators, and furthermore for the application towards adaptive constrained motion control. In this paper, after reviewing the effective implementation, we discuss experimental results performed on a 4-degrees-of-freedom high-performance articulated robot arm. Of particular interest are the excellent performance during high speed joint space control, as well as a comparison of direct and composite adaptive controllers.

Following a brief review of the direct adaptive controller of [Slotine and Li, 1986], Section 2 summarizes the recursive algorithm necessary for an effective implementation thereof. The experimental results are then reported in Section 3, which also includes the description of additional switching control terms. Section 4 offers brief concluding remarks.

2. RECURSIVE IMPLEMENTATION OF ADAPTIVE MANIPULATOR CONTROLLERS

In this section, we briefly review the computationally efficient and effective implementation of direct adaptive manipulator controllers. Section 2.1 describes the direct adaptive controller of [Slotine and Li, 1986, 1987a]. The recursive implementation of the algorithm, as detailed in [Niemeyer and Slotine, 1988, 1989], is summarized in section 2.2.

2.1 Direct Adaptive Control

In the following, we briefly summarize the direct adaptive manipulator control algorithm of [Slotine and Li, 1986, 1987a] (to which the reader is referred for details). In the absence of friction and other disturbances, the dynamics of a rigid manipulator (with the load considered as part of the last link) can be written as

$$H(q)\,\ddot{q}\,+\,C(q,\dot{q})\,\dot{q}\,+\,G(q)\,=\,\tau$$

where q is the $n{\times}1$ vector of joint displacements, τ is the $n{\times}1$ vector of applied joint torques (or forces), $H(q)$ is the $n{\times}n$ symmetric positive definite manipulator inertia matrix, $C(q,\dot{q})\,\dot{q}$ is the $n{\times}1$ vector of centripetal and Coriolis torques, and $G(q)$ is the $n{\times}1$ vector of gravitational torques. In addition, a friction model can be added using, for example, $D_s(\dot{q})$ as the $n{\times}1$ vector of static friction torques, and $D_v\,\dot{q}$ as the $n{\times}1$ vector of viscous friction torques. The adaptive controller design problem is as follows: given the desired trajectory $q_d(t)$, with some or all of the components of the (properly defined) $m{\times}1$ manipulator dynamic parameter vector a being unknown, and with the joint position and velocity measured, derive a control law for the actuator torques, and an adaptation (or estimation) law for the unknown parameters, such that the manipulator joint position $q(t)$ closely track the desired trajectory $q_d(t)$.

Define the tracking error measures

$$\tilde{q}\,=\,q\,-\,q_d \qquad , \qquad s\,=\,\dot{\tilde{q}}\,+\,\Lambda\,\tilde{q}\,=\,\dot{q}\,-\,\dot{q}_r$$

where $\dot{q}_r = \dot{q}_d - \Lambda\,\tilde{q}$ and Λ is a symmetric positive definite (s.p.d.) matrix (or, more generally, a matrix whose eigenvalues are strictly in the right-half complex plane). Also, define $\hat{a}(t)$ as the current estimate of the constant parameter vector a, with $\tilde{a} = \hat{a} - a$. Then, using the Lyapunov-like function

$$V(t)\,=\,\tfrac{1}{2}\,[\,s^T H\,s\,+\,\tilde{a}^T P^{-1}\,\tilde{a}\,]$$

and exploiting fundamental physical properties of the system such as conservation of energy and the positive-definiteness of the inertia matrix H, it can be easily shown that the control and adaptation laws

$$\tau\,=\,Y\,\hat{a}\,-\,K_D\,s \qquad , \qquad \dot{\hat{a}}(t)\,=\,-\,P\,Y^T s$$

where K_D and P are s.p.d. gain matrices and the matrix $Y = Y(q,\dot{q},\dot{q}_r,\ddot{q}_r)$ is defined by

$$Y(q,\dot{q},\dot{q}_r,\ddot{q}_r)\,\hat{a}\,=\,\hat{H}(q)\,\ddot{q}_r\,+\,\hat{C}(q,\dot{q})\,\dot{q}_r\,+\,\hat{G}(q)\,+\,\hat{D}_s(\dot{q})\,+\,\hat{D}_v\,\dot{q}_r$$

yield $\dot{V}(t) = -\,s^T(\,D_v + K_D\,)\,s \leq 0$. Using simple functional analysis arguments, this result can be shown to imply that $s \to 0$ as $t \to \infty$, which in turn implies that $\tilde{q}$ and $\dot{\tilde{q}}$ both tend to 0, and therefore guarantees the global tracking convergence of the algorithm, independently of the initial parameter estimates. Note that the control law is composed of a "P.D." term $-\,K_D\,s$ and a particular type of "feedforward" $Y\,\hat{a}$, which uses Coriolis and viscous friction torques to assist in the convergence process.

2.2 Recursive Implementation

As is the case for all manipulator controllers using the full non-linear dynamics, the computational complexity of the control laws prevents their immediate implementation for multi-d.o.f. systems. Many control algorithms solve this problem by using the physically motivated recursive Newton-Euler algorithm to calculate the dynamic forces and thus achieve an efficient implementation. Adaptive sliding controllers, however, modify the natural dynamics by introducing a reference velocity and therefore can not make use of the standard Newton-Euler algorithm.

Nevertheless it is possible to generalize the Newton-Euler algorithm such that it can calculate modified manipulator dynamics. [Walker, 1988] suggests a particular generalization and achieves a recursive, hence efficient, control algorithm. This algorithm, however, can no longer be represented by joint-space equations and thus differs from the original algorithm of [Slotine and Li, 1986].

In the following, we review a different generalization, as presented in [Niemeyer and Slotine, 1988, 1989], which gives the true implementation of the original algorithm. It is not only physically motivated, as is the standard Newton-Euler algorithm, but can also be derived through substitutions in the original joint-space equations. It is applicable to both open and closed kinematic chains, though detailed here for open chains only. The standard Denavit-Hartenberg convention is used, thus numbering the links from zero (base) to n (tip) with joint i connecting link $i-1$ to link i.

initialize:

$$\dot{w}_0 = -g$$

upward:

$$v_k = v_{k-1} + d_k \dot{q}_k$$
$$w_k = w_{k-1} + d_k \dot{q}_{r_k}$$
$$\dot{w}_k = \dot{w}_{k-1} + d_k \ddot{q}_{r_k} + v_{k-1} \times d_k \dot{q}_{r_k}$$
$$e_k = v_k - w_k$$

downward:

$$f_k^i = \frac{1}{2} v_k \times R_i\, w_k + \frac{1}{2} w_k \times R_i\, v_k + \frac{1}{2} R_i\, w_k \times v_k + R_i\, \dot{w}_k$$

$$F_k = F_{k+1} + \sum_{i=1}^{10} f_k^i\, a_k^i$$

$$\tau_k = d_k^T F_k + (\hat{D}_s + \hat{D}_v\, \dot{q}_r - K_D\, s)_k$$

$$\hat{\dot{a}}_k^i = - P_k^i\, e_k^T\, f_k^i$$

$$\hat{\dot{D}}_{s_k} = - P_k^s\, s_k$$

$$\hat{\dot{D}}_{v_k} = - P_k^v\, s_k\, \dot{q}_{r_k}$$

Table 1: Equations for the recursive implementation of a direct adaptive controller

As other recursive algorithms, this particular algorithm can be partitioned into an upward and downward section. During the first phase the kinematic structure is traversed in an upward fashion starting from the base and finally reaching the end-effector. Throughout this phase all the joint information is used to compute the cartesian velocities and acceleration of each individual link. Given such data the dynamic equations are then applied to compute cartesian forces acting locally on each link. Throughout the second phase, which traverses the kinematic structure downwards, these forces are then combined and mapped to the different actuators. This then determines the control torques, which are to be applied.

Table 1 gives a summary of all involved equations. All cartesian quantities are expressed in the spatial vector notation of [Featherstone, 1987], which considerably reduces the complexity and number of equations necessary. The sparse placement matrices R give the location of different parameters in the appropriate spatial inertia matrix, which itself includes all ten inertial parameters of any given link. The vectors d describe the direction of the appropriate joint axes.

The gravitational forces are implemented as a vertical acceleration in gravity-free space, thus eliminating their explicit calculation. For best efficiency the following reference frames should be used: for velocities, accelerations, inertias, and local force components the frame attached to the given link, while for the joint-axes and summed forces the frame attached to the link directly below is best suited. Furthermore it is important to customize the algorithm, including the analytical evaluation of cross products in the individual force equations, and to use minimal parameter sets.

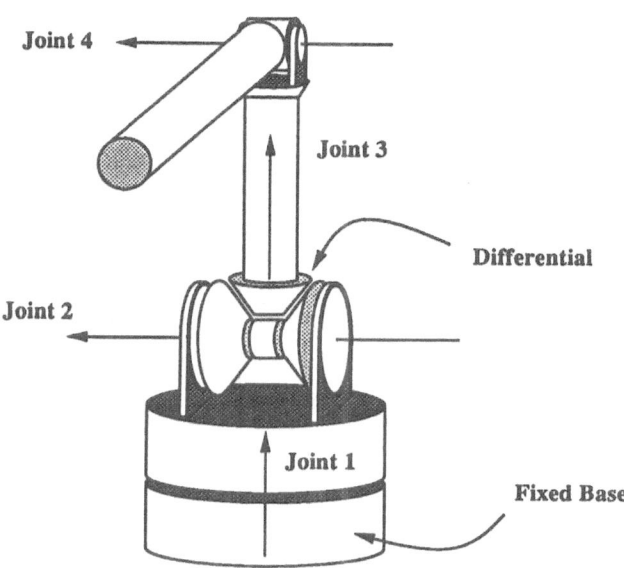

Figure 1: The 4 d.o.f. whole arm manipulator

3. EXPERIMENTAL RESULTS

The above developments were tested on a 4 d.o.f. cable-driven "whole-arm" manipulator designed at the M.I.T. Artificial Intelligence Laboratory [Townsend, 1988; Salisbury, Townsend, *et al.*, 1988]. The manipulator, shown in Figure 1, has a configuration similar to that of a human arm, with an extended length of 1 meter. The joint-torques are generated by four pulse-width modulated motors, capable of delivering a maximum of 1.5 Nm each. The motors are located close to the base and connected to the lightweight links via cables, introducing a transmission-ratio varying between 1:20 and 1:30. Reduction is done at the joints. Position measurements are made at the motor shafts using resolvers. In order to exploit the wide dynamic range that the manipulator can potentially achieve, the arm is connected to a VME-Bus based multiprocessor system [Narasimhan, *et al.*, 1988], consisting of up to six 68020 based processor boards, a D/A and A/D board, a parallel interface, as well as other boards. This system is interfaced with the network via a Sun-3 Workstation of Sun Microsystems, Inc. The different processors are used to implement high-speed input/output routines, the basic controller algorithm, the adaptation algorithm, the inverse kinematics, and the trajectory generation as well as other higher level tasks. Input acquisition and filtering is performed at 4 KHz, while the control, adaptation, and inverse kinematics algorithms are executed on separate processors at 125 Hz. P.D. torques are calculated at a rate of 250 Hz and additional switching terms computed at

1.5 KHz. The velocity signals are created by filtered differentiation of the 16 bit position signals (12 bit resolver signals plus rotation count). The integration in the adaptation algorithm is performed using a 2nd order Adams-Bashforth scheme. All programs are written in C.

The friction model used for the manipulator consists of viscous and Coulomb friction, where Coulomb friction is allowed to be direction-dependent. We therefore deal with 12 friction parameters in addition to the 24 inertial parameters, and thus we are adapting to 36 unknown values. While dealing with unknown loads only requires in principle to adapt to 10 unknown parameters, the capability of effectively adapting to all 36 parameters allows the range of application of the algorithm to be considerably extended, as discussed later. Also, closed chains are used in the kinematics, as the cable-transmission involves a differential introducing two extra drums.

3.1 Switching Control Term

The adaptive controller detailed so far will cause the tracking error to converge to zero in the absence of disturbances. In practice this assumption is obviously not valid and as a result a residual tracking error will remain, due in particular to imperfect friction models, a 5% torque ripple, and measurement noise. To further reduce this residual error, we use of an additional switching control term. In the absence of a detailed model of disturbance sources, we simply add to the torques τ^{adapt} commanded by the adaptive controller a torque τ^{switch} of constant magnitude, and of sign (component-wise) opposite to the tracking error measure s.

$$\tau_i = \tau_i^{adapt} - \tau_i^{switch} \, \text{sgn} \, (s_i)$$

Ideally, each the switching torque magnitudes should be larger or equal to an upper bound on the corresponding disturbance component. Note that the extra "dither" control term is in s rather than in velocity, which has the major advantage of coming with a Lyapunov performance proof. Indeed, assume that each disturbance component $d_i(t)$ is bounded in magnitude by D_i. The Lyapunov-like function used for proving stability and convergence of the adaptive controller remains unchanged. Its derivative, however, is changed from $\dot{V} = - s^T K_D s$ as follows, with sgn(.) switching each individual element [Slotine and Li, 1987a]

$$\dot{V} = s^T \, (- K_D s + d - \tau^{switch} \, \text{sgn}(s)) \; \leq \; - s^T K_D s \; \leq 0$$

$$\text{as} \quad |d_i| \leq D_i \leq \tau_i^{switch} \quad => \quad |s^T d| \leq s^T \tau^{switch} \, \text{sgn}(s)$$

The complete controller now guarantees convergence even in the presence of bounded disturbances. Another way of viewing the switching is to interpret it as a locally infinitely stiff P.D. Yet, even in the case of electrical motors, which can naturally deal with such high-frequency inputs (an even more efficient implementation would be to use switching in s directly in place of pulse-width modulation, but this would require redesign of the motor amplifiers), the well-known difficulty with switching controllers is their tendency to cause vibration and high-frequency excitation. Ideally, the frequency of the switching should be chosen well *beyond* that of significant structural vibration modes, while remaining *below* the actuators' bandwidth and the frequency of the pulse-width modulation: these requirements were compatible in the available hardware, and the switching frequency was chosen to be 1.5Khz, or 1/3rd of the (relatively low) PWM frequency (and very close to the actuators' bandwidth). Roughly speaking, given the structural resonance properties of the arm, the allowable amplitude of the switching (and therefore its disturbance rejection capabilities) varies as the inverse of the switching frequency (using a standard Filippov approximation argument). This assumes that meaningful state estimates can be be provided at the selected switching frequency, and indeed, another important performance limit is due to measurement noise in velocity (recall that the only measurements are of joint positions). Given all these limitations, the use of a switching controller is only effective when the remaining disturbances are quite small. This is the case for our adaptive controller, since all inertial and gravitational effects (as well as friction to a significant extent) are adaptively compensated for. The experiments verify this, as the simple additional switching control reduces the residual tracking error by 50% without causing noticeable vibrations. Also, it considerably improves the quality of low-speed behavior, which without is extremely sensitive to torque ripple, stiction, and other modeling imperfections.

3.2 Discussion

Figures 2a and 2b show data obtained while tracking a sinusoidal trajectory of period 1 second with an amplitude of $\pm 45^o$ per joint. During this trajectory the tip travels approximately 5 meters per second, with maximum tip velocities and accelerations of 7m/s and 5g. This was done (i) with a simple well-tuned P.D. (whose performance is considerably helped by the presence of the transmission ratios), and (ii) with the adaptive scheme starting as a P.D., i.e. with parameter estimates initialized at zero. The plots clearly demonstrate a factor 10 to 20 improvement in tracking error after about 1/2 second when adapting to all 36 parameters. Furthermore, although both controller start identical, the maximum tracking error of the adaptive controller during transients remains 2 to 10 times smaller than that of the P.D. . Nevertheless the generated joint torques are very similar in both smoothness and amplitude. Adaptation dead-

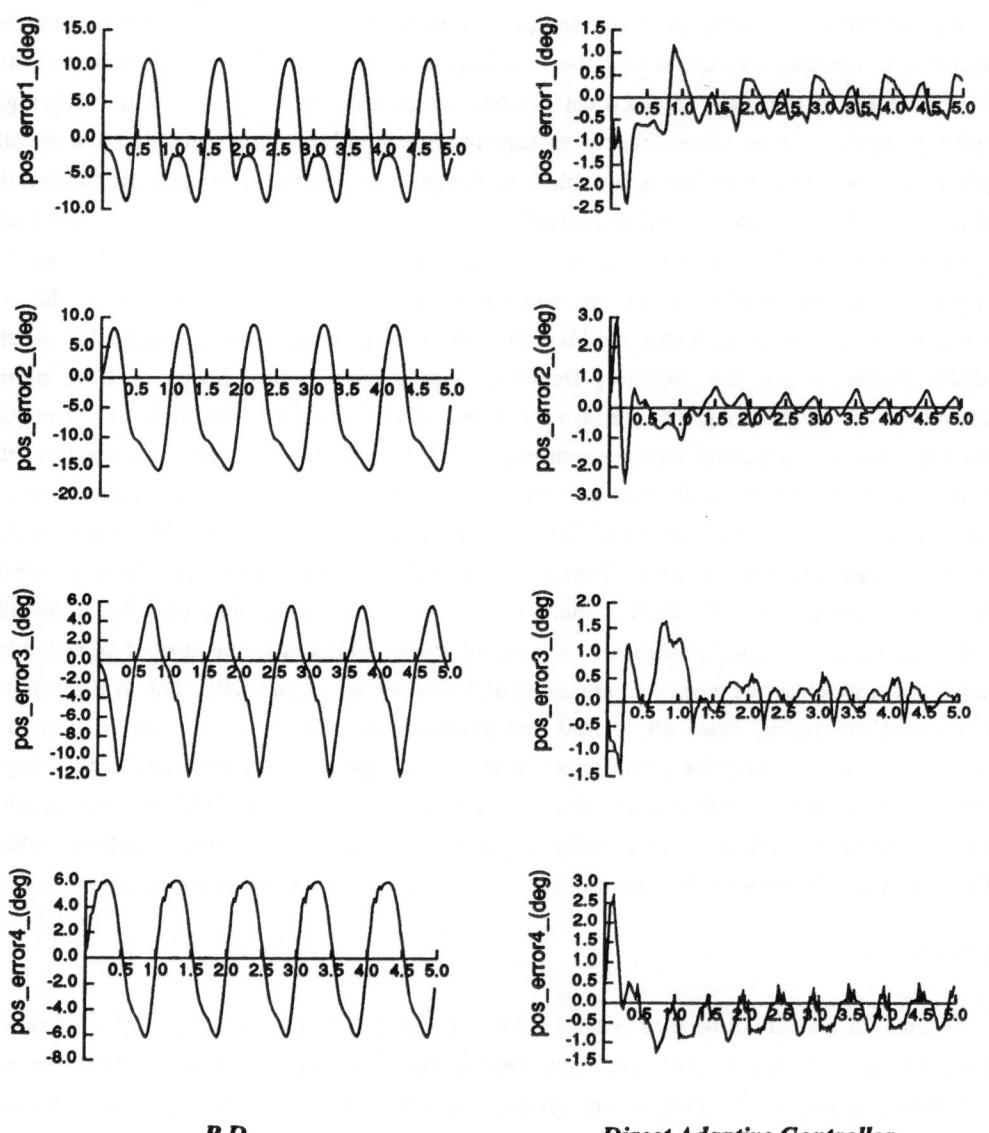

P.D.   Direct Adaptive Controller

Figure 2.a: Tracking Errors $\tilde{q}$ (in degrees)

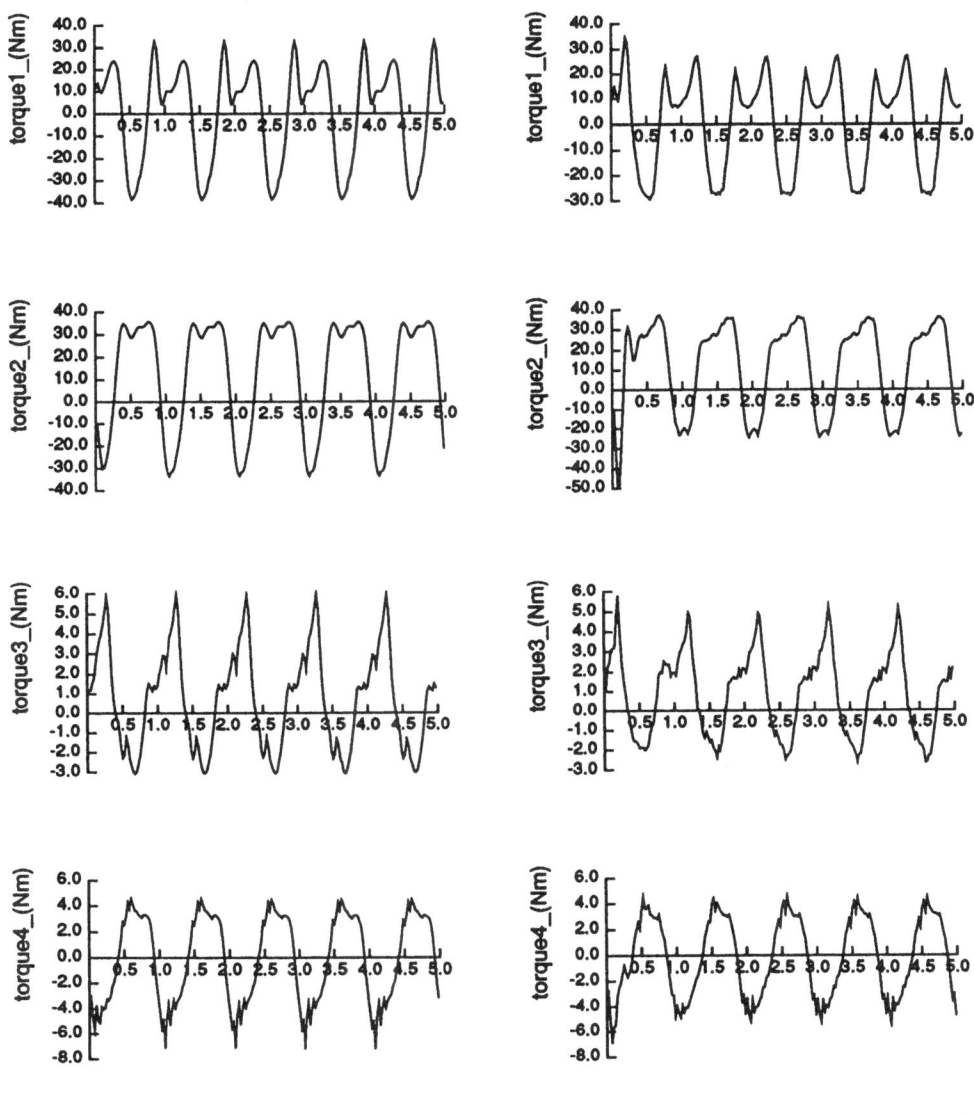

P.D.   Direct Adaptive Controller

Figure 2.b: Joint Torques τ   (in N.m.)

zones of 0.15 rad/s in the components of s (or about 1°, with $\lambda = 10$) are used in order to account for residual parametric uncertainty (e.g., inaccurate friction models, torque ripple) and enhance robustness to noise and unmodeled dynamics.

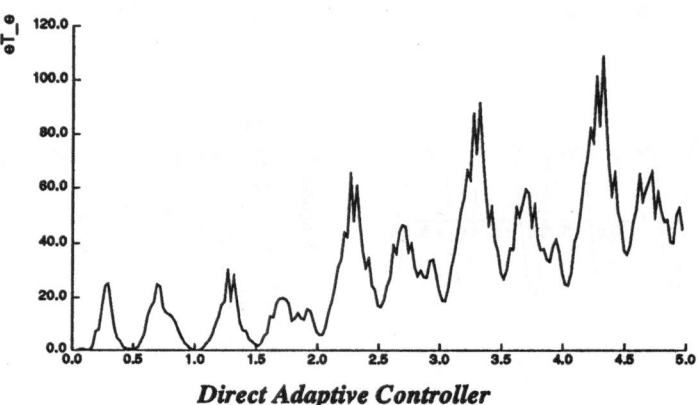

Direct Adaptive Controller

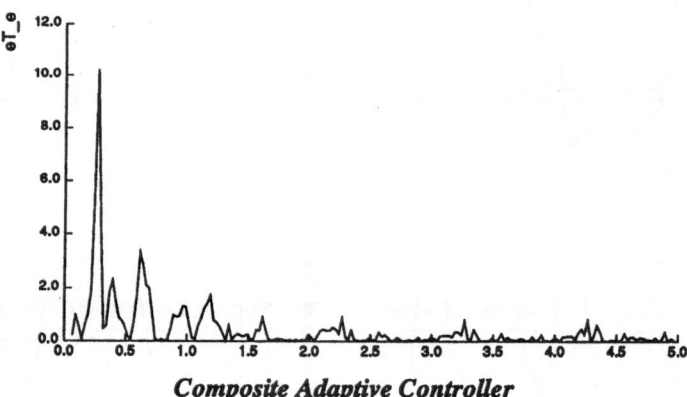

Composite Adaptive Controller

Figure 3: Prediction Error

In other experiments, the adaptation was kept on while using the last link to push a large box (of weight about ten times that of the last link, and of volume about 0.2 m^3) located on the floor, at speeds of about .75 m/s. This exploited the kinematic redundancy of the arm to maintain line contact with the box while pushing. Despite the fact that accurate models of the relative motion of the box with respect to the arm or of the friction between the box and the floor were not developed, this strategy allowed accurate "whole-arm" manipulation of the box, the presence of the box being in this case largely interpreted by the algorithm as a change in the inertial and friction coefficients of the base link. We believe that this robustness is quite remarkable for a reasonably complex high-performance algorithm, and that it should extend the range of applications of adaptive tracking manipulator controllers well beyond adaptation to grasped loads. Also, the accuracy of the manipulator controller allows us to make full use of the "geometric" stability properties of the task ([Mason, 1987] in the case of pushing), thus allowing large uncertainties on the initial position of the box to be tolerated, and high speed performance to be achieved without slowing down the desired motion during the transition (making contact) phase.

Further experiments compared the direct adaptive controller to the composite adaptive controller of [Slotine and Li, 1987e, 1988b]. In this extension both output (i.e. tracking) error, as well as input (i.e. predicted torque) error drive the adaptation process. The experiments verify that while the introduction of the composite adaptation has little effect on tracking performance it guarantees the additional convergence of the prediction error. This allows the controller to handle variations in trajectory excitation more effectively. Figure 3 compares the prediction error of direct and composite controller operating on the same trajectory. Notice that the prediction of the direct controller is completely dominated by the actual behavior of the filtered power input.

4. CONCLUDING REMARKS

We believe that advanced control concepts have exceptional potential in the development of robust, reliable, high-performance robotic systems. We hope studies such as the one presented here will accelerate such implementations in common industrial and scientific practice, and thus allow the full potential of ever increasing computational capabilities to be exploited effectively.

Acknowledgements: We are grateful to Ken Salisbury and Bill Townsend for allowing us to test the above results on the fine manipulator they developed, and for stimulating discussions. We would also like to thank Brian Eberman and Sundar Narasimhan for their help in the implementation. This work was supported in part by a student Grant from the Perry Foundation, and in part by Grant 8803767-MSM from the National Science Foundation. Development of the arm and control hardware was supported by Grants N00014-86-K-0685 and N00015-85-K-0214 from the Office of Naval Research.

REFERENCES

An, C.H., Atkeson, C.G. and Hollerbach, J.M., 1985. Estimation of inertial parameters of rigid body links of manipulators, *I.E.E.E. Conf. Decision and Control*, Fort Lauderdale.

Asada, H., and Slotine, J.J.E., 1986. Robot Analysis and Control, *Wiley*.

Baillieul, J., 1985. Kinematic Programming Alternatives for Redundant Manipulators, *IEEE Int. Conf. on Robotics and Automation*, St. Louis

Baker, D.R., and Wampler, C.W.,II, 1988. On the Inverse Kinematics of Redundant Manipulators, *Int. J. Robotics Res.*, Vol. 7, No.2

Bayard, D.S., and Wen, J.T., 1987. Simple Adaptive Control Laws for Robotic Manipulators, *Proceedings of the Fifth Yale Workshop on the Applications of Adaptive Systems Theory*.

Craig, J.J., Hsu, P. and Sastry, S., 1986. Adaptive Control of Mechanical Manipulators, *I.E.E.E. Int. Conf. Robotics and Automation*, San Francisco.

Featherstone, R., 1987. Robot Dynamics Algorithms, *Kluwer Academic Publishers*.

Hsu, P., Sastry, S. , Bodson, M. and Paden, B. 1987. Adaptive Identification and Control of Manipulators Without Joint Acceleration Measurements, *I.E.E.E. Int. Conf. Robotics and Automation*, Raleigh, NC.

Khosla, P.,and Kanade, T., 1985. Parameter Identification of Robot Dynamics, *I.E.E.E. Conf. Decision and Control*, Fort Lauderdale.

Koditschek, D.E., 1987. Adaptive Techniques for Mechanical Systems, *Proceedings of the Fifth Yale Workshop on the Applications of Adaptive Systems Theory*.

Landau, I., and Horowitz, R., 1988. *I.E.E.E. Int. Conf. Robotics and Automation*, Philadelphia, PA.

Li, W. and Slotine, J.J.E. 1987. Parameter Estimation Strategies for Robotic Applications. *A.S.M.E. Winter Annual Meeting*, Boston, MA.

Li, W. and Slotine, J.J.E. 1988a. Indirect Adaptive Robot Control, *5th I.E.E.E. Int. Conf. Robotics and Automation*, Philadelphia, PA.

Mayeda, H., Yoshida, K., and Osuka, K., 1988. Base Parameters of Manipulator Dynamic Models, *5th I.E.E.E. Int. Conf. Robotics and Automation*, Philadelphia, PA.

Middleton, R.H. and Goodwin, G.C. , 1986. Adaptive Computed Torque Control for Rigid Link Manipulators, 25th *I.E.E.E. Conf. on Dec. and Contr.*, Athens, Greece.

Narasimhan, S., Siegel, D., and Hollerbach, J., 1988. *I.E.E.E. Int. Conf. Robotics and Automation*, Philadelphia, PA.

Niemeyer, G., and Slotine, J.J.E., 1988. Performance in Adaptive Manipulator Control, *I.E.E.E. Conf. on Decision and Control*, Austin, TX.

Niemeyer, G., and Slotine, J.J.E., 1989. Computational Algorithms for Adaptive Compliant Motion, *I.E.E.E. Int. Conf. Robotics and Automation*, Scottsdale, AZ.

Ortega, R., and Spong, M., 1988. *I.E.E.E. Int. Conf. Decision and Control*, Austin, TX.

Sadegh, N., and Horowitz, R., 1987. Stability Analysis of an Adaptive Controller for Robotic Manipulators. *I.E.E.E. Int. Conf. Robotics and Automation*, Raleigh, NC.

Salisbury, J.K., et al., 1988. Preliminary Design of a Whole-Arm Manipulator System, *I.E.E.E. Int. Conf. Robotics and Automation*, Philadelphia, PA.

Slotine, J.J.E., and Li, W., 1986. On The Adaptive Control of Robot Manipulators, *A.S.M.E. Winter Annual Meeting*, Anaheim, CA.

Slotine, J.J.E., and Li, W., 1987a. On the Adaptive Control of Robot Manipulators, *Int. J. Robotics Res.*, vol. No.3

Slotine, J.J.E., and Li, W., 1987b. Adaptive Robot Control, A Case Study, *I.E.E.E. Int. Conf. Robotics and Automation*, Raleigh, NC.

Slotine, J.J.E., and Li, W., 1987c. Adaptive Strategies in Constrained Manipulation, *I.E.E.E. Int. Conf. Robotics and Automation*, Raleigh, NC.

Slotine, J.J.E. and Li, W., 1987d. Theoretical Issues In Adaptive Manipulator Control. In *the Proceedings of the Fifth Yale Workshop on Applications of Adaptive Systems Theory*.

Slotine, J.J.E. and Li, W., 1987e. Adaptive Robot Control - A New Perspective, *I.E.E.E. Conf. Decision and Control* L.A., CA.

Slotine, J.J.E., and Li, W., 1988a. Adaptive Manipulator Control: A Case Study, *I.E.E.E. Trans. Autom. Control, 33, 11*.

Slotine, J.J.E., and Li, W., 1988b. Composite Adaptive Robot Control, MIT-NSL-880501, submitted to Automatica.

Slotine, J.J.E., and Ydtsie, B.E., 1988. Control of Nonlinear Chemical Processes: A Physically-Motivated Approach, submitted to *Automatica*.

Townsend, W., 1988. "The Effect of Transmission Design on Force-Controlled Manipulator Performance", Ph.D. Thesis, M.I.T., Department of Mechanical Engineering.

Walker, M.W., 1988. An Efficient Algorithm for the Adaptive Control of a Manipulator, *5th I.E.E.E. Int. Conf. Robotics and Automation*, Philadelphia, PA.

CONTROL OF MACHINES WITH NON-LINEAR, LOW-VELOCITY FRICTION: A DIMENSIONAL ANALYSIS

Brian S. R. Armstrong
University of Wisconsin - Milwaukee
P.O. 784, Milwaukee, Wisconsin 53201 U.S.A.

Abstract

Machines at low velocity exhibit stick-slip motion. This phenomena is of particular importance in force control of robot manipulators because much of the force correcting action can occur in the low velocity regime of stick-slip. Experimental work has mapped out the structure of Stribeck friction, a non-linear low-velocity friction effect that contributes to and perhaps dominates stick-slip. In this paper I explore the implications of Stribeck friction for feedback control, in particular, for a system with Stribeck friction the following are examined:

- The minimum velocity below which stick-slip will occur;
- Accuracy of sensing required to eliminate stick-slip by feedback control;
- The jerk size during stick-slip motion;
- Mechanical attributes that will affect Stribeck friction.

A 7 parameter friction plus control model is proposed; dimensional analysis is employed to reduce the degrees of freedom of this model to 3. Results obtained from the study of this lower order model may be scaled to apply to all physical systems describable by the 7 parameter model.

1   Introduction

In fluid or grease lubricated mechanisms, friction decreases as the velocity increases away from zero. In general terms this effect is understood. It is due to the transition from boundary lubrication to fluid lubrication. In boundary lubrication, metal parts are separated by extremely thin, perhaps mono-molecular, layers of materials that adhere to the metal surfaces. These materials are chosen to have low shear strength, so as to reduce friction, good bonding and a variety of other properties such as stability, corrosion resistance or solubility in the bulk lubricant. Boundary lubricants are standard in greases and oils specified for precision machine applications. It should be noted that in virtually all applications, it is the prevention of wear during start and stop which dominates the choice of boundary lubricant. With the exception of whey lubricants, the friction properties of boundary lubricants have been a secondary consideration [Rabinowicz 78].

When a machine is operating at sufficient speed, the bulk lubricant, be it oil or grease, is held in the load bearing interface by its own viscosity and the friction is governed by hydrodynamics. When the load bearing parts are conformal, that is to say that their curvatures are nearly equal, the load bearing area is large and the stresses are relatively low, so that the deformation of the loaded parts is small in relation to the film thickness. This situation is called hydrodynamic lubrication. It arises in flat machine guides and journal bearings.

The far more common situation is that of non-conformal components, such as ball bearings or gear teeth. In these cases the contact area is governed by part deformation, normally modeled as hertizian contact, and local stresses are extremely high. This situation is called elasto-hydrodynamic lubrication (EHL). The study of elasto-hydrodynamic lubrication is quite active. It is, generally speaking, pursued numerically because Reynolds equation of fluid flow and the elastic deformation equation can not be simultaneously satisfied analytically. The theory of EHL gives rise to some rather counter intuitive results, for example, the film thickness between load bearing parts can increase with increased load [Pan and Hamrock 89]. It is interesting to note that in many cases EHL functions because lubricant viscosity increases exponentially with pressure and may be hundreds of times higher in an EHL interface than at ambient pressures. Dowson and Higginson, [66], is widely cited on the subject of EHL.

The tribological problem of central concern in the fine control of mechanisms is the dynamic of the transition from boundry to fluid lubrication. The problem is a difficult one theoretically because the details of surface roughness strongly influence the fluid flow. As yet a complete picture of the transition process seems unavailable, but work is underway [Zhu and Cheng 88, Sadeghi and Sui 89].

Experiments reported in [Armstrong 88] show a friction structure that is quite different from that previously studied in the controls literature. The structure, Stribeck friction, allows that friction is constant at extremely low velocities and makes a smooth transition from the higher static friction to the lower kinetic or kinetic plus viscous friction. Experiments reported in [Gassenfeit and Soom 88] show a similar structure. A model based on Stribeck friction will predict steady motion at extremely low velocities, instability through a range of low velocities and stable motion above a threshold velocity. A plot of friction as a function of velocity for joint 1 of a PUMA 560 manipulator is presented in figure 1.

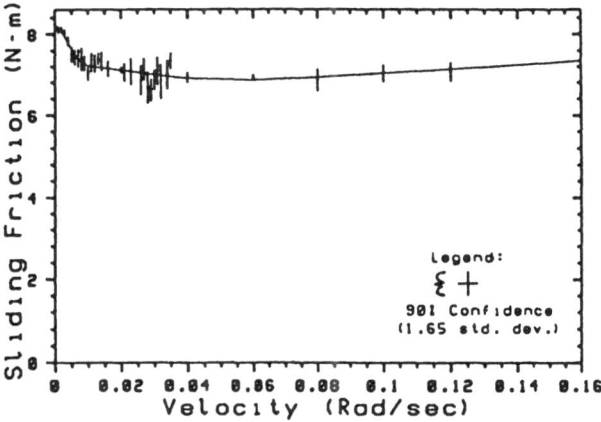

Figure 1.   Friction as Function of Velocity for Joint 1 of a PUMA 560 Manipulator. Stribeck Friction is demonstrated in the transition from Static to Kinetic plus Viscous Friction (From [Armstrong 88]).

The friction data presented in figure 1 were obtained with an apparatus shown schematically in figure 2. Torque applied at joint 1 resulted in a force applied at the contact, where springs of various stiffness were used. A torque applied was a triangle wave: a ramp up and a ramp down. Ideally, the motion would have a triangular shape with the rate proportional to the torque rate and inversely proportional to the spring stiffness. Actual motions are shown in figures 3a, 3b and 3c. Note the different scales of the plots of figure 3, the motion of figure 3a is extremely small, that of 3b larger, and that of 3c larger yet. Here we see the three regions of the Stribeck effect: steady motion at extremely low velocities, stick-slip in evidence at low velocities and steady motion again above a threshold velocity.

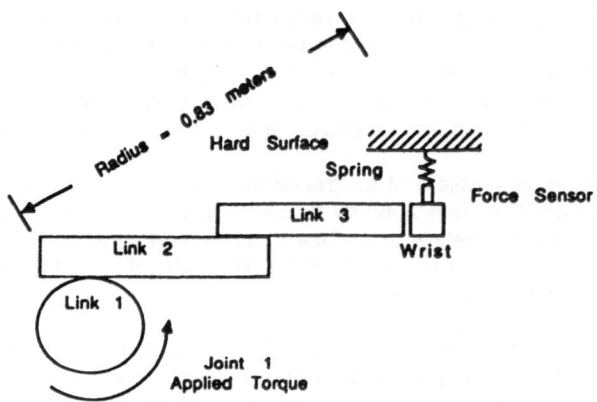

Figure 2.   Schematic of the Apparatus for Measuring Stribeck Friction.

To study the Stribeck friction, precise velocity measurements were required. These were made using a rotational accelerometer and the force sensors. The integral of the accelerometer signal provided one velocity estimate, the derivative of the force signal, smoothed off-line with a non-causal filter, provided a second velocity estimate. With careful attention to noise sources, the two independent velocity estimates agreed to within 70 microradians per second. They were merged by a cross-over filter that took advantage of the high frequency sensitivity of the accelerometer data and the low frequency sensitivity of the force rate data.

Friction was measured by subtracting inertial and contact torques from the applied torque, according to:

$$Friction\_Torque = Applied\_Torque - Contact\_Torque - Inertial\_Torque \qquad (1)$$

The contact force was measured directly, and converted to torque. The inertial torque was computed from the accelerometer signal and the known inertia of the arm. No correction was made for the bending dynamics of the arm.

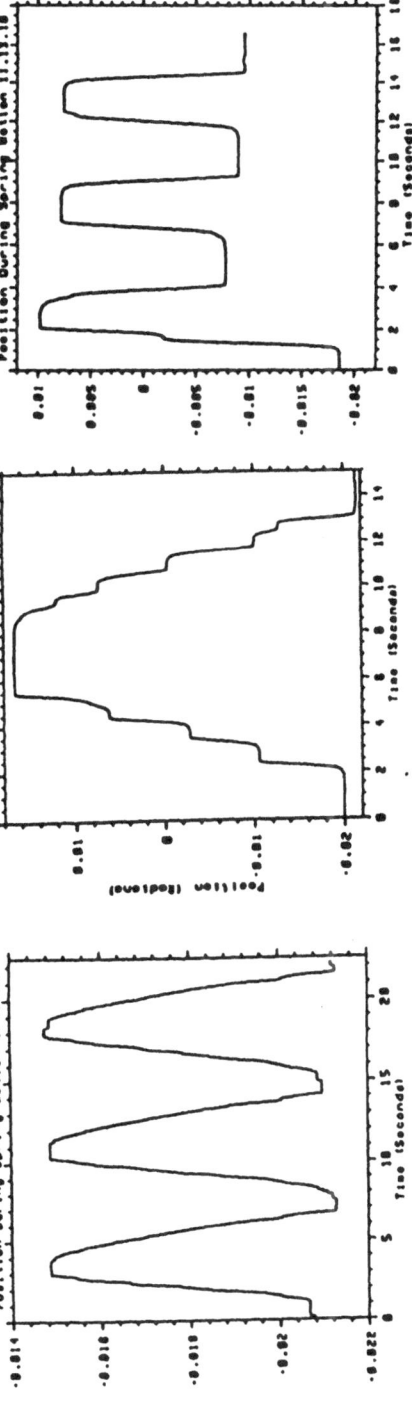

Figure 3a.   Mean Vel.:   0.003   Figure 3b.   Mean Vel.:   0.012   Figure 3c.   Mean Vel.:   0.020
                  Spring Stiff.: 3296   Spring Stiff.: 554   Spring   Stiff.: 770

Figure 3.   Three Plots of Position versus Time for Motions of the Apparatus of Figure 2.   Mean
                  Velocity in [Rad/Sec], Spring Stiffness in [N·m/Rad].

184

2   Dimensional Analysis

The Stribeck friction may be represented by:

$$F(x) = Fo\ sgn(\dot{x}) + Fs\ (exp(-(\dot{x}/Vc)^2) - 1)\ sgn(\dot{x}) \qquad (2)$$

Where Fo   is the static friction,
        Fs   is the magnitude of the Stribeck effect,
        Vc   is the critical velocity of the Stribeck effect,
        $\dot{x}$ is the velocity.

This model is empirical, but fits the data better than a ninth order polynomial or other exponential models. The model friction curve of figure 1 is drawn with two friction breaks, represented by two exponential terms; it is believed that these correspond to two different rubbing interfaces with two different effective radii, so that the two different critical velocities expressed in radians per second correspond to similar linear rubbing velocity.

We will investigate a system with the above friction plus proportional and derivative feedback (or perhaps inherent stiffness and damping) so that the dynamic model is:

$$M\ x = - K(x - x_d) - Kv(\dot{x} - \dot{x}_d) - Fo\ sgn(\dot{x}) - Fs\ (exp(-(\dot{x}/Vc)^2) - 1)\ sgn(\dot{x}) \qquad (3)$$

Where   M   is the mass,
          K   is the proportional feedback gain (stiffness),
          Kv   is the derivative feedback gain (damping),
          $\dot{x}_d$ is the desired velocity.

The situation under investigation is steady, low-velocity motion. By making the assumption that the velocity does not reverse (a fairly mild requirement on the size of the slips), the $sgn(\dot{x})$ functions may be eliminated. By redefining position, the Fo term may be eliminated:

Defining: $x = x - x_d + Fo\ /\ K,$ (4)
then: $\dot{x} = \dot{x} - \dot{x}_d,$
               $\ddot{x} = \ddot{x};$

and equation (3) becomes:

$$M\ \ddot{x} = - K(x) - Kv(\dot{x}) - Fs\ (exp(-(\dot{x}/Vc + \dot{x}_d/Vc)^2) - 1) \qquad (5)$$

Rescaling position by a factor α to get a new position value, y, and time by a factor β to get a new time value, s, gives:

$$\alpha\ y = x \qquad and \qquad s = \beta\ t \qquad (6)$$

We find that

$$\dot{x} = \alpha\ \beta\ \dot{y}, \qquad\qquad \ddot{x} = \alpha\ \beta^2\ \ddot{y}. \qquad (7)$$

In what follows with dimensionless coordinates, all derivatives are taken w.r.t dimensionless time, s. Expressing the system dynamic, (4), in terms of y:

$$\alpha \beta^2 M \ddot{y} = -K(\alpha y) - Kv(\alpha \beta \dot{y}) - Fs (exp(-(\alpha \beta \dot{y}/Vc + \alpha \beta \dot{y}d/Vc)^2) - 1) \qquad (8)$$

where $\dot{y}d$ is the scaled desired velocity, $\alpha \beta \dot{y}d = \dot{x}d$.

Now, choosing α and β to satisfy two constraints - that dividing through by $\alpha \beta^2 M$ will cancel Fs, and that dividing through by $\alpha \beta$ will cancel Vc - gives :

$$\alpha \beta^2 = Fs / M, \qquad \alpha \beta = Vc.$$

Solving for α and β gives:

$$\alpha = Vc^2 /(Fs/M), \qquad \beta = (Fs/M)/Vc. \qquad (9)$$

Looking at the units (m: meters, s: seconds, kg: kilo-grams),

$$\alpha = [m^2 s^{-2} / (kg\ m\ s^{-2} / kg)] = [m]$$
$$\beta = [(kg\ m\ s^{-2} / kg) / (m\ s^{-1})] = [s^{-1}].$$

Thus y is dimensionless distance and s dimensionless time. Incorporating the scaling factors into K* and Kv*, the dynamic model, (8), becomes

$$\ddot{y} = -K^* y - Kv^* \dot{y} - (exp(-(\dot{y} + \dot{y}d)^2) - 1) \qquad (10)$$

Equation (10) is written in 3 parameters, K*, Kv* and yd but can represent any system in the original space of 7 parameters, through the scaling of equation (6) and the offset of equation (4). Note that K* and Kv* are themselves dimensionless:

$$K^* = K / \beta^2 M = K Vc^2 M / Fs^2 : \quad [(kg\ m\ s^{-2} / m)/(s^{-2}\ kg)] \quad or\ [\ ] \qquad (11)$$
$$Kv^* = Kv / \beta M = K Vc / Fs: \quad [(kg\ m\ s^{-2} / m\ s^{-1})/(s\ kg)] \quad or\ [\ ]$$
$$\dot{y}d = \dot{x}d / \alpha \beta = \dot{x}d / Vc \quad [(m\ s^{-1}) / (m\ s^{-1})] \quad or\ [\ ]$$

Equations (10) and (11) provide a dimensionless model of a proportional-derivative controlled system with Stribeck friction. It can be used answer questions of interest, such as what desired velocity, yd, is the most difficult to stabilize; what damping is required to stabilize that velocity, and what sensor noise may be allowed while state error is maintained within bounds. The original model, (3), may be modified to incorporate force control or joint torque feedback, and the same questions may be answered for these configurations. By integrating (10) for various values of K*, Kv* and yd, the stick-slip motion may be studied and properties such as the Kv* required to prevent stick and the expected slip distance may be determined.

3 Stabilizing $\ddot{y} = -K^* y - Kv^* \dot{y} - (exp(-(\dot{y} + \dot{y}d)^2) - 1)$

Asymptotic stability will be achieved if $d\ddot{y}/d\dot{y} < 0$ for all $\dot{y}$.

$$d\ddot{y}/d\dot{y} = -Kv + 2(\dot{y} + \dot{y}d) exp(-(\dot{y} + \dot{y}d)^2) \qquad (12)$$

The value of $(\dot{y}+\dot{y}_d)$ for which $2\,(\dot{y}+\dot{y}_d)\,exp(-(\dot{y}+\dot{y}_d)^2$ takes on the greatest value is $(\dot{y}+\dot{y}_d)$ = $1/\sqrt{2}$. Choosing yd so that we consider the case of trying to achieve steady motion at this velocity and evaluating (12), we find that for stability

$$Kv > 2 * 1/\sqrt{2}\ exp(-1/2) = .858 \qquad (13)$$

The desired velocity most difficult to stabilize is $\dot{y}_d = 1/\sqrt{2}$; we must have a dimensionless damping of .858 to achieve *asymptotically* stable motion at that velocity.

4 The Required Velocity Measurement to Achieve Allowable Variance in the State

If we imagine that the damping term of equation (10) is implemented by derivative feedback, then noise in the estimate of velocity will give rise to state error. Stochastic analysis may be applied to the dynamic of equation (10) to relate an anticipated noise in the estimate of velocity to the expected deviation from desired state. There is one possible deviation that is of particular interest: a swing from the desired velocity to the stuck condition. It is of little use to stabilize a system by derivative feedback if the noise thereby injected drives the system into sticktion.

If we restrict ourselves to the least stable value of $\dot{y}_d$, $1/\sqrt{2}$, and the minimum stabilizing value of Kv*, .858, then (10) becomes a system in 1 parameter: K*. We will find K* to be independent of what we seek to discover, namely maximum variance in the estimate of velocity is that will provide an acceptable level of probability that the system will be driven into the stuck condition.

Equation (10) is a non-linear dynamic. Thus the probability distribution of the state will not be Gaussian, even if the driving noise is Gaussian. But we may follow the standard quadratic analysis to produce an integral equation for the probability distribution, which may be integrated for a particular boundry condition. The boundry condition will represent a chosen threshold of probability that the system arrives at the stuck state due to noise induced deviation.

By writing (10) in matrix form, so that Y represents a two-vector of y and $\dot{y}$, the dynamic equation becomes:

$$Y = (F - GC)\,Y\ +\ \begin{bmatrix} 0 \\ fs \end{bmatrix}\ -GCY \qquad (14)$$

Where Y   is the state vector $[\ y\ \ \dot{y}\ ]'$,
    F   is the coupling of state to the derivative of state,
    G   is the coupling of control to the derivative of state,
    C   is the control vector, [K*   Kv*],
    fs   is the Stribeck friction, $(exp(-(\dot{y}+\dot{y}_d)^2) - 1)$,
    Y   is the vector of noise in the state estimate.

Here we have considered only noise in the velocity estimate, and we shall be concerned only with error in the velocity itself and the probability that velocity goes to zero due to this error. Multiplying (14) on the right by Y', and adding the result to it's own transpose gives:

$$Y\,Y' + Y\,Y' = (F - G\,C)\,Y\,Y' + Y\,Y'\,(F - G\,C) + \begin{bmatrix} 0 \\ fs \end{bmatrix} Y' + \dot{Y}\,[\,0\ fs\,] - G\,C\,Y\,Y' - Y\,Y'\,G'C' \tag{15}$$

Taking the expectation value of (15) and noting that

$$E\{Y\,Y'\} = Y \quad \text{and} \quad E\{Y\,Y' + y\,y'\} = \dot{Y}$$

Where Y is the covariance matrix of Y,
$\dot{Y}$ is the derivative of the covariance matrix of Y.

gives a dynamic equation in the variance of Y:

$$\dot{Y} = (F - G\,C)\,Y + Y\,(F - G\,C)' + E\left\{\begin{bmatrix} 0 & 0 \\ 0 & 2\,fs\,\dot{y} \end{bmatrix}\right\} - E\{G\,C\,Y\,Y' + Y\,Y'\,G'C'\} \tag{16}$$

Because of the coupling between control and state, $E\{G\,C\,Y\,Y' + Y\,Y'\,G'C'\} = -G\,C\,Y\,C'\,G'$. (see Bryson and Ho 75, p. 335). In steady state $Y = 0$, giving:

$$0 = (F - G\,C)\,Y + Y\,(F - G\,C)' + E\left\{\begin{bmatrix} 0 & 0 \\ 0 & 2\,fs\,\dot{y} \end{bmatrix}\right\} + G\,C\,Y\,C'\,G' \tag{17}$$

This matrix ricatti equation allows us to solve for the elements of Y. Solving for y22 gives

$$2(K^{*}\,y12 + Kv^{*}\,y22) = (K^{*2}\,\sigma^2 p + Kv^{*2}\,\sigma^2 v) + E\{2\,fs\,\dot{y}\} \tag{18}$$

where $\sigma^2 p$ is the variance in the position measurement (scaled to dimen'less units)
$\sigma^2 v$ is the variance in the velocity measurement (scaled to dimen'less units)

Solving for y12 gives y12 = 0, taking $\sigma^2 p \ll \sigma^2 v$ (a condition that must be checked with the scaling relations and sensor performance of a particular situation, but it would be extraordinary if not true) (18) gives y22:

$$y22 = (Kv^{*2}\,\sigma^2 v + E\{2\,fs\,\dot{y}\})\,/\,(2\,Kv^{*})\,. \tag{19}$$

Evaluating the expected value requires knowing the probability distribution on $\dot{y}$. This can be evaluated as the solution to the differential equation giving a probability distribution satisfying (18) and in steady state:

$$1/2\ Kv^2\ \sigma^2 v\ p'(\dot{y}) = p(\dot{y})\,(-Kv^{*}\,\dot{y} - fs\,) \tag{20}$$

where $p'(\dot{y})$ is the derivative of the probability distribution, $p(\dot{y})$, w.r.t. $\dot{y}$.

Taking as a boundry condition that $p(\dot{y} = -1/\sqrt{2}) = 0.0046$, equation (20) may be integrated. This initial condition corresponds to the probability level at 3 sigma in a gaussian distribution and the expectation that the system will arrive in the stuck condition roughly once every 800 time constants. Note that $\dot{y} = -1/\sqrt{2}$ is the stuck condition, that is $\dot{y}$ is expressed with respect to the desired velocity. The second constraint on the probability distribution of $\dot{y}$ is that the mean velocity must be 0. (20) may be integrated for various values of $\sigma^2 v$ until mean($\dot{y}$) = 0. This exercise gives:

For Kv = .858 and becoming stuck is a three sigma event, $\sigma^2 v = 0.0180$ (21)

Scaling from dimensionless units to the original problem: $\sigma x <= 0.134\ Vc$, where σx is the standard deviation of the velocity measurement in the original coordinates.

Thus a velocity estimate with a standard deviation of about 1/8th the critical velocity of the Stribeck effect is necessary to allow a negative velocity feedback large enough to stabilize the system and hold noise to a level where jittering to the stuck condition is a 3 sigma event. The PUMA 560 manipulator described in [Armstrong 88] had the following parameters:

$$
\begin{aligned}
Fs &= 1.02 &&\text{Newton - meters} \\
Vc &= 0.0061 &&\text{radians per second} \\
M &= 5.0 &&\text{kilogram - meters}^2
\end{aligned}
\qquad (22)
$$

To achieve stable motion throughout the low velocity spectrum by feedback control, a velocity estimate with a standard deviation of 0.000817 [rad/sec] would be required. For the PUMA, a first difference of position at 200 hertz results in a standard deviation of 0.0082 [r/s], inadequate by an order of magnitude. Velocity estimate noise levels of $\sigma v = 0.0057$ and 0.0032 [r/s] are reported in [Armstrong 88] for experiments done by Craig [86] on an Adept 1 arm, and An, Atkeson and Hollerbach [86] on the MIT/Asada arm. These figures too are substantially higher than that required to implement a stabilizing Kv. Note that all of these figures are based on quantization alone and neglect other noise processes present in the sensors.

The value of 0.134 Vc is confirmed by solving (19) in an approximate way. If the velocity distribution is assumed to be gaussian then it is known, because the mean is zero and the deviation is such as to make arriving in the stuck condition a 3 sigma event. Thus, the unknown expected value in equation [19] may be computed directly, giving $\sigma x <= 0.128\ Vc$. This approximate calculation and the exact calculation above are relatively close, indicating that the actual distribution of $\dot{y}$ is not far from Gaussian.

5   Sensor Accuracy Required: Force or Joint Torque Sensing

The appropriate choice of model is less clear when force sensing is employed because of the number of possible formulations and the important role of mechanism bending dynamics. Two limiting cases are considered: force control through a known stiffness and control of free motion where torque sensing is used to compensate for friction.

When a mechanism operates through a stiffness, ke, that will depend upon both the mechanism and the environment, and the contact force is sensed, then actuator velocity may be calculated by:

$$\dot{x} = ke \cdot \dot{fe}; \qquad \text{where } \dot{fe} \text{ is the force measurement,} \qquad (23)$$
$$\text{ke is the environment/mechanism stiffness.}$$

If the mechanism flexibility lies between the source of Stribeck friction and the end effector, $\dot{x}$ will be an estimate of the velocity difference between the actuator velocity and the velocity required to track the surface. Assuming that the surface is stationary or slowly varying, the high frequency information in $\dot{f}e$ will provide an estimate of the high frequency component of $\dot{x}$. Repeating the noise analysis above, but replacing $Kv \cdot \dot{y}$ with $Kv\ ke\ \dot{f}e\ /Vc$ (to scale to dimensionless units), yields the result that:

$$\sigma \dot{f}e <= .134\ Vc\ ke. \tag{24}$$

In practice the force rate information may be currupted by noise sources in the force sensing process. But reasonable rate information can be achieved, figure 4 is a plot of velocity estimated during motions of a PUMA arm. There are two curves: velocity estimated by integrating acceleration and velocity estimated by differentiating measured contact force. As mentioned with the experimental description above, they agree quite well. Several things

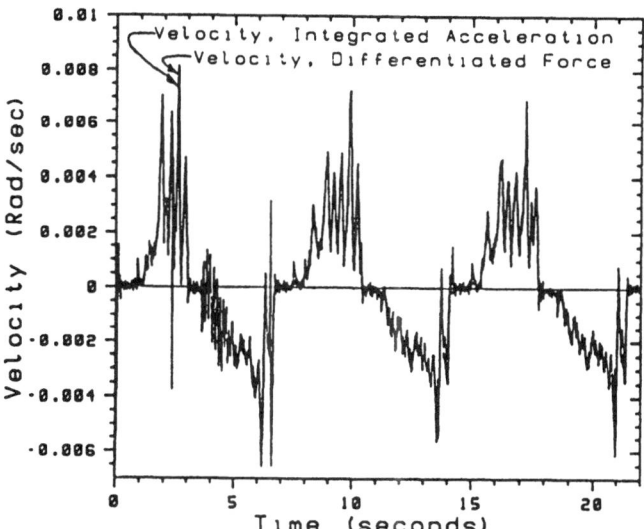

Figure 4. Velocity Estimated Through Force and Acceleration Sensors.

assisted the acquisition of an accurate force rate signal in this case: motions in contact were extremely slow and there was no tracking of a surface, the data were processed off-line, and smoothed with a non-causal filter. None the less, the noise level of 70 micro-radians per second is substantially better than the level required to stabilize the PUMA arm, suggesting that force rate feedback may be useful.

Joint torque sensing may be employed to compensate for unknown friction [Pfeffer, Khatib and Hake 86]. If the control law were:

$$u = -K\ (x - x_d)\ - Kv(\dot{x} - \dot{x}_d)\ +\ Ff$$
$$Ff = Fo\ sgn(\dot{x}) + Fs\ (exp(-(\dot{x}/Vc)^2 - 1)sgn(\dot{x}) + w$$

where   Ff   is the sensed friction;

           w   is a noise signal in the friction sensing process.

The noise analysis may be completed in the standard way, using w as a noise signal injected on the control. To retain our requirement that to reach the stuck state through noise injection be a 3 sigma event, we require that

$$.23 <= 1/(2 \; Kv^*) \; \sigma^2 w^*$$

where $\sigma^2\_w^*$ is the variance of the force signal in dimensionless units.

                    Note: Force scales as $(1 / (M \; \alpha \; \beta^2)) = (1 / Fs)$

Solving for σ_w in the original coordinates gives

$$\sigma\_w <= .23 \; \sqrt{2 \; Kv \; Vc \; Fs} \qquad\qquad (25)$$

For the PUMA arm, with values as above, and Kv = 16 N-m / (rad sec-1) (dimensioned coordinates), σ_w must be less than 0.10 (N-m), again a sensitive but achievable level of sensing.

6   Integration of the Dimensionless Dynamic

6.1   Evaluation of Kv* Required to Prevent Stick-Slip

The dimensionless damping required for asymptotic stability, $Kv^* = 0.858$, will provide steady motion, even at the velocity of greatest instability, $\dot{y}_d = 1/\sqrt{2}$. To prevent stick-slip, the damping must be sufficient to prevent a trajectory which starts in the stuck condition from arriving again at the stuck condition. In general, K*, Kv* and $\dot{y}_d$ will interact to determine whether stick-slip motion occures, for any value of K* and $\dot{y}_d$ there will be a minimum value of Kv* (Kv*_min) which will prevent a limit cycle from reaching the stuck condition. Figure 5 is a plot of iso-Kv*_min curves as a function of K* and $\dot{y}_d$, these data were obtained by integrating equation (10). As one might intuit, Kv*_min is highest when y_d is near to the most unstable value, $\dot{y}_d = 1/\sqrt{2}$.

The highest Kv*_min values discovered, Kv*_min = 0.71, are near to the 0.858 value required for asymptotic stability. As $\dot{y}_d$ increases, the required damping decreases until a practically achievable value of Kv*_min is reached.

Stick-slip trials, as described above, were run with 17 combinations of spring stiffness and desired velocity. These are shown as X's and O'x in figure 5; X's where stick-slip occurred and O's where it did not. Damping in the experiment was provided by the viscous friction of the machine. The viscous friction value away from zero velocity gives a dimensionless damping of $Kv^* = 0.03$, which agrees reasonably well with the required value determined by numerical integration for the high velocity cases. The low-velocity, high-stiffness values that do not exhibit stick-slip require substantially higher damping for steady motion. The physical process providing this damping is not understood; however, it may relate to break-away and to a time lag between changing velocity and changing friction.

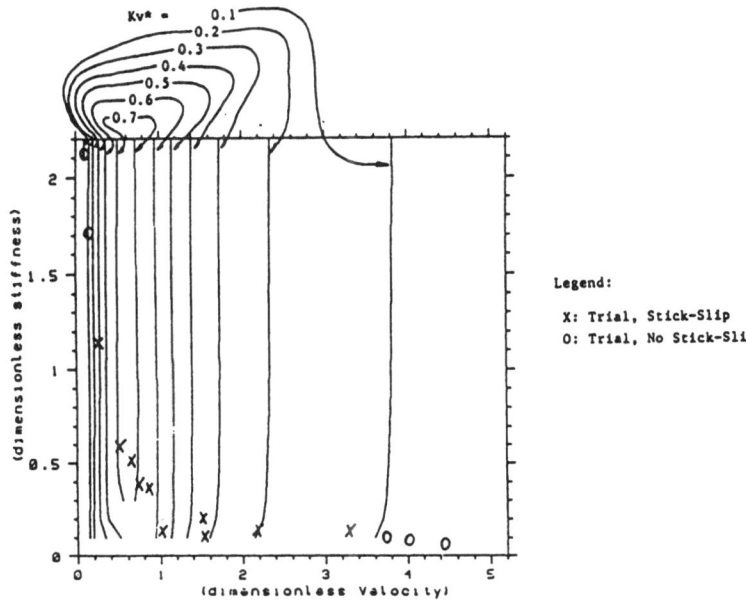

Figure 5. Contours of Damping Required to Prevent Stick-Slip versus a Range of K* and yd.

6.2 Evaluation of the Slip Distance

In table 1, the slip distances measured in the 11 trials that exhibited slip are reported, along with slip distances predicted by the model. The predicted distances are consistently smaller than the measured values. No explanation is available to explain this effect, but an empirical fitting of the measured to the predicted data yields equation (26).

$$Jm = 1.32 \, Jp + 2.54 \qquad\qquad (26)$$

Where   Jm   is the measured slip distance;
         Jp   is the predicted slip distance.

Equation (26) predicts the observed slip distances (in dimensionless distance) to +- 7%, a figure consistent with the observed variance in the data.

Table 1. Measured and Predicted Slip Distances

Trial	Dimensionless Parameters		Slip Distance [milli-radians]		Dimensionless Slip [No Dimensions]		Slip Predicted By [milli-Radians] Equation (28)
	K	$\dot{y}_d$	Mean	Standard Deviation	Experimental	Predicted	
11.13.2	0.63	0.48	1.53	0.22	8.5	4.5	1.66
11.13.4	0.36	0.74	2.27	0.50	12.6	8.5	2.59
11.13.5	0.19	1.57	7.68	0.42	42.3	25.0	6.21
11.19.2	0.60	0.50	1.01	0.20	5.6	4.5	1.73
11.19.3	0.36	0.83	2.62	0.40	14.6	9.6	2.81
11.19.4	0.36	0.83	2.41	0.73	13.4	9.6	2.81
11.19.5	0.36	0.34	1.67	0.85	9.3	4.2	1.75
11.19.6	0.10	1.50	8.73	1.07	48.5	35.0	8.00
11.19.8	0.14	3.31	13.37	0.66	74.3	21.5	14.48
11.19.9	0.14	2.16	9.91	1.94	55.1	39.0	9.61
11.19.10	0.14	1.08	5.63	1.62	31.3	21.5	5.14

An effort was made to establish a relationship between slip distance and the friction model parameters; the hope was to establish a straight forward experiment, involving only the measurement of slip distance, which could identify the Stribeck friction. The slip distance, however, proved to be independent of the friction model. This counter-intuitive result obtains because the system is dominated by the mass-spring characteristic, with the de-stabilizing friction and stabilizing damping biasing the trajectory, but making only a small impact on the time in motion during a cycle. The slip distance is the difference between initial and final position plus the travel of the reference point during the slip. For the range of K^* and $\dot{y}_d$ explored, travel of the reference point dominates, so that slip distance may be approximated by:

$$Jp' = 2\,\Pi\,\sqrt{(1/K^*)}\;\dot{y}_d \tag{27}$$

Equation (27) is evaluated and presented in table 1; it is presented in radians (dimensioned coordinates). With a new empirical fit, equation (27) is found to fit the data to +- 10%. The fit of equation (27) to the data is given by:

$$Jm = 1.35\;2\,\Pi\,\sqrt{(M/K)}\;\dot{x}_d\;+\;730\;[\text{micro-radians}]. \tag{28}$$

7   Discussion

The situation of interest is a mechanism with Stribeck friction being called upon to move steadily at low velocity. The direct dynamic model of such a system, which has 7 parameters, was rewritten as a dimensionless model with only 3 independent variables. In addition to reducing the space of systems that must be explored numerically, the dimensional analysis reveals the scaling effect of the friction model parameters, Fs and Vc. With the dimensionless system it is straight forward to determine the velocity most de-stabilized by Stribeck friction and the damping necessary for asympototic stability at that velocity. This is a result of practical importance because fixing the velocity and damping permits the determination of the required sensor performance for stabilization by feedback control.

The tolerable noise in the velocity estimate proved to be within a feasible range - about an order of magnitude better than that available in experiments reported. The experimental noise figures discussed consider only direct velocity measurement; improvement will be achievable through estimation. Estimation is not expected to be a panacea, however, because the system itself is unstable in this regime, so the introduction of a small amount of process noise will rapidly degrade the estimate quality. The analysis also neglects problems of controller lag or systematic error sources (such as position dependent friction) which may substantially increase the damping required to prevent stick-slip.

The required sensor sensitivity was also investigated for force control and joint torque sensing configurations. A PUMA like system with a reasonable value of Kv was found to require a standard deviation in joint torque measurement equal to about one tenth the magnitude of the Stribeck friction. For the PUMA, the Stribeck friction is itself about 1% as large as full torque at the joint, so a joint toque sensor with a signal to noise ratio of 1000 would be required: again, feasible, but challenging.

Integration of the dimensionless model permitted evaluating the required damping by the standard of eliminating stick-slip, rather than that of asympototic stability at the most de-stabilized velocity. The required damping was found to strongly dependent on desired velocity and weakly dependent on the stiffness. The dimensionless damping evaluated for experimental trails that were just barley stable (made stable by increasing the desired velocity) agrees roughly with the predicted value, suggesting the possibility of estimating the minimum stabilizable velocity, given the parameters of a friction model and damping.

Slip distances were also computed by integrating the model and compared to measured values. The process revealed that slip distance is very weakly coupled to the friction model, dashing all hopes of constructing a procedure for identifying the friction parameters using measured slip distances. There appears to be some possibility of identifying friction parameters based on an experimentally measured boundry of stable velocity vs. damping. Such an experiment would require variable damping. The damping gain, Kv, required to achieve modest dimensionless gains, Kv*, may be extremely high, suggesting that it may not be possible to fully explore this curve with damping provided with feedback control.

8 Discussion of Mechanical Considerations

The relationship between issues of mechanical design and stick-slip induced by Stribeck friction is an interesting one. At the friction site itself, the friction will be a linear force, dependent upon the linear velocity of the contact. In rotating machinery, the translation of these parameters from their linear coordinates to the rotational joint coordinates will depend upon the radius at which they act. The scaling effects of radius on the dimensionless parameters will permit comparisons of the likely stick-slip behavior of different mechanisms, and may provide guidlines in placing the friction interfaces.

As the radius to a rubbing interface increases, the rotational velocity corresponding to a given linear velocity decreases (Vc decreases) and the torque corresponding to a given linear friction increases (Fs increases). Examining the relationship between these parameters and dimensionless damping and stiffness, equation (11), shows that dimensionless damping, Kv*, decreases as the square of radius and that dimensionless stiffness, K*, decreases as the fourth power of radius. Assuming the friction itself is not modified by the mechanical change, as the friction point is moved outward, the system becomes more difficult to stabilize. It may however be less necessary to stabilize the system across all velocities: the dimensionless desired velocity increases with decreasing Vc and thus with increasing radius. The damping required for stable motion drops off rapidly as the dimensionless desired velocity increases above 2. Thus, it is possible that if a specific minimum motion velocity is required, moving the friction contacts outward (by use of a larger bull gear, or one with a higher inter-tooth slip ratio!) might make the specified motion achievable, but slower motions more difficult to control.

Acknowledgements

The experimental work reported here was supported by the Stanford Institute for Manufacturing Automation and the Hewlett Packard Corporation. A high precision rotational accelerometer was lent by the Systron Donner Corporation. Subsequent work was supported by the University of Wisconsin - Milwaukee. The author is indebted to professors Tom Binford and Gene Franklin, for ideas that are still coming to light.

References

Atkeson, C.G., An, C.H. and Hollervach, J.M. 1986.
"Estimation of Inertial Parameters of Manipulators Loads and Links," Inter. Journal of Robotic Research 5(3)101:119.

Armstrong, B. 1988.
"Dynamics for Robot Control: Friction Modeling and Ensuring Excitation During Parameter Identification," PhD Thesis, Electrical Engineering Dept., Stanford University. (Computer Science Dept Report No. STAN-CS-88-1025.)

Armstrong, B. 1988 (April).
"Friction: Experimental Determination, Modeling and Compensation," Proc. 1988 Conf. on Robotics and Automation, IEEE: Philadelphia.

Bryson, A.E. and Ho, Y. 1975.
Applied Optimal Control, New York: Hemisphere Publishing Co.

Craig, J.J. 1986.
"Adaptive Control of Mechanical Manipulators," PhD Thesis, Electrical Engineering Dept., Stanford University.

Dowson, D. and Higginson, G.R. 1965.
"Elasto-Hydrodynamic Lubrication," New York: Pergamon Press.

Gassenfeit, E.H. and Soom, A. 1988 (July).
"Friction Coefficients Measured at Lubricated Planar Contacts During Start-Up," ASME Journal of Tribology vol.110, pp.533-8.

Pan, P. and Hamrock, B.J. 1989 (April).
"Simple Formulas for Performance Parameters Used in Elastohydrodynamically Lubricated Line Contacts," ASME Journal of Tribology vol.111, pp.246-9.

Pfefer, L., Khatib, O. and Hake, J. 1986 .
"Joint Torque Sensory Feedback in the Control of a PUMA Manipulator," Proc. 1986 American Control Conference: Seattle.

Rabinowicz, E. 1978.
"Friction - Especially Low Friction," Fundamentals of Tribology, Suh and Saka, eds., Cambridge: MIT Press.

Sadeghi, F. and Sui, P.C. 1989 (January).
"Compressible Elastohydrodynamic Lubrication of Rough Surfaces," ASME Journal of Tribology vol.111, pp.56-62.

Zhu, D. and Cheng, H.S. 1988 (January).
"Effect of Surface ROughness on Point Contact EHL," ASME Journal of Tribology vol.99, pp.32-7.

Experimental Results on Adaptive Friction Compensation in Robot Manipulators: low velocities

C. Canudas de Wit

LABORATOIRE D 'AUTOMATIQUE DE GRENOBLE (C.N.R.S. UA 228)
GRECO (69) "Systèmes Adaptatifs" ENSIEG-INPG
BP 46, 38402 Saint-Martin d'Hères, FRANCE

Abstract

The paper analyzes the problem of modelling and compensation of friction at velocities close to zero. A new model, linear in parameters, which captures the downward bends at low velocity is used to adaptively compensate for friction. The need for this type of models is mainly motivated by instability phenomena that can be caused by overcompensation when simple models (such as Coulomb friction models) are used as a basis for the friction compensation. This model, in combination with an adaptive computed torque method were tested experimentally in a robot manipulator.

1 Introduction

Inaccuracies in servo-systems are often caused by the presence of friction in the motor shaft, mechanical links, gears, etc. Typical errors caused by friction are: steady-state errors and tracking lags. The former are mainly due to the static (or dry) friction which is proportional to the velocity sign. The latter are generated by viscouse friction which increases the damping of the system. To deal with friction, it is first necessary to have a good characterization of the structure of the friction model and then to resort to the appropriate compensation technique.

Friction models have been widely studied. It has been well established that friction depends on the direction of rotation, but the character of the function, in particular for low-velocities, gives rise to some disagreements. Phenomenas such as stiction (torque needed to start the motion), as well as a downward bend at low-velocities, have been identified. Recent detailed experiments [1] realized at low-velocities have confirmed Tustin's model [14], which includes a decaying exponential term in the friction model. This downward bend appears after has surmounted the break-away (stiction) torque. This phenomena is however not new, it has been detected in connection with flat surfaces, see measurements collected by Mc Kee [15], and explained by the empiric Buckinham's exponential formula [15]. In fact, friction is a complicated combination of all the force components opposing the motion which are distributed along the mechanical links : flat surfaces, gear boxes, pulleys, bearings, etc. These forces may vary not only as a function of temperature changes but also depend on the forces orthogonal to the motions.

In this paper we propose a simple model, which captures the exponential friction characterization at low-frequencies and is linear in parameters. This type of model has two purposes : on one hand, parameter linearity is suitable for on-line identification. On the other hand, this type of model structure avoids instability problems resulting from overcompensation when simple models, such as Coulomb friction, are used as a basis for friction compensation. Futhermore, friction compensation can be realized adaptively because of the model parameter linearity [4].

Although friction compensation is a good candidate among the strategies for beating friction, attention must be paid when the compensation is performed on the basis of an inexact model, in particular when friction is overcompensated. The paper analyzes the existence of possible limit-cycles caused by the overcompensation. We draw conditions needed to avoid this oscillation in terms of compensation gains and friction model accuracy. This analysis justifies the use of more complex friction models and the adaptive techniques. These ideas were experimentally tested in a robot manipulator ACMA-S58 in connexion with the computed torque method.

The paper organization is as follows : Section 2 discusses the main components of the friction models and proposes one which is linear in parameters and includes the downward bend at low velocities. Section 3 analyzes the phenomena of overcompensation. Section 4 gives an adaptive on-line control schema and the algorithm used to estimate friction under possible inaccuracies (bounded) on the friction model. Section 5 shows the experimental results obtained in an industrial robot ACMA-S58. Part of this work was previouly presented in [18].

2 Friction Models

Friction models have been extensively discussed in the literature, see for example [1], [3], [4], [6], [7], [8], [11], [14] and [15]. These models established that the friction torque is a function of the angular velocity, there is, however, disagreement as far as the character of the function is concerned. The aim of this section is to describe the main friction component torque and to discuss its relative importance in robot arm configurations.

(i) Coulomb/Stiction :   In classical Coulomb friction models, there is a constant friction torque opposing the motion when the velocity is different from zero. For zero velocity, the stiction will oppose all motions as long as the torques are smaller in magnitude than the stiction torque.

(ii) Asymmetries : Imperfection in the motor mechanics and unbalances on the motor shaft yield asymmetric behaviour of the motor dynamics. The model proposed in [4] includes Coulomb and viscouse friction and accounts for friction asymmetries. Some experiments realized with this model showed that the asymmetry in the Coulomb friction components were dominant. This observation was also corroborated by the results presented in [1].

(iii) Position dependence : In some mechanical set-ups, the friction torques also depend on the angular position. In motor-drives, the position dependence can be considered as a consequence of imperfections on the shaft and reductor centers. These imperfections will generate oscillations with a period equal to the reductor ratio. Experiments, carried out in industrial manipulators [1], [10], had shown that this dependence is relatively weak (it modifies no more than 5 % of the maximum absolute value of friction) so that it can be neglected for most purposes. It is worthwhile to noting that in other processes, such as in the rolling mill or

in paper factories, small imperfections in the rollers generate important changes in the contact forces and hence induce non negligable friction torque variations.

(iv) Downwards bends : It has been experimentally verified that after the stiction torque has been surmounted the friction torque decreases exponentially reaching approximately 60 % of the break-away torque and then increases proportionally to the velocity. These bends occur at velocities close to zero. This type of friction structure, sometimes addressed as a slip-stick friction, was already used in connection with an adaptive compensation scheme for a tracking telescope [7]. Before that it was identified in connection with flat surfaces by measurements collected by Mc Kee [15]. In mechanical gears, it was established empirically by Buckingham's formula [15]. Recently, detailed experiments [1], [5] carried out in industrial manipulators have confirmed this negative velocity dependence at low velocity.

The following expressions summarize the main friction components (i) - (iv)

(i) $\tau_f(\dot{q}) = \alpha\,\text{sgn}(\dot{q})$ (2.1a)

(ii) $\tau_f(\dot{q}) = \alpha_i\,\text{sgn}(\dot{q}) + \beta_i\,\dot{q}$ (2.1b)

(iii) $\tau_f(\dot{q}) = K_f\sin(\omega_o\,\dot{q} + \varphi)$ (2.1c)

(iv) $\tau_f(\dot{q}) = [\alpha_o + \alpha_1\,e^{-\beta|\dot{q}|}]\,\text{sgn}(\dot{q})$ (2.1d)

where α in (i) represent the Coulomb friction, α_i and β_i in (ii) the asymmetric model of Coulomb and viscouse friction, K_f, ω_o and φ, in (iii), the amplitude, frequency and phase of the oscillations due to the position dependence. In (iv) the sum, $\alpha_o + \alpha_1$, represents the break-away torque and β the slip constant. Each of these components are present to some extent, but its relative importance shall be evaluated by engineer intuition or by observing collected measurements. As mentioned previously, in robotic applications (Industrial robots), the position dependence seems not to have a fundamental importance while the negative velocity dependence may induce a variation of 40 % of the maximum friction torques. The total friction model, in robot manipulators, is in fact a complicated function resulting from the combination of the components (i)-(iv), distributed along the robot mechanical links; ball bearings, gear boxes, flat surfaces, pulleys, etc. A friction model covering most of these components can be expressed as follows :

$$\tau_f(q) = [\,\alpha_o + \alpha_1\,e^{-\beta_1|\dot{q}|} + \alpha_2\,(1 - e^{-\beta_2|\dot{q}|})\,]\,\text{sgn}(\dot{q}) \qquad (2.2)$$

where the α_i's and the β_i's are positive constant. Note that we have substitued the term

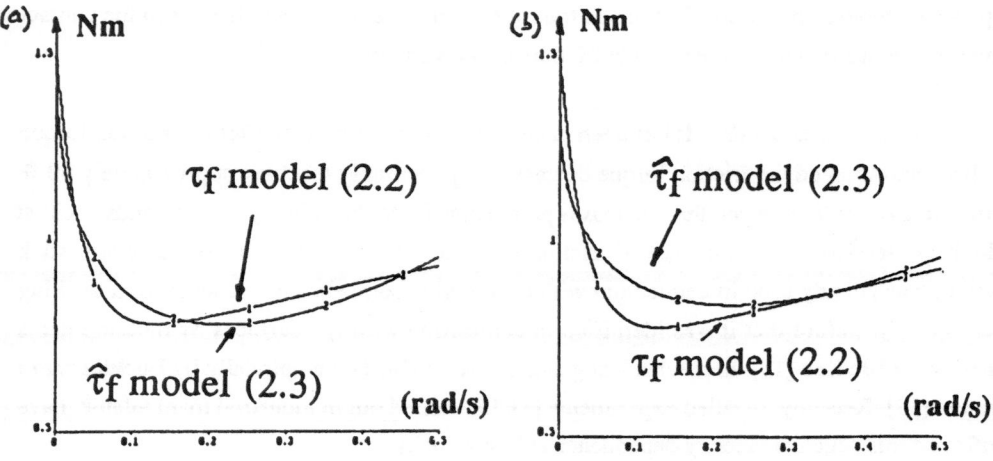

Figure 1. Comparison of the exponential friction model (2.2), τ_f, and the estimated linear model (2.3), $\hat{\tau}_f$. (a) First link, (b) Second link.

proportional to the velocity by an exponential function The measurement data motivated the use of model (2.2) instead of a combination of (2.1b) and (2.1d). Asymmetries can be included in (2.2) by letting the α_i's be different for different rotation directions. The β_i's can be maintained constant, however. In spite of the completeness of model (2.2), the nonlinearity of parameters α_i and β_i restricts its utility for on-line identification, (linear predictors require a model expression which is linear in parameters). It is then important to modify (2.2) such that the simplified model is still capturing the downward bends and possible asymmetries while remaining linear in the unknown parameters. Such a model is the following :

$$\tau_f(q) = [\; \alpha_0 + \alpha_1 \,|\, \dot{q} \,|^{1/2} + \alpha_2 \,|\, \dot{q}| \;]\; \text{sgn}(\dot{q}) \tag{2.3}$$

The model (2.2) can thus be used for simulating the real friction forces whereas the simplified model (2.3) is used as a basis for compensation. To evaluate the precision that can be achieved with this reduction, we have estimated the parameters α_i of model (2.3) by minimizing a least-square estimation algorithm. Figure 1 shows the comparison at low velocities, of the friction obtained directly from model (2.2) and the estimated friction (2.3). These curves indicate that a good precision is achieved with (2.3). This precision achieved depends, of course, on the spectral frequency distribution of the applied torque and more precisely on the spectral frequency distribution of the induced velocity. Remark also, that the uniqueness of the α_i's does not exist. Indeed, there may exist several sets of parameters α_i leading to an equivalent approximation.

3 Stability issues

This section discusses some of the strategies for dealing with friction and analyses the stability problems caused by friction overcompensation. Consider for simplicity's sake a servo-motor with friction, i.e.

$$J \ddot{q} = \tau_m - \tau_f(\dot{q}) \tag{3.1}$$

where J is the nominal inertia (motor plus charge) q the angular position and τ_f the friction torque function of the angular velocity. Let the applied torque, τ_m, be composed of a linear compensator by position and velocity feedback plus an additional term compensating for friction :

$$\tau_m = H_1 (q_r - q) - H_2 \dot{q} + \hat{\tau}_f(\dot{q}) \tag{3.2}$$

where H_1 and H_2 are linear compensators and $\hat{\tau}_f$ is an estimate of τ_f. The desired reference trajectory is q_r. Closed-loop stability can be studied by introducing (3.2) in (3.1) and by approximating the difference of $\hat{\tau}_f$ and τ_f by its describing function N(A). Let H (jω) be a linear portion of the closed-loop system which defines the characteristic equation $1 + H (j\omega) N(A) = 0$. The intersection the Nyquist of these two curves, H (jω) and -1/N(A), if any, detects the existence of limit cycles, with an amplitude given by A.

3.1 Strategies for beating friction

High gain controllers : The influence of the nonlinearities can to some extent be reduced by high gain linear feedback [16]. This approach has, however, some severe limitations because the nonlinearities will dominate any compensation for small errors. Limit cycles may appear as a consequence of the dynamic interaction of the friction forces and the high gain controllers [3].

High-frequency bias signal injection : Limit cycles can only be removed by reshaping either $H(j\omega)$ or $-1/N(A)$ such as to avoid their intersection. Modifing $H(j\omega)$ implies a new control design which can become complicated and does not necessarily solve the original problem. An alternative is then to modify $N(A)$ by adding a dither to the torque input, τ, i.e. :

$$\tau^0 = \tau + A_0 \sin \omega_0 t.$$

where ω_0 should be chosen greater than the system bandwidth and A_0 exceeding the static friction levels. The oscillation can be decreased in magnitude but increased in frequency. Although the injection of an additional high-frequency noise partially alleviates the friction effects, it may alsoyield undesirable consequences : the excitation of the high frequency harmonics (resulting from the elasticity of the motor-load coupling) and the fatigue of the system actuators.

Friction compensation: :Another way to reshape $N(A)$, is by nonlinear compensation. Indeed, the key idea is not to modify $N(A)$ but rather totally nullify it. The selection of the adequate technique to compensate for friction torques depends on the choice of the friction model. For the dynamic model proposed in [11] it is possible to predict the friction behaviour and to compensate it by feedforward. An alternative approach is to cancel the friction effects by feedback compensation, as described by the control law (3.2). Indeed, the nonlinear compensation (feedback) acts as a linearization loop which renders the system linear at any operating point; for further details see [3]. The success of the compensation clearly depends on the accuracy of the estimated friction.

3.2 Inexact friction compensation

Friction compensation requires a high degree of accuracy in the predicted friction. In practice, friction can be estimated with a precision of approximately 80-90 % [1]. The remaining errors can thus induce a friction under or over compensation. The problem of an inexact friction compensation and the influence of the downward bends can be studied assuming that the real friction is given by :

$$\tau_f(\dot{q}) = \begin{cases} \dfrac{a-b}{\tau}\dot{q} + b\,\text{sgn}(\dot{q}) & \text{if}\ |\dot{q}| \le \tau \\[4mm] a\,\text{sgn}(\dot{q}) & \text{if}\ |\dot{q}| > \tau \end{cases} \tag{3.3}$$

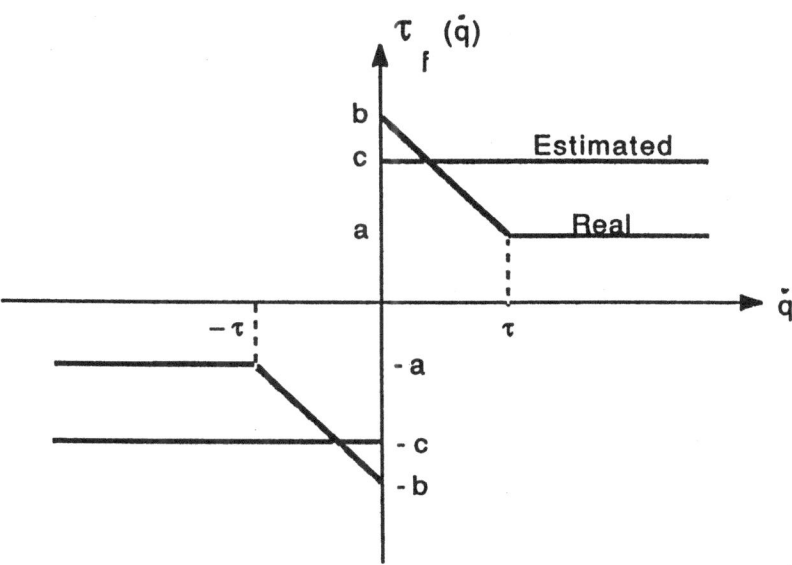

Figure 2 : Real and estimated friction.

and that the estimated friction is,

$$\hat{\tau}_f(\dot{q}) = c \, \text{sgn}(\dot{q}) \qquad (3.4)$$

where we have omitted the viscous friction and used, in expression (3.3), straight lines instead of exponential curves for simplicity's sake. Generality is, however, not lost. Figure 2 shows these friction curves.

In connection with the servo (3.1) and the control law (3.2), we can now analyze the closed-loop stability when the compensation is performed on the basis of the simplified model (3.4). Introducing N(A) as the describing function of the difference $\tilde{\tau}_f(\dot{q}) = \tau_f(\dot{q}) - \hat{\tau}_f(\dot{q})$, the characteristic equation of the closed-loop transfer function obtained by this approximation, is given by $1 + H(j\omega) N(A) = 0$. Where $H(j\omega)$ and $-1/N(A)$ are given as :

$$H(s) = \frac{s}{Js^2 + H_2(s)\,s + H_1(s)} \qquad (3.5)$$

$$-\frac{1}{N(A)} = \begin{cases} \dfrac{-A}{A\,\eta_o + \eta_1} & \text{if } A < \tau \\[4mm] \dfrac{-A}{\dfrac{2}{\pi}\,\eta_o\,[\gamma_o(A) + \gamma_1(A)\,] + \eta_1} & \text{if } A \geq \tau \end{cases} \qquad (3.6)$$

where,

$$\eta_o = (a - b)/\tau \qquad\qquad \gamma_o(A) = A \text{ arc sin } (\tau/A)$$
$$\eta_1 = 4\,(b - c)/\pi \qquad\qquad \gamma_1(A) = \tau\,(1 - \tau^2/A^2)^{1/2}$$

Further details of the derivation of the function $N(A)$ can be found in [16]. Limit-cycles can now be studied by analyzing possible intersections of the Nyquist of $H(j\omega)$ with $-1/N(A)$.

Undercompensation ($c = a$). Undercompensation occurs when the estimated friction is smaller in magnitude than the real friction torque. It is simple to see that oscillations can not be caused by under compensation when PD ($H_1 = k_p$, $H_2 = k_v$) or PID ($H_1(s) = k_p + k_i/s$, $H_2 = k_v$) compensators are used. First, note that static nonlinearities without hysteresis do not cause phase lags. Therefore, the curve of $-1/N(A)$ can only lie on the real axis. Secondly, the Nyquist of $H_{PD}(j\omega)$ evolves within the first and fourth quadrant whereas $H_{PD}(j\omega)$ evolves from the second to the fourth quadrant in the clockwise. The negative real axis is thus not crossed either by $H_{PD}(j\omega)$ or by $H_{PID}(j\omega)$. A detailed analysis of the variation of $-1/N(A)$ with respect to A, see [16], shows that the maximum of $-1/N(A)$ is always negative for any $c \in (0,a)$. Intersection of $H(j\omega)$ with $-1/N(A)$ is then not possible. Figure 3(a) shows this.

Overcompensation ($c = b$). Overcompensation occurs if friction is overestimated, for example when the downward bends are not modelled or if compensation is performed on the basis of the stiction torque. Limit cycles are, in this case, avoided provided that the compensator gains k_p, k_i, k_d verify the following:

$$(i)_{PD} \quad \frac{1}{k_d} < \frac{\tau}{b - a}$$

$$(ii)_{PID} \quad \frac{1}{k_d - Jk_i/k_p} < \frac{\tau}{b - a}$$

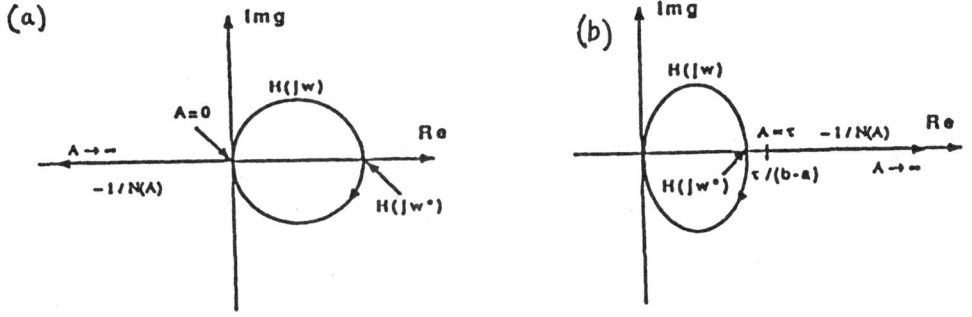

Figure 3 : Nyquist of H_{PD} (jω) and -1/N(A). (a) Friction undercompensation. (b) Friction overcompensation.

which are obtained from the following observation: the intersection of H(jω) and -1/N(A), if any, can only occur on the positive real axis at the frequency, $ω^*$, for which the phase of H(jω) is zero. For both cases, PD and PID compensators, Img{H(jω)} cancels at $ω^* = (k_p/J)^{1/2}$. The respective magnitudes are $|H_{PD}(j\;ω^*)|=1/k_d$ and $|H_{PID}\;(ω^*)| = 1/(k_d - J\;k_i/k_p)$. This gives the left terms of inegalities (i) and (ii). The function -1/N(A) has a unique minimum at A = τ which is τ/(b - a), and then increases monotonically as A →∞, see Figure 3(b). It is then sufficient that the magnitude of $H(jω_0)$ be smaller than the minimum of -1/N(A), as indicated by the above conditions, in order to avoid limit cycles, see [16] for further details.

Comments : (1) For the PID controller, the gains k_p, k_d and k_i shall also verify $k_p\;k_d > Jk_i$ such that the left hand side of inequality (ii) remain positive. Hence k_i can not be increased arbitrarily. (2)The previous analysis addressed only the problem of a single servo-system with a simplified real friction model. It gives however a good insight into the choice of the compensator gains. Friction overcompensation for τ closed to zero (the model (3.3) becomes very close to the stiction/Coulomb model) requires gains yielding very small values for $|H_{PD}(jω^*)|$ and $|H_{PID}\;(ω^*)|$. In practice overcompensation may produce oscillations because the compensator's gains can not be arbitrarely modified (saturation and measurement noise impose the major limitations). Therefore, the amplitude of the oscillations can be reduced by increasing k_d. The frequency can be decreased by reducing k_d. (3)The previous analysis did not include in the motor model dynamics, high-frequency modes resulting from the elasticity in the motor mechanic links. These unmodelled modes may increase the lag of H(jω). In such cases, both under and over friction compensation may limit the compensator gains.

4 Adaptive compensation

This section presents the model parametrization and the algorithm used to estimate the unknown parameters under possible model structural inaccuracies.

4.1 Estimation framework

Without loss of generality, consider only the last link of a robot manipulator placed on the xy plane see figure 4 :

$$(mr^2 + I)\ \ddot{q} + mg\ r\ \cos(q) + \tau_f(\dot{q}) = \tau \tag{4.1}$$

where q describes the link displacement, I is the inertia associated to the motor shaft and the mechanical transmissions, τ is applied motor torque and g is the gravity constant. A payload, considered as a punctual mass m, is located at distance r from the center of the coordinate q. The first two terms of the above expression, derive from the potential and kinetic energy and are well characterized, the friction forces, $\tau_f(q)$ are however difficult to model exactly. In order to use linear parameter algorithms for estimate I, m and τ_f, it is necessary to use friction models such as the one given by Equation (2.3), which preserves the parameter linearity i.e.

$$\tau_f(\dot{q}) = [\ \alpha_o + \alpha_1\ |\dot{q}|^{1/2} + \alpha_2\ |\dot{q}|\ ]\ \text{sgn}(\dot{q}) \tag{4.2}$$

where the α_i parameters are not unique and are function of the operating velocity. These parameters may also change with time and temperature variations. We can assume that Model (4.2) can capture around 90 % of the friction components. The remaining structural friction error is then assumed to be smaller than 10 % of the break-away friction value (Stiction). A rough estimation of stiction torque can be simply performed by placing the payload in such a position that friction torque equals the gravity components. The angle resulting from this position is then used to compute $\tau_f(0)$, i.e. $mgr\ \cos(q) = -\ \tau_f(0)$. With payload mass of 11.5 Kg, calculation yields $\tau_f(0) = 8$ Nm. An upperbound on the structural friction errors can therefore be fixed at 0.8 Nm. In the following discussion this upperbound is called δ.

4.2 Parametrization

The model (4.1) together with the friction model (4.2), can be reparametrized as follows :

$$\tau(t) = \theta^T\ \Phi(t) \tag{4.3}$$

$\theta^T = [m, I, \alpha_o, \alpha_1, \alpha_2]$

$\Phi = [(\ddot{q} + g\, r \cos (q)) , \ \ddot{q}, \ \text{sgn} (q), \ |\dot{q}|^{1/2} \, \text{sgn} (\dot{q}), \ q]^T$

The measurement of $\ddot{q}$ can be avoided by filtering the equation (4.3) by a low pass filter, $F(s) = 1/(\zeta s + 1)$, with $\zeta > 0$. Defining $\widetilde{\Phi}$ as $\{F(s)\Phi\}$ and $\widetilde{\tau}$ as $\{F(s)\tau\}$, the filtered equation (4.3) becomes : $\widetilde{\tau} = \theta^T \widetilde{\Phi}(t)$. Now sampling at $t = k$, the estimation error, e_k is defined as :

$$e_k = \widetilde{\tau}_k - \theta^T \widetilde{\Phi}_k \qquad (4.4)$$

which, according to the above discussion, abs (e_k) is upperbounded by $\delta = 0.8$ Nm.

4.3 Estimation algorithm

The real process (model 4.1 with the complete friction model (2.2)) is assumed to be reparametrized as :

$$\widetilde{\tau}_k = \theta^T \widetilde{\Phi}_k + \varepsilon_k \qquad (4.5)$$

where θ and $\widetilde{\Phi}_k$ follow the same friction definition (4.3) and ε_k captures friction structural inaccuracies (differences between model (2.2) and (2.3)) and other possible torques unexplained by kinetic and potential forces. The estimation problem is then to estimate I, m and τ_f in presence of the noise ε_k, which is assumed to have the known upperbound δ (which can also be time varying i.e. $|\varepsilon_k| \leq |\delta_k|$). Several estimation algorithms formulated within this framework have already been proposed in the literature. One of them is the following, see [13] and references therein :

$$\theta_k = \theta_{k-1} + \lambda_k P_k \widetilde{\Phi}_k e_k \qquad (4.6)$$

$$P_k = \frac{1}{k}\left[P_{k-1} - \frac{\lambda_k P_{k-1} \widetilde{\Phi}_k \widetilde{\Phi}_k^T P_{k-1}}{\lambda + \lambda_k \gamma_k} \right] \qquad (4.7)$$

with

$$e_k = \widetilde{\tau}_k - \theta^T_{k-1} \widetilde{\Phi}_k \quad , \quad \gamma_k = \widetilde{\Phi}_k^T P_{k-1} \widetilde{\Phi}_k \qquad (4.8)$$

$$\lambda_k = \frac{a_k \lambda}{\gamma_k} \left[\left| \frac{e_k}{\delta_k} \right| - 1 \right] \tag{4.9}$$

$$a_k = \begin{cases} 0 & \text{if } \gamma_k = 0 \text{ or } |e_k| < |\delta_k| \\ 1 & \text{otherwise} \end{cases} \tag{4.10}$$

where the estimate of θ, θ_k, is obtained by minimizing an exponentially weighted least-squares cost function under the noise constraint $|e_k| < |\delta_k|$. The weight $\lambda \in [0, 1]$, is fixed by the user and exponentially weights the arrival information. The time-varying weight, $\lambda_k \geq 0$, is function of the collected measurements. The former weights the importance attached to the past of the data while the latter provides an instantaneous weight to the last measurement. The fundamental role λ_k is to ensure the boundedness of $\{\theta_k\}$ faced with the noise presence. Note that the updating of θ_k and P_k is stopped ($a_k = 0$) if either the estimation error is within a "dead zone" ($|e_k| \leq |\delta_{kp}|$), or the arrival data are meanless ($\gamma_k = 0$). Further details about the derivation and the properties of the algorithm can be found in [3], [13].

4.4 Friction compensation

Friction compensation was realized in combination with a computed-torque law with integral action, i.e.

$$\tau_m = (\hat{m}\, r + \hat{I})\, u + \hat{m}\, g\, r \cos(q) + \rho\, \hat{\tau}_f(\dot{q}) \tag{4.11}$$

$$u = \ddot{q}_d + K_v (\dot{q}_d - \dot{q}) + k_p (q_d - q) + K_i \int_0^T (q_d - q)\, dt \tag{4.12}$$

where $\hat{m}$, $\hat{I}$ and $\hat{\tau}_f$ are the corresponding estimates of m, I and τ_f. The desired position, velocity and acceleration trajectories are q_d, $\dot{q}_d$ and $\ddot{q}_d$ respectively. The parameter $\rho \in (0,1]$ is included to avoid friction, overcompensation and represents the degree of confidence associated to the estimated friction torque. In view of our previous discussion ρ can be chosen within [0.8, 0.9].

There is, however, another potential source of bad compensation: the noise of the measured velocity. At low velocities, the measurement noise can have important consequences during the friction compensation. For instance, noise peaks may produce a change of the sign of the measured velocity. Compensation may then acts in the wrong direction creating instabilities. The problem can,

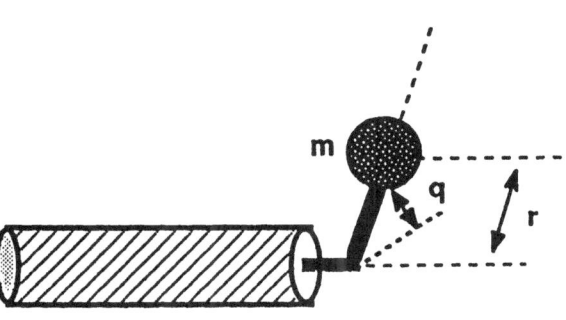

Figure 4 . Last link of a robot manipulator.

in this case, be alleviated by replacing $\dot{q}$ by $\dot{q}_d$ in the computation of $\hat{\tau}_f(\dot{q})$, i.e.

$$\hat{\tau}_f(q_d) = [\hat{\alpha}_0 + \hat{\alpha}_1 |\dot{q}_d|^{1/2} + \hat{\alpha}_2 |\dot{q}_d|] \, sgn(\dot{q}_d) \qquad (4.13)$$

This corresponds to a feedforward friction compensation. It is sometimes possible to combine a *feedback* with a *feedforward* compensation. This is particularily suitable when the desired trajectories include a combination of low and high velocities. Then, $\hat{\tau}_f$ is computed as :

$$\hat{\tau}_f = \begin{cases} \hat{\tau}_f(\dot{q}_d) & \text{if} & \dot{q}_d \leq \dot{q}_0 \\ \hat{\tau}_f(\dot{q}) & \text{if} & \dot{q}_d > \dot{q}_0 \end{cases} \qquad (4.14)$$

where $\dot{q}_0$ is a pre-defined threshold which defines the limits between high and low velocities.

5 Experimentation

The algorithm was tested experimentally on a robot manipulator ACMA-S18. The experiments were performed on an IBM-PC with 8087 floating point chip and data translation AD and DA converters. The major part of the software was written in C-Langage in order to improve the computation speed. The sampling time was 13 ms. Two types of experiments were realized.

First, a *pre-identification* sequence with high-excited signal was implemented. Figure 5 shows the velocity trajectory. The objective was to estimate the parameters I and m , which are invariant during the robot task operation, in order to reduce the total number of parameters to be estimated on-line (m can also be treated as time-varying if that is the case). This sequence was realized in

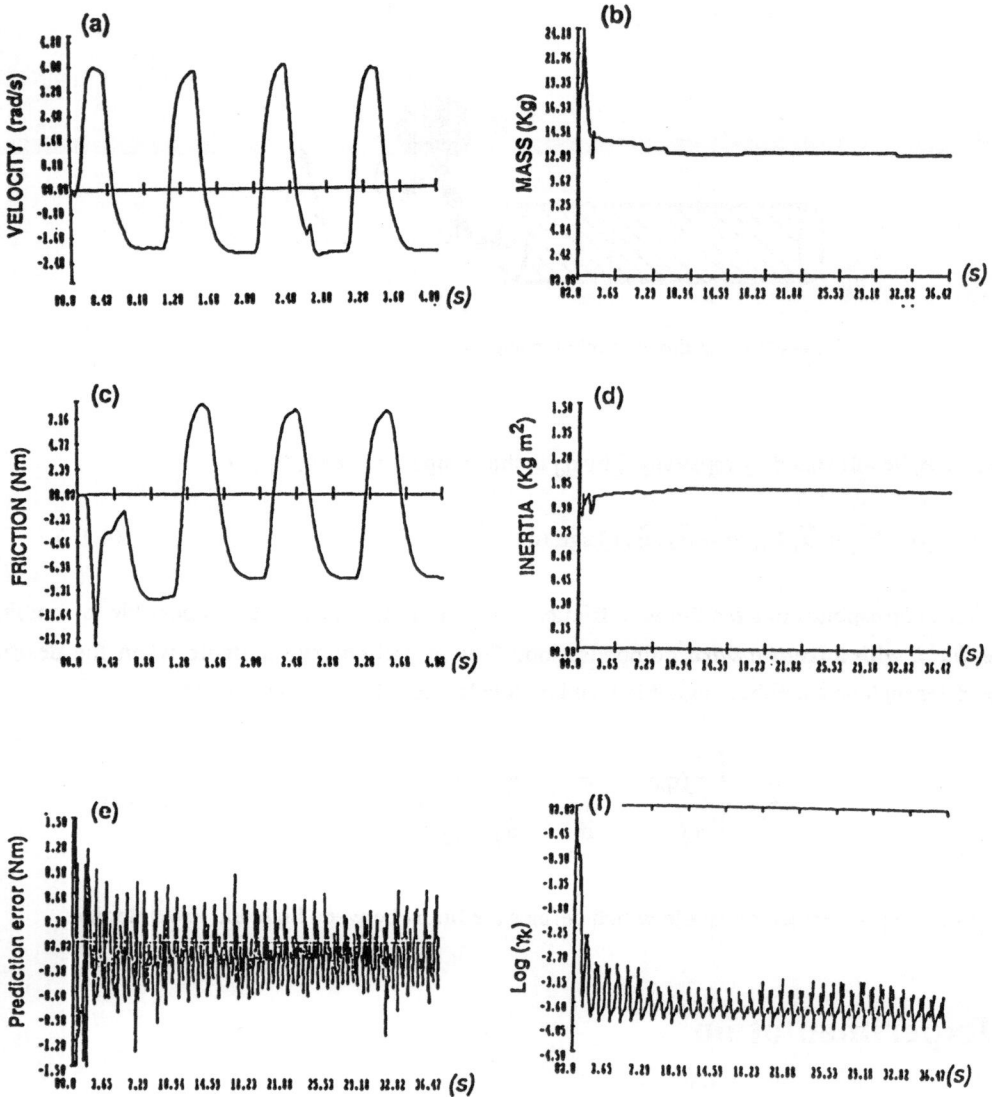

Figure 5 . Results obtained during the pre-identification sequence : (a). Measured velocity, (b) Payload mass m. (c) Estimated friction $\hat{\tau}_f$. (d) Inertia I . (e) Prediction error. (f) Measurement of arrival information ; γ_k.

closed-loop under a simple PID control loop. Figure 5 shows the time-evolution of the estimates. Inertia and payload mass were accurately estimated (the measured physical values were : $m = 11.5$ Kg, $I \approx 1 \text{Kg/m}^2$). The estimated friction was computed from the estimated $\widehat{\alpha}_i$ and the measured velocity $\dot{q}$. Its accuracy can be determined from the estimation error, see Figure 5(e), which reflects the remaining structural errors of the estimated friction forces. This figure validates our initial assumptions on the upperbound on e_k (the threshold previously defined was $\pm$ 0.8 Nm) since the prediction error seldom overcomes this bound. The exponentially weighting factor λ was 0.97. This choice is enough to keep γ_k bounded from zero, as shown by figure 5(f).

The second sequence uses the previous estimates of I and m.. Therefore, the parameter vector θ contains only the parameters associated with friction, i.e. the α_i's. This sequence involves low-speed trajectory motions ($\pm$ 0.12 Rad/sec) for which friction is difficult to estimate. The estimation of τ_f was realized in combination with a computed-torque law with integral action and adaptive friction compensation.

$$\tau_m = (m_0 \, r + I_0) \, u + m_0 \, g \, r \cos (q) + \rho \; \theta_k \, \Phi_k$$

$$u = \ddot{q}_d + k_v \, (\dot{q}_d - \dot{q}) + k_p \, (q_d - q) + k_i \int_0^T (q_d - q) \, dt$$

where $\theta_k = (\widehat{\alpha}_0, \widehat{\alpha}_1, \widehat{\alpha}_2)_k{}^T$, are the current estimates and the estimated friction forces are given by $\theta_k \, \Phi_k$. The pre-identified values of m and I are m_0 and I_0 respectively. The confidence degree of $\widehat{\tau}_f$ was chosen as $\rho = 0.9$ and friction for these low trajectories was compensated by feedforward. Figure 6(a) shows the desired velocity and the measured velocity. The peaks at the points where motion starts, represent the break-away points due to stiction friction. Figure 6(b) shows the time-evolution of the estimates $\widehat{\alpha}_i$. The variation of these parameters can be explained from its non-uniqueness. They vary when the velocity traverses the valley created by the negative exponential friction characterization and the zero-cross points. The estimated friction forces are displayed by Figure 6(c). As before we can quantify the accuracy of the friction estimation by observing the prediction error shown by Figure 6(e). The difficulties in having a good friction estimate at low-velocities become clear when comparing these curves with the ones obtained at high-velocities. The variation of the prediction error is more important in this case. Despite these differences the error e_k remains, in mean, within the pre-defined threshold $\pm$ 0.8 Nm. Unlike the previous case, the zero set-up of γ_k was not trivial. For instance, γ_k has decreased up to 1×10^{-10}, as is shown in Figure 6(f), leading to important numerical problems. To counteract this difficulty a lower bound on γ_k has been defined, below which a resetting on the P_k matrix is carried out, i.e. P_k

212

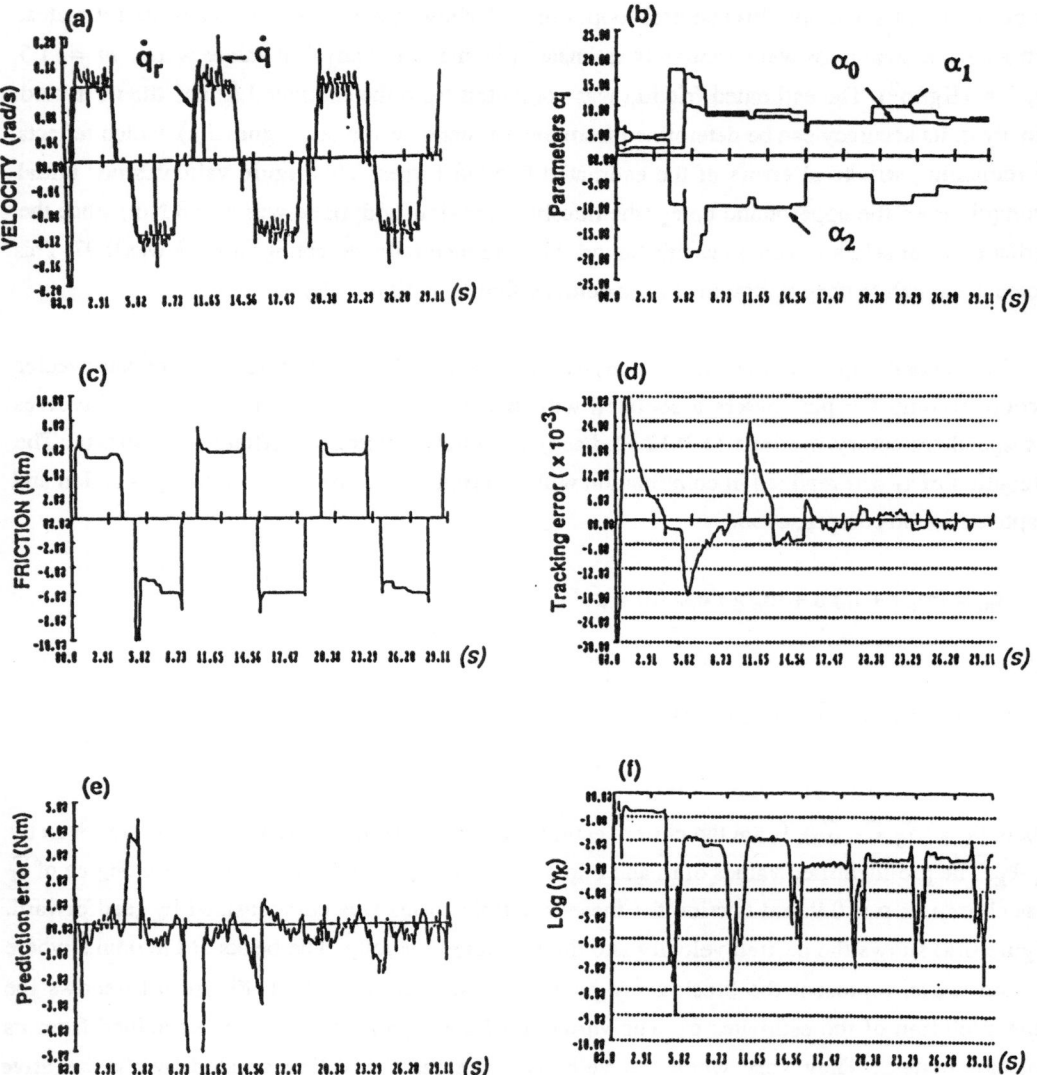

Figure 6. Results obtained during the second sequence: (a) Measured velocity *vs* Desired velocity. (b) Friction parameters $\hat{\alpha}_i$. (c) Estimated friction $\hat{\tau}_f$. (d) Tracking error. (e) Prediction error. (f) Measurement of excitation; γ_k

$= P_k + 1 \times 10^{-6}$. This permits the estimation algorithm to follow time-variations of θ while γ_k remains bounded. A final evaluation of the friction estimates can be observed from figure 6(d) which shows the improvements obtained by the adaptive feedback friction compensation realized at t = 11 sec.

6 Conclusions

In this paper we have addressed the problem of adaptive friction compensation at low-velocitiesand we have proposed a simple model, linear in parameter, which captures the downward bends at low velocities. It is important to have this type of model when friction compensation is realized, because it avoids the possibility of friction overcompensation when simplest models are used. We had shown that friction overcompensation may create limit-cycles. For this reason we have introduced a confidence degree in the predicted friction. The validity of this degree of confidence was corroborated during the experiments. The prediction error e_k, which directly gives the accuracy of the predicted friction remained within the pre-established threshold of ± 0.8 Nm. Adaptive techniques are necessary because friction not only depends on the temperature changes but also on the forces orthogonal to the motions. Experiments on an industrial robot showed that improvements on the tracking errors were obtained after the adaptive friction compensation was effectuated. For low velocities, where the velocity noise ratio is low, it is advisable to use adaptive feedforward instead of feedback friction compensation. The experiments were realized in simple PC computer leading to a sampling time of 13 ms. Additional improvements in the tracking errors will be expected if the sampling period can be reduced. For this a new processor architecture is probably requided.

References

[1] *Armstrong, B. (1988)* "Friction : Experimental Determination, Modeling and Compensation" IEEE International Conference on Robotics and Automation, Philadelphia, U.S.A, Vol. 3, pp. 1422-1427.

[2] *Braun K. (1985)* "Implementation of an Adaptive Friction Compensation", Report CODEN : LUFTD2/TFRT-7156/1-10 (1985), Dept. of Auto. Control, LTH, Lund, Sweden.

[3] *Canudas de Wit, C (1988)* "Adaptive Control for Partially Known Systems : Theory and Applications", To appear : Elsevier Science Publishers.

[4] *Canudas de Wit, C., K.J. Åström and K. Braun (1986)* "Adaptive Friction Compensation in DC Motor Drives", *IEEE Conferences on Robotics and Automation*, San

Francisco, Cal., USA. Also *IEEE journal of Robotics and Automation,* vol. RA-3, N°6, Dec.

[5] *Canudas de Wit, C. and Noel, P (1988)* " Adaptive Friction Compensation: Application to Industrial Robot". Internal rapport (in French).Laboratory of Automatic Control, ENSIEG, Grenoble.

[6] *Dahl, P.R. (1977)* "Measurement of Solid Friction Parameters of Ball Bearings", *Proc. of 6th Annual Symposium on Incremental Motion Control Systems and Devices,* University of Illinois.

[7] *Gilbart, J.W. and G.C. Winston (1974)* "Adaptive Compensation for an Optical Tracking Telescope", *Automatica,* vol. 10, pp. 125-131.

[8] *Handlykken, M. and T. Turner (1980)* "Control System Analysis and Synthesis for a Six Degree-of-freedom Universal Force Reflecting Hand Controller", *9th IEEE Conf. on Decision and Control,* vol. 1.

[9] *Kubo T., G. Anwar and M. Tomizuka (1986)* "Application of Nonlinear Friction Compensation to Robot Arm Control", *Proceedings of the 1986, IEEE Conference of an Robotics and Automation,* San Francisco, April.

[10] *Maiaux, L. (1986)* "Compensation adaptive des frottements pour les asservissements avec moteur à courant continu", Rapport Interne (in French), Laboratoire d'Automatique de Grenoble, ENSIEG, France.

[11] *Walrath, C.D. (1984)* "Adaptive bearing friction compensation based on recent knowledge of dynamic friction", *Automatica ,* Vol. 20, pp. 717-727.

[12] *Paul, P. (1972)* "Modeling trajectory calculation and servoing of a computer controller arm", Stanford Artificial Intelligence Lab., Stanford Univ., Stanford, CA, A.I. Memo, 177.

[13] *Canudas de Wit, C. and Carillo, J. (1988)* "A Modified EW-RLS algorithm for systems with Bounded Disturbances", IFAC Conference on Identification and Parameter Estimation", Beijing, China.

[14] *Tustin , A. (1947)* "The effects of Backlash and of Speed-Dependent Friction on the Stability of Closed-Cycle Control Systems", *J. of the Institution of Electrical Engineers,* 94(2A) : 143-51.

[15] *Hersey, M.D. (1966)* " Theory and Research in Lubrification", John Wiley and Sons, INC New-York.

[16] *Brogliato, B. Aubin, A. and Canudas de Wit, C.(1988)* "Stability of Servo-motors under inexact friction compensation", Internal Raport.

[17] *Wu, C.H. and Paul, P. (1980)* "Manipulator compliance based on joint torque control", *Proc. 9th IEEE Conf. Decision and Control,* Vol. 1.

[18] *Canudas de Wit, C. , P. Noel, A. Aubin, B. Brogliato, and P. Drevet (1988)* "Adaptive Friction Compensation in Robots Manipulators: low velocities",IEEE-Conference on Robotics and Automation, Poenix Arizona, USA.

Modeling and Analysis of a High-Torque, Hydrostatic Actuator for Robotic Applications

JAMES E. BOBROW AND JAYESH DESAI
DEPARTMENT OF MECHANICAL ENGINEERING
UNIVERSITY OF CALIFORNIA, IRVINE
IRVINE, CALIFORNIA 92717

ABSTRACT

The goal of this research is to develop a robot actuator capable of producing and controlling large output torques. Because of friction and backlash, it is difficult to control large output torques if they are obtained from an electric motor through a gear train. If no gearing is used, it is possible to accurately control torque output, but large torques are not possible unless heavy direct-drive motors and high-powered current amplifiers are used. In this paper, we describe a pressure-controlled hydrostatic transmission which can be used as an alternative to a gear train. It uses a fixed-displacement hydraulic pump and rotary actuator to eliminate problems due to backlash, and enables large output forces or torques of an actuator to be accurately measured and controlled. An analog control system is used to achieve a desired force output, and a digital compensator is used to obtain position control. Modeling, simulations, and experiments are presented to describe the system and its capabilities.

This research was supported by Parker-Hannifin, Parker-Bertea Aerospace Division.

INTRODUCTION

In many of today's applications of robotics, there is a pervasive need for a system with a large load carrying capacity, that is light-weight, that has the ability to exert a specified force on its environment, and is relatively inexpensive. The key to improving robot performance is to improve the performance of their actuators. Several kinds of actuators are presently being used, including electric motors and hydraulic actuators.

Electric motors are used to drive most robot systems, using either direct-drive designs [Asada and Kanade, '87] or gear trains. Direct drive systems are the most straight-forward to control, but they are also very inefficient unless they are used with a mechanical linkage such as a parallel drive mechanism [Asada and Youcef-Toumi, '84]. The main problem is that motors deliver their peak power output at 1/2 their maximum angular velocity, and they are seldom used at this speed. They are usually operated a relatively slow speeds. Hence, direct-drive motors must be sized to handle large static torques, and then are seldom operated at their peak power output levels. Also the weight of the motor's themselves may introduce large loads for a manipulator to support.

One solution is to use a large speed reducer or gear train which allows the motor to be sized so that it operates more often at the speed that it is capable of producing its maximum power. Unfortunately, the nonlinearities of friction and backlash arising from speed reducers create major obstacles for control system designers. Even if the speed reducer has a minimum amount of friction and backlash there is still another problem. In many robot applications it is desirable to control the force that the actuator exerts on its environment. This requires that a strain gage and flexure be mounted to sense the output force. This increases the expense and complicates the design, and the added dynamics due to the flexure complicates the control system design. Example applications and designs of torque control systems for robots are given in [Luh, et. al. '81, Paul and Shimano '76, Pfeffer, et. al. '87, and Wu '85]. The use of force control for robot hands is discussed in Salisbury and Craig, '82.

Hydraulic servoactuators have been used for a number of years for both position and force control applications in aircraft, machine tools, and robots [Maskrey and Thayer '78, Blackburn et. al. '60]. The main advantage of these systems is the high power and force output levels they are capable of achieving. The main disadvantages are that they require a large external accumulator and pump, they require extensive plumbing from the accumulator to the actuators, and they use servovalves for control which are relatively expensive. However, even with these substantial disadvantages, they are prevalent in nearly every application where large forces or torques are required.

Recently, a self-contained hydraulic system has been developed termed an "electrohydrostatic actuator" [Parker Bertea Aerospace '88] that accomplishes the speed reduction of a gear train and has high position accuracy. It is capable of producing forces as large as 50000 lb and weighs a total of 30 lb. It requires no external accumulator and pump as needed in conventional servovalve controlled systems. In this paper we present experiments with a similar system which uses a rotary actuator to produce large output torques. It is fitted with pressure transducers on each side of the actuator, and this enables direct control of the output torque. A dynamic model of the nonlinear system is presented, and a pressure control system is simulated and tested experimentally. A position control design is also presented.

SYSTEM DESCRIPTION AND MODEL

Description. The system shown in Figure 1 consists of a brushless DC motor, a fixed displacement hydraulic pump, a small accumulator, and a rotary actuator. A rotary actuator was used rather than a linear one in order to compare its performance with that obtained from motors and gear trains.

A pulse-width-modulated current amplifier drives the brushless DC motor. The motor drives the pump, and depending on its direction of rotation, the pump forces fluid from side 1 to side 2 of the actuator or vice-versa. This causes motion of the rotary actuator in direct proportion to rotation of the pump, except for a small error due to leakage which occurs between the two sides of the pump and the two sides of the actuator. The fluid in the accumulator is pressurized with a small piston and spring to approximately 100 psi. When the pressure on either side of the actuator drops below the pressure in the accumulator, a one-way valve opens and fluid from the accumulator enters the system.

The key to successful operation of this system is this small accumulator and check valves. When the actuator forces a large, a sizable volume of fluid is compressed on the high pressure side of the actuator. If there were no accumulator, a void in the fluid would be created on the low pressure side of the pump, and it would no longer function properly. The accumulator and one-way valves provide fluid to the lines to make up for this compressibility and hence prevent cavitation.

In order to obtain light weight, a small displacement pump is used in conjunction with a motor capable of rotating at high angular velocities. If a high speed motor is used, a lower torque is required for a given power output. Hence, a smaller and lighter motor can be used. We experimented with several combinations of motors and pumps, and found that a 18000 rpm motor worked excellent, but needed large cooling fins to dissipate excess heat generated after prolonged use. For the experiments reported here, a larger 6000 rpm motor was used. Its peak torque output is 40 in-lb, and is geared down though this hydraulic transmission with a ratio of 484.7:1 to produce a theoretical peak torque of 19388 in-lb and a maximum speed of 12.4 rpm. The actual peak torque was limited to about 7300 in-lb (600 ft-lb) because of pressure limitations (2000 psi) on the internal seals of the rotary actuator used in the experiments (it was a standard rotary actuator, [Parker-Hannifin '85]).

Dynamic Model. To develop a force (pressure) control system, a dynamic model of the hydraulic system is required. Consider the dynamics of the motor-amplifier combination. Because the amplifier senses and controls the motor current, the bandwidth (response time) of the motor is much higher than that of the overall system. Therefore, we assume that the torque output from the motor is directly proportional to the voltage input to the amplifier,

$$(1) \qquad\qquad T_p(t) = k_1 v(t)$$

where T_p is the torque from the motor windings applied to the pump and motor armature, v is the voltage input to the amplifier, and k_1 is a proportionality constant.

Next consider the dynamics of the pump. If we assume that there is no friction or leakage in the pump, the power delivered to the pump from the motor is $T_p\dot\theta$, where θ is the angular displacement of the motor, and the dot indicates derivative with respect to time. The power delivered by the pump is Δpq, where $\Delta p = p_2 - p_1$ is the pressure difference across the lines, and q is the fluid flow rate through the pump. For the ideal system which is not accelerating and has no power loss,

$$(2) \qquad\qquad T_p\dot\theta = \Delta pq.$$

The volumetric displacement D, which is the ratio of the flow rate through the pump to the angular velocity of the pump, is given by $D = \frac{q}{\dot\theta}$. Hence, for an ideal system that is not accelerating and has no leakage

$$(3) \qquad\qquad T_p = \Delta pD.$$

For a realistic model, leakage and friction must be accounted for. The torque exerted on the motor armature and pump must overcome friction and the torque due to the pressure difference across the pistons in (3). The excess torque, if any, will accelerate or decelerate the pump. A good model for the dynamics of the motor and pump assembly is [Merrit '67]

$$(4) \qquad T_p = D\Delta p + b\dot{\theta} + (T_s + f(p_1 + p_2))\frac{\dot{\theta}}{|\dot{\theta}|} + J\ddot{\theta}.$$

where J is the inertia of the coupled pump and motor rotors, b is the coefficient of viscous friction, T_s is the Coulomb friction due to the seals and bearings, and $f(p_1 + p_2)$ is a Coulomb friction term which increases linearly with pressure.

Because of the high pressures and forces obtained from the pump, the compressibility of the fluid must be modeled. Compressibility is defined as the change in volume V per unit volume for a unit change in pressure p. Bulk modulus β, which is the reciprocal of compressibility, is given by $\beta = -V\frac{dp}{dV}$. Hence, the rate of change of pressure is related to the bulk modulus by

$$(5) \qquad \dot{p} = -\frac{\beta}{V}\dot{V},$$

where we have assumed the pressure increases uniformly throughout V.

The volume of fluid on either side of the pump is

$$(6) \qquad V_1(\alpha) = V_{ms} - \frac{r^2 d}{2}\alpha,$$

$$(7) \qquad V_2(\alpha) = V_{ms} + \frac{r^2 d}{2}\alpha,$$

where r is the radius of the actuator, d is the depth of the actuator, V_{ms} is the volume of fluid in the actuator and in the lines in the mid-stroke position, and α is angular position of the rotary actuator measured from the mid-stroke position. The change in fluid volume $\dot{V}$ in (5) is due to flow from the pump and to motion of the actuator. For side one, $\dot{V} = \dot{V}_1 + q_1$, where q_1 is the flow exiting the actuator (entering the pump). Using (5) and (6), the pressure equation for side one is

$$(9) \qquad \dot{p}_1 = -\frac{\beta}{V_1(\alpha)}(q_1 - \frac{r^2 d}{2}\dot{\alpha}).$$

Similarly for side 2

$$(10) \qquad \dot{p}_2 = -\frac{\beta}{V_2(\alpha)}(q_2 + \frac{r^2 d}{2}\dot{\alpha}).$$

For the ideal case, the fluid flow entering the pump equals that leaving it, or $D\dot{\theta} = q_1 = -q_2$. However, some leakage occurs in the pump which can be modeled by [Merrit '67]

$$(11) \qquad q_1 = D\dot{\theta} - c_I(p_2 - p_1) + c_E(p1 - p3)$$

$$(12) \qquad q_2 = -D\dot{\theta} + c_I(p_2 - p_1) + c_E(p2 - p3),$$

where c_I determines the internal cross-port leakage from side one to side two, and c_E gives the external leakage to the accumulator. Substituting these flow equations into (9) and (10) gives

(13)
$$\dot{p}_1 = -\frac{\beta}{(V_{m\bullet} - \frac{r^2 d}{2}\alpha)}(D\dot{\theta} - c_I(p_2 - p_1) + c_E(p1 - p3) - \frac{r^2 d}{2}\dot{\alpha})$$

and

(14)
$$\dot{p}_2 = -\frac{\beta}{(V_{m\bullet} + \frac{r^2 d}{2}\alpha)}(-D\dot{\theta} + c_I(p_2 - p_1) + c_E(p2 - p3) + \frac{r^2 d}{2}\dot{\alpha}).$$

The rotary actuator drives a one-link arm, with the equation of motion

(15)
$$J_a \ddot{\alpha} = k_2(p_2 - p_1) - mgl\cos\alpha,$$

where J_a is the moment of inertia of the arm about the fixed pivot, m is the mass of the arm, l is the distance from the pivot to the center of mass, and the term $k_2(p_2 - p_1)$ gives the torque on the arm from the pressure difference Δp across the rotary actuator.

The nonlinear equations given by (4), (13), (14), and (15) represent the dominant dynamics for the motor/piston/actuator system. Because (4) and (15) are second order, a total of six states are needed to simulate the system. In the simulations of these equations, the six states were chosen as $\theta, \dot{\theta}, \alpha, \dot{\alpha}, p_1$, and p_2. We note that the three main assumptions used to obtain this model are: a) the DC motor dynamics are negligible, or, the motor torque is directly proportional the the amplifier voltage; b) the fluid compresses uniformly on either side of the actuator, which means that the pressure is constant throughout the volume; and c) idealized models were used for friction and leakage in (4), (13), and (14).

Determining the System Parameters. Before simulating the pressure control system, the constants used in the equations of motion are needed. We determined the constants either experimentally or from manufacturer's specifications as listed in Table 1. To determine the Coulomb friction in the pump T_s, the minimum voltage required to turn the motor was found experimentally. To determine b in (4), a valve connecting the two pressure sides was opened to make the pressure on both the sides equal at all times. The amplifier was given a constant voltage, and time was allowed for the system to reach a steady-state so that $\ddot{\theta} = 0$ and $\dot{\theta}$ is constant. Since $p_2 - p_1 = 0$, and assuming f is small, (4) shows that the torque supplied overcomes Coulomb friction and viscous friction. A number of readings were taken for different speeds and a plot of T_p versus $\dot{\theta}$ was made. The slope gives the constant b and the intercept is the value of T_s, which is a check for the previously calculated value. Explanation as to how we obtained the constants c_E and c_I is given in the following discussion of the open loop response.

OPEN AND CLOSED-LOOP EXPERIMENTS AND SIMULATIONS

By applying a step input voltage v to the motor amplifier in (1), the response of the open-loop (no feedback) system can be obtained. To investigate the behavior of the fluid model, the output shaft was clamped fixed at the midstroke position $\alpha = 0$. A constant voltage was input to the motor amplifier which creates a constant torque on the pump rotor. The resulting pressure Δp was input to the oscilloscope, and interesting behavior was observed for low input voltages. Erratic fluctuations in pressure occurred on the high pressure side of the actuator. The fluctuations are caused by alternating leakage an rotation of the pump. After a peak in pressure, the friction in the pump and seals holds the rotor stationary until the pressure drops low enough for the constant input torque to cause the pump to rotate once again.

To simulate the behavior, we adjusted the values of c_E and c_I in (11) and (12) until our numerical solution agreed fairly well with the experimental response. An example simulation is shown in Figure 2. At higher input torques applied to the pump, the stick-slip phenomena described above did not occur. Figure 3 shows the experimental response and the simulated response to an input torque of 10 in-lb. The noise in the experimental response is due to problems with the pressure transducer amplification electronics. The behavior is analogous to the response of a mass-spring-damper due to a constant force input. The effective spring constant is due to the hydraulic fluid compressibility.

Analog pressure control. The fluctuations in the actuator output torque due to variations in pressure described in the last section are undesirable if the actuator is to be used for robotic applications. In many instances it is necessary to have control of the output torque from the actuator. For example, torque control is needed to exert a given force for an assembly operation or to cause a robot arm to follow a prescribed trajectory. To achieve torque control, an analog circuit was designed. An analog controller was used rather than a digital one because it allows for a natural separation between the position controller and the force controller. In addition, the bandwidth of the pressure controller is an order of magnitude higher than the position controller so a higher sampling rate would have been needed for digital control.

We tested several controllers with experiments and simulations, and found that in order to obtain a fast response, a feedback compensator which performed well is

$$(16) \qquad v = k_p\big((\Delta p_{ref} - \Delta p) - k_v\dot{\theta},$$

where v is the voltage to the motor-amplifier in (1), Δp_{ref} is the desired pressure difference across the actuator, and the gains k_p, k_v were chosen experimentally as described below. Note that this compensator uses the state variables $\dot{\theta}, p_1$, and p_2, which corresponds to full state feedback if one does not control θ in (4), and α is held fixed. With this control law, the desired torque corresponding to Δp_{ref} is input to the system, and the actual output torque given by Δp is quickly driven to the desired value.

If the system given by (4), (13), and (14) were linearized for a given α, and no friction were assumed, full state feedback would allow arbitrary placement of the closed-loop poles for this system. Hence, any desired response time could be obtained. However, for the actual system, several limitations exist. The two most prominent factors which limit the performance of the pressure controller are the current (torque) limit of the motor and amplifier, and the unmodeled dynamics of pressure waves in the hydraulic fluid. In (13) and (14), we assumed the pressure remains uniform throughout sides one and two. In reality, pressure waves travel through the lines at the speed of sound in oil when the pump operates. The natural frequency of the pressure waves was measured to be about 300 hz, and these waves are the source of some of the noise in our response plots.

Figures 4 and 5 show both experimental and simulated plots of the actual pressure output Δp and simulated output to a square wave and a sine wave command pressure Δp_{ref} in (16). To obtain these results, the pressure control law (16) was implemented on the experiment with analog electronics. The simulation included the motor saturation nonlinearity given by the constraint $\|T_p\| \leq 40.0$ in-lb. The values used for k_p and k_v were chosen experimentally to give a fast, stable response. The steady-state error could be eliminated using a small feed-forward term which changes linearly with Δp_{ref}.

Figures 6 and 7 show experimentally measured frequency response plots of the closed loop pressure controller for two fixed positions α of the rotary actuator. Although the system is nonlinear, the frequency response plots indicate that the dominant dynamics are approximately second order. The bandwidth of the pressure controller appears to be about $200/(2\pi) = 32$ hz.

POSITION CONTROL

Given the ability to control the torque applied to the actuator T_a with a relatively high bandwidth using the analog pressure controller of the last section, it is straight-forward to design a position controller. We found that a simple PD digital compensator performed well. The control law was implemented digitally as

(17) $$\Delta p_{ref}(t) = -c_1 e(t) - c_2/T(e(t) - e(t - T))$$

where $e(t) = \alpha(t) - \alpha_{ref}(t)$ is the position error at time t, T is the sampling interval, α_{ref}, is the desired position of the arm, and the constants c_1, c_2 are the feedback gains for the position loop.

The control computer (an IBM PC) inputs the position α at time t, computes the desired command Δp_{ref} and outputs this value through a D/A converter to the analog pressure controller of (16). It is helpful to have an approximate linear model for this system in order to choose the gains c_1 and c_2. Figure 8 shows the model we used. For clarity, a model for an approximate continuous-time system is shown, even though the control loop was implemented digitally. In the figure, the closed-loop pressure control system is modeled as a second order linear system with $\omega_n = 150$ rad/sec and $\varsigma = .5$ which are roughly the values which match the results of the response curves in Figures 6 and 7 at low frequencies.

A root-locus plot is shown for this system in Figure 9, where it was assumed that the ratio $c_1/c_2 = 30$, and a variable gain g multiplies c_1 and c_2 to produce the plot. Note that the system is stable until $g = 770$, at which point the locus crosses the imaginary axis. More sophisticated controllers could be used to obtain a faster response, but this controller performed well for our application. Figure 10 shows the step response for the actual system with $g = 350$.

CONCLUSION

An alternative actuator to a motor and gear train has been presented which uses a hydrostatic transmission to enable large speed reductions and corresponding torque amplification. Problems due to friction and backlash in conventional gear trains are eliminated with the closed hydraulic system presented. A dynamic model of the system was developed, and it was shown that the output torque of the actuator can be controlled by sensing the pressure of the working fluid. Simulations and experiments were used to test the system and model.

REFERENCES

1 H. Asada and T. Kanade, "Design of Direct-Drive Arms," ASME Journal of Vibration, Stress, and Reliability in Design, Vol 105, No 3, 1983.

2 H. Asada and K. Youcef-Toumi, "Analysis and Design of a Direct-Drive Arm With a Five Bar Link Parallel Drive Mechanism," ASME Journal of Dynamic Systems, Measurement, and Control, Vol 106, No 3, pp 225-230, 1984.

3 J.F. Blackburn, G. Reethof, and J.J. Shearer, Fluid Power Control, Wiley, 1960.

4 J.Y.S. Luh, W.D. Fisher, and R.P. Paul, "Joint Torque Control by Direct Feedback for Industrial Robots," Proc. IEEE Conference on Decision and Control, pp 265-271, San Diego, Ca, 1981

5 R.H. Maskrey and W.J. Thayer, "A Brief History of Electrohydraulic Servomechanisms," ASME Journal of Dynamic Systems, Measurement, and Control, Vol 100, No 2, pp 110-116, June 1978.

6 E.H. Merrit, Hydraulic Control Systems, Wiley, 1967.

7 Parker Hannifin Inc, Hydraulic and Pneumatic Rotary Actuators, Catalog 1800-1, Cylinder Division, Des Plaines IL, 1985.

8 Parker Hannifin Inc, Parker Bertea Aerospace, Electro Hydrostatic Actuator, Control Systems Division, Irvine CA, 1988.

9 R. Paul and B. Shimano, "Compliance and Control," Proc. IEEE Joint Automatic Control Conference, pp 694-699, San Francisco, Ca, 1976

10 L. Pfeffer, O. Khatib, and J. Hake, "Joint Torque Sensory Feedback in the Control of a PUMA Manipulator," Proc. IEEE International Conference on Robotics and Automation, pp 1966-1971, Raleigh, NC, 1987.

11 J.K. Salisbury and J.J. Craig, "Articulated Hands: Force Control and Kinematic Issues," International Journal of Robotics Research, Vol 1, No 1, 1982.

12 C. H. Wu, "Compliance Control of a Robot Manipulator Based on Joint Torque Servo," International Journal of Robotics Research, Vol 4, No 3, pp 55-71, 1985.

Table 1: List of parameters used in the simulation.

Parameter	Value	Units	Method Determined
β	247,000	psi	from the specifications for the oil used
r	1.83	in	from actuator data
d	2.20	in	from actuator data
V_{me}	3.16	in^3	from actuator data
D	.00796	in^3/rad	pump data
p_3	100.0	psi	preset
k_1	4.0	in-lb/volt	set by adjusting the amplifier gain
k_2	3.68	in-lb/psi	from the transmission ratio
J	0.03	lb-in-sec^2	from motor and pump data
J_a	23.3	lb-in-sec^2	from arm mass properties
T_s	4.0	in-lb	experimentally
b	0.01455	in-lb-sec	experimentally
c_E	6.5×10^{-6}	in^3/(sec-psi)	compared experiments to simulation
c_I	1.0×10^{-8}	in^3/(sec-psi)	compared experiments to simulation
T_{pmax}	40.0	in-lb	set by current limiting in amplifier
f	0.0	in-lb/psi	compared experiments to simulation

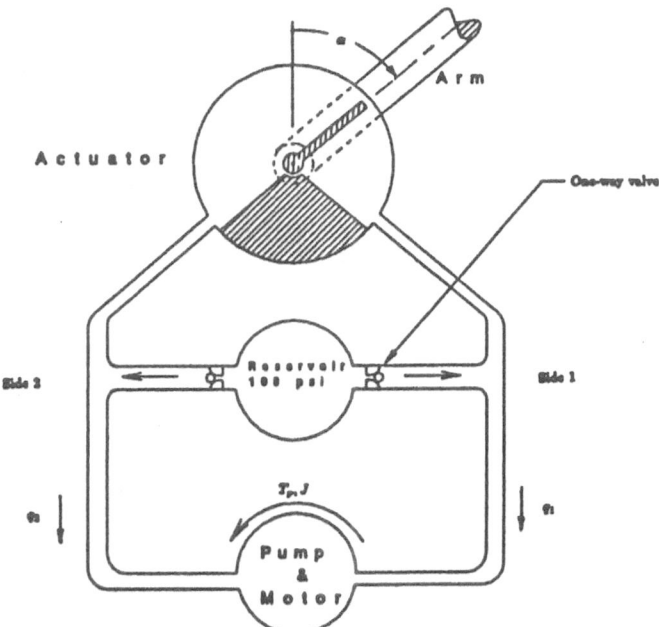

Figure 1: Test apparatus consisting of a DC motor, pump, small accumulator, and rotary actuator.

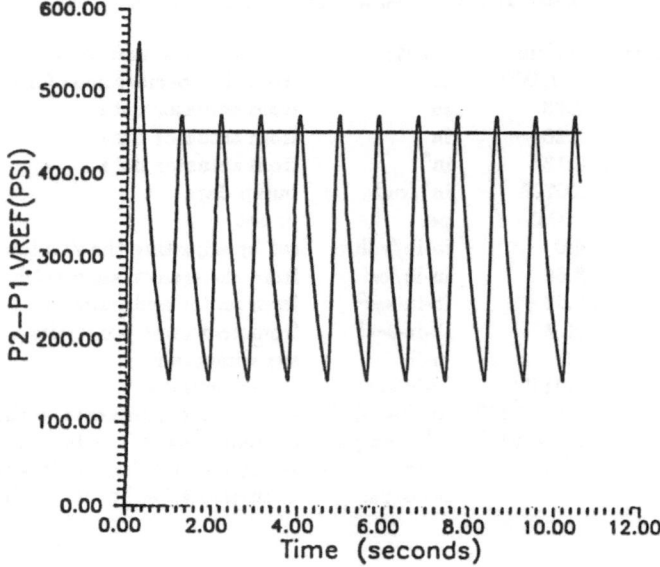

Figure 2: Simulation of open-loop stick-slip phenomenon at low pressures.

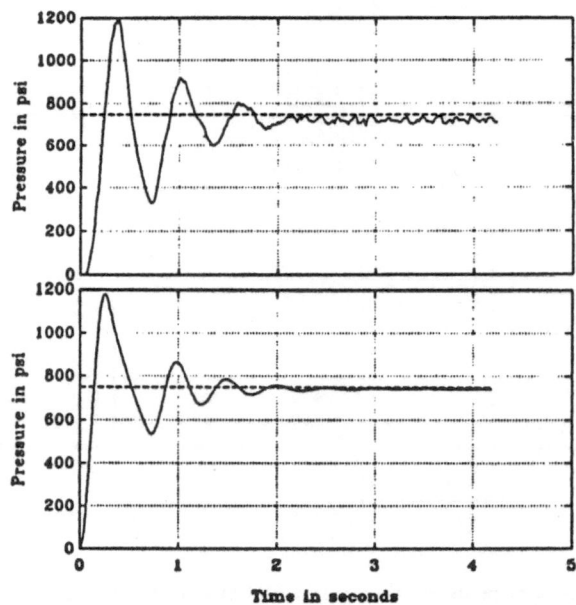

Figure 3: Experimental (upper plot) and simulated (lower plot) response of the open-loop system at higher pressures.

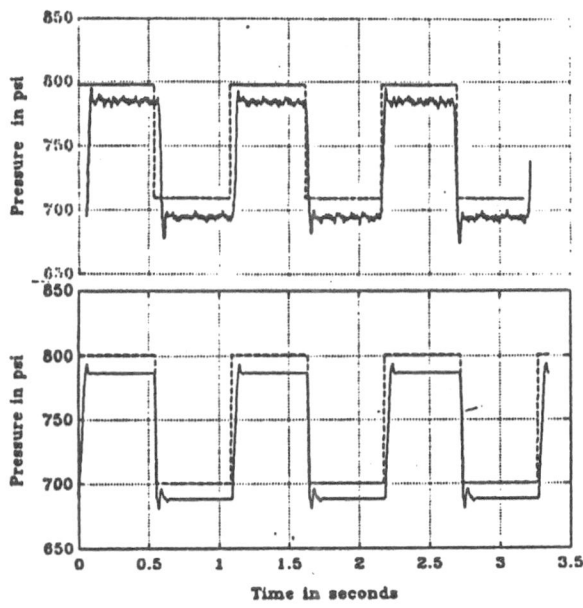

Figure 4: Experimental (upper) and simulated (lower) pressure response to a square wave reference input.

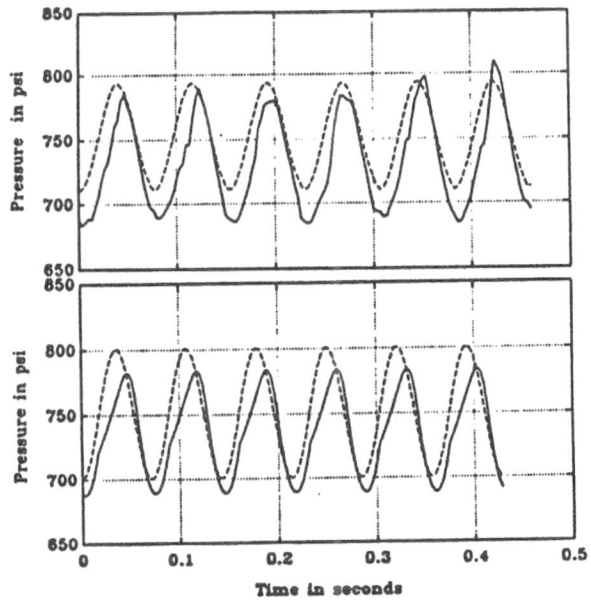

Figure 5: Experimental (upper) and simulated (lower) pressure response to a sine wave reference input.

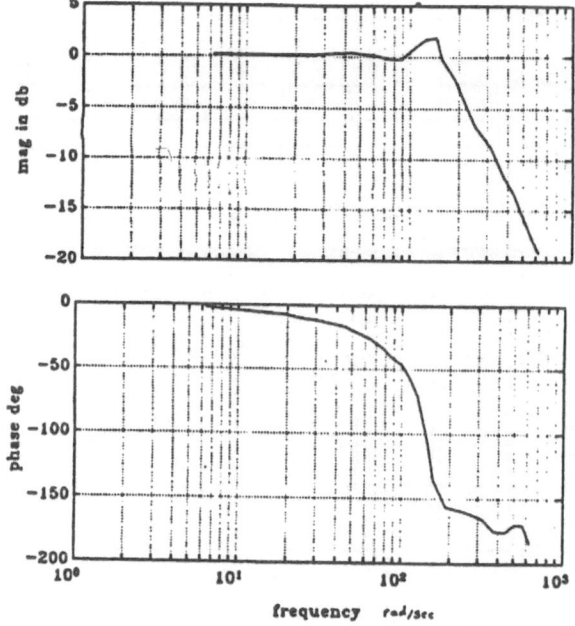

Figure 6: Experimental frequency response of the pressure control system when the actuator is in the midstroke position $\alpha = 0$.

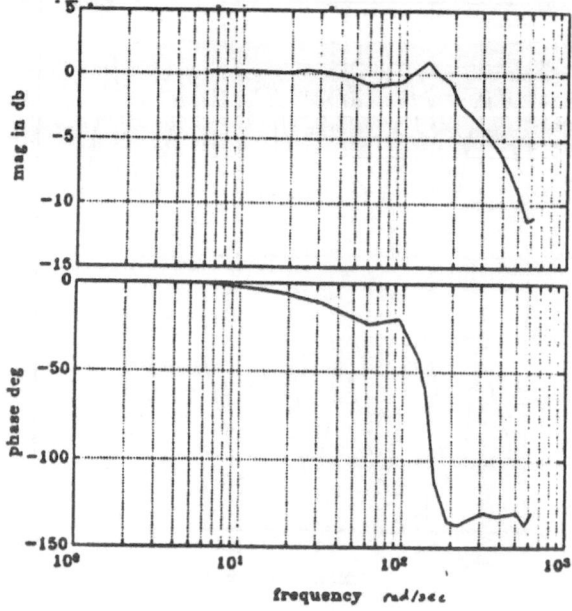

Figure 7: Experimental frequency response of the pressure control system when the actuator is moved 90 degrees $\alpha = \pi/2$.

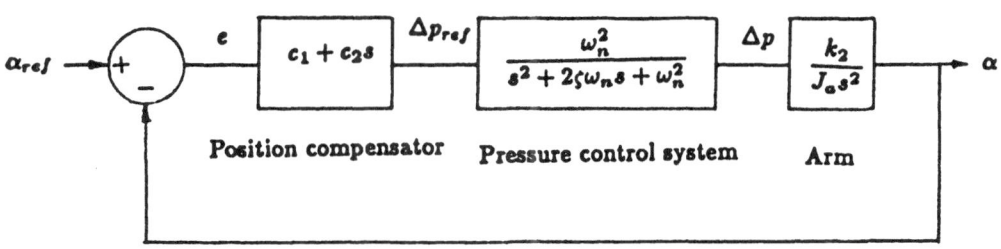

Figure 8: Block diagram of the overall position control system.

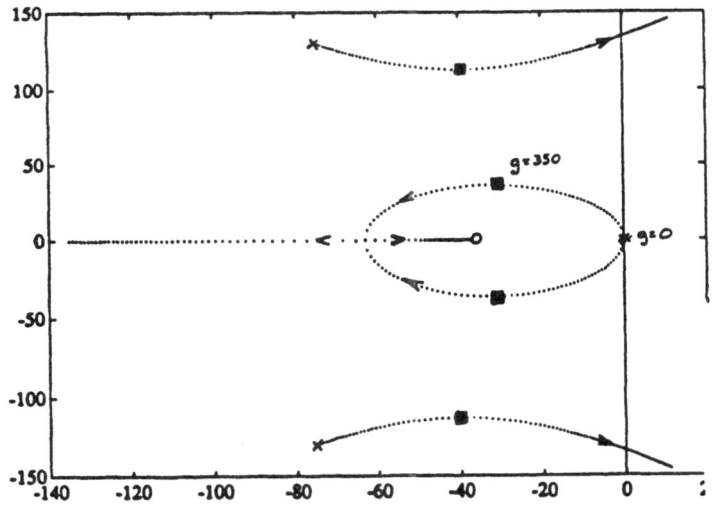

Figure 9: Root-Locus of the position control system with $c_1/c_2 = 30$.

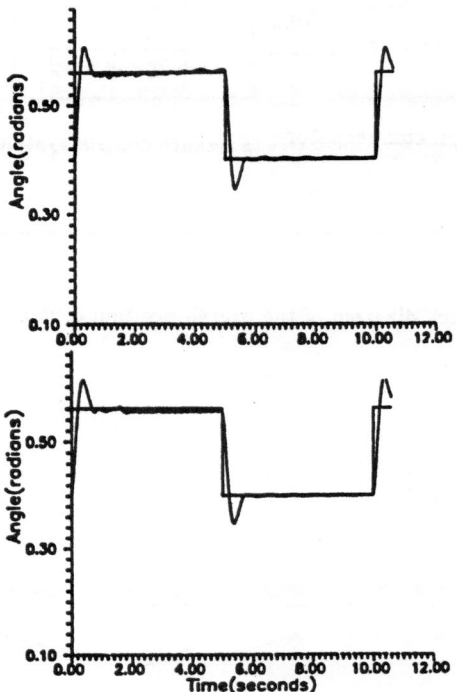

Figure 10: Experimental and simulated step response of the position control system.

High-Speed Digital Controller for Magnetic Servo Levitation of Robot Mechanisms

Masahiro TSUDA *, ** †, Yoshihiko NAKAMURA **, Toshiro HIGUCHI*

* Institute of Industrial Science, University of Tokyo
Roppongi, Minato-ku, Tokyo 106, Japan
** Center for Robotic Systems in Microelectronics, University of California, Santa Barbara
Santa Barbara, CA93106, U.S.A.

Magnetic levitation has the strong potential to be a precision force/position-controllable actuator with inherent force-sensing capability in itself. We call such a system *magnetic servo levitation* to distinguish it from conventional magnetic suspension-only levitation. Magnetic levitation inherently has strong instability and needs high-speed closed-loop control. There were few applications of magnetic servo levitation because high-speed digital controllers were not available. The authors developed a digital controller using a digital signal processor, and experimentally verified that magnetic servo levitation is promising and realizable by the state-of-the-art device technology.

1. Introduction

1.1. Force/Position Control and Servo Levitation for Robots

We propose to apply magnetic levitation to robot mechanisms and to control the levitated object by using levitating force in order to perform precision force/position-control for robots. Many researchers have been asserting that precision force/position-control is necessary for robots to achieve advanced tasks, and have been developing sophisticated force/position control schemes. Those control schemes have been implemented in robots but have limitations in precision, response and so on owing to unmodeled factors such as friction and backlash dogging conventional robot-mechanisms. Levitation is expected to dynamically reduce such mechanical uncertainties in control and to bring out full performance of force/position-control schemes.

Levitation has many advantages, such as: mechanical-friction-free, backlash-free, wear-free, maintenance-free, lubrication-free, and so on. Those advantages have found significant applications in suspension mechanisms, such as bearing and levitated vehicles. Consequently levitation has been associated with ideal suspension mechanisms

However, in addition to the above advantages, levitation permits precision control of position and attitude of the levitated object by controlling levitating-force. Levitation can work as multi-degree-of-freedom actuators. We refer to such levitation as *"Servo Levitation."*

---

† M. Tsuda, a Ph.D. candidate at the Institute of Industrial Science, University of Tokyo, is presently a visiting researcher at the Center for Robotic Systems in Microelectronics, University of California, Santa Barbara, for collaborative research of the two institutions.

1.2. Magnetic Servo Levitation

There are several types of levitation: magnetic, electromagnetic, electrostatic, pneumatic and so on. Electrostatic levitation does not have enough force available in reality. Pneumatic levitation has a controlled system too complicated to control the levitated object. Magnetic levitation and electromagnetic levitation have great possibilities for servo levitation. Magnetic levitation can further be classified as possessing both passive and active characteristics. Passive magnetic levitation uses either repulsive or restoring force or both by means of permanent magnets and/or electromagnets for the levitating-force, and is stable without control. Active magnetic levitation uses controllable attractive force of electromagnets and is unstable without control. Electromagnetic levitation uses Lorentz-force for the levitating-force.

Table 1 shows advantages and disadvantages of the three types of levitation. Passive magnetic levitation is suitable for suspension-only mechanisms, but not for servo levitation due to the difficulty of controlling levitating-force. Active magnetic levitation and electromagnetic levitation are both suitable for servo levitation and already have been applied to antenna pointing systems, robot end-effectors, and robot joints [1 to 5].

Both have advantages and disadvantages. Permanent magnets are essential to electromagnetic levitation, but are not desirable from the standpoint of control and mechanism design, because they have characteristics varied by heat and time and also decrease freedom of design of mechanisms. In addition, electromagnetic levitation requires the levitated object to have coils and/or permanent magnets, which structure also produces constraints in designing mechanisms. On the other hand, active levitation, which requires only that the object be steel, has more flexibility in mechanism design. It can be combined easily with other mechanisms such as induction motors and/or stepping motors like magnetic bearings [5, 6]. This feature is a notable advantage in applying servo levitation to robot mechanisms. However, active magnetic levitation has nonlinearity in the control system, since attractive force generated by electromagnets has nonlinear characteristics. Also, active magnetic levitation inherently has strong instability and requires high-speed closed-loop control to be stable. The nonlinearity and the inherent instability make control difficult.

We consider that active magnetic levitation is more applicable to various kinds of robot-mechanisms due to the advantages derived from its simple mechanical-structure. Its disadvantages can be avoided by applying advanced control schemes. We call active magnetic levitation used for servo levitation *"magnetic servo levitation."*

Table 1. Comparison of three types of levitation.

type of levitation	active magnetic levitation	passive magnetic levitation	electromagnetic levitation
stability	×	O	Δ
low power consumption	Δ	O	Δ
control of force	Δ	×	O
invariability of system	O	Δ	Δ
largeness of force	O	Δ	Δ
movable range	Δ	O	O
freedom in mechanism	O	Δ	×

Magnetic servo levitation can work as multi-axis force-sensors as well as multi-DOF actuators, because the friction-free and backlash-free structure enables us to solve the motion equation precisely for external force with known levitating-force. In addition, magnetic servo levitation generally produces redundant DOF in robots and allows micro/macro-manipulator structures, which are also suitable for force/position control [7, 8]. Magnetic servo levitation is certainly expected to improve performance of robots with force/position control. Furthermore, magnetic levitation, which is lubrication-free, can be used in special environments such as space, vacuum, and super-clean environments.

1.3. Goals of This Paper

The idea of magnetic servo levitation is not new but has not been utilized for a long time due to the lack of economic digital controllers with high-speed control ability. In this paper, we synthesize a high-speed digital controller in order to provide the required technology for magnetic servo levitation. High-speed digital controllers are needed because:

1) To stabilize levitation, which has inherent instability, high-speed closed-loop control is necessary. Up to now, analog regulation controllers have been used to stabilize active magnetic levitation in conventional systems, such as magnetic bearings and magnetic suspension systems.

2) The friction-free and backlash-free structure of magnetic servo levitation is expected to accommodate precision force/position control using advanced control schemes for robots. Implementation of such advanced control schemes requires a digital controller.

3) The control system of magnetic servo levitation involves nonlinearity of magnetic force. In order to make the most of magnetic servo levitation, the controller should be a digital controller to deal with the nonlinearity.

For the above reasons, a high-speed digital controller is an essential device technology for the realization of magnetic servo levitation. We have developed a high-speed digital controller using a digital signal processor (DSP). DSP's are relatively cheap and have high computational ability. In the digital controller, the DSP is not used as a co-processor of a computer, but works as a CPU which can directly control the controller-units, such as A/D and D/A converters, with tight hardware connections. This enables the DSP to display its full performance so as to execute closed-loop control as fast as possible. Using the controller we have demonstrated that magnetic servo levitation is realizable by state-of-the-art device technology.

This paper describes the basic concept and principle of magnetic servo levitation, and shows the structure of the high-speed digital controller. The results of experiments using the controller are shown to examine basic characteristics and capabilities of magnetic servo levitation.

2. Principle of Magnetic Servo Levitation

Magnetic servo levitation has two main control problems: nonlinearity of magnetic force; and inherent instability. Those problems are inherent in all types of magnetic servo levitation, although magnetic servo levitation may be classified according to the number of controllable DOF. An n-DOF-controllable magnetic servo levitation is discussed for generality. For simplicity, we first discuss 1-DOF-controllable magnetic servo levitation. It does not lose any generality for n-DOF types.

In the next section, we will illustrate a servo-control system of magnetic servo levitation in order to clarify the principle and problems. Note that the servo-control system shown is a typical example but other servo-control systems also may allow magnetic servo levitation.

2.1. 1-DOF-Controllable Magnetic Servo Levitation

Figure 1 shows a 1-DOF-controllable magnetic servo levitation. We assume that the levitated object can move only in the horizontal direction. All the other DOF is ideally constrained. An electromagnet can apply only attractive force to the object. Therefore, two electromagnets are necessary to apply both positive and negative force to the object. We will refer to the electromagnet generating plus (or minus) force as electromagnet P (or M). If constant currents are supplied to the pair of electromagnets, the levitation is obviously unstable. To stabilize it, we must control the currents, measuring the position of the object

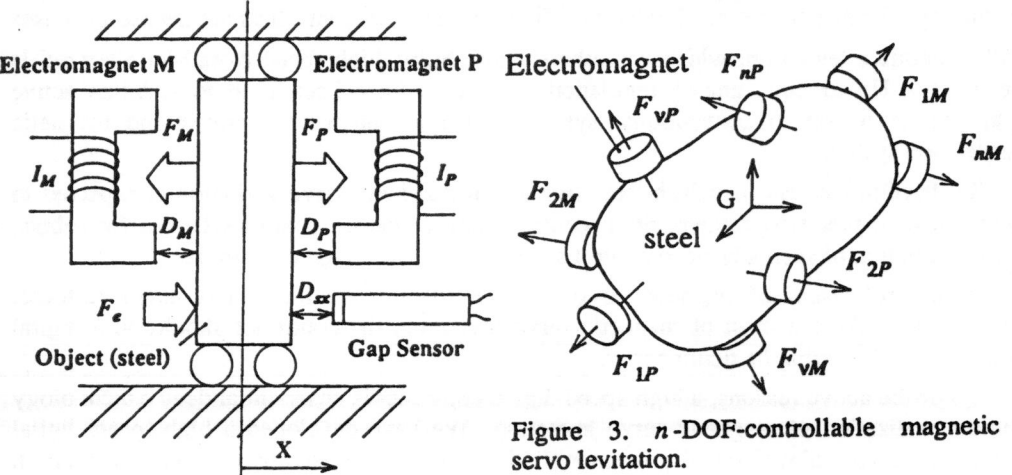

Figure 1. 1-DOF-controllable magnetic servo levitation.

Figure 3. n-DOF-controllable magnetic servo levitation.

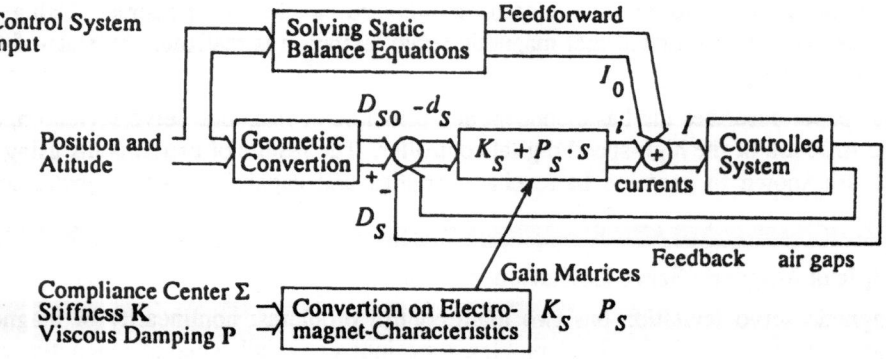

Figure 2. Block diagram of control system.

by a gap-sensor.

In the following, we construct a servo-control system, shown in figure 2, by modifying the conventional control system of active magnetic levitation [9]. Let us design a control system provided with the following closed-loop characteristics.

$$m\ (\ddot{x} - \ddot{x}_0) = -K\ (x - x_0) - P\ (\dot{x} - \dot{x}_0)\ , \tag{1}$$

where m is a mass of the object; x and x_0 are respectively actual and desired positions of the object; K and P are respectively support-stiffness and viscous-damping coefficients; K, P and x_0 are given as input; and x is measured as output. The control system is stable if and only if K and P are both positive.

We can describe the motion equation of the object as:

$$m\ \ddot{x} = F_P - F_M + F_e\ , \tag{2}$$

where F_P, F_M and F_e denote attractive forces of electromagnets P and M, and external force, respectively. Magnetic force $F_\alpha\ (\alpha = P, M)$ can be considered a nonlinear function of current I_α and air-gap D_α:

$$F_\alpha = F_\alpha\ (I_\alpha, D_\alpha)\ . \tag{3}$$

Note that it is current I_α that we directly control. We control current I_α as the sum of feedforward current $I_{0\alpha}$ and feedback current i_α, where I_α must be nonnegative.

Let $D_{0\alpha}$ indicate the air-gap D_α when the object is located at desired position x_0. The known external force, such as gravity, is denoted by F_{e0}, while the unknown external force $(F_e - F_{e0})$ is treated as disturbance. We can determine feedforward currents I_{0P} and I_{0M} to satisfy:

$$-m\ \ddot{x}_0 + F_{0P} - F_{0M} + F_{e0} = 0\ , \quad \text{where}\ F_{0\alpha} \equiv F_\alpha\ (I_{0\alpha}, D_{0\alpha})\ , \tag{4}$$

$$I_{0\alpha} \gg |\ i_\alpha\ |\ . \tag{5}$$

Then we can approximately linearize equation (3) to:

$$F_\alpha\ (I_{0\alpha} + i_\alpha, D_{0\alpha} + d_\alpha) \approx F_{0\alpha} + A_{I\alpha}\ i_\alpha - A_{D\alpha}\ d_\alpha\ , \tag{6}$$

where

$$d_\alpha \equiv D_\alpha - D_{0\alpha},\quad |\ d_\alpha\ | \ll |\ D_{0\alpha}\ |\ ,$$

$$A_{I\alpha} \equiv \frac{\partial F_\alpha\ (I_{0\alpha}, D_{0\alpha})}{\partial I_\alpha}\ ,\quad A_{D\alpha} \equiv -\frac{\partial F_\alpha\ (I_{0\alpha}, D_{0\alpha})}{\partial D_\alpha}\ .$$

On the other hand, we geometrically obtain the following condition:

$$x - x_0 = -d_P = d_M \equiv d\ . \tag{7}$$

Substituting equations (4), (6) and (7) into (2), we obtain:

$$m\ \ddot{d} = A_I\ i + A_D\ d\ , \tag{8}$$

where

$$i \equiv i_P = -i_M\ ,\quad A_I \equiv A_{IP} + A_{IM}\ ,\quad A_D \equiv A_{DP} + A_{DM}\ . \tag{9}$$

From equations (1) and (9), feedback current i is determined as:

$$i = K_s\, d + P_s\, \dot{d}\,, \tag{11}$$

where

$$K_s \equiv A_I^{-1}\,(K + A_D)\,, \quad P_s \equiv A_I^{-1}\, P\,. \tag{12}$$

This establishes a servo–control system represented by equation (1) as a typical example.

Magnetic servo levitation inherently possesses force-sensing capability. Unknown external force F_e can be measured as the following algorithm: Every magnetic force F_α can be calculated from current I_α and air-gap D_α. Using the F_α's, motion equation (2) can be solved for F_e.

2.2. n-DOF Servo levitation

In order to control n DOF of a levitation object, at least $n+1$ electromagnets are necessary and sufficient [11]. In this paper, however, we use n pairs of electromagnets, that is, $2n$ electromagnets to simplify the control system. Every pair of electromagnets confront each other so that we can obtain both positive and negative forces, as shown in figure 3. In that case, we can easily generalize the 1-DOF control system to a n-DOF one, where variables i, d, D_0 and I_0 are vectors. Gains K_s and P_s are matrices. Input x_0 is a vector and the other inputs K and P are matrices (See figure 2.)

We designed and developed the Magnetically Supported Intelligent Hand (MSIH) by applying 5-DOF-controllable magnetic servo levitation to a robot end-effector for automatic precision assembly [3, 10] †. Its schematic structure and cross section are shown in figures 4 and 5 respectively. The detail of the control system of the MSIH was explained in [3, 10].

3. Digital Controller

A high-speed digital controller is essential for magnetic servo levitation. Several digital controllers have been developed for magnetic levitation. However, most of them, using a microcomputer, were only able to regulate 1 DOF [12, 13]. Their reported sampling periods were longer than 400 [μsec]. They succeeded in stabilizing levitation, but did not have sufficient ability to accommodate high stiffness, quick response or servo levitation of multi-DOF.

Compared with regulated levitation, magnetic servo levitation requires controllers to have still more ability for fast control, because precision force/position control needs more computational capability than the conventional regulation. Therefore, ordinary microcomputer controllers can hardly provide for magnetic servo levitation, although their performance has been improving rapidly.

Nowadays, digital signal processors (DSP) are available with cost-and-performance efficiency. We devised a digital controller using a DSP for 5-DOF-controllable servo levitation. There are several contrivances in hardware for the controller to work at sufficiently high speed to stabilize 5-DOF levitation with high stiffness and rapid response. The DSP directly controls the control-units, such as A/D- and D/A- converters, with tight hardware connections so as to display its full talent of fast operation. It can execute the closed-loop control of 5 DOF, described in chapter 3, in 54 [μsec], including all calculation,

---

† The MSIH was developed at Institute of Industrial Science, University of Tokyo.

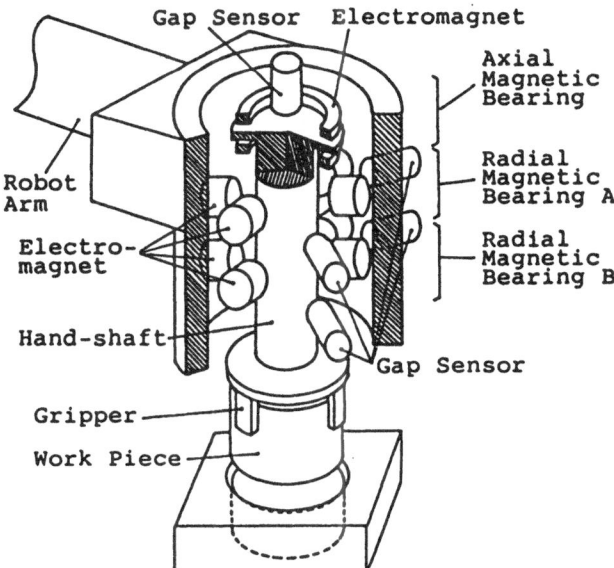

Figure 4. Schematic of MSIH.

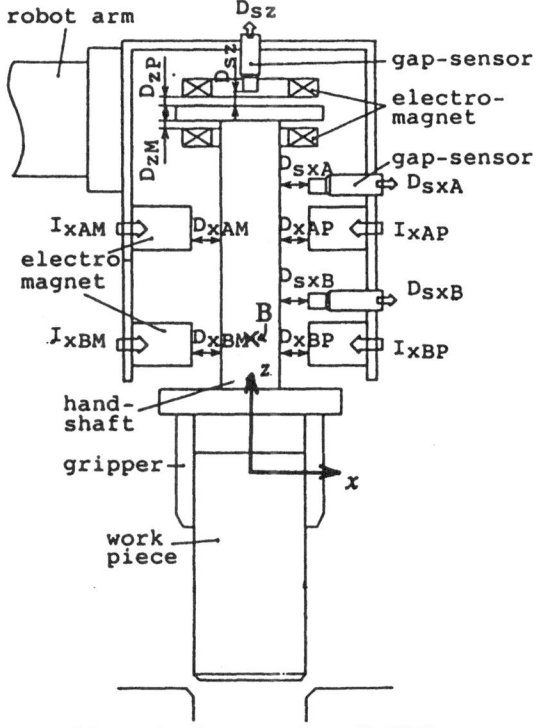

Figure 5. Cross section of MSIH.

data-conversion and others. It has succeeded in stabilizing 5-DOF levitation with stiffness of 3.0×10^5 [N/m] and 500 [N m/rad] (which does not indicate the limit of stiffness.)

The structure of the controller is as follows: The controller includes a DSP μPD77230 (NEC) and a personal-computer with an i80286 (Intel), and has a hierarchy structure of the dual CPU's as shown in figure 6. The DSP is tightly coupled with five A/D converters and ten D/A converters so as to provide direct control of them. Otherwise communication of control-information would be necessary between the personal computer and the DSP, increasing time-delay in control which may cause instability. The DSP can execute each operation in 150 [nsec] , such as floating-point multiplication-and-addition. Each D/A converter is connected to a current-driver circuit for an electromagnet. Each A/D converter receives output of a gap-sensor as input through a sample/hold device. The DSP and the control units work mainly for the stabilizing closed-loop. The DSP can be interrupted by a programmable timer controlled by the personal computer to keep the accurate sampling period. The complicated calculations for outer feedback-loop and feedforward are mostly done by the personal computer. The personal computer is connected with the DSP through two-port RAM and several input/output-ports. It takes 64 [μsec] to transmit all control data between the two CPU's, such as gap-sensor outputs D_s, coil-currents I and control gains K_s, P_s. The personal computer works at relatively low speed, but has advantages of low cost and easy maintenance for both software and hardware. The personal computer can also be used for recording/analyzing experimental data and for developing programs of the DSP.

Note again that the above structure was designed to accomplish high-speed-control at a low cost. The controller can be easily modified to control 6- or more-DOF levitation.

4. Experiments

We carried out experiments for three purposes: (1) to verify that the controller has sufficient ability to give high stiffness and rapid response for magnetic servo levitation; (2) to evaluate the accuracy of the mathematical model by comparing experimental results with computer simulation, which will indicate the simple dynamics of magnetic servo levitation; and (3) to demonstrate that magnetic servo levitation is realizable by the current device technology and has promising capability.

The experiments were done with the MSIH experimental system, shown in figure 7.

4.1. Mathematical Model of Electromagnets

Magnetic levitation can be exactly modeled if we know the characteristics of electromagnets shown as equation (3), since it involves no mechanical uncertainties. We can, in theory, model the static magnetic force by the following equation, if we assume constant permeability and no leak of magnetic flux:

$$F_\alpha = Q_\alpha \frac{I_\alpha^2}{D_\alpha^2} . \tag{13}$$

We experimentally confirmed that the above model is not accurate enough for magnetic servo levitation. We have introduced the following empirical model:

$$F_\alpha = \sum_{n=1}^{14} Q_{\alpha n} \frac{I_\alpha^{\rho_n}}{D_\alpha^{\delta_n}} , \tag{14}$$

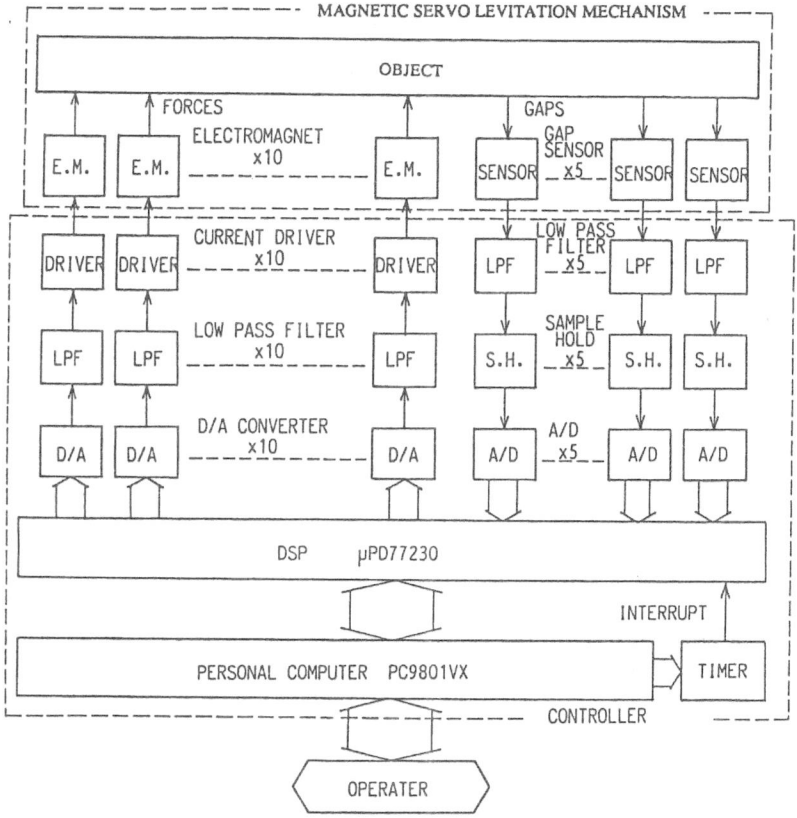

Figure 6. Structure of high-speed digital controller.

Figure 7. Photo of the experimental system.

where $Q_{\alpha n}$'s are coefficients; ρ_n's are 1 and 2; and δ_n's are 0.5, 1.0, 1.5,..., 3.0, 4.0. We refer to the ten electromagnets as α such as xAP, xAM, xBP,..., zM.(See figure 5.) Equation (14) is a second-order polynominal of current I_α and can be easily solved for I_α with F_α and D_α given, which helps calculating feedforward current I_0.

We measured 76,173 sets of $(F_\alpha, I_\alpha, D_\alpha)$ under various static-conditions, and obtained $Q_{\alpha n}$'s from those data sets with the regression analysis. We used equation (14) as the mathematical model of the electromagnets in the experiments of this paper.

4.2. Transient Response of Dynamic System

Transient response was measured when a stepped disturbance was given to coil-current I_{xBP}. The experiments were compared with computer simulations based on the mathematical model. The experimental conditions are as follows: (1) The mass and inertia-moment around the center of gravity of the levitated object are respectively 1.42 [kg] and 3.12×10^{-3} [kg m^2]. (2) Position D_z is measured by the gap-sensors every 120 [μsec] and is plotted versus time in a graph. (3) It takes 55 [μsec] to settle the coil-currents I after the outputs D_z of the gap-sensors are sampled and held. In other words, the lag of the control system is 55 [μsec]. (4) Sampling period T and gain matrices K_s and P_s are variable as experimental parameters. (5) The A/D- and D/A-converters have 12-bit resolution. Their 1 bits respectively represent 1 [μm] and 0.5 [mA]. (6) Gain matrices P_s should be multiplied by velocity of the levitated object. However, we have no sensors capable of measuring the velocity directly. As a pseudo velocity, we used the difference between a current gap and a 4-sampling-period-previous gap divided by a 4-sampling-period. The difference period (4-sampling-period) was decided by experimental consideration to minimize vibration in the steady state. In that case, the vibration is about ± 1 [μm]. (7) A stepped disturbance is given to feedforward coil-current I_{0xBP}. (See figure 5.) I_{0xBP} is usually equal to 0.6 [A], while it becomes 1.0 [A] for 1.2 [msec] with the disturbance given. (8) The desired air-gaps D_{z0} are equal to $(2.0, 2.0, 1.1, 1.1, 2.0, 2.0, 1.1, 1.1, 1.6, 1.6)^T$ [mm]. (9) Every current I_α is nonnegative and is less than 1.8 [A].

First of all, we experimented with T changed and with K_s and P_s constant as

$$K_s = \begin{bmatrix} K_{sx} & 0 & 0 \\ 0 & K_{sy} & 0 \\ 0 & 0 & K_{sz} \end{bmatrix}, \quad P_s = \begin{bmatrix} P_{sx} & 0 & 0 \\ 0 & P_{sy} & 0 \\ 0 & 0 & P_{sz} \end{bmatrix} \tag{15}$$

where

$$K_{sx} = K_{sy} = \begin{bmatrix} 11.53 & 0.45 \\ -11.53 & -0.45 \\ -2.93 & 5.57 \\ 2.93 & -5.57 \end{bmatrix}, \quad K_z = \begin{bmatrix} 9.47 \\ -9.47 \end{bmatrix} \tag{16}$$

$$P_{sx} = P_{sy} = \begin{bmatrix} 5.04 & 2.16 \\ -5.04 & -2.16 \\ -3.86 & 11.06 \\ 3.86 & -11.06 \end{bmatrix}, \quad P_z = \begin{bmatrix} 17.28 \\ -17.28 \end{bmatrix} \tag{17}$$

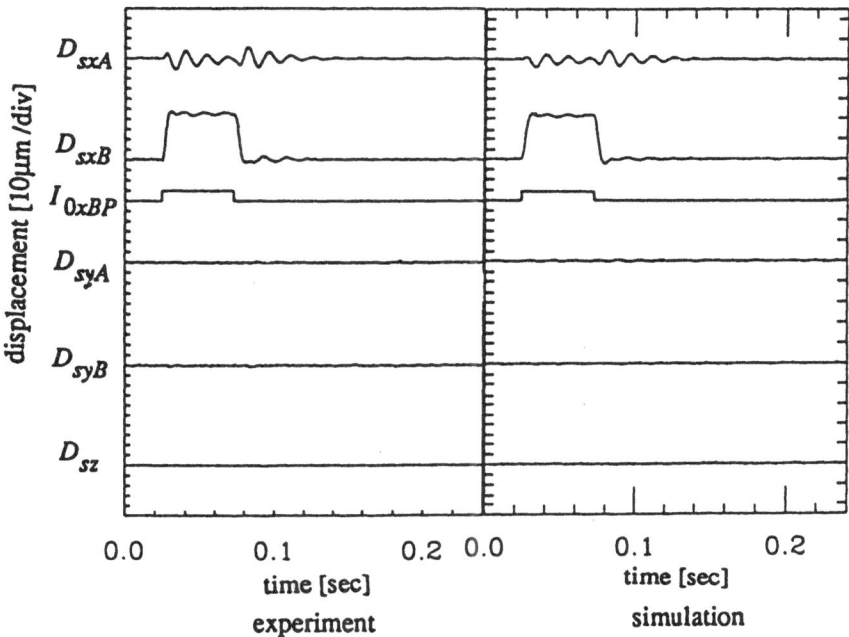

Figure 8. Transient response with sampling period 120 μsec.

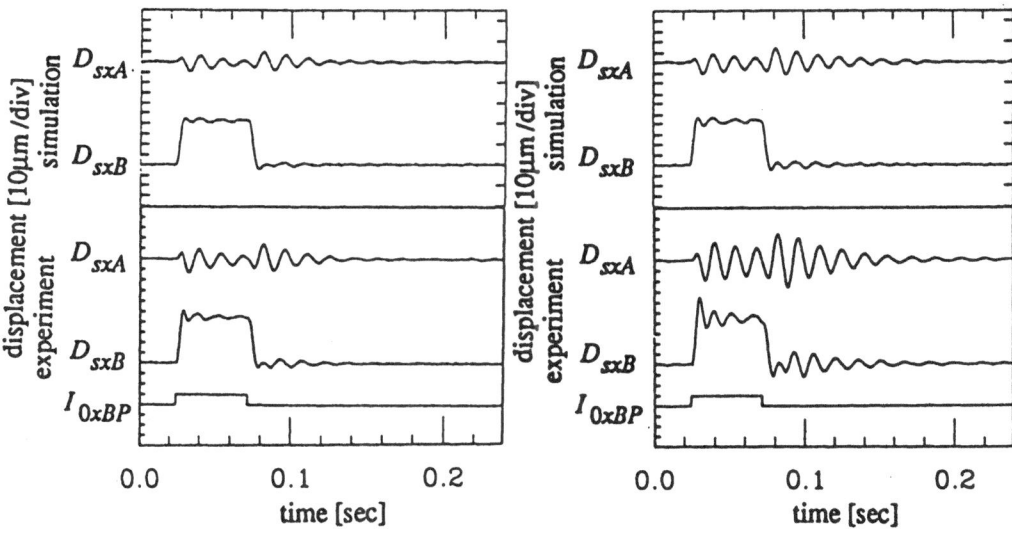

Figure 9. Transient response with sampling period 240 μsec.

Figure 10. Transient response with sampling period 360 μsec.

The above value of K_s corresponds with translational stiffness of 300 [N/mm] and rotational stiffness of 500 [N m/rad] around the center of gravity. We measured transient responses with sampling periods of 120, 240, 360 and 480 [μsec]. Figures 8, 9 and 10 show experimental and simulation results of sampling periods of 120, 240 and 360 [μsec] respectively. In the case of 480 [μsec], the levitation is unstable in both the experiment and the simulation. In the graphs, measured outputs of the A/D converts are plotted versus time. The simulations include all of the above experimental conditions and also assume that the outputs of the A/D converters have a random noise of ±1 bit (= ±1 [μm]) with the possibility of once per 10-sampling-periods. Figure 11 shows a simulation result of the continuous-time control system with all other conditions the same, while figure 12 shows a simulation result of the continuous-time linearized control system which does not include nonlinearity of the electromagnets.

Figure 13 shows a set of data on the same experiment of sampling period 240 [μsec] with different damping matrix P_s which is twice as large as the previous one . Figure 14 shows a result with a different stiffness matrix K_s which is

$$K_{sx} = K_{sy} = \begin{bmatrix} 7.22 & -6.26 \\ -7.22 & 6.26 \\ -7.44 & 15.41 \\ 7.44 & -15.41 \end{bmatrix}, \quad K_z = \begin{bmatrix} 9.47 \\ -9.47 \end{bmatrix}. \tag{18}$$

P_s is specified in equation (19). This K_s can be considered physically a stiffness with the same magnitude as the previous K_s and the different coordinate origin of point B (shown in figure 5.) In this control system, when external lateral force acts at point B, the levitated-unit, in the static state, should have only translational displacement without rotational displacement.

This series of experiments verified that the digital controller can control magnetic servo levitation with high stiffness and rapid response in the sufficiently short sampling period. In addition, the mathematical model has proven to correspond accurately with the practical system, as predicted. It is also shown that magnetic force can be modeled accurately enough by experimental data obtained in static conditions. Those results suggest good implementation of precision force/position control to magnetic servo levitation for robot mechanisms.

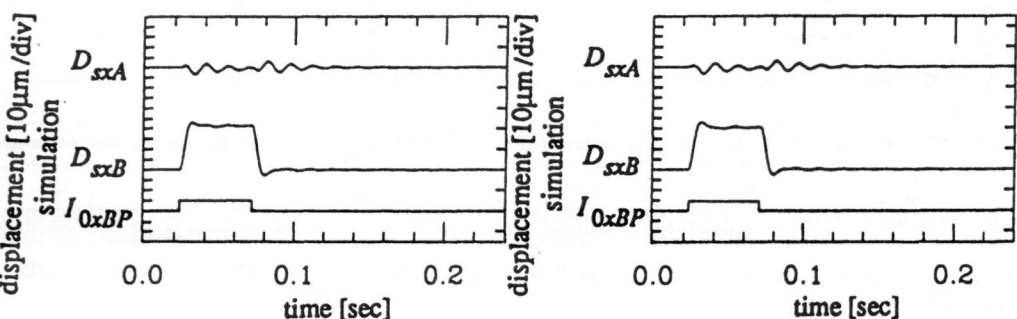

Figure 11. Transient response in continuous time control with nonlinearity.

Figure 12. Transient response in continuous time control of linear model.

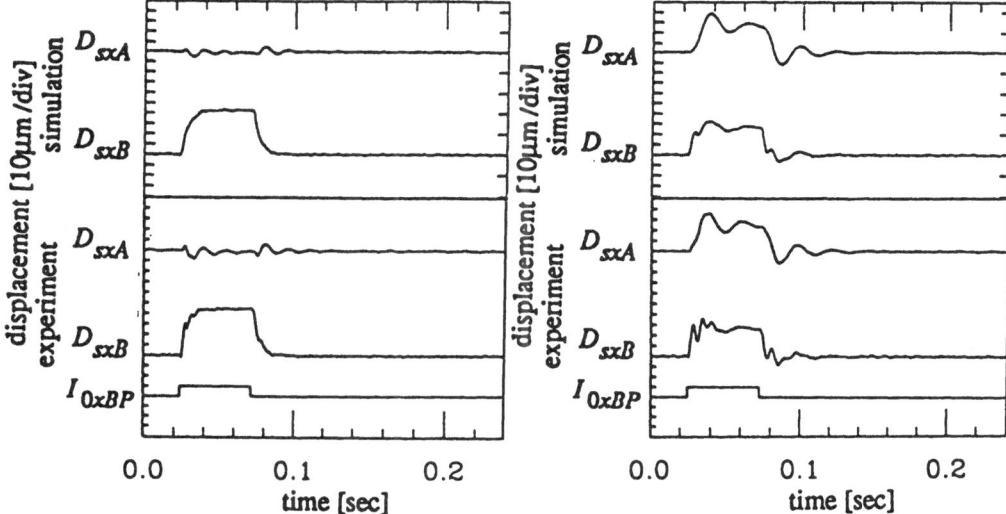

Figure 13. Transient response with twice as large damping gain in sampling period 240 μsec.

Figure 14. Transient response with the different coordinate origin of stiffness.

4.3. Experimental Verifications of Magnetic Servo Levitation

We verified, by using the MSIH, that magnetic servo levitation can be a precision multi-DOF actuator with programmable compliance and simultaneously can be a multi-axis sensor of force and moment [10, 14]. The resolutions of the sensing have proven to be less than 0.05 [N] and 0.001 [N m]. The resolution of the actuation is almost equal to 1 [μm]. As a typical experimental result, figure 15 shows output of the force/moment-sensing with several known loads applied to the levitated object. The solid line represents the predicted outputs. The O's and the □'s show actual output of force-and-moment-sensing respectively.

The experiments have demonstrated that magnetic servo levitation is achievable by a state-of-the-art digital controller.

5. Conclusion

(1) Active magnetic levitation, having a friction-free and backlash-free structure, has experimentally been proven to work as a precision force/position-controllable actuator with force-sensing capability. We call such magnetic levitation *"Magnetic Servo Levitation."*

(2) A high-speed digital controller is essential to realize magnetic servo levitation. We developed a digital controller using a digital signal processor. We experimentally verified that the digital controller has sufficient ability of high-speed control for magnetic servo levitation with high stiffness and rapid response needed for robot applications.

(3) We experimentally demonstrated that magnetic servo levitation involves no mechanical uncertainties which interfere with precision force/position control and has a strong potential to improve the performance of robots.

Acknowledgement

The experiments shown in this paper were done at Institute of Industrial Science, University of Tokyo. The theoretical analysis was done at both the Center for Robotic Systems in Microelectronics, University of California, Santa Barbara, and the IIS. This material was partially supported by the National Science Foundation under Cooperative Agreement number 8421415. Any opinions, findings, conclusions, or recommendations expressed in this publication are those of the authors and do not necessarily reflect the views of the Foundation.

The authors would like to thank Saku Egawa of IIS and Shigeki Fujiwara of Matsushita Electric Works, Co., Ltd. for their valuable support during the development of this work.

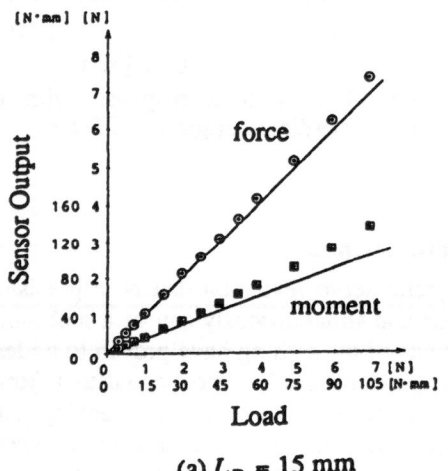

(a) $L_P = 15$ mm

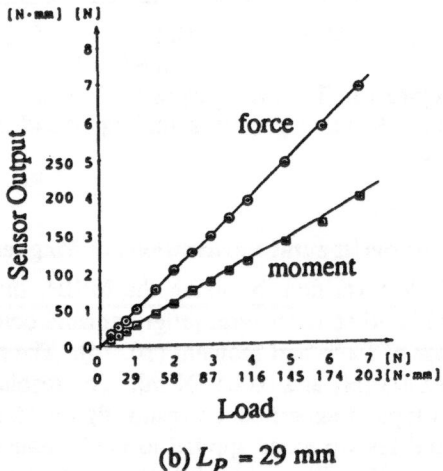

(b) $L_P = 29$ mm

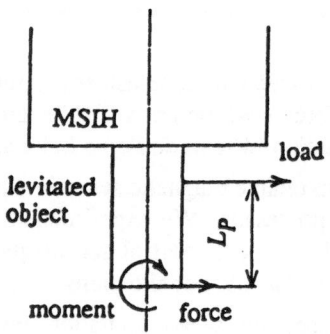

Figure 15. Force/moment sensing by magnetic servo levitation. O's and □'s show measured force and moment respectively. Solid lines indicate the predicted outputs of the sensing.

References

[1] K. Takahara et al., "Development of a Magnetically Suspended Tetrahedron-Shaped Antenna Pointing System," NASA CP-2506. 22nd Aerospace Mechanisms Symp. 1988, pp. 133-147.

[2] S. Iwaki and R. Matsuda, "Testing and Investigation of Magnetically Suspended APM," Proc. of Int. Space Technology and Science 86, 1986, pp. 787.

[3] T. Higuchi, M. Tsuda and S. Fujiwara, "Magnetic Supported Intelligent Hand for Automated Precise Assembly," Proc. of 1987 Int. Conf. on Industrial Electronics, Control, and Instrumentation, SPIE Vol. 857, 1987, pp. 926-933.

[4] R. L. Hollis, A. P. Allan, and S. Salcudean, "A Six Degree-of-freedom Magnetically Levitated Variable Compliance Fine Motion Wrist," Proc. of 4th Int. Symp. on Robotics Research, 1987, pp. 241-249.

[5] T. Higuchi, K. Oka and H. Sugawara, "Development of Clean Room Robot with Contact-less Joints using Magnetic Bearings," Proc. of USA-Japan Symp. on Flexible Automation, 1988.

[6] T. Higuchi and H. Kawakatsu, "Development of Super-Clean Actuator for Machines and Robots," Proc. of 12th MOTOR-CON'88, 1987, pp. 322-334.

[7] O. Khatib, "Augmented Object and Reduced Effective Inertia in Robot Systems," Proc. of 1988 American Control Conference, 1988, pp. 2140-2147.

[8] S. Salcudean and Chae An, "On the Control of Redundant Coarse-Fine Manipulators," Proc. of 1989 IEEE Int. Conf. on Robotics and Automation, 1989, pp. 1834-1840.

[9] H. Ulbrich and G. Schwetzer, "A Rotor Supported Without Contract - Theory and Application," Proc. of 5th World Congress on Theory of Machines and Mechanisms, 1979, pp. 181-184.

[10] M. Tsuda, "Development of Magnetically Supported Intelligent Hand," MS thesis, Department of Precision Mechanical Engineering, University of Tokyo, 1988, (in Japanese.)

[11] M. Tsuda, Y. Nakamura and T. Higuchi, "Design of Magnetic Servo Levitation for Robot Mechanisms," Proc. of 20th Int. Symp. on Industrial Robots, 1989.

[12] T. Chikada and K. Furuta, "Computer Control of a Magnetic Levitation System," Trans. Society of Instrument and Control of Engineers, Vol. 17, No. 7, 1981, pp. 713-720, (in Japanese.)

[13] M. Hisatani Y. Inoue and J. Mitsui, "Development of Digitally Controlled Magnetic Bearing," Trans. Japan Society of Mechanical Engineers, Vol. 51 No. 465, 1985, pp. 1095-1100, (in Japanese.)

[14] M. Tsuda, T. Higuchi, S. Fujiwara, "Functions of Magnetically Supported Intelligent Hand for Automatic Precision Assembly," Proc. of 20th Int. Symp. on Industrial Robots, 1989.

HYBRID POSITION FORCE CONTROL OF ROBOT MANIPULATOR WITH AN INSTRUMENTED COMPLIANT WRIST

Yangsheng Xu

Richard P. Paul

Peter I. Corke **

General Robotics Active Sensory Perception Laboratory
Department of Computer and Information Science
University of Pennsylvania
Philadelphia, PA 19104

ABSTRACT

A six DOF compliant wrist which combines passive compliance and active sensing has been developed to provide the necessary flexibility for force and contact control as well as being accurately controllable in position. This paper describes the compliant wrist and sensing mechanism design. Utilizing the sensed information from this wrist allows the apparent stiffness of the end-effector to be increased in unconstrained mode and decreased in constrained modes where the contact force is controlled. An active feedback control scheme for both position and force control is presented and the dynamic behavior of the system is analyzed. As the manipulator is partially constrained by the environment, a new hybrid control algorithm with consideration of passive compliance is proposed. The applicability of the method is demonstrated by experimental results. The entire system performance is analyzed under conditions with variable environment characteristics, controller parameters, and contact force, etc. A sinusoid surface tracking experiment was performed as an application of the hybrid control scheme and some useful results were obtained. The velocity discontinuity as the robot makes contact with, or breaks from, the environment is accommodated by the passive compliance of the wrist. The method shown is simple, economical, and applicable in industry.

This material is based on work supported by the National Science Foundation under Grant No. DMC-8512838. Any opinions, findings, and conclusions or recommendations expressed in this publication are those of the authors and do not necessarily reflect the views of the National Science Foundation.

** Research Scientist, CSIRO Division of Manufacturing Technology, Melbourne, Australia.

1. INTRODUCTION

When robots are used in operations where end effectors contact the environment, compliance is beneficial in allowing external constraints to modify the trajectory. Considerable attention has been directed to compliant motion of robot manipulators in this decade. We may categorize currently available compliant motion control techniques into two basic types. Firstly, active compliance is specified in the joint servo either by setting a linear relation between the force and displacement (or velocity and displacement) such as impedance control [14], damping control [24], stiffness control [12], or by controlling force in certain degrees while controlling position in the remaining degrees, such as compliance control [19], compliance and force control [11], hybrid control [15]. Secondly, passive compliance is provided by a compliant element near the end-effector, usually incorporated into a wrist, hand, or fingers. The most well known example is the Remote Center of Compliance (RCC) [4] although many different versions have been developed in Japan [20][21], France [16][22], West Germany [18] and USA [13][23].

There are fundamental problems for both techniques. For active compliance, an instability problem in a stiff environment is observed, thus a passive compliance installed in the end-effector is desirable to reduce the overall system stiffness. Passive compliance also possesses other advantages such as accommodating geometric uncertainties and dimensional tolerances, reducing the high forces or moments usually caused by assembly or other contact operations, and avoiding costly electronic instruments normally required in precision manufacturing. Using passive compliance alone, however, the positioning capability of robot is degraded. Based on these two main problems, many papers have been published recently [8][9][10][17], but a simple, economical and reliable method is still demanded so that compliant motion of the robot manipulators can be finally implemented in industry.

In this paper, we propose to use a passive compliance mechanism with six DOF compliance which is also capable of measuring the six DOF deflection within the device, that is, between the end-effector and robot wrist. Passive compliance can correct the position error automatically and allow relaxed tolerances, as well as accommodating the transition between the position and force control modes. The sensed deflection in the wrist can be used for feedback control such that the entire system is controllable.

Such a device, with both a passive compliance and active sensing mechanism, was developed in our laboratory. The passive compliance consists of a compliant rubber element yielding compliance along, and about, all axes. The device is instrumented by providing a simple six joint serial linkage with potentiometer sensors at its joints. The sensing information is used in two ways. In position control, the sensed information is utilized to compensate deflection of the wrist, due to the load or external forces, in such a way as to increase the apparent stiffness of the manipulator wrist system. In force control, the wrist is used as a force sensor by which means the manipulator is driven in the same direction as the sensed force allowing the desired contact force to be maintained.

The rest of the paper is organized as follows. In the first section, the compliant wrist device combining passive compliance and a sensing mechanism is described and the kinematics of the mechanism is presented. In the second section, position compensation of the robotic system in the presence of a passive compliance is investigated. Two control algorithms, operating in Cartesian space and joint space, are proposed. The analytical and experimental results are illustrated. The third section presents a force control scheme which is demonstrated by experiment and analysis. The fourth section develops a hybrid control strategy based on the previously discussed position and force control schemes. The fifth section describes a sinusoid surface tracking experiment as an application of the hybrid control scheme and some useful results are obtained. The sixth section analyzes the system performance and the effect of controller parameters such as gain and the digital filter on the entire system. Finally we present our conclusions.

2. PASSIVE COMPLIANCE AND SENSING MECHANISM

The compliance device includes two parts: a special rubber element acting as a damped compliance, and a sensing mechanism to measure the deflection of the device during end-effector motion or contact. The sensor mechanism is capable of measuring six DOF motions of the upper plate of the device relative to the lower one. The mechanism is a serial linkage with a potentiometer sensor at each of its six joints, instead of six actuators as for a manipulator. The task of computing from the sensed joint data is simply the direct kinematics so that the end-point motion is identified. We at first, intended to use a parallel mechanism as in paper [6], and use LVDTs as displacement sensors. However, the direct kinematics is difficult for a parallel mechanism, while inverse kinematics is easy. On the contrary, for a serial mechanism the direct kinematics is much easier than the inverse [7], so this type of mechanism was chosen. Additionally the serial linkage was easier to fabricate than the parallel one. A disadvantage is the error accumulation of a serial mechanism, while a parallel mechanism compensates for error. We, however, carefully calibrate the mechanism and potentiometers and filter the data, so the designed precision of the device is obtained. The device features are shown in Figure 1(a) and the mechanism's kinematic skeleton is shown in Figure 1(b) with coordinate frame's assigned to the links.

$$T_w = Trans\,(-l_7, l_3, l_1,)\;Rot\,(z\,, \theta_1)\;Trans\,(-l_2, 0, 0)\;Rot\,(x\,, \theta_2)\;Trans\,(0, -l_3, 0)\;Rot\,(x\,, \theta_3)$$

$$Trans\,(l_4, -l_5, 0)\;Rot\,(z\,, \theta_4)\;Trans\,(0, 0, l_6)\;Rot\,(y\,, \theta_5)\;Trans\,(0, l_5, 0)\;Rot\,(z\,, \theta_6)\;Trans\,(l_7, 0, l_8)$$

$$(1)$$

where T_w is the transformation of the wrist from lower plate to top plate. The A transformation matrices for the wrist device are as follows.

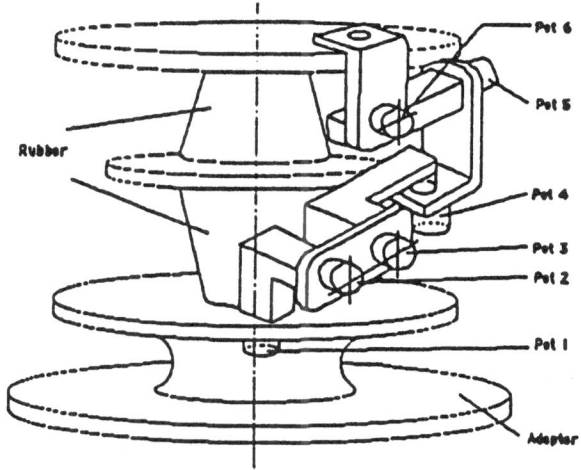

Fig. 1(a)   Mechanical structure of the compliant wrist

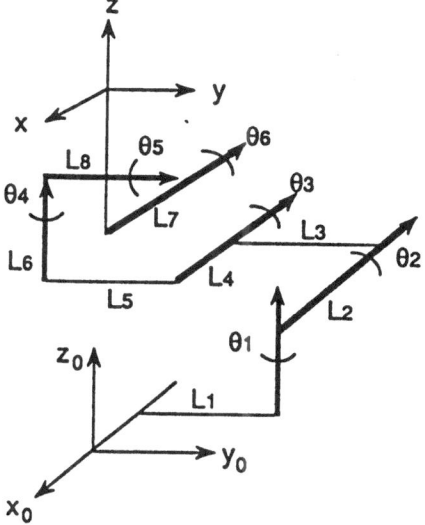

Fig. 1(b)   Kinematic skeleton of the wrist sensing mechanism

$$A_1 = Trans(-l_7, l_3, l_1)\, Rot(z, \theta_1) = \begin{bmatrix} C_1 & -S_1 & 0 & -l_7 \\ S_1 & C_1 & 0 & l_3 \\ 0 & 0 & 1 & l_1 \\ 0 & 0 & 0 & 1 \end{bmatrix}$$

$$A_2 = Trans(-l_2, 0, 0)\, Rot(x, \theta_2) = \begin{bmatrix} 1 & 0 & 0 & -l_2 \\ 0 & C_2 & -S_2 & 0 \\ 0 & S_2 & C_2 & 0 \\ 0 & 0 & 0 & 1 \end{bmatrix}$$

$$A_3 = Trans(0, -l_3, 0)\, Rot(x, \theta_3) = \begin{bmatrix} 1 & 0 & 0 & 0 \\ 0 & C_3 & -S_3 & -l_3 \\ 0 & S_3 & C_3 & 0 \\ 0 & 0 & 0 & 1 \end{bmatrix}$$

$$A_4 = Trans(l_4, -l_5, 0)\, Rot(z, \theta_4) = \begin{bmatrix} C_4 & -S_4 & 0 & l_4 \\ S_4 & C_4 & 0 & -l_5 \\ 0 & 0 & 1 & 0 \\ 0 & 0 & 0 & 1 \end{bmatrix}$$

$$A_5 = Trans(0, 0, l_6)\, Rot(y, \theta_5) = \begin{bmatrix} C_5 & 0 & S_5 & 0 \\ 0 & 1 & 0 & 0 \\ -S_5 & 0 & C_5 & l_6 \\ 0 & 0 & 0 & 1 \end{bmatrix}$$

$$A_6 = Trans(0, l_5, 0)\, Rot(z, \theta_6) = \begin{bmatrix} 1 & 0 & 0 & 0 \\ 0 & C_6 & -S_6 & l_5 \\ 0 & S_6 & C_6 & 0 \\ 1 & 0 & 0 & 1 \end{bmatrix}$$

$$A_7 = Trans(l_7, 0, l_8) = \begin{bmatrix} 1 & 0 & 0 & l_7 \\ 0 & 1 & 0 & 0 \\ 0 & 0 & 1 & l_8 \\ 0 & 0 & 0 & 1 \end{bmatrix}$$

The products of the A transformations for the the device can be evaluated starting at the upper plate and working back to the robot end-plate. The determinant of the Jacobian matrix was calculated to investigate singularities. The results show that there is no singularity around the home position of the linkage.

The rubber element is chosen because the stiffness of rubber and its shape is such as to yield reasonable stiffness in each direction. Also, from a stability analysis, some damping in the device is necessary as the damping ratio of the system is critical for system performance [1]. The rubber material in the device provides significant inherent damping. The stiffness in each direction was measured, and the results are listed in the Table below. The stiffness of the device can be represented in matrix form as

$$K_w = \begin{bmatrix} K_{ll} & 0 & 0 & 0 & K_{bl} & 0 \\ 0 & K_{ll} & 0 & -K_{bl} & 0 & 0 \\ 0 & 0 & K_{aa} & 0 & 0 & 0 \\ 0 & -K_{lb} & 0 & K_{bb} & 0 & 0 \\ K_{lb} & 0 & 0 & 0 & K_{bb} & 0 \\ 0 & 0 & 0 & 0 & 0 & K_{tt} \end{bmatrix}$$

where

K_{ll}: Lateral force/lateral displacement;

K_{aa}: Axial force/axial displacement;

K_{tt}: Torsional torque/torsional angle;

K_{bb}: Bending torque/bending angle;

K_{bl}: Bending torque/lateral displacement;

K_{lb}: Lateral force/bending angle.

Compliant Wrist Stiffness Characteristics					
K_{ll}	K_{aa}	K_{tt}	K_{bb}	K_{bl}	K_{lb}
(N/m)	(N/m)	(N-m/rad)	(N-m/rad)	(N)	(N/rad)
441.00	5512.5	0.056	0.34	43.09	1.19

We introduced passive compliance in each of six directions instead of the three or five directions of most passive compliance devices. The reason is that the device is used not only to correct lateral and torsional errors in assembly operations but also to absorb kinetic energy when the robot tool stops suddenly upon making contact with the environment. In this way the transition between force and position control is accommodated. The positioning capability will not be degraded because of active sensing and compensation in the feedback loop.

3. POSITION CONTROL

Experiments with the compliant wrist shown in Figure 2 were performed on a PUMA 560. Before the experiment, all six potentiometers were adjusted in a proper range and the compliant wrist sensor was calibrated carefully. The control was executed on a Microvax 2 using the RCI primitives of RCCL [5], which allowed the software to directly command robot joint angles. The software package allowed various parameters to be set, and also allowed trajectory and wrist displacement data to be logged to a file for subsequent analysis.

The objective of position control of the robotic system, including a compliant wrist, is to increase the apparent stiffness of the device by moving the manipulator in the opposite direction to the sensed displacement. In other words, the deflection of the compliant wrist due to load or external forces is compensated and the static positioning accuracy of robot manipulator is maintained. A detailed analysis for selecting controller parameters, as well as the effect of the compliant wrist on the system performance, can be found in our earlier papers [1][2].

Two different control schemes, control in Cartesian coordinates and control in joint coordinates, are investigated for position control.

We first discuss the control scheme in Cartesian coordinates. We define the transformation from the base coordinates to the lower plate of the compliant wrist device as T_6, that from the lower plate to the upper plate of the compliant wrist as T_w, and that from the base to the upper plate of the compliant device as B which is considered as the task coordinate transformation. The kinematic relation at the initial state is

$$T_6 T_w = B \qquad (2)$$

Suppose at the current state the compliant wrist coordinate frame T_w is changed to T_w' due to load or other external force. The task coordinate transformation is thus changed to B' and the kinematic relation becomes

$$T_6 T_w' = B'$$

In order that the positioning ability of the robotic system is retained, it is our aim that the robot coordinate transformation T_6 be modified to T_6' such that the task coordinate transformation B remaining unchanged. Therefore, the control goal is

$$T_6' T_w' = B \qquad (3)$$

Equating (2) and (3) yields

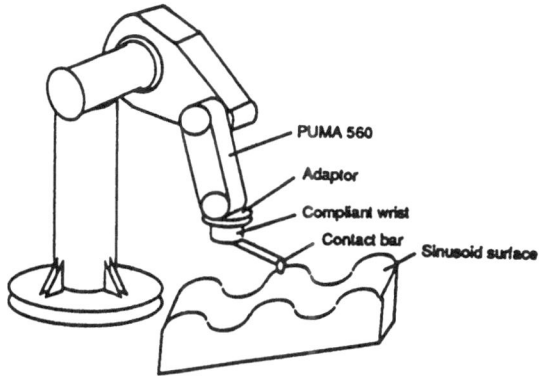

Fig. 2   Experiment with the compliant wrist on PUMA 560

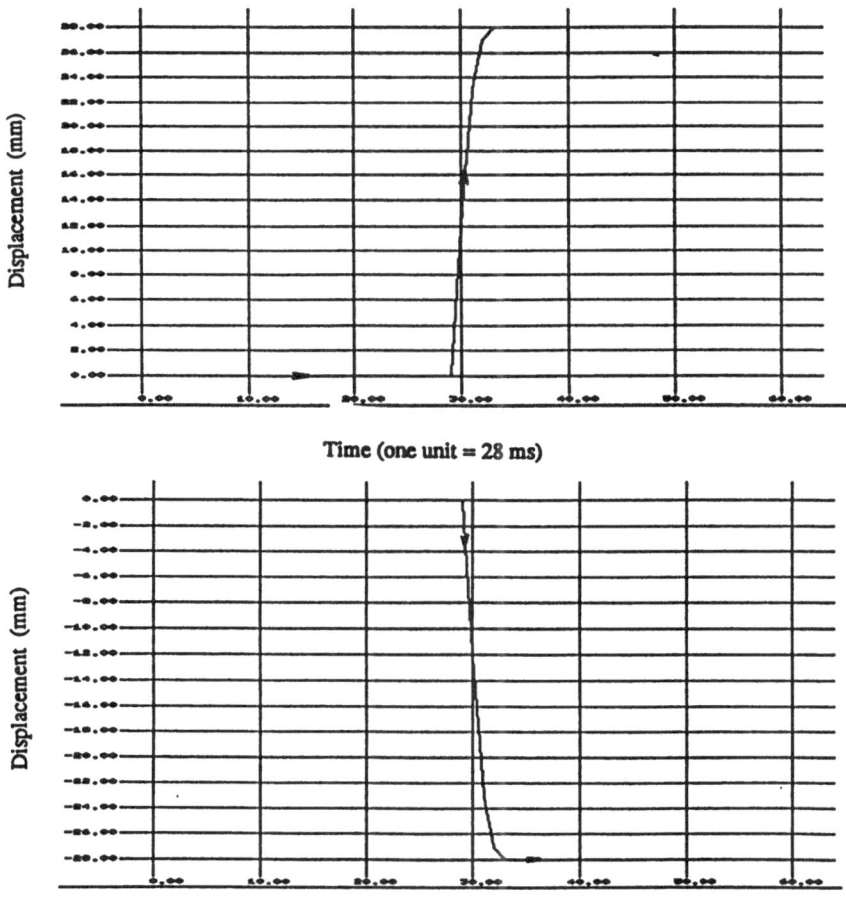

Fig. 3   Deflection reading of the wrist and manipulator end-point response in position compensation

$$T_6' T_w' = T_6 T_w$$

or,

$$T_6' = T_6 T_w (T_w')^{-1} \tag{4}$$

An alternative, which we use, is joint rate control utilizing the differential displacement of the compliant wrist. There are two ways to obtain a six components generalized differential displacement vector ΔX_w, i.e., three position displacements and three orientations (relative to the initial position where the deflection is zero) from the sensing information. Firstly, ΔX_w may be extracted from the updated transformation matrix of the compliant wrist T_w. Secondly, these six differential displacements may also be calculated from the wrist mechanism Jacobian matrix J_w and $\Delta\theta_w$ which both depend upon sensor joint angles.

$$\Delta X_w = J_w \Delta\theta_w \tag{5}$$

Joint rate control of the manipulator is defined as

$$\Delta X = J_m \Delta\theta_m \tag{6}$$

where J_m and $\Delta\theta_m$ are the Jacobian matrix and joint differential change of the manipulator respectively. Since compensation is desired in position control so that the manipulator moves in the opposite direction to the wrist displacement, ΔX in (6) must be the same as ΔX_w in (5) but with the opposite sign if complete compensation is desired. Therefore,

$$\Delta\theta_m = -J_m^{-1}\Delta X_w \tag{7}$$

More generally we may introduce a gain matrix K_P so that the desired joint command vector θ_{des} becomes

$$\theta_{des} = \theta_{traj} + \Delta\theta_m = \theta_{traj} - J_m^{-1} K_P \Delta X_w \tag{8}$$

where θ_{traj} is the desired joint angles, supplied by a trajectory generator function.

An experiment consisted of applying an incremental change of load in or around a certain direction. The six DOF wrist deflections and six DOF manipulator endpoint were recorded. The position response shown in Figure 3 are the results as the load is applied around the X direction, with a gain K_p of 1.0. The upper curve is the observed wrist angle deflection around the X direction θ_x and the lower curve is a recording of the manipulator endpoint response in the same direction. From these curves, we can see that the manipulator moves in the opposite direction but by the same amount as the wrist deflection so that the absolute endpoint location is maintained.

The experimental results indicate that the use of active sensing of the compliant wrist makes it possible to retain the original static stiffness characteristics of the manipulator in spite of the presence of a passive compliance in the wrist.

We may analyze the system dynamic performance with a simple single degree of freedom system as shown in Figure 4. The end-effector is represented as a mass m and the compliant wrist is represented as a spring with stiffness K_w and a viscous damping C_w which are attached to the manipulator end-point. The manipulator is assumed to be rigid. The end-effector motion is X_2 and the manipulator end-point motion is X_1. The compliant sensor records the difference of the motions $\Delta X = X_2 - X_1$. The controller uses both the error information ΔX with the end-point command X_{2c} to drive the end-point motion X_1 in such a way that the end-effector motion X_2 is not affected by the external force F_{ext} in position control.

This simple system and controller is shown in block diagram form in Figure 5. The feedback loop can be switched to a negative gain for position control or a positive gain for force control (which we will discuss later). The transfer function from X_1 to X_2, G_1, and from the external force F_{ext} to X_2, G_2, may be obtained from the system of Figure 4.

$$G_1(s) = \frac{X_2(s)}{X_1(s)} = \frac{C_w s + K_w}{ms^2 + C_w s + K_w} \tag{9}$$

$$G_2(s) = \frac{X_2(s)}{F_{ext}(s)} = \frac{1}{ms^2 + C_w s + K_w} \tag{10}$$

In position control, the system function is

$$(1 - K_P) G_2 F_{ext} + G_1 X_{2c} = (1 - K_P + K_P G_1) X_2 \tag{11}$$

To make a comparison, the system function for Figure 4 without active feedback control may also be written as

$$G_2 F_{ext} + G_1 X_{2c} = X_2 \tag{12}$$

From (11), it is clear that the effect of the external force is scaled by $(1 - K_P)$. If $K_P = 1$, the effect of an external force on end-point position can be eliminated.

The system's dynamic stiffness $Z(s)$ can be derived from (11) if the second term is not considered.

$$Z(s) = \frac{F_{ext}(s)}{X_2(s)} = ms^2 + \frac{C_w s}{1 - K_P} + \frac{K_w}{1 - K_P} \tag{13}$$

For the case without feedback control (12), it is

$$Z(s) = \frac{F_{ext}(s)}{X_2(s)} = ms^2 + C_w s + K_w \tag{14}$$

Therefore the virtual stiffness in steady state is increased by $1/(1 - K_P)$ due to introduction of active feedback, thus providing a better positioning ability despite the presence of passive compliance.

The system's characteristic equation from (11) for position control mode is

$$(1-K_P)ms^2 + C_w s + K_w = 0 \tag{15}$$

Constraining the feedback gain $K_P < 1$ in position control ensures the system stability from (15).

4. FORCE CONTROL

In the force control case, where the end-effector is partially constrained by the workpiece, we use the compliant wrist as a sensor to detect the force exerted on the end-effector. The sensed deflection is used to drive the manipulator in the same direction as the deflection of the wrist so that the apparent stiffness is decreased and the desired contact force is obtained.

We again utilize a joint rate control scheme for force control. As discussed above, we can obtain the six component generalized displacement of the wrist ΔX_w from either the wrist Jacobian matrix or the transformation matrix.

The manipulator rate control scheme from (6) is

$$\Delta\theta_m = J_m^{-1}\Delta X_m \tag{16}$$

Since in our method, the manipulator is driven to a certain displacement in response to the sensed force, this force control scheme is actually controls the compliance or stiffness of the system, rather than force. The stiffness to be controlled is determined by the ratio of the sensed force to the displacement response of the system. This displacement ΔX relates to the exerted force F_w and the desired stiffness K_d

$$F_w = K_d\,\Delta X \tag{17}$$

whereas measured displacement ΔX_w relates to the exerted force F_w by

$$F_w = K_w\,\Delta X_w \tag{18}$$

where K_w is the actual stiffness of the wrist. Substituting yields

$$\Delta X = K_F\,\Delta X_w \tag{19}$$

where K_F is a dimensionless stiffness ratio

$$K_F = K_d^{-1}K_w \tag{20}$$

Substituting Equation (19) to Equation (16) results in

$$\Delta\theta_m = J_m^{-1}K_F\Delta X_w \tag{21}$$

The desired joint angles thus are

$$\theta_{des} = \theta_{curr} + \Delta\theta_m = \theta_{curr} + J_m^{-1}K_F\Delta X_w \qquad (22)$$

where θ_{curr} is the current joint angles.

Comparing the control algorithm in position control (8) with that in force control (22), it is worthwhile noting the following points.

(1) The control algorithms are very similar, and both contain a dimensionless gain matrix K_P or K_F, but with a different sign in front. The former represents the gain controlling how much of the deflection is to be compensated in position control. The latter represents the gain relating the natural stiffness to the effective stiffness of the system which may be determined based on different force control tasks. If complete compensation in and around each direction is required in position control, the gain K_P is an identity matrix. If the desired compliance level is as same as the natural compliance K_w, which is mainly contributed by the passive compliant mechanism of the wrist, the gain matrix K_F is again an identity matrix.

(2) In position control, the end-effector must be driven in the opposite direction to the displacement measured, but in the same direction for force control. Therefore, the updated differential displacement $\Delta\theta_m$ is negative in Equation (8) while positive in Equation (22). As a result, the overall stiffness of the system is increased in position control mode and decreased in force control mode.

(3) Provided the end-effector is in steady-state and a constant deflection (i.e, constant force) exists in the compliant wrist, the manipulator should keep moving in the force control mode till the specified contact force is obtained, while it should stop if a constant compensation has been achieved to in position control mode. Therefore, the desired joint angles θ_{des} should be based on the specified joint angles θ_{traj} in position control (8), but based on the current joint angles θ_{curr} in force control (22).

In experiments, a constant force was applied suddenly, simulating the end-effector coming into contacted with the environment. Figure 6 shows the deflection of the wrist and the endpoint response of the manipulator, provided the desired force level is zero. When the force was applied, the wrist deformed, and the manipulator end-point moved in the same direction as the measured force. As the manipulator reached the desired force level, specified as zero, the wrist deflection becomes zero and the manipulator stops.

In force control, we also can analyze the system dynamics as we did for the position control case. The system block diagram is shown in Figure 5 where the feedback loop is switched to "force control".

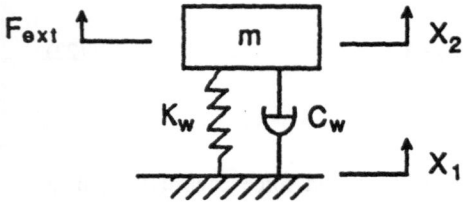

Fig. 4   Simple model of the wrist system

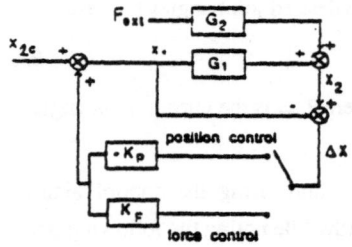

Fig. 5   Block diagram of the active feedback control

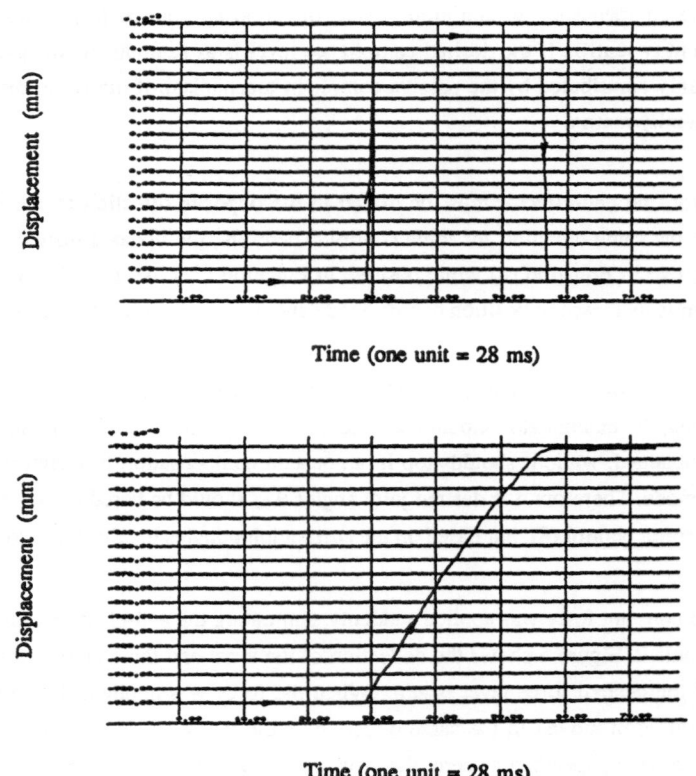

Time (one unit = 28 ms)

Time (one unit = 28 ms)

Fig. 6   Deflection reading of the wrist and manipulator end-point response in force control

(the desired force is zero)

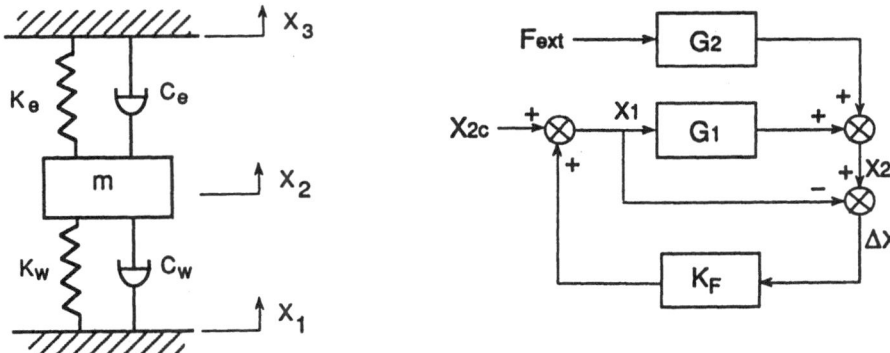

Fig. 7   Simple model of the wrist and onvironment system   Fig. 8   Block diagram of force control scheme.

The system function is

$$(1+K_F)G_2 F_{ext} + G_1 X_{2c} = (1+K_F - K_F G_1)X_2 \tag{23}$$

Here it is clear that feedback control causes the external force to dominate the system. The dynamic stiffness is

$$Z(s) = \frac{F_{ext}(s)}{X_2(s)} = ms^2 + \frac{C_w s}{1+K_F} + \frac{K_w}{1+K_F} \tag{24}$$

Comparing (24) to (14), the virtual stiffness of the system is decreased by $1/(1+K_F)$ times. A high gain K_F causes a decrease of stiffness of the closed-loop system. The system's characteristic equation is

$$(1+K_F)ms^2 + C_w s + K_w = 0 \tag{25}$$

which indicates that the system is invariantly stable in force control. However, as the gain K_F is increased, the effective damping ratio ζ is decreased

$$\zeta = \frac{2C_w}{[(1+K_F)mK_w]^{1/2}}$$

affecting the relative stability and system performance. Using different gains K_F, surface tracking experiments have been performed. The experiments showed that the force gain has to be limited so that the effective damping is adequate and oscillation is prevented.

Force gain is also dependent on the desired pole assignment of the closed-loop system. Since the physical and desired stiffness in each direction or around each axis is different, the force gains in the experiment were set differently for each axis.

If the environmental characteristics must be considered, the system can be modeled as a spring and a damper between the end-effector X_2 and a rigid hypothetic surface X_3 as shown in Figure 7. The stiffness and damping of the environment are represented as K_e and C_e. The problem can be interpreted as controlling the compliance between X_2 and X_3 so that the manipulator can move the end-effector X_1 and then drive the compliant wrist X_2 to respond to variation of X_3 produced by errors of surface geometry and material elasticity. In this way the compliance between X_2 and X_3 is maintained at the desired level. The control block diagram can be modified as shown in Figure 8 and the system transfer functions can be derived as follows.

$$G_1 = \frac{C_w s + K_w}{ms^2 + (C_e + C_w)s + (K_e + K_w)} \tag{26}$$

$$G_3 = \frac{C_e s + K_e}{ms^2 + (C_e + C_w)s + (K_e + K_w)} \tag{27}$$

$$\frac{X_2}{X_3} = \frac{(1+K_F)G_3}{1+K_F-K_F G_1} = \frac{C_e s + K_e}{ms^2 + (C_e + \frac{C_w}{1+K_F})s + (K_e + \frac{K_w}{1+K_F})} \tag{28}$$

Without feedback, the transfer function from X_3 to X_2 is actually G_3. Comparing G_3 in Equation (27) with Equation (28), it is clear that the system becomes more compliant than that without feedback. The feedback gain K_F modifies the stiffness and damping of the compliant wrist, i.e., K_w and C_w, thus modifying the effective stiffness and damping of the entire system. It is also apparent that, for different environmental stiffness, control gain may be selected differently. For a soft environment, a large gain can be selected, while for a stiff environment, only small gain is allowable for an equal output behavior. This conclusion verifies our primal rationale to install a passive compliance in a rigid robotic system, especially for a rigid environment.

5. HYBRID CONTROL

It is well known that every manipulator task can be broken down into elemental components that are defined by a particular set of contacting surfaces. A generalized surface can be defined in a constraint space having six degree of freedom, with position constraints along the normal to this surface and force constraints along the tangents. These two types of constraints, force and position, partition the degrees of freedom of possible end-effector motion into two orthogonal sets, that must be controlled according to different control strategies (8) and (22). Since the desired angles θ_{des} are based on θ_{traj} in position control (8), but based on θ_{curr} in force control (22), the hybrid control can not simply combine Equation (8) and (22). Here we present a hybrid control scheme for a robot system including an instrumented passive compliance.

At first, we partition ΔX_w which is the Cartesian error determined from the wrist sensor into two sets; ΔX_w^F corresponding to the direction in which force control is required, and ΔX_w^P in the remaining directions in which the position control is required. For example, if the force in the Z direction and the torques around the X and Y directions are controlled and the remaining directions are position controlled, we partition

$$\Delta X_w = \left[\Delta x \ \Delta y \ \Delta z \ \Delta \theta_x \ \Delta \theta_y \ \Delta \theta_z \right]^T \tag{29}$$

into

$$\Delta X_w^F = \left[0 \ 0 \ \Delta z \ \Delta \theta_x \ \Delta \theta_y \ 0 \right]^T \tag{30}$$

$$\Delta X_w^P = \left[\Delta x \ \Delta y \ 0 \ 0 \ 0 \ \Delta \theta_z \right]^T$$

If the desired force $\mathbf{F}_d$ is given, the desired displacement ΔX_d can be computed by

$$\Delta X_d = K_w^{-1} F_d \tag{31}$$

Multiplying by a gain matrix, the desired differential motion of the end-effector corresponding to position and force control schemes can be obtained.

$$\Delta X_P = K_P \ \Delta X_w^P \tag{32}$$

$$\Delta X_F = K_F \ (\Delta X_w^F - \Delta X_d) \tag{33}$$

to achieve the required position and force control.

From differential motions ΔX_P and ΔX_F we can form 4×4 differential transform matrices $T_{\Delta X_P}$ and $T_{\Delta X_F}$ respectively. When force control is considered in Cartesian space, the desired motion of the end-effector is $T_d^{(j)}$, where the superscript j refers to time.

$$T_d^{(j)} = T_d^{(j-1)} * T_{\Delta X_F} \tag{34}$$

Since we must consider the deflection of the end-effector in the presence of the passive compliance, the desired motion $T_d^{(j)}$ has to be modified by the differential motion $T_{\Delta X_P}$ which represents the deflection of the compliant wrist in (32). Therefore, the required motion $T_r^{(j)}$ to yield the desired motion $T_d^{(j)}$ is

$$T_r^{(j)} = T_d^{(j)} * T_{\Delta X_P}^{-1} \tag{35}$$

Thus, the end-effector not only provides the desired compliance in the specified degrees of freedom, but compensates simultaneously for the deflection of the compliant wrist.

We can also perform hybrid control in joint space. The joint rate control scheme from Equation (6) is

$$\Delta \theta = J_m^{-1} \Delta X \tag{36}$$

where J_m is the manipulator Jacobian, and $\Delta \theta$ and ΔX are the joint differential motions and the corresponding Cartesian differential motion of end-effector respectively. The desired joint angles of end-effector must move in the same direction as the differential joint angles caused by ΔX_F, based on the current desired joint angles as in Cartesian space control (34).

$$(\theta_{des})_j = (\theta_{des})_{j-1} + J_m^{-1} \Delta X_F \tag{37}$$

Also, the end-effector motion must be modified by the differential joint angles which represents the deflection of passive compliance ΔX_P in (32),

$$(\theta_{req})_j = (\theta_{des})_j - J_m^{-1} \Delta X_P \tag{38}$$

Equation (37) and (38) represents hybrid position and force control in joint space in correspondence with Equation (34) and (35) for Cartesian space.

An experiment based on this control scheme was performed. We specified the force along the Z axis and the torques around the X and Y axes while the remaining directions were position controlled. Force in an arbitrary direction was applied and the wrist deflection as well as the end-effector motion was recorded. The experimental curves in force control mode are identical with the curves shown in Figure 6 and those in position control mode are identical with the curves shown in Figure 3. The results were satisfactory and demonstrated that the control scheme is stable.

6. SURFACE TRACKING

A surface tracking experiment was performed as an application of the hybrid control strategy of the robotic compliant wrist system. The surface used is a sinusoid, curved as shown in Figure 2. The control software has no a priori knowledge of the surface and the end-effector trajectory is modified by sensed contact forces. The tool descends at a fixed velocity until it makes contact with the surface, upon which the controller is switched from full position control (i.e., six DOF position control) to the hybrid control scheme. The force normal to the surface is set to be controlled, while in other directions position is controlled. Desired contact forces, force feedback gain, position gain, and other parameters can be specified interactively.

Experiments have showed applicability of the method. Based on the experimental results, the effects of different parameters on the system performance have been investigated and summarized as follows.

(1) The force gain K_F must be selected carefully. As we expected, a high gain may be selected when the physical stiffness of the system decreases because of the environmental compliance and the wrist compliant mechanism. The primary rationale to introduce passive compliance in the robotic system is so that a much larger force gain can be used than that without. However, the force gain still has to be carefully selected according to the physical compliance of the wrist as well as that of environment. Too large a force gain may still cause instability. Figure 9 are the curves recorded for two translational motions Y and Z of the end effector as the end-effector is tracking a sinusoidal surface. We may see that in the direction of the force control, Z axis, the record is a sinusoid. Since the tracking is achieved by moving along the Y direction at a fixed velocity, the recording of Y direction motion is a line with a constant slope. The

upper curves are for force gain of 0.2, while the lower curves are for force gain of 0.6. It is clear that in the latter case, tracking does not accurately follow the surface.

(2) As the compliance of the system including robot and environment increases, so to does the maximum allowable gain. Since the stiffness component in the Z direction of the wrist is much larger than that in the X direction, the critical value of gain in the Z direction is much smaller than that in the X direction. This conclusion is confirmed from experimental results of tracking operations where force is controlled in the Z direction or in the X direction respectively.

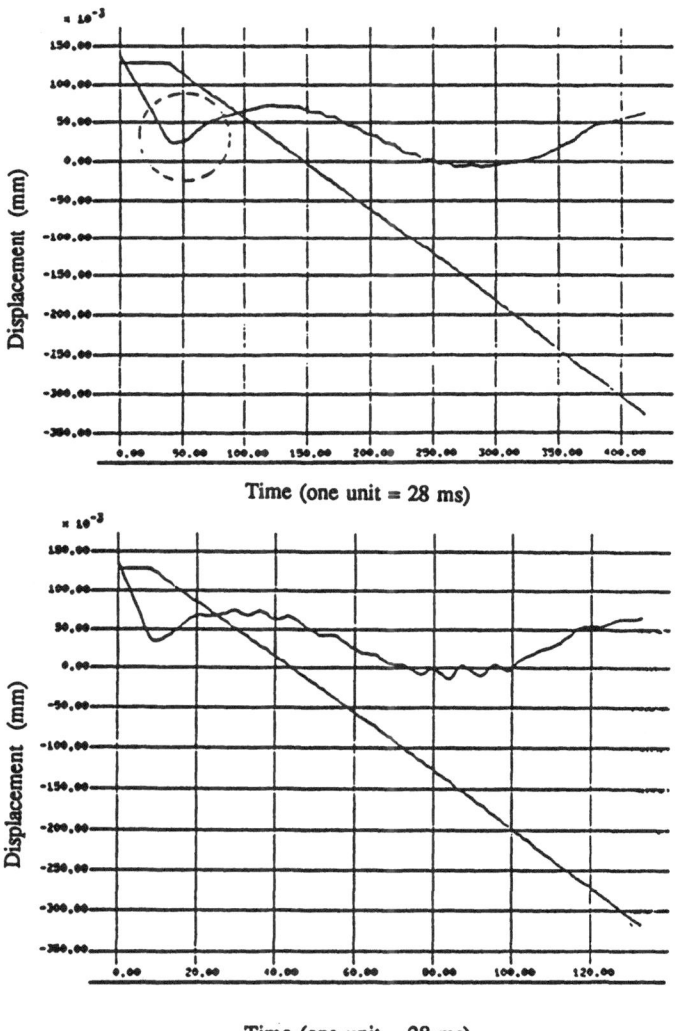

Fig. 9   Sinusoid surface tracking reading of the end-effector (in y and z direction)

(3) Transition as the robot makes or breaks contact with the surface can be accommodated. The passive compliance provides a mechanism to absorb the kinetic energy of impulsive force as the robot, moving at significant velocity, suddenly stops on the workpiece. In experiments, we used various velocities prior to contact and contact transition force is limited. The recording curve circled in Figure 9 shows a smooth contact between robot and workpiece before tracking.

(4) A reasonably large contact force is desirable. The experiment shows as specified contact force is increased, contact becomes smoother, as long as the deflection of the wrist is within a permitted range.

7. SYSTEM PERFORMANCE

We have found it is necessary to use a digital filter in the closed-loop of the system because of mechanical backlash in the potentiometers and other electronic noise. A first order digital filter with a unity-gain is utilized in the closed-loop system. The digital filter, in Z transform, is

$$F(z) = \frac{Y(z)}{U(z)} = \frac{1-a}{z-a} \tag{39}$$

The low pass corner frequency is controlled by the pole of the digital filter, a

$$a = e^{-\omega_- T} \tag{40}$$

where T is the sample interval, in our case 28 ms.

The system performance is dominated by the digital filter. The system block diagram without the filter in Figure 5 can be modified as in the diagram of Figure 10, where F denotes the digital filter. Using the bilinear relation

$$s = \frac{2}{T}(\frac{1-z}{1+z}) \tag{41}$$

and substituting (41) to (39) yields the transfer function of the filter in Laplace transform form

$$F(s) = \frac{(\frac{1-a}{1+a})(\frac{2}{T}-s)}{s + \frac{2}{T}\frac{(1-a)}{(1+a)}} \tag{42}$$

Let

$$a_0 = \frac{(1-a)}{(1+a)}, \qquad f_0 = \frac{2}{T} \tag{43}$$

thus

$$F(s) = \frac{a_0 f_o - a_0 s}{s + a_0 f_0} \tag{44}$$

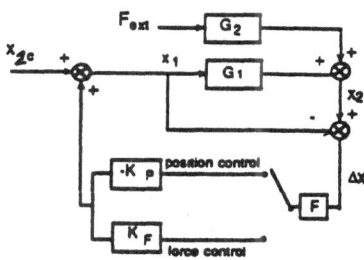

Fig. 10 Block diagram of the active feedback control with a digital filter

As the pole of the digital filter a becomes zero, a_0 goes to unity, and vice versa. The system transfer function from X_{2c} to X_2 in position control mode (11) becomes

$$\frac{G_1}{1-K_P F+K_P G_1 F} = \frac{C_w s+K_w}{(ms^2+C_w s+K_w)-K_P ms^2 F}$$

$$= \frac{(C_w s+K_w)(s+a_0 f_0)}{(ms^2+C_w s+K_w)(s+a_0 f_0)-K_P ms^2(a_0 f_0-a_0 s)} \tag{45}$$

The characteristic function is

$$(m+K_P a_0 m)s^3+(ma_0 f_0-K_P a_0 mf_0+C_w)s^2+(C_w a_0 f_0+K_w)s+K_w a_0 f_0 = 0 \tag{46}$$

and the stability conditions for position control mode is

$$ma_0 f_0-K_P ma_0 f_0+C_w > 0 \tag{47}$$

$$[ma_0 f_0(1-K_P)+C_w](C_w a_0 f_0+K_w) > K_w a_0 f_0(1+K_P a_0)m \tag{48}$$

From the stability condition, we may see:

(i) Passive damper is of significance. If the damping $C_w=0$, the second condition and first condition yield respectively

$$-1 > a_0 \qquad \text{and} \qquad K_p < 1$$

The former is contradictory and the latter means that full compensation is impossible.

(ii) The pole of the digital filter a should be selected as close to unity as possible. As a goes to unity, a_0 become zero, and the first and second conditions yield

$$C_w > 0, \qquad \text{and} \qquad C_w K_w > 0$$

which are invariantly true. However, as a goes to zero, i.e., a_0 goes to unity, the condition (48) results in

$$K_P < \frac{C_w(C_w + mf_0)}{m(2K_w + f_0 C_w)}$$

which restricts the gain selection of feedback control.

(iii) The gain K_p also dominates the system. As K_p is selected close to unity, both conditions are critical, and the relative stability condition is affected, especially when the system including environment is stiff and poorly damped. The condition (48), as K_P goes to unity, can be written as

$$K_w < \frac{C_w^2 a_0 f_0}{ma_0 f_0(1+a_0) - C_w}$$

Therefore, passive compliance is necessary to stabilize the system in this case.

Since the mass of the end-effector in our experiment is small and parameters K_P and a_0 are both less than 1, we neglect the third order term in (46), so that the system can be viewed as a second order model and its natural frequency ω_n can be obtained

$$\omega_n = \sqrt{\frac{K_w a_0 f_0}{(1 - K_P)a_0 mf_0 + C_w}} \qquad (49)$$

Compared to the natural frequency of the original system ω_{n0}

$$\omega_{n0} = \sqrt{\frac{K_w}{m}}$$

Equation(49) can be written in the form of

$$\omega_n = P_0 \omega_{n0} \qquad (50)$$

where the coefficient P_0 is the ratio of ω_n / ω_{n0}

$$P_0 = \sqrt{\frac{1}{(1 - K_P) + \frac{C_w}{a_0 mf_0}}} \qquad (51)$$

In general, the frequency coefficient P_0 is less than unity. As the damping coefficient C_w and the pole of digital filter a are decreased, or the gain K_P and mass m are increased, P_0 is increased. If P_0 approaches unity, oscillation may occur, while if P_0 approaches zero, the system response becomes slow. Therefore, the frequency coefficient P_0 represents the system performance to a certain extent.

To demonstrate the significance of the filter and passive damping on the system performance, let us suppose $K_P = 0.8$, $f_0 = 2000/28 \; s^{-1}$, and $m = 0.1kg$. The frequency coefficient P_0 can be computed in terms of the pole of the digital filter a and the damping coefficient C_w as shown in Figure 11 and Figure 12. From the plot in Figure 11, it is clear that the filter pole must be assigned close to unity, which results in a slow response. As the pole approaches the origin, the system becomes unstable. Therefore, we may conclude that the system bandwidth is limited by the natural frequency of the passive compliant mechanism

ω_{n0}. In other words, a soft passive compliance results in a slow system response. A compromise between the response time and the system compliance must be carefully made with respect to different operations. For most of experiments in position control, the frequency coefficient is set as around 0.25, and the first order natural frequency of the compliant wrist system is approximately 8 Hz, thus the closed-loop effective frequency is near 2 Hz, i.e., the response time is 0.5 sec.

Figure 12 depicts that the significance of passive damping to the system performance. Less damping may cause instability, while higher damping produces a slow response. Therefore, a proper selection of the damping in the device is important.

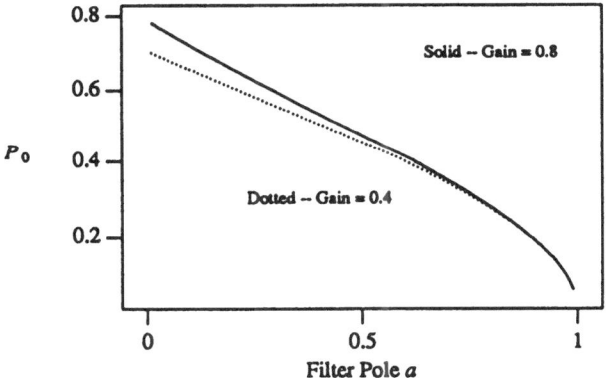

Fig. 11 The frequency coefficient P_0 varies with the pole of digital filter a

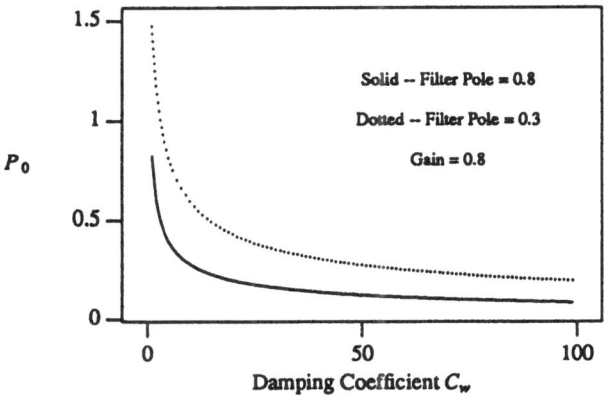

Fig. 12 The frequency coefficient P_0 varies with the wrist damping coefficient C_w

The surface tracking experiment demonstrated that for position control mode if the pole of the digital filter is lowered to a certain value, oscillation may occur. Using the same parameters, but changing the filter pole in position control mode, the experimental results are shown in Figure 13. The upper curve is for the case in which the pole is 0.95, while the lower curve is for the case in which the pole is 0.6 where oscillation is evident.

We also can investigate the system performance for force control mode with a similar analysis as shown above, provided a zero contact force is desired. For simplicity, we only give a stability condition.

$$K_F \, a_0 < 1$$

It is clear that if the force gain K_F and the pole of the digital filter are both selected to be less than unity, the system will be stable. The natural frequency in force control mode is

$$\omega_n = \sqrt{\frac{K_w a_0 f_0}{m a_0 f_0 (1+K_F) + C_w}} = Q_0 \omega_{n0} \tag{52}$$

where

$$Q_0 = \sqrt{\frac{1}{(1+K_F) + \dfrac{C_w}{m a_0 f_0}}} \tag{53}$$

Compare the frequency coefficient P_0 in position control (51) with Q_0 in force control (53). We may see that the condition is much improved in force control since the denominator in (53) always larger than unity, thus Q_0 cannot approach unity. However, in this case the gain is much smaller than unity ($K_F \ll 1$), and as the pole of filter a approaches unity ($a_0 = 0$), the second term of the denominator in (53) vanishes, and resonance may be induced. Therefore, the gain K_F must be selected as small as possible.

8. CONCLUSIONS

(1) The compliant wrist performs successfully both in passive compliance and active sensing. Using such a device makes it possible to provide the system with a flexibility which simplifies both contact force control and transient state control, and compensates the end-effector deflection due to the external forces.

(2) The applicability of the position and force control schemes is shown by experiment. In position control, the manipulator moves in the opposite direction to the deflection of the wrist so that the apparent stiffness of the end-effector is increased. In force control, the manipulator is driven in the same direction as the sensed force so as to decrease stiffness.

267

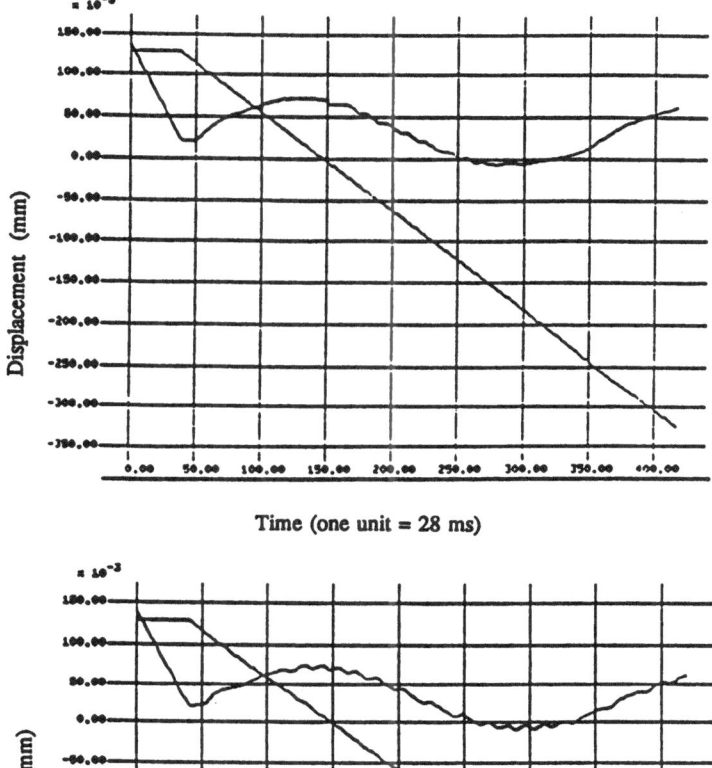

Time (one unit = 28 ms)

Time (one unit = 28 ms)

Fig. 13 Sinusoid surface tracking reading of the end-effector (in y and z direction)

(3) Hybrid position and force control for the robot system including passive compliance is possible and the control scheme's feasibility is demonstrated by the experimental results. The scheme can also be utilized as a free joint control scheme, as well as a conventional hybrid control without a passive compliance.

(4) The digital filter dynamics dominate the systems' response. In position control, a high pole of the filter is desirable, while in force control, a low pole is required to stabilize the system. A compromise has to be made between fast response and a large compliance of the wrist device.

(5) Force gains should be selected carefully according to the stiffness of the passive compliance device, the mass of the end-effector, and the desired compliance of the system. The different force gains are necessary for each direction and around each axis. If the passive compliance device or environment is low, a relatively high force gain can be used.

9. REFERENCES

[1] Y. Xu and R.P. Paul, "On position compensation and force control stability of a robot with a compliant wrist", *Proceedings of the IEEE International Conference on Robotics and Automation*, P.1173-1178, 1988

[2] R.P. Paul, Y. Xu and X. Yun, "Terminal link force and position control of a robot manipulator", *Seventh CISM and IFToMM International Symposium on Theory and Practice of Robots and Manipulators*, 1988

[3] R. K. Roberts, R. P. Paul, and B. M. Hillberry, "The effect of wrist force sensor stiffness on the control of robot manipulators", *Proceedings of the IEEE International Conference on Robotics and Automation*, P.269-274, 1985

[4] D. E. Whitney and J. M. Rourke, "Mechanical behavior and design equations for elastomer shear pad remote center compliance", *ASME Journal of Dynamic System, Measurement, and Control*, Vol. 108, P.223-232, 1986

[5] V. Hayward, *RCCL User's Guide*, edited for CVaRL by John Lloyd, 1984

[6] H. Inoue, Y. Tsusaka, and T. Fukuizumi, "Parallel manipulator", *Proceedings of Third International Symposium of Robotics Research*, P.321-327, 1986

[7]   K.J. Waldron and K.H. Hunt, "Serial-parallel dualities in actively coordinated mechanisms", *Fourth International Symposium on Robotics Research*, Santa Cruz, Sept. 1987

[8]   D. S. Seltzer, "Compliant robot wrist sensing for precision assembly", *Robotics: Theory and Application*, P.161-168, 1986

[9]   H. Kazerooni, and J. Guo, "Direct-drive, active compliant end-effector" *Proceedings of the IEEE International Conference on Robotics and Automation*, P.758-766, 1987

[10] J.De Schutter, "Compliant robot motion control methods for rigid manipulators based on a generic scheme", *Proceedings of the IEEE International Conference on Robotics and Automation*, P.1060-1065, 1987

[11] M. Mason, "Compliance and force control for computer controlled manipulators", in *Robot Motion Planning and Control*, M. Brady et al, ed., The MIT Press, Cambridge, MA, 1982, P.373-404. ch.5.

[12] K. Salisbury, "Active stiffness control of a manipulator in cartesian coordinates", *Proc. 19th IEEE Conference on Decision and Control*, Albuquerque, NM, December 1980, P.87-97.

[13] M.R. Cutkosky and P.K. Wright, "Position Sensing Wrists for Industrial Manipulators", *12th International Symposium on Industrial Robots*, Paris, France, June 1982.

[14] N. Hogan, "Impedance control of industrial robots", *Robotics and Computer Integrated Manufacturing*, Vol.1, No.1, 1984, P.97-113.

[15] M.H. Raibert and J.J. Craig, "Hybrid position/force control of manipulators", *ASME Journal of Dynamics system, Measurement, and Control*, Vol. 102, 1981, P.126-133

[16] C. Reboulet and A. Robert, "Hybrid control of a manipulator with an active compliant wrist", *Proceedings of the Third International Symposium on Robotics Research*, 1985

[17] H.V. Brussel, H. Thielman and J. Simons, "Further developments of the active adaptive compliant wrist (AACW) for robot assembly", *11th International Symposium on Industrial Robots*, Tokyo, Japan, October 1981, P.377-384.

[18] R. Dillmann, "A Sensor Controlled gripper with tactile and non-tactile sensor environment", *Proceeding of the 2nd International Conference on Robot Vision and Sensory Controls*. Stuffgart, Germany, November, 1982, P.159-170

[19] R.P.C. Paul and B. Shimano, "Compliance and control", in *Robot Motion Planning and Control*, M. Brady et al, ed., The MIT Press, Cambridge, MA, 1982, P.404-418, ch.5.

[20] K. Takuse, H. Inoue, K. Sato and S. Hagiuara, "The design of the articulated manipulator with torque control ability", *Fourth International Symposium on Industrial Robots*, Nov. 1974

[21] H. Asada and K. Ogawa, "On the dynamic analysis of a manipulator and its end effector interacting with the environment", *Proceedings of the IEEE International Conference on Robotics and Automation*, P.751-756, 1987

[22] J-P. Merlet, "C-surface applied to the design of an hybrid force-position robot controller", *Proceedings of the IEEE International Conference on Robotics and Automation*, 1987

[23] J. J. Bausch, B. M. Kramer and H. Kazerooni, "The development of compliant tool holders for robotic deburring", *Robotics: Theory and Application*, P.79-89,

[24] D.E. Whitney, "Force feedback control of manipulator fine motions", *ASME Journal of Dynamic Systems, Measurement and Control*, June 1977, P.91-97.

Design and Development of Torque-Controlled Joints

Dieter Vischer and Oussama Khatib

Robotics Laboratory
Computer Science Department
Stanford University

Abstract

This paper discusses the effect of basic manipulator characteristics upon the implementation of high performance joint torque control. Two manipulators with very different characteristics (high and low gear ratios) are used in this analysis: The PUMA 560 manipulator and ARTISAN, a ten degree-of-freedom manipulator currently under development at Stanford. The experimental results obtained with a prototype link of ARTISAN are presented and compared to those previously obtained with the PUMA. This paper also describes conceptually a new type of torque sensor, developed during the course of this project. With this new sensor, using inductive contactless transducers, torques are evaluated by distance measurements of deflections in the sensor's structure. The new sensor provides a substantial increase in accuracy over conventional strain gauge sensors, achieves higher mechanical robustness, and presents lower sensitivity to electrical noise.

1 Introduction

The work reported in this paper is part of a larger research effort concerned with the development of a high-performance, force-controlled, ten degree-of-freedom manipulator, ARTISAN, currently under development at Stanford University (Roth et al. 1988). Force control has emerged as one of the basic means to extend robot capabilities in performing advanced tasks in complex environments. A prerequisite to force control implementation is the manipulator's ability to achieve precise control of joint torques. This ability, however, is considerably restricted by the nonlinearities and friction inherent in the actuator-transmission systems generally found in industrial robots.

The list of desirable properties of a force controlled manipulator includes: high backdrivability, low friction, minimal effects of ripple torques and dynamic forces, high ratio of force capacity to force accuracy (Townsend 1988), little backlash, and negligible distributed elasticities (Colgate and Hogan 1989, Eppinger 1986 and 1987).

While avoiding transmission nonlinearities, direct drive manipulators become increasingly massive and bulky with increases in the number of degrees of freedom. The solution

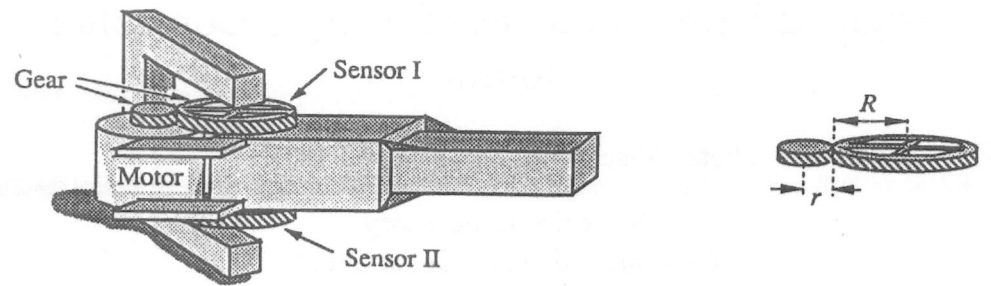

Figure 1: Prototype Link

we have adopted for the actuation of ARTISAN has been to use brushless (permanent magnet) motors and a single stage, low gear reduction (evoloid) system with torque sensing. Joint torque feedback is aimed at reducing friction and transmission effects, thus providing high performance joint torque control. The goal is to design, for each joint, an independent, high-bandwidth, robust torque servo controller.

The first part of the following description deals with the basic characteristics of torque controlled joints, where high-geared and low-geared manipulators will be compared. The second part focuses on the design and development of the new torque sensor.

2   Torque Control

Joint torque performance depends strongly on the gains achievable by the controller. A better understanding of the limitations placed on these gains has been one of the major objectives in our experiments with the prototype link.

2.1   Prototype Link

The prototype link for ARTISAN is shown in Figure 1. It uses a brushless dc motor, mounted at the base of the link in order to counterbalance the link's mass. The motor torque is transmitted in parallel through two single-stage low gear-reduction transmissions. The torque sensor is integrated with the gear. A shaft encoder is located on the motor axis to measure the relative position between the link and the motor. The parameters of the prototype link are given in Table 1.

The equations of motion of the link involve additional dynamic forces resulting from the rotation of the motor relative to the link. During a full revolution of the link, the motor makes $N + 1$ turns

$$T_m - \frac{T_s}{N} = J_m(\ddot{\theta}_m + \ddot{\theta}_l) + d_m\dot{\theta}_m; \tag{1}$$

$$T_s(1 + \frac{1}{N}) - T_m = (\tilde{J}_l + m(R+r)^2)\ddot{\theta}_l + d_l\dot{\theta}_l - d_m\dot{\theta}_m. \tag{2}$$

Table 1: Link Parameters

N	Gear Ratio	m	Motor Mass
$\breve{J}_l$	Link Inertia	J_m	Motor Inertia
d_l	Link Friction	d_m	Motor Friction
T_s	Sensor Torque	T_m	Motor Torque
θ_l	Link Angle	θ_m	Motor Angle

The torque sensor can be modeled as a linear spring, where k_s is the total stiffness of sensors I and II

$$T_s = k_s \theta_s; \tag{3}$$

and θ_s is the angle of sensor deflection

$$\theta_s = \frac{\theta_m}{N} - \theta_l. \tag{4}$$

2.2 Torque Transfer Function

The transfer function of motor torque T_m to sensed torque T_s (which corresponds to the joint torque) can be obtained from equations (1), (2), (3) and (4) as

$$\frac{T_s}{T_m} = \frac{(\frac{k_s}{NJ_m} + \frac{k_s}{NJ_l} + \frac{k_s}{J_l})s + \frac{d_l k_s}{NJ_m J_l}}{s^3 + (\frac{d_m}{J_m} + \frac{d_l}{J_l} + \frac{d_m}{J_l})s^2 + (\frac{k_s}{N^2 J_m} + \frac{k_s}{J_l} + \frac{2k_s}{NJ_l} + \frac{k_s}{N^2 J_l} + \frac{d_l d_m}{J_m J_l})s + \frac{k_s}{J_m J_l}(\frac{d_l}{N^2} + d_m)}; \tag{5}$$

where

$$J_l = m(R+r)^2 + \breve{J}_l.$$

Experimental frequency and step responses have shown the system to have a dominant second order behavior with weak damping. Thus, a first approximation of the transfer function can be obtained by neglecting the damping terms d_m and d_l. By setting $(d_m = d_l = 0)$, equation (5) is reduced to

$$\frac{T_s}{N T_m} = \frac{k_0 \omega_0^2}{s^2 + \omega_0^2} \quad \left\{ \begin{array}{l} k_0 = \frac{(N+1)J_m + J_l}{(N+1)^2 J_m + J_l} \le 1; \\ \omega_0^2 = k_s \frac{(N+1)^2 J_m + J_l}{N^2 J_m J_l}; \end{array} \right. \tag{6}$$

where k_0 is the open-loop gain and $\omega_0/2\pi$ is the open-loop frequency, f_0. Observing that $N \gg 1$ and $J_l \gg N J_m$, the parameters k_0 and ω_0 can be further simplified to

$$k_0 = \frac{1}{1 + \frac{N^2 J_m}{J_l}}; \tag{7}$$

$$\omega_0^2 = k_s(\frac{1}{J_l} + \frac{1}{N^2 J_m}). \tag{8}$$

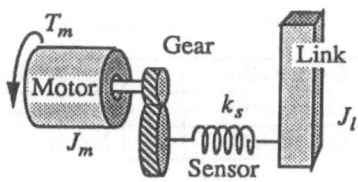

Figure 2: Approximated Model

Equations (7) and (8) correspond to a link with an inertia J_l driven by a motor, which is fixed in the inertial frame, as shown in Figure 2.

The transfer function for the constrained link can be found from equation (5) by letting J_l go to infinity

$$\frac{T_s}{NT_m} = \frac{k_0\omega_0^2}{s^2 + 2\xi_0\omega_0 s + \omega_0^2}; \quad \text{with} \quad k_0 = 1; \quad \xi_0 = \frac{d_m}{2\omega_0 J_m}; \quad \text{and} \quad \omega_0^2 = \frac{k_s}{N^2 J_m}. \quad (9)$$

2.3 Disturbances

Static friction (Coulomb friction), dynamic friction, and motor ripple torques can be regarded as disturbance torques acting on the motor. The sensed torque is also effected by other disturbances originating at the link (friction in the gears and dynamic forces resulting from the action of other links or from the link interaction with the environment).

Some of these effects can be modeled, identified, and compensated for in the control algorithm. Armstrong (1988) has shown the effectiveness of feedforward compensations using look up tables. Our goal is to couple feedforward compensations with joint torque feedback to provide high reduction of unmodeled disturbances and robustness to model errors. Unknown disturbances and model errors can be grouped into two classes:

- T_{d1} : the sum of disturbances originating at the motor;

- T_{d2} : the sum of disturbances acting at the link.

The transfer function (with $d_l = d_m = 0$) then becomes

$$T_s = \frac{k_0\omega_0^2}{s^2 + \omega_0^2}(NT_m + T_d); \quad \text{with} \quad T_d = NT_{d1} + \frac{N^2 J_m}{J_l}T_{d2}. \quad (10)$$

Equation (10) describes the influence of disturbances on the open-loop system (see Figure 3). One of the advantages of low geared robots such as ARTISAN is that their open-loop characteristics, k_0 and f_0, are less sensitive to changes in the link inertia, J_l. In addition, sensed torques are less sensitive to the disturbance torque, T_{d2}, acting at the link (equation 10). This is illustrated in the following two examples.

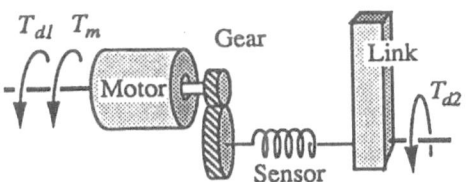

Figure 3: Disturbance Torques

Table 2: Prototype and PUMA Parameters

	Prototype	PUMA
Gear Ratio, N	6	53.7
$\frac{N^2 J_m}{J_l}$	0.12	2.5
Motor Inertia, J_m	0.000755 kg m^2	0.000288 kg m^2
Link Inertia, J_l	0.23 kg m^2	0.336 kg m^2
Spring Constant, k_s	452000 Nm/rad	16000 Nm/rad

2.4 Examples

Pfeffer, Khatib, and Hake (1986) have developed a joint torque sensory feedback controller for the third link of a PUMA 560 manipulator. The PUMA 560 is an example of a relatively high geared manipulator. Here, it is used to provide a basis for comparison with the low geared prototype link of ARTISAN.

The relevant parameters for the ARTISAN prototype and the PUMA are shown in Table 2. The open-loop characteristic for the two examples, obtained from equations (6), (9), and (10), are summarized in Table 3.

The data in Table 3 confirms the advantages of low over high gear-ratio manipulators discussed above. The dominance of a second order behavior in the dynamics of the prototype link has been confirmed by experimental open-loop step responses (see Figure 9). These experiments have shown the resonant frequency to be at 230 Hz for both the

Table 3: Open-Loop Characteristics

		Prototype	PUMA
Open-Loop Gain, k_0	Link Free	0.88	0.29
	Link Fixed	1	1
Open-Loop Frequency, f_0	Link Free	699 Hz	41 Hz
	Link Fixed	650 Hz	22 Hz
Disturbance (Link Free)	$\frac{T_l}{T_{d1}}\big|_{s=0}$	5.28	15.6
	$\frac{T_l}{T_{d2}}\big|_{s=0}$	0.18	0.73

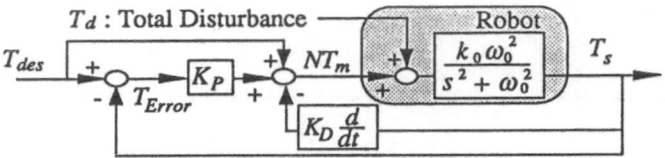

Figure 4: PD Controller

free and constrained link cases. This value is significantly lower than the frequencies of 699 Hz and 650 Hz estimated using the model. This discrepancy is due to significant flexibility in the transmission system that was neglected in the model. In effect, the experimental data have shown the overall stiffness to be much lower than the stiffness of the sensor.

2.5 Lead Controllers

Disturbance rejection and robustness characteristics of lead-type controllers can best be analyzed by using a simple PD controller. The control law (see Figure 4) is

$$NT_m = T_{des} - K_P(T_s - T_{des}) - K_D\dot{T}_s;$$ (11)

where K_P and K_D are the proportional and derivative feedback gains, respectively. The input feedforward allows to reduce static errors.

The closed-loop transfer function can be obtained from equations (10) and (11) as

$$T_s = \frac{k_c\omega_c^2}{s^2 + 2\xi_c\omega_c s + \omega_c^2}(T_{des} + \frac{T_d}{K_P + 1});$$ (12)

where the closed-loop gain k_c, damping ξ_c and frequency f_c are

$$k_c = \frac{1}{1 + \frac{J_m^*}{J_l}}; \quad \xi_c = \frac{K_D k_0 \omega_0^2}{2\omega_c}; \quad \text{and} \quad \omega_c^2 = k_s(\frac{1}{J_m^*} + \frac{1}{J_l});$$ (13)

with

$$J_m^* = \frac{N^2 J_m}{K_P + 1}.$$ (14)

The transfer function associated with the closed-loop system is similar to that of the open-loop. However, the link is now driven by a "motor," whose inertia is $K_P + 1$ times smaller. Disturbance torques are also $K_P + 1$ times smaller. A mechanical equivalent to the closed-loop system is shown in Figure 5, where J_m^* is the equivalent motor inertia. Given that $J_l \gg J_m^*$, the closed-loop parameters can be approximated as

$$k_c \approx 1; \quad \text{and} \quad \omega_c^2 \approx \frac{k_s}{J_m^*} = k_s \frac{K_P + 1}{N^2 J_m}.$$ (15)

With $K_D > 0$, equations (12), (13) and (14) show the closed-loop system to be stable for all gains $K_P > -1$. For gains between -1 and 0 disturbances are amplified. Thus, in

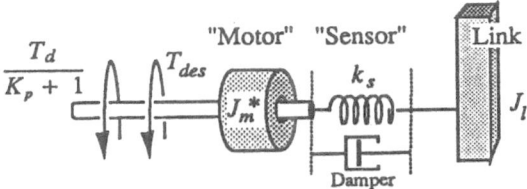

Figure 5: Closed-Loop System

order to achieve an improvement over the open-loop system, these gains must be larger than zero. This also means that the closed-loop frequency f_c must be larger than the open-loop frequency f_0.

Equation (15) provides a basis for trade-offs between achievable disturbance rejection $K_P + 1$, closed-loop bandwidth f_c, and sensor stiffness k_s. This equation (15) also shows that for a given achievable bandwidth, a lower sensor stiffness corresponds to a higher disturbance rejection, as observed by Whitney (1985) and Roberts (1985).

2.6 System Limitations

The closed-loop bandwidth, f_c, and the achievable gain, K_P, are limited by the following considerations:

- The sampling frequency of the digital controller must be at least five times higher than the closed-loop frequency f_c;

- The overall deadlag time in the measurements should not exceed a third of $1/f_c$;

- The first unmodeled resonant mode should be about three times higher then the closed-loop frequency;

- There are other limitations which result from noise in measurements and limit cycles due to backlash and nonlinear friction (Luh et al. 1983).

While a low sensor stiffness is desirable (equation 15) to achieve larger gains and better disturbance rejection, high gains result in increased motor activities. Higher motor torques will then be needed to counter the lower rigidity of the system. The transfer function of desired torque to motor torque is

$$NT_m = \frac{(K_P + 1)(s^2 + \omega_0^2)}{s^2 + 2\xi_c\omega_c s + \omega_c^2} T_{des}.$$ (16)

The Bode diagram of this transfer function (see Figure 6) shows that changes in the desired torque at frequencies higher than f_0 will require up to $K_P + 1$ times higher actuator torques. Below f_0 the required actuator torques are comparable to those needed for a rigid link.

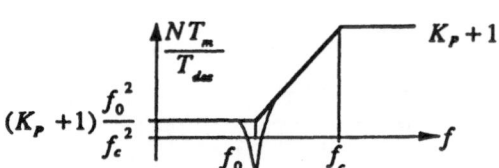

Figure 6: Required Motor Torque

PUMA Link

The limiting factor for the bandwidth in the PUMA link experiment has been the sampling frequency. Computer hardware has allowed a maximum sampling frequency of 500 Hz. The achievable closed-loop frequency can be estimated at $f_c = 100$ Hz which corresponds to a gain, K_P, of 19 (equation 15). This results in a 95% reduction of the effective friction $(1 - \frac{1}{K_P+1})$. In the actual experiment a factor of 97% has been achieved with a second order digital controller.

Torque changes with frequencies above the open-loop frequency (22 Hz for the fixed link and 41 Hz for the free link) require motor torques with very high amplitudes (up to 20 times (K_P+1) the desired torque). In practice, desired torques are much below the open-loop frequency. Under a lead-type controller, the high geared PUMA $(N^2 J_m / J_l = 2.5)$ becomes equivalent to a low geared manipulator with $J_m^* / J_l = 0.13$.

ARTISAN Link Prototype

For a low geared robot $(N^2 J_m \ll J_l)$, the achievable disturbance rejection $K_P + 1$ can be estimated as

$$K_P + 1 \approx (\frac{f_c}{f_0})^2. \tag{17}$$

The open-loop frequency of the link prototype is 230 Hz. A reduction factor of 95% of the effective friction corresponds to a closed-loop frequency around 1000 Hz. This is clearly much too high, considering the bandwidth limitations discussed above.

2.7 Lag Controller

When the dynamics of the open-loop system are too fast to be effectively controlled, one alternative is to ignore the high frequency of the open-loop system and to design a controller operating at lower frequency. The open-loop transfer function can then be approximated by

$$(k_s \to \infty) \quad T_s \approx k_0 (N T_m + T_d). \tag{18}$$

Clearly, a lag controller is the most suitable for this zero-order system. A simple integral controller (see Figure 7) is selected for the following investigation of bandwidth limitations and disturbance rejection with lag-type controllers. The control law is

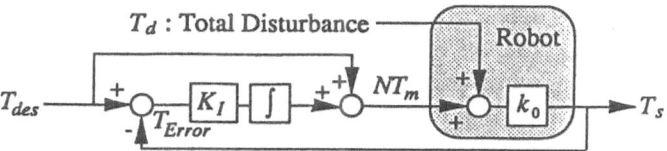

Figure 7: Integral Controller

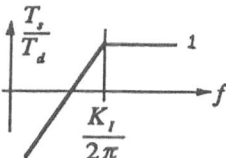

Figure 8: Disturbance Rejection

$$NT_m = T_{des} - K_I \int (T_s - T_{des})dt; \qquad (19)$$

where K_I is the integral gain. The closed-loop transfer function can be obtained from equations (18) and (19) as

$$T_s = \frac{k_0(s + K_I)}{s + k_0 K_I} T_{des} + \frac{k_0 s}{s + k_0 K_I} T_d. \qquad (20)$$

For low geared manipulators ($N^2 J_m \ll J_l$) the transfer function can be simplified to

$$T_s \approx T_{des} + \frac{s}{s + K_I} T_d. \qquad (21)$$

With the above controller, disturbances are reduced for frequencies below $K_I/2\pi$ (see Figure 8). This type of controller is independent of the stiffness of the sensor. The achievable disturbance rejection will only depend on bandwidth limitations as described above. The behavior of the closed-loop system is similar to that of a direct drive manipulator ($T_s \approx T_{des}$).

2.8 Experimental Results

Figure 9 shows different step responses for the prototype link. The same digital controller with an overall lag characteristic was used to control the free link and the constrained link. The schematic of the controller is given in Figure 10.

The ARTISAN prototype has little natural damping and its open-loop bandwidth is 230 Hz as shown in Figure 9.a. The disturbance rejection bandwidth, $K_I/2\pi$, is 30 Hz. The 10-90% rise time for the closed-loop system (see Figure 9.b) is less than 10ms. Desired torque inputs, which are typically below the 30 Hz disturbance rejection bandwidth, are accurately controlled. A second order low-pass filter (150 Hz, $\xi = 1$) for the input is used to prevent the command torque from exiting the first resonant mode of 230 Hz.

In the case where the link is free (see Figures 9.c and 9.d), the resonant mode is excited by cogging in the transmission system. While the controller is unable to prevent this behavior, it provides a significant reduction of overshoot and static errors.

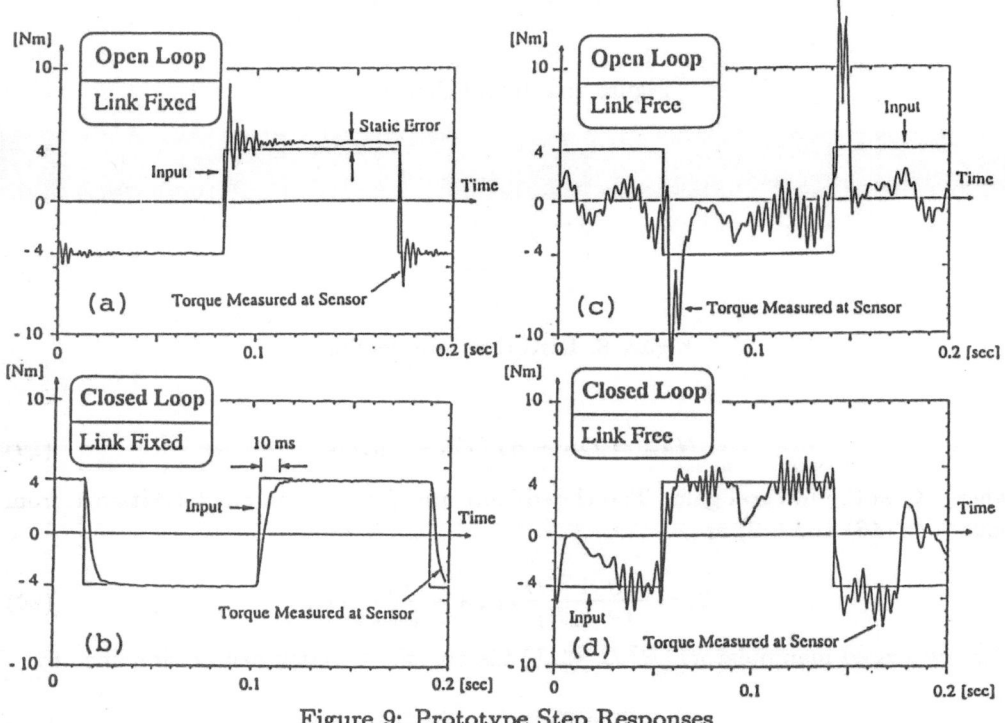

Figure 9: Prototype Step Responses

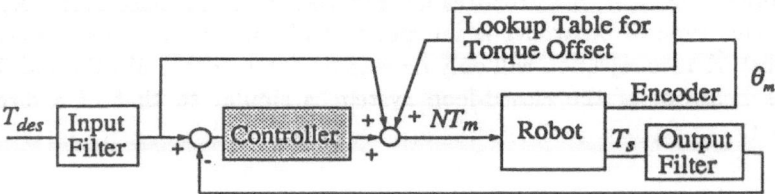

Figure 10: Controller Schematic

3 New Torque Sensor Design

Breakaway friction for manipulators with brush type servo-motors, e.g. the PUMA, is much higher (one order of magnitude) than the friction resulting from brushless motors. To bring about a reduction of the already low friction, high torque accuracy is needed for manipulators with brushless motors. For instance, the resolution needed

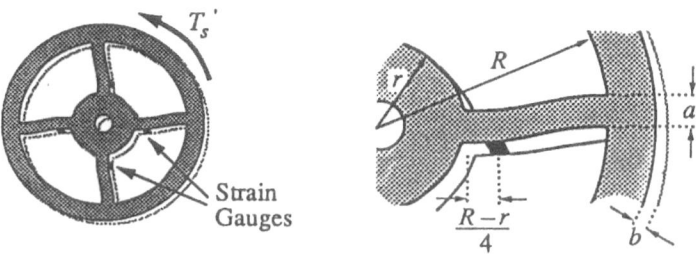

Figure 11: Strain Gauge Sensor

(torque accuracy to maximum torque) for the sensor used in the prototype link has been estimated at 0.03%. Building a sensor with such a resolution, in practice, is not an easy task.

3.1 Strain Gauge Sensors

As a first attempt at torque sensing, we have used the four beam torque sensor shown in Figure 11. Beam deflection is measured by semiconductor strain gauges. The torsional stiffness k'_s and the strain ε at the gauge location can be found as

$$k'_s = \frac{4Ea^3b(R^2 + Rr + r^2)}{3(R-r)^3}; \quad \text{and} \quad \varepsilon = \frac{3(R-r)(5R+r)}{16Ea^2b(R^2 + Rr + r^2)}T'_s; \quad (22)$$

where E is the Young's modulus. Eight strain gauges arranged in Wheatstone bridges were used for each of the two sensors integrated in the two gears.

With this type of sensors, gear eccentricities have been shown to result in a position-dependent torque offset, which was 5 to 10 times higher than the required accuracy. The use of lookup tables to compensate for this dependency resulted in significant improvements but did not allow us to obtain the required resolution of 0.03%.

3.2 Sensor Placement

In the initial design of the prototype link, the sensor was placed in the gear as shown in Figure 12.a. With this arrangement, the sensor is not sensitive to those dynamic forces acting on the link, which do not contribute to the joint torque. At that location, however, the sensor is exposed to large radial forces caused by the transmission system. We have found and experimentally verified that very small eccentricities in the gear (10^{-5} to 10^{-6} meters) could result in radial forces high enough to saturate (and sometimes break) the strain gauges.

An alternative to this initial design (Figure 12.a) is shown in Figure 12.b. In this configuration the sensors are exposed to dynamic forces that do not contribute to the joint torque but are protected from the much larger radial forces due to eccentricities of the gear.

This second design was used for the new prototype. Further improvements in sensor disturbance protection can be achieved by using the configuration shown in Figure 12.c.

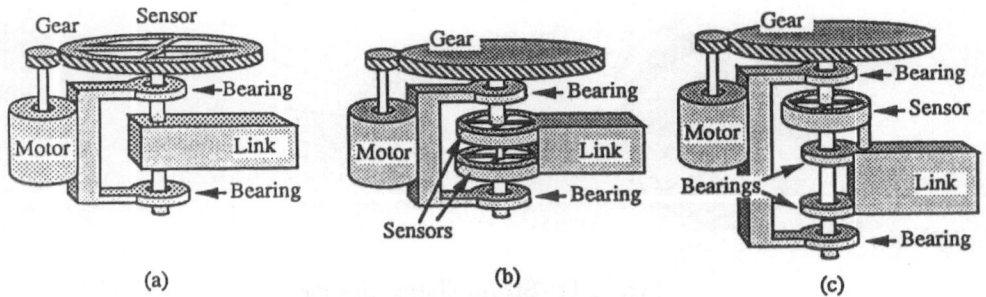

Figure 12: Sensor Placement

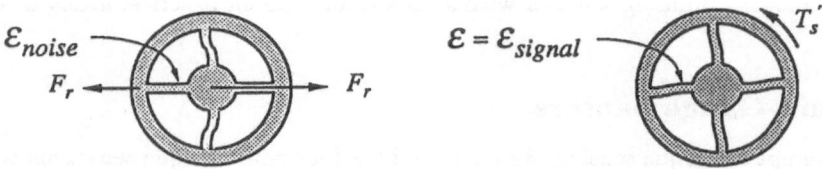

Figure 13: Signal-to-Noise Ratio

A second set of bearings is used to better protect the sensor from disturbances. Although significant improvements can be obtained by careful sensor design, radial forces cannot entirely be suppressed.

3.3 Modeling of Radial Forces

The effect of radial forces acting on each individual gauge can be formulated as a mechanical signal-to-noise ratio as shown in Figure 13.

The strain ε_{noise} caused by the radial force F_r depends on the direction of F_r. It can be shown that ε_{noise} reaches its maximum when F_r is parallel to the beam. The minimal signal-to-noise ratio (worst case) can be calculated as

$$\frac{\varepsilon_{signal}}{\varepsilon_{noise}} = \gamma\,\frac{T_s'}{F_r}\,; \quad \text{with} \quad \gamma = \frac{3(5R+r)(R^2 - 2Rr + r^2 + a^2)}{8a(R-r)(R^2 + Rr + r^2)}. \tag{23}$$

The factor γ depends only on the geometry of the sensor. Since a 10 to 20 times increase of γ would provide the needed accuracy, we focused our effort on designs that would maximize the value of γ. We have also analyzed the signal-to-noise ratio characteristics for other sensor configurations, as shown in Figure 14.

Figure 14.a shows a torsional sensor as describe by Wu and Paul (1980). Those sensors have high robustness to radial forces, but are quite sensitive to torque disturbances in other directions. Figure 14.b shows an eight beam sensor. Although the addition of beams improves the signal-to-noise ratio, the manufacturing difficulty and sensor cost rapidly increase.

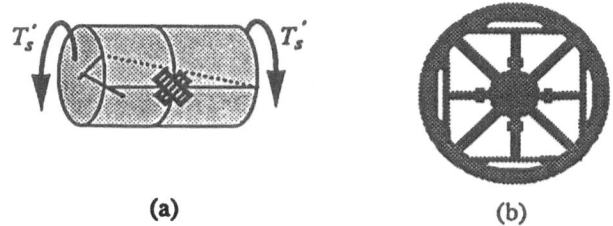

(a)   (b)

Figure 14: Torque Sensor Designs

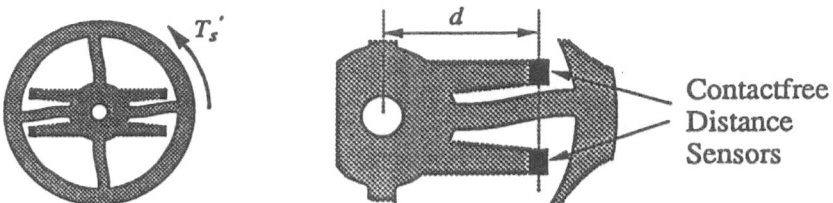

Contactfree
Distance
Sensors

Figure 15: New Torque Sensor

3.4   New Torque Sensor

The design concept of the new sensor is illustrated in Figure 15. Torques are obtained from measurements of the beam deflections using four contact-free distance sensors.

The deflection at the sensor location can be evaluated as

$$\Delta x = \Delta x_{signal} \;=\; d\,\frac{T_s'}{k_s'}. \tag{24}$$

A radial force, F_r, will result in a deflection $\Delta x = \Delta x_{noise}$, which depends on the direction of F_r. Δx_{noise} reaches its maximum when F_r is parallel to the measured deflection. The minimal signal-to-noise ratio (worst case) is

$$\frac{\Delta x_{signal}}{\Delta x_{noise}} = \gamma\,\frac{T_s'}{F_r}\;; \quad \text{with} \quad \gamma = \frac{3d(R^2 - 2Rr + r^2 + a^2)}{2a^2(R^2 + Rr + r^2)}. \tag{25}$$

A comparison of equations (23) and (25) clearly shows the advantages of the contact-free sensor over strain gauge based sensing.

The final layout of the new sensor is shown in Figure 16 and uses a six beam structure. Four inductive transducers are arranged in a Wheatstone bridge configuration. The signal-to-noise ratio of the new sensor is 24 times higher than the value obtained with the initial sensor ($\gamma = 1730$ m^{-1}).

3.5   Mechanical Robustness and Electrical Noise

The signal-to-noise ratio described above is an important parameter in the design of a torque sensor. However, the new torque sensor has other advantages, which for some applications might be more useful than its high resolution.

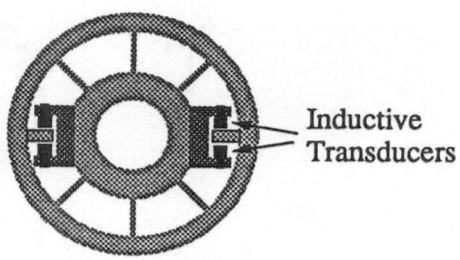

Figure 16: Torque Sensor Layout

Strain gauges are quite fragile and tend to break easily. Their maximum allowable strain is also fairly close to the strain at which they break, making failure prevention difficult. The inductive sensors are housed in steel cases and can withstand torques that are at least one order of magnitude higher than the maximum measurable torque.

Inductive sensors are also easy to mount and easily replaced, whereas mounting strain gauges (especially semiconductor ones) involves a lengthy procedure that requires expertise. Another shortcoming of strain gauges is their sensitivity to electrical noise. Their resolution (ratio of minimal to maximal measurable strain) is very high (up to 0.003% for semiconductors) but this accuracy can only be reached in an environment free of noise. With the new sensor, the inductive bridge is modulated with a carrier frequency of 5 kHz, which significantly reduces the sensitivity to electrical noise.

3.6   Conclusion

The issue of joint torque feedback has been discussed in the context of two manipulators with very different characteristics. The PUMA 560 is a high gear-ratio manipulator with a low open-loop resonant frequency, a characteristic which was shown to be compatible with a lead-type joint torque controller. For the low gear-ratio prototype link with high open-loop resonant frequency, a lag-type controller was used, resulting in a closed-loop behavior similar to that of a direct-drive system. These two controllers represent the two extremes in a wide spectrum of possible lead-lag controllers.

This paper also presented the design concept of a new type of torque sensor based on contactless distance measurements using inductive transducers. The signal-to-noise ratio achieved with the new sensor has been shown to be 24 times higher than the ratio obtained with conventional strain gauge sensors.

In addition to the conclusive results on the impact of joint torque feedback, these experiments have resulted in a better understanding of many theoretical and practical issues associated with the design and control of actuator/transmission systems using joint torque sensory feedback.

Acknowledgments

The financial support of SIMA, the Swiss NSF, and DARPA (contract DAAA21-89-C0002) are acknowledged. We are thankful to Professors Bernard Roth and Kenneth Waldron and to Richard Voyles, Larry Pfeffer and David Williams, who have made valuable contributions to the development of this work.

4   References

Armstrong, B., "Dynamics for Robot Control: Friction Modeling and Ensuring Excitation during Parameter Identification," Ph.D thesis, Department of Electrical Engineering, Stanford University, Stanford, CA, June 1988.

Asada, H., Youcef-Toumi, K., and Lim, S.K., "Joint Torque Measurement of a Direct Drive Arm," 23d IEEE Conference on Decision and Control, December 1984, pp. 1332-1337.

Colgate, E., and Hogan, N., "An Analysis of Contact Instability in Terms of Passive Physical Environments," Proc. IEEE International Conference on Robotics and Automation, Scottsdale, Arizona, 1989, pp. 404-409.

Eppinger S.D., and Seering W. P., "On Dynamics of Robot Force Control," Proc. IEEE International Conference on Robotics and Automation, San Francisco, CA, 1986.

Eppinger, S.D., and Seering, W. P., "Understanding Bandwidth Limitations in Robot Force Control," Proc. 1987 IEEE International Conference on Robotics and Automation, Raleigh, N.C., CA, April 1987.

Khatib, O., Burdick, J., "Motion and Force Control of Robot Manipulators," Proc. International Conference on Robotics and Automation, San Francisco, CA, April 1986, pp. 1381-1386.

Luh, J.Y.S., Fisher, W.D., and Paul, R.P., "Joint Torque Control by a Direct Feedback for Industrial Robots," IEEE Transactions on Automatic Control, AC-28, 1983, pp. 153-160.

Pfeffer, L., Khatib, O., and Hake, J., "Joint Torque Sensory Feedback in the Control of a PUMA Manipulator," Proc. American Control Conference, Seattle, Washington, June 1986, pp. 818-824.

Roberts, R.K. Paul, R.P., and Hillberry, B.M., "The Effect of Wrist Sensor Stiffness on the Control of Robot Manipulators," Proc. IEEE International Conference on Robotics and Automation, St. Louis, MO, March 1985, pp. 269-274.

Roth, B., Raghavan, M., Khatib, O., and Waldron, K., "Kinematic Structure for a Force Controlled Redundant Manipulator," Proc. International Meeting on Advances in Robot Kinematics, Ljubljana, Yugoslavia, September 1988, pp. 62-66.

Townsend, W., "The Effect of Transmission Design on Force-Controlled Manipulator Performance," Ph.D thesis, Artificial Intelligence Laboratory, Massachusetts Institute of Technology, Cambridge, MA, 1988.

Whitney, D.E., "Historical Perspective and State of the Art in Robot Force Control," Proc. IEEE International Conference on Robotics and Automation, St. Louis, MO, March 1985, pp. 883-889.

Wu, C.H., and Paul, R.P., "Manipulator Compliance Based on Joint Torque Control," 19th IEEE Conference on Decision and Control, December 1980, pp. 88-94.

On a Unified Concept for a New Generation of Light–Weight Robots

J. Dietrich, G. Hirzinger, B. Gombert, J. Schott

D L R

German Aerospace Research Establishment

Institute for Flight Systems Dynamics

D-8031 Oberpfaffenhofen, FRG

Abstract

The paper outlines a concept for the design of a new generation of light weight robots using an integrated and unified approach for all relevant elements. A light multisensory gripper as the critical end-mass of any robot is presented that contains different-type redundant force-torque sensors, tactile arrays, 9 laser-range-finders including a scanner, and a tiny stereo-camera. A new electrical gripper drive has been developed, the weight to grasp force ratio of which aimed at an improvement factor of 5-10 compared to hitherto known systems. The same order of improvement has been the design goal for a new modular system of light-weight arms, the joint drives being integrated into ultra-light carbon-fibre-grid structures. These joint drives - as well as the gripper drives - are based on a complete redesign of the control system of commercially available stepping motors that turns them into electronically commutated high-speed and highly dynamics dc-motors, and a new type of light, compact gearing with reduction rates of 600 and more easily achievable.

1 Introduction

Present-day robots as they are working in our factories have often been called "dino-saurs" as they are moving around heavy arm and joint masses with modest bandwidth while carrying small tools only. The argument has always been that robots have to guarantee a certain end point precision. But indeed robot manufacturers today only guarantee some repetitive positioning accuracy, that is a robot repeating a path under a certain load should reach the previous endpoint say in a neighbourhood of ± 0,1 mm. Absolute positioning accuray has been approached by kinematic calibration schemes using e.g. external optical measurements [1], but kinematic errors caused by fabrication imprecision represent only part of the problem, the rest is due to position-dependant, unavoidable compliance and backlash in the joints, and uncertainty in the environment. What we really need is **relative precision with respect to the environment**, and the only solution we can see here is multisensory perception of the environment, similar to the amazing performance of the light, fragile human arm that does the most delicate, fast manipulations by using sensory feedback.

Now assuming that by extensive use of sensors integrated into a light, multisensory gripper with high controllable grasp force capabilities we do no longer need hundreds of kilopounds of steel or aluminum to assure some dubious positioning accuracy, light-weight arms increasing the speed and bandwidth of motion might even be flexible in the links. However from the present viewpoint we feel that flexible arm control – due to its complexity – will have its main applications in large space manipulators, while for earth-bound robots we prefer light-structure kinematically redundant, but stiff robots made of composite fibres with dedicated active or passive compliance concentrated in the wrist or fingers. Clearly adding a "micromanipulator" at the end of a "macromanipulator" may be the ultimate solution, but leaves the arguments as given here for the macromanipulator untouched. What about the joints? Apparently making the links out of composite material, while keeping the old joints that weigh nearly as much as the original links, is not a real solution to the light-weight problem. Discussion about joints is strongly affected by control aspects; the most promising schemes implying wrist force feedback (see e.g. [1] or [2]) are based on commanding torques at the joint level. But present robot joints e.g. using harmonic drives or other gearings show up friction of 40 % or more, so gearless direct drives seemed to be the most reasonable solution in the past. On the other side we know mean-while [3] that even with direct–drives torque sensing and feedback in the joints seems necessary (e.g. to compensate for torque ripple characteristics); and as one cannot circumvent physics, direct drives tend to be voluminous and heavy.

As a consequence of the above remarks our goal has been to design new drive systems for electrical grippers as well as for the robot joints that are characterized by

- small and light, highly dynamic, high–speed motors (with relatively high torque)

- highly compact and light, stiff, high reduction, low friction, low–backlash gearings which are easily complemented by joint torque or grasp force measurement and feedback schemes to compensate the remaining friction effects if necessary.

By integrating these kind of drives into light carbon fibre links as well as into new light multisensory gripper as described in the next section, we tried to arrive at an overall optimal design for an advanced robot generation.

2 Multisensory integration in a robot gripper

There are a number of reasons why sensorcontrolled and thus more intelligent robots are still hardly to find in industrial applications despite of all the predictions that have been made already years ago.

Sensors in general are

- too expensive

- too voluminous for an elegant integration into smart grippers

- not reliable enough, with a "black box" and a considerable amount of wiring coming with each sensor, so that multisensory integration became more a pure laboratory concept than a practical item.

- difficult to integrate into robot control systems, from the data technical view point as well as from the algorithmic one.

In our laboratory these observations have led to the design of a multisensory gripper which (as critical end-mass of a robot) had to be of light-weight and in which we tried to avoid all the above-mentioned problems; i.e. the design goals were:

- minimal size and weight for all sensors, with overload protection - as necessary in case of force sensors - integrated, i.e. not introducing extra weight.

- fabrication costs considerably lower than was usual in the past

- all analog electronics and digital preprocessing integrated in the sensors or at least in the wrist

- sensors exchangeable and mountable mechanically in a simple way; but more important they should be easily integrable electrically, that is from the hardware point of view, as well as from the software point of view in a completely modular way.

Before describing the different systems in a more detailed way, let us briefly focus on the above topics of modular integration. Fig. 1 shows that each sensor as well as the gripper drive contains analog electronics, digital electronics and power supply on tiny SMD–boards inside the sensor or as an additional circular slice in the wrist. The one important item is power supply, as in general the sensors need different voltages each and they have to be galvanically decoupled - main reasons for the excessive wirings in the past. In our design there is only one 20 kHz – 50 V power supply providing a rectangularly alternating voltage source; each of the sensors has its own tiny transformer via which it derives the individually needed voltages. Thus galvanic decoupling is assured automatically. Concerning data transmission in the past most sensors were provided with slow serial RS232 interfaces and the robot control system hat to care for the data protocols with each sensor. In our new concept there is now a fas. serial bus (RS485) with 375 kBaud signal rate, to which each sensor is connected; a bus master requests data from the active sensors (thus we have a master slave system) and collects the data into a dualport-RAM, from where the higher–level robot control system can simply read them without caring for protocols or interrupt handling. This concept meanwhile is going to influence German robot industry towards creating much more efficient sensor interfaces than were available in the past.

The modularity concept is pertained in the software of all sensors (fig. 2). There is a measurement module that differs with the physical principles used in the different sensors but uniquely supplies digitized voltages into a switch buffer from which the preprocessing module (e.g. using calibration laws) issues physically relevant values. From a second switch buffer the most recent values are fetched by a bus master request.

Fig. 3 shows a block diagram, fig. 4 the real arrangement of the different components integrated into our prototype gripper; prototype here means that the developments on one side were supported by the German ministery of technology in a major national research and development project aiming at the design of "multisensory intelligent robot grippers for future assembly automation", on the other side they became the basis for our ROTEX project [4], which provides a small sensor–controlled robot to fly with the next German spacelab–mission D2 (in 1991). The robot is supposed to perform different tasks like assembly of a truss structure, plug/unplug electrical connectors and grasp a floating object; the operational modes are automatic, teleoperation on board, by astronauts and teleoperation from ground.

The gripper's sensory components are (see fig. 3):

a) an array of 9 very small laser range finders based on triangulation, one of them being somewhat bigger (half the size of a match box, switchable into a rotational scanning mode) for the medium range of $\approx$ 5–30 cm, and 4 tiny ones in each finger for lower ranges of 0–3 cm. The range finders are the result of many years development aiming at a precise performance over a remarkable range, independent of the slant angle and surface of the measured object. One of the main problems to be solved in this context was to design a nonlinear digital control system that adapts the light transmitter's intensity depending on the reflected light intensity in a range of 1 to 4000 within 10 μsec, for each measuring pulse independently. This indeed enables the sensors to measure distances with respect to surfaces that show up strongly and quickly changing reflection characteristics.

The light source for all range finders is a laser diode with integrated power monitor diode. The divergent laser light is focused by a lens to form a parallel light beam, the diameter of which is $\approx$ 1 mm with divergence in the range of 2 mrad. The laser's output power is 3 mW at 790 mm wavelength.

All receiver systems are built up with linear position sensitive detectors (PSD). The received environment light is suppressed by different techniques, e.g. by optical daylight cut-off filters, filtering amplifiers adapted to the laser pulse characteristics, and sample-hold techniques that registrate and subtract the environment signal from the reflected light pulse measurements.

Resolution for the medium range/scanner sensor as well as for the finger sensors varies from 0,1 % (near) to 3 % (far) of measurement valve.

The medium distance range finder weighs only 0,15 N. So in order to operate it optionally as a scanner we decided to rotate the whole system using a tiny stepper motor instead of implying some oscillating mirror that would yield smaller scan angles and might induce vibrations onto the instrumented compliance. The problem of wiring power and data to such a rotating sensor head is solved by using the scanner's ball-bearings as conductive transducers.

b) a tactile array of 4 x 8 sensing elements (conductive rubber "taxels") in each finger. The dimensions of the tactile area is 32 x 16 mm. The binary state of each taxel is serially transmitted through the analog multiplexers without additional wiring.

c) a "stiff" 6 axis force–torque sensor based on strain–gauge measurements and a compliant optical one. Originally it seemed necessary to make a final decision between these two principles, but as indicated in fig. 3 and fig. 4 they finally were combined into a ring–shaped system around the gripper drive, the instrumented compliance being lockable and unlockable electrically. Shaping these sensors as rings around the gripper drive shows up different advantages:

- it does not prolong the axial wrist length

- it brings the measurement system nearer to the origin of forces–torques and yields a better ratio of torque range to force range than achievable with a compact form.

The "compliant" optical force–torque sensor consists of an inner and an outer part (fig. 5). The basic measuring arrangement in the inner ring is composed of a LED, a slit and perpendicular to it a linear position sensitive detector (PSD). The slit/LED combination is mobile against the remaining system. Six of such systems (rotated by 60 degrees each) are mounted in a plane, whereby the slits alternatingly are vertical and parallel to the plane. The ring with PSD's is fixed inside the outer part and connected via springs with the LED–slit–basis. The springs bring the inner part back to neutral position when no forces/torques are exerted. There exists a particularly simple and unique transformation from PSD–signals $U_1...U_6$ to the unknown displacements.

The stiff, strain–gauged force–torque sensor is a completely new design and will be described in another paper. It performs automatic temperature compensation based on the temperature characteristic as stored during the calibration process and continues operating reliably with reduced accuracy if one of the strain–gauges is damaged.

d) A pair of tiny stereo CCD-cameras, the CCD's plus optics plus minimum electronics in a first realization was taking a volume smaller than a match–box, too. Of course the camera's wiring is independent of the rest of sensors and signal processing is not done in the gripper.

e) An electrical gripper drive, the motor of which is treated like a sensor with respect to the data bus and the 20 kHz power supply connections. The design criteria for this drive are outlined in the next chapter.

3 Electrical drive systems in our light—weight concept

3.1 General remarks

From the present state of technology we mainly see two promising techniques for electrical drive systems for joints and grippers in light—weight robots of the future:

- tendons with drive systems in the robot's base

- light, high torque, highly dynamic motors integrated into the joints or grippers in an optimized combination with new light, high reduction, low—friction gears tailored to these applications.

We have been focussing on the latter approach and tried to solve the problem for the rotational—translational transmission of motion in a gripper (as in a prismatic joint) as well as for the rotational—rotational transmission in a revolute joint. Surely any design in some form represents a tradeoff, nevertheless our goal was to achieve an improvement of at least a factor of 5 compared to presently available designs.

3.2 The motor concept

In both the joint and the gripper drives we provide using the same motor concept (although different sizes), based on our own redesign of commercially available two—phase stepping motors of the ESCAP type (see fig. 4). These motors were basically attractive for our light—weight goals because they show up

- very low rotor inertia (flat disc)

- small size, low weight (1,5 N) but relatively high torque (0,33 Nm), the rotor permanent magnets being integrated into a rare earth disc with **large** diameter

- high positioning accuracy (25 pole pairs).

What we wanted to arrive at however was not really a stepping motor, but an electronically commutated dc motor with the option of switching into a micro—step mode for fine—positioning. Due to the motor's mechanical construction it was fairly easy to integrate hall sensors that allow for the electronic commutation and at the same time serve as position sensors; thus extra resolvers were not needed. Some efforts have been made to raise the torque—speed characteristics at higher speeds (e.g. 6000 – 16000 r.p.m) by a speed—dependent phase—lead circuitry.

3.3 Rotational–translational gearing

In the gripper the problem is to transform the motor's high–speed rotational motion into a fairly slow axial motion to move the fingers (fig. 4). For this type of transmission a new mechanical spindle concept has been designed as shown in fig. 6 [5]. The spindle fixed to the motor axis wears a very fine thread with extremely small pitch (e.g. 0,2 mm/revolution). Six planetary rollers arranged concentrically between spindle and the finger–driving nut show up fine grooves (no pitch) that fit precisely into the grooves of the spindle thread. However they also show up much coarser grooves that mate with corresponding coarse grooves inside the nut, so the rollers do not move axially relative to the nut nor do they change their mutual distance. Why not? The real trick has been to provide the rollers' fine grooves with a "phase shift" from one roller to the next (fig. 6), so that without using a multiple thread (not realizeable with this pitch anyway) they simulate a nut with the desired low pitch. And, as the other crucial factor, by using rollers of this type sliding friction as prevailing in a pure spindle–nut concept has been replaced by much smaller rolling friction (1 % as the presently envisioned goal).

What we have gained with this motor–gearing combination is a small prismatic drive (applicable also in a robot joint), which used as gripper drive allows to exert grasp forces of more than 400 N with a gripper weight of 5 N and a grasp speed of about 15 cm/sec; without measuring and feeding back grasp forces we arrived at a feedforward grasp force control resolution of $\approx$ 4 N (1 % of max force) with high repeatability. Reduction rate referring to the finger rotation is $\approx$ 1 : 1000.

3.4 Rotational–rotational gearing

The "phase–shift" ideas as outlined in the last chapter have meanwhile been transferred to pure rotational gearings, too [6]. The basic concept of our design is schematically depicted in fig. 7, consisting of two simple planetary gearings (similar to the socalled WOLFROM – gearing) that are connected via the planetary wheels e.g. via friction so that a friction clutch (very useful in case of joint overload) is inherent in the system. The phase–shift idea now is realized by letting the number of teeth in the second (moving) hollow tooth wheel z_5 differ by one compared to the first (fixed) one z_3. To assure, that despite of this both wheels of a planetary wheel block simultaneously mate with their corresponding hollow wheels, they must show up a certain "phase shift"

$$\Delta\varphi = \frac{360^o}{z_4 \cdot n_p} \cdot i \quad , \ i = 1, \ldots n_p$$

of their teeth (see fig. 7), where z_4 denotes the number of teeth in the planetary wheels and n_p is the number of revolving blocks. Due to the friction clutch the phase shift is adopted automatically. Gear transmission ratio is

$$r = \frac{z_2 z_5 (z_3 + z_1)}{z_1 (z_2 z_5 - z_3 z_4)}$$

A reduction rate of $\approx 1 : 600$ turned out to be easily realizable (a factor of $1 : 100$ being contributed by hollow wheels with 100 and 101 teeth), but higher reductions are no problem, too. Note that the gearing is very compact, has more teeth in contact (i.e. better force transmission) and is considerably stiffer than a harmonic drive. Integrating joint torque measurement and control to compensate for the inevitable friction effects is one of the remaining final design steps now.

3.5 Integration into carbon fibre grid links

Presently we are developing a 6 axis carbon fibre light–weight robot as a ground training device for the above–mentioned space robot technology experiment ROTEX. The flight robot as realized in conventional technique cannot sustain itself in a 1 g–environment. However the integration of the above described joint drives into carbon fibre grid structure links as developed by our colleagues from the institute "for structural research and development" in Stuttgart (see fig. 8) is not bound to a special robot system, but envisioned in a completely modular way. Typically the link shown in fig. 8 with approximately 0,5 m length weighs $\approx 2,5$ N, while the joint (including motor + electronics + gearing) is about 3,5 N with a torque capability of ≈ 150 Nm and 2 rad/sec speed. We feel that with designs of this type robots might be 5 – 10 times lighter than they are today without any loss in performance but higher bandwidth of reaction.

4 Conclusion

The paper tries to emphasize, that on the way to smarter, especially lighter and more intelligent robots, all components of a robot arm have to be optimized. In the work presented here, the first goal has been to design a light, really multisensory gripper with electrical gripper drive as the critical endpoint–mass of a robot. 9 range–finders including a scanner, a stiff and a lockable compliant wrist–force–torque sensor, tactile sensors and stereo–cameras are integrated following a completely modular, high–integration hard– and software–concept. The complete gripper with grasp force capability of more than 400 N weighs only ≈ 12 N. The same motor concept (small, light, high–dynamics, high–torque, electronic commutation derived from stepper motor characteristics) as realized in the gripper has been transferred into our new joint drives. The same basic ideas resembling some mechanical "phase shift" have been applied to the design of compact, light–weight high–reduction prismatic gearings (e.g. for the gripper drive in our context) as well as revolute gearings in the joints. Integration of the joint drives into modular grid–structure carbon fibre links was a final step of this unified approach, in which we tried to avoid solving isolated problems only. The main activities will now concentrate on realizing fast digital controllers on the joint level (sampling rates $\prec 1$ msec) to make use of the high dynamics, integrate output–torque sensing and feedback in the joint gears, and build several multisensory redundant and nonredundant light–weight robots, the first of which will be a ground training system for our space robot experiment ROTEX. It is supposed to be nearly 10 times lighter than the kinematically similar commercial robot MANUTEC R2.

References

[1] Khatib, O., *A unified approach for motion and force control of robot manipulators: the operational space formulation.* IEEE J. Robotics and Automation. RA-3; pp. 43-53.

[2] An, Ch.H., Atkeson, Ch.G., Hollerbach, J.M., *Model-based control of a robot manipulator*, MIT Press 1988.

[3] Nsada, H. and Lim, S.K., *The joint torque feedback control of a direct drive arm* in Proc. IFAC World Congress, Munich, 1987.

[4] Hirzinger G., Heindl, H., Landzettel, K., *Predictive and Knowledge-based Telerobotic Control Concepts*, IEEE Conf. on Robotics and Automation, Scottsdale, 1989.

[5] Dietrich, J., Gombert, B., *Vorrichtung zur Umwandlung einer Drehbewegung in eine Axialbewegung*, European Patent No. 0320621 pending.

[6] Dietrich, J., Gombert, B., *Getriebeanordnung in Form eines Umlaufraedergetriebes*, Patent No. 39011070 pending.

[7] Hirzinger, G., Landzettel, K., *Sensory feedback structures for robots with supervised learning*, IEEE Int. Conf. on Robotics and Automation, St. Louis, Missouri, March 1985.

[8] Hirzinger, G., Dietrich, J., *Multisensory robots and sensorbased path generation*, IEEE Int. Conference on Robotics and Automation, San Francisco, April 7-10, 1986.

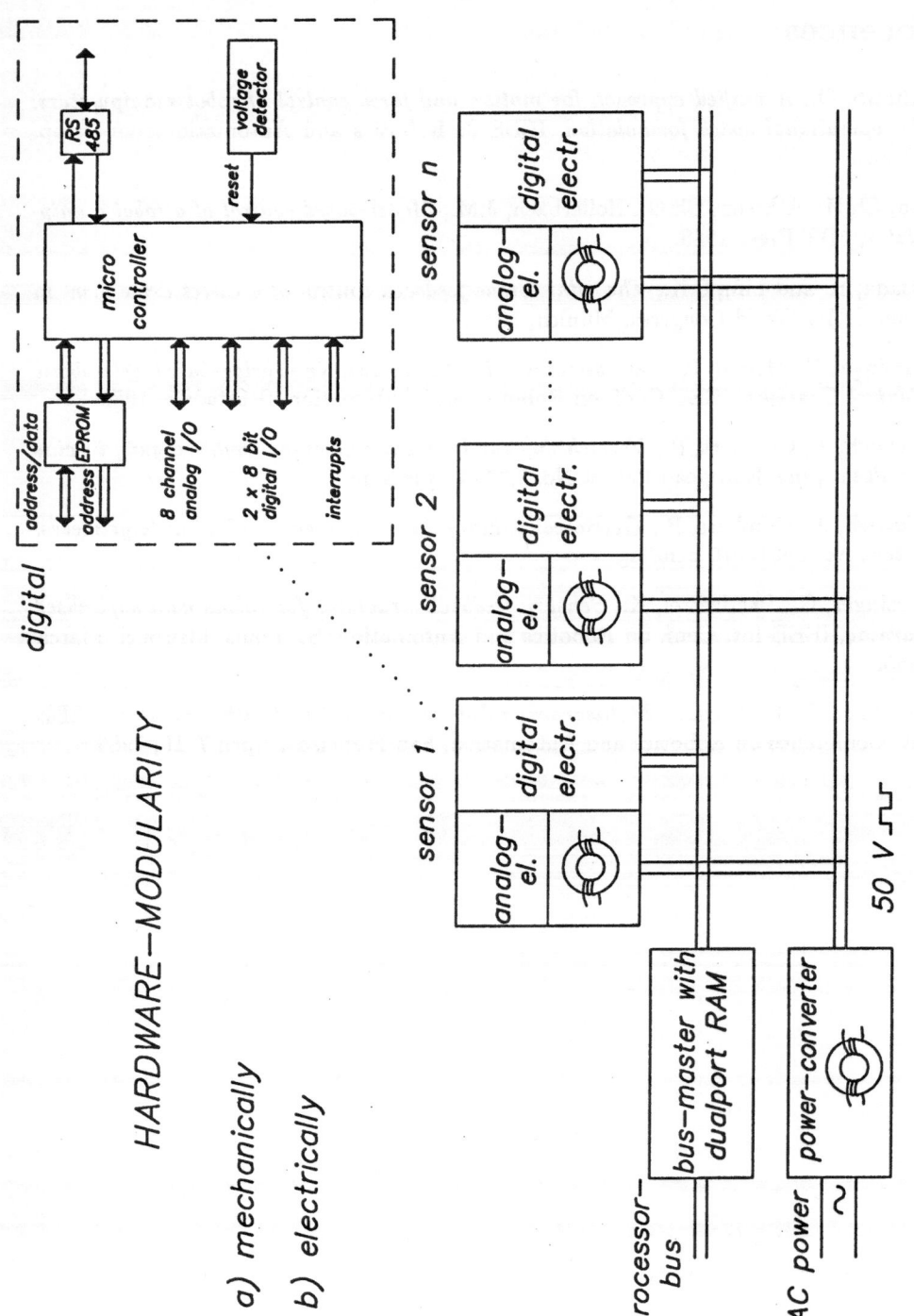

Fig. 1 Hardware modularity in our multisensory gripper

Generation of Data   Generation of Status Data

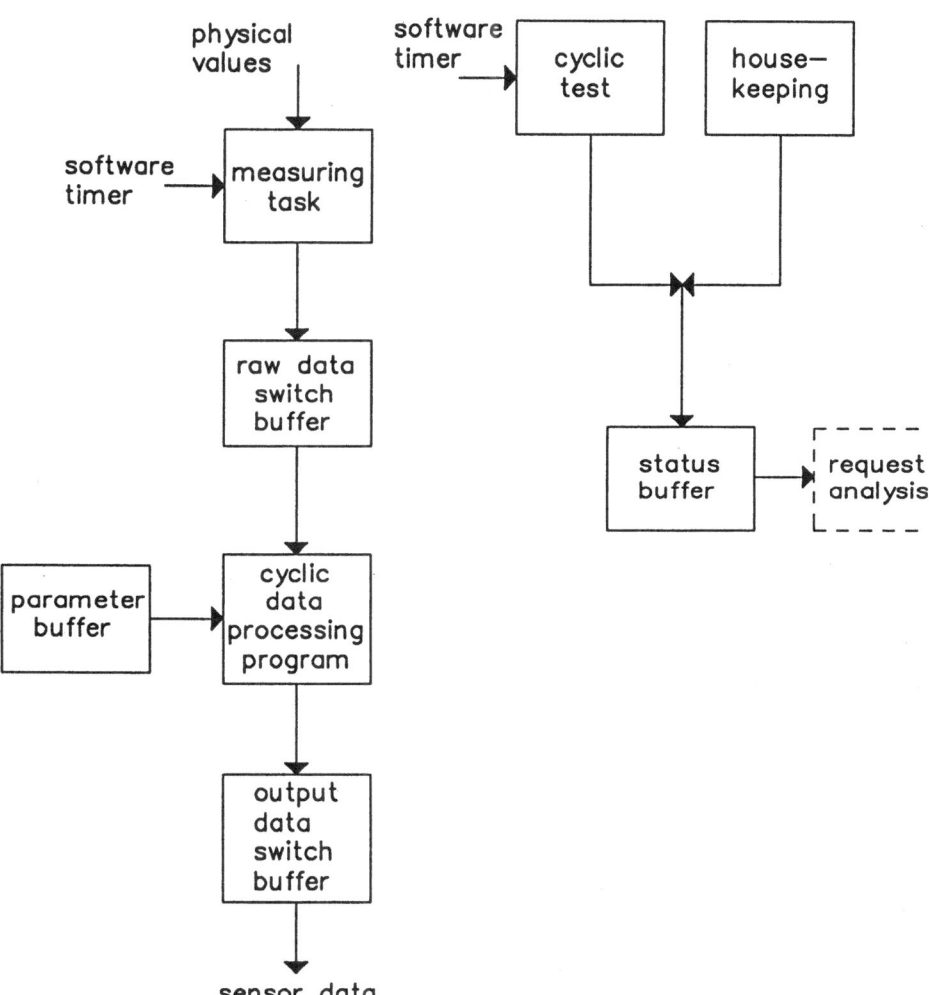

Fig. 2 Software modularity in the sensor processors

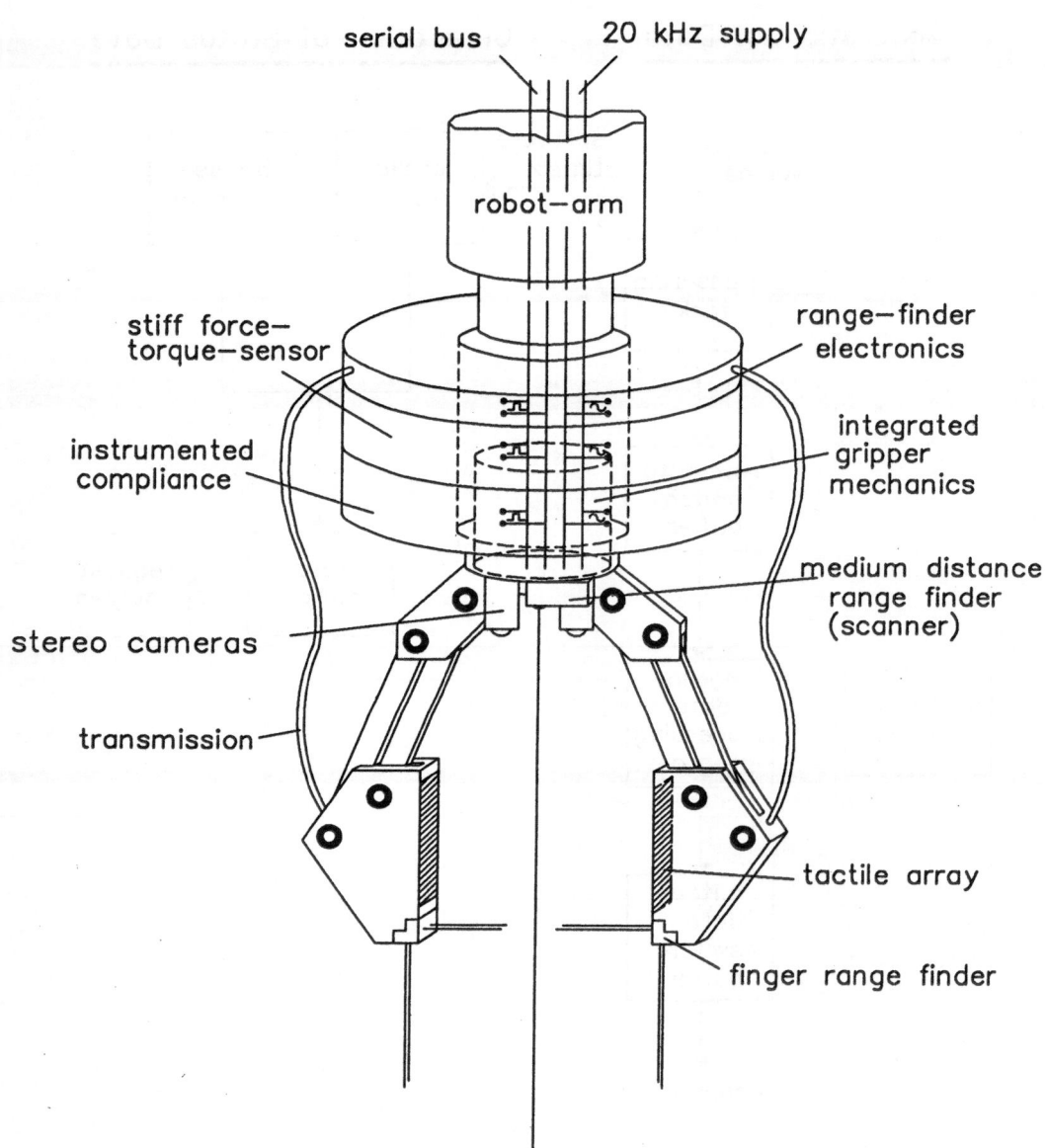

serial bus

20 kHz supply

robot—arm

stiff force—torque—sensor

instrumented compliance

stereo cameras

transmission

range—finder electronics

integrated gripper mechanics

medium distance range finder (scanner)

tactile array

finger range finder

*Fig. 3 Schematic arrangement of sensors
in our prototype gripper*

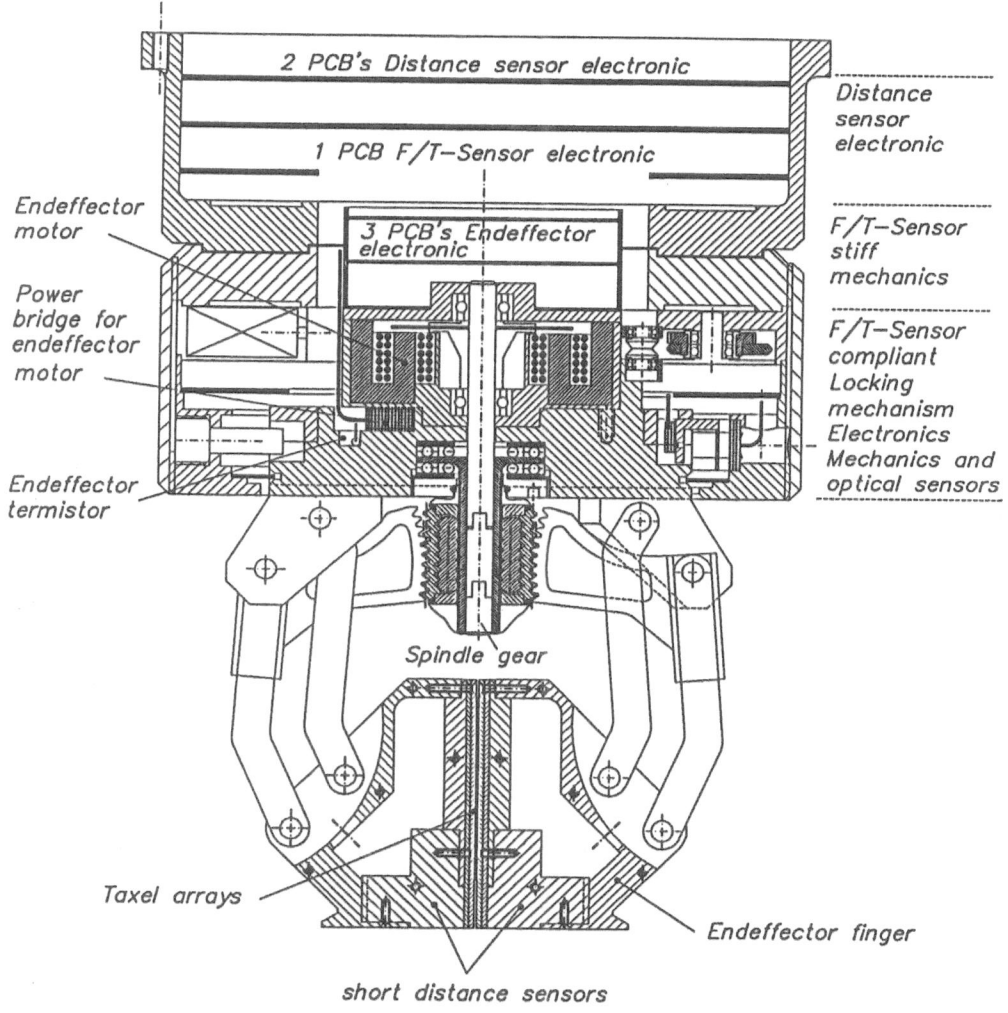

Fig. 4 Gripper (endeffector) section drawing

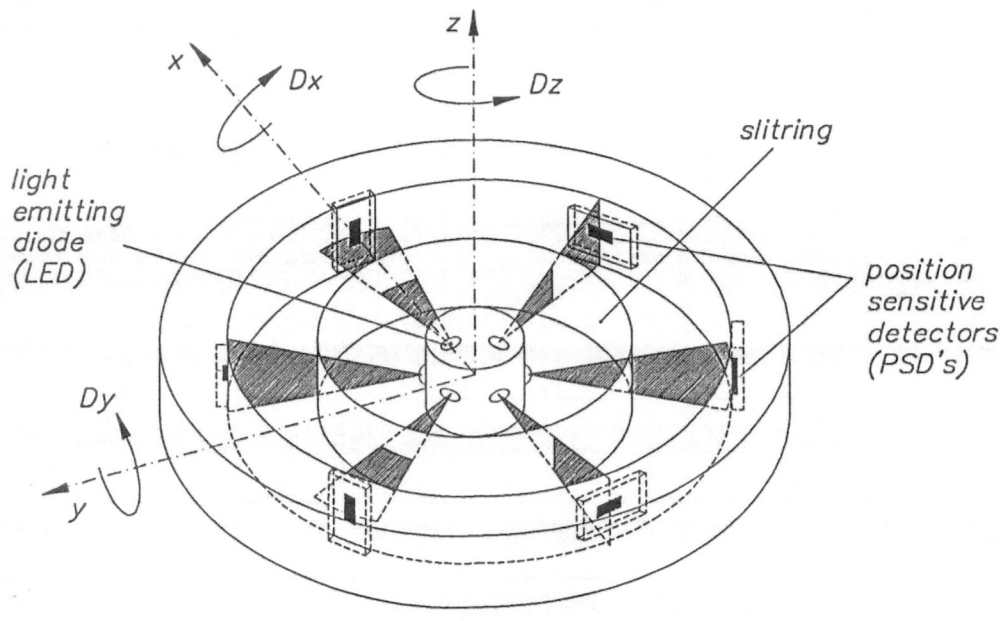

Fig. 5   Schematic diagram and ring—shape
realization of the instrumented compliance

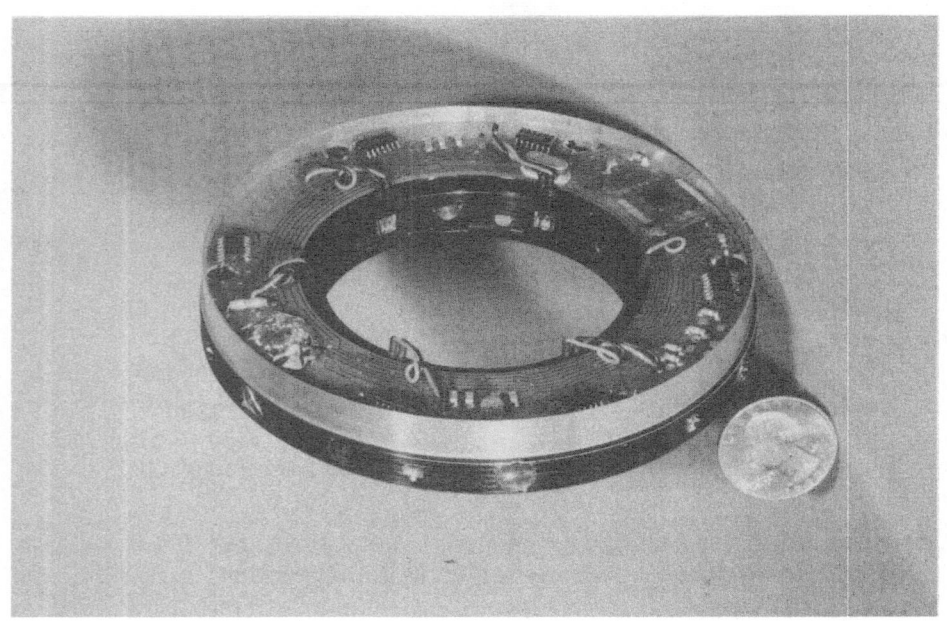

301

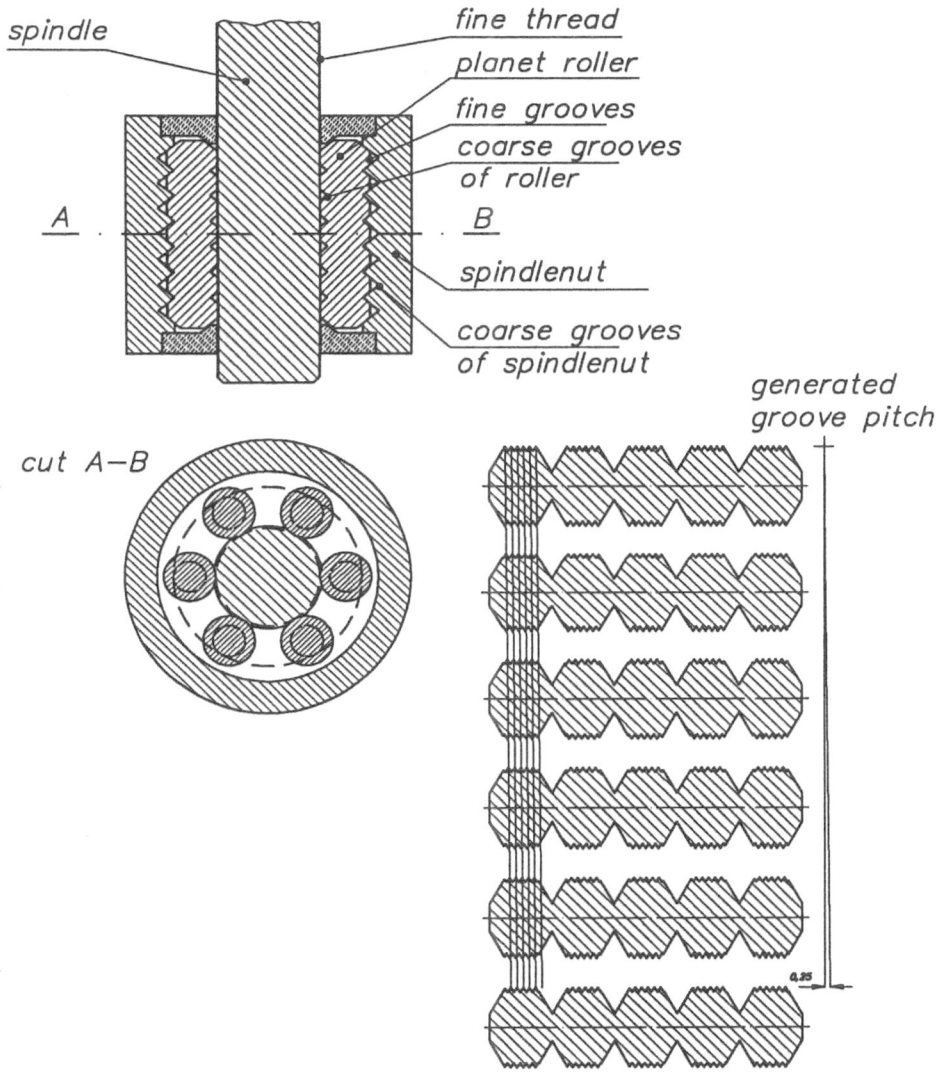

Fig. 6   Our new spindle concept for the gripper drive

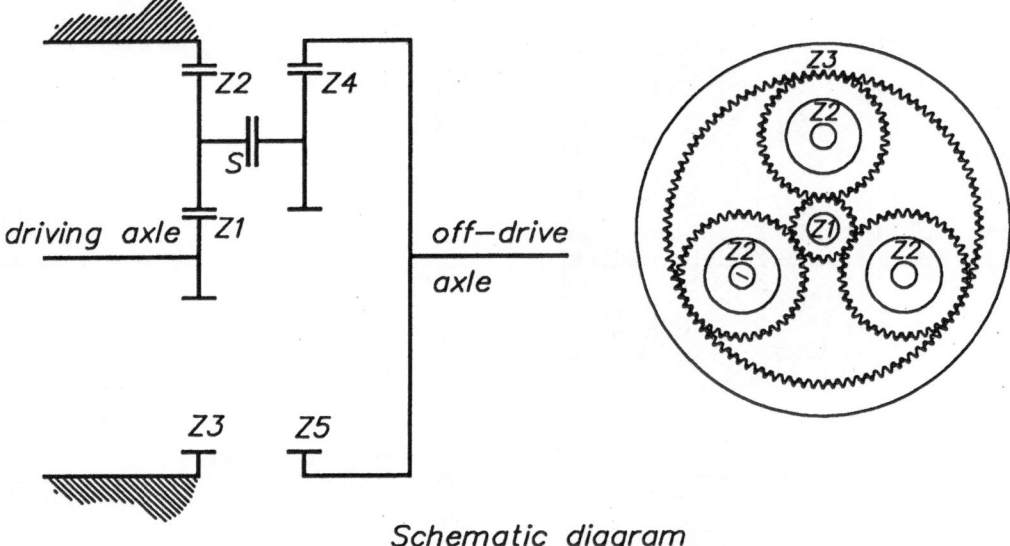

Schematic diagram

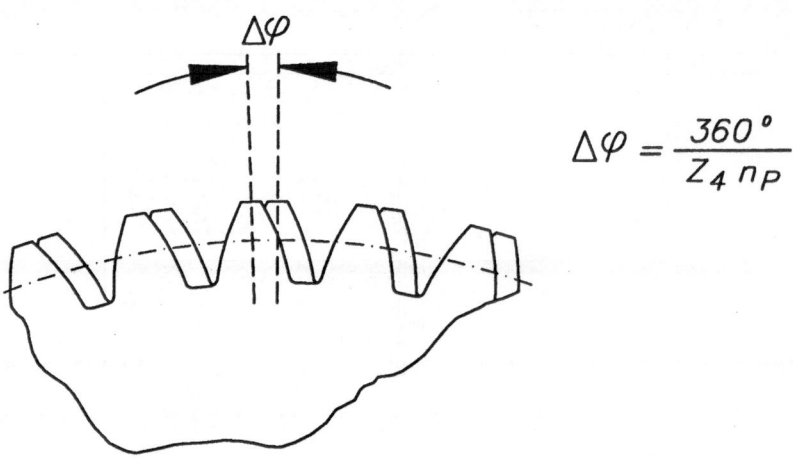

$$\Delta\varphi = \frac{360°}{Z_4 \, n_P}$$

"Phase shift between planetary wheels Z2 and Z4

Fig. 7   Our new compact, high reduction gearing

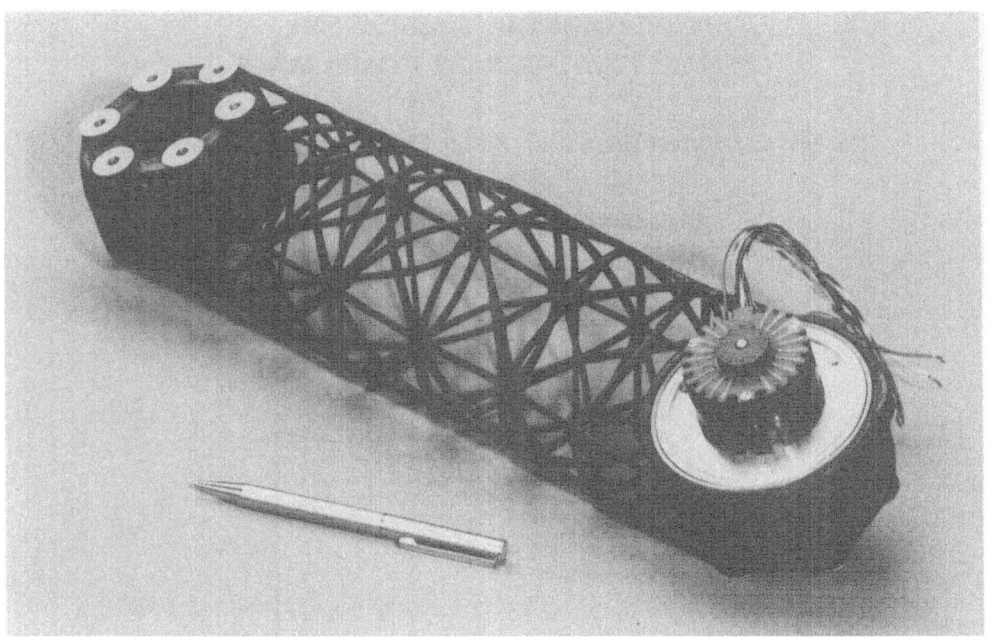

*Fig. 8   First prototype of new joint integrated
into a carbon fibre grid structure link*

EXPERIMENTAL SIMULATION
OF MANIPULATOR BASE COMPLIANCE

Harry West, Norbert Hootsmans, Steven Dubowsky, Nathan Stelman

Department of Mechanical Engineering
Massachusetts Institute of Technology
Cambridge, MA 02139, USA

Abstract

Many future applications of robotic systems will require manipulators to operate from moving vehicles. Such vehicles will be compliant in comparison to the rigid bases on which most manipulators are mounted today. Base compliance can seriously degrade system performance. Statically base compliance may lead to error in the position of the end effector, and dynamically base compliance may interact with the motion of the manipulator and impair the stability of the system. The accuracy of the manipulator may be improved by modelling the base compliance and compensating for its deflection. Further improvement in accuracy may be achieved by endpoint feedback control of the position of the end effector relative to the task frame. A Vehicle Emulator System (VES) has been developed for experimental investigation of the static and dynamic behavior of manipulators mounted on compliant bases. The VES operates under admittance control and can experimentally simulate a wide variety of linear and non-linear six-degree-of-freedom compliances. A series of experiments are described that use modelling of the base compliance and end point control to achieve precise positioning of the end effector of a manipulator manipulator mounted on a compliant base.

1 Introduction

Many future applications of robotic systems will require manipulators to operate from moving vehicles. Such vehicles will be compliant in comparison to the rigid bases on which most manipulators are mounted today. Examples of manipulators on complaint bases are mobile robotic systems for nuclear environments, and *Field Material Handling robots* such as the one shown in Figure 1. An extreme example of a compliant base is a manipulator free floating in space, which is a base compliance with no stiffness or damping.

The compliance at the base of the robot may seriously degrade the performance of the manipulator. Statically the base compliance may lead to error in the position of the end effector, and dynamically the base compliance may interact with the motion of the manipulator and impair the stability of the system. It is possible to improve the accuracy of the manipulator by modelling the base compliance and compensating for the deflection at the end effector. However, it may be difficult to obtain an accurate model of the base compliance. Further improvement in the accuracy of the manipulator can be achieved by directly measuring the position of the end effector relative to the task frame, and controlling the motion of the end effector relative to the task.

Figure 1. Mobile Robot

A *Vehicle Emulator System* has been developed for experimental investigation of the behavior of manipulators mounted on compliant or moving bases. The VES operates under admittance control and can experimentally simulate a wide variety of linear and non-linear six-degree-of-freedom compliances. A description of the VES is given in the Appendix. The goal in developing this apparatus was to provide a general purpose vehicle emulation system with the capabilities to:

(a)   Impose an arbitrary trajectory on the base of a manipulator
(b)   Emulate the weightlessness of a manipulator free floating in space
(c)   Emulate a vehicle suspension including its non-linear characteristics

Experimental work on (a) and (b) has been described in [West, Papadopoulos, Dubowsky and Cheah 1989; Dubowky, Paul and West 1988; Dubowsky and Tanner 1987; Lynch 1987].

This paper describes a series of experiments using the vehicle emulator system to simulate the behavior of a manipulator on a compliant base. The experiments investigate the potential of compliant modelling and end point control to compensate for the error in the position of the end effector caused by base compliance. This work is motivated by the challenge of achieving precise positioning of a large field material handling robot such as the one shown in Figure 1.

2   End Effector Deflection due to Base Compliance

A. One Degree of Freedom Example

An analysis of the deflection of the end effector due to compliance at the base will be developed with reference to a simple one-degree-of-freedom example before the more general case is presented. The manipulator shown in Figure 2 has a single joint, about its z_1 axis, and is mounted on a base with compliance about the y-axis, represented by the vehicle emulator system. The reference frame is attached to the center of the force sensor which is also the center of compliance. The compliance has been chosen to be attached to the inertial frame rather than the moving frame, but for the one degree of freedom example this choice has no effect.

The moment at the center of compliance is given by:

$$\mathbf{m} = {}^{B}\mathbf{r} \times \mathbf{f} + \mathbf{m}_0 \tag{1}$$

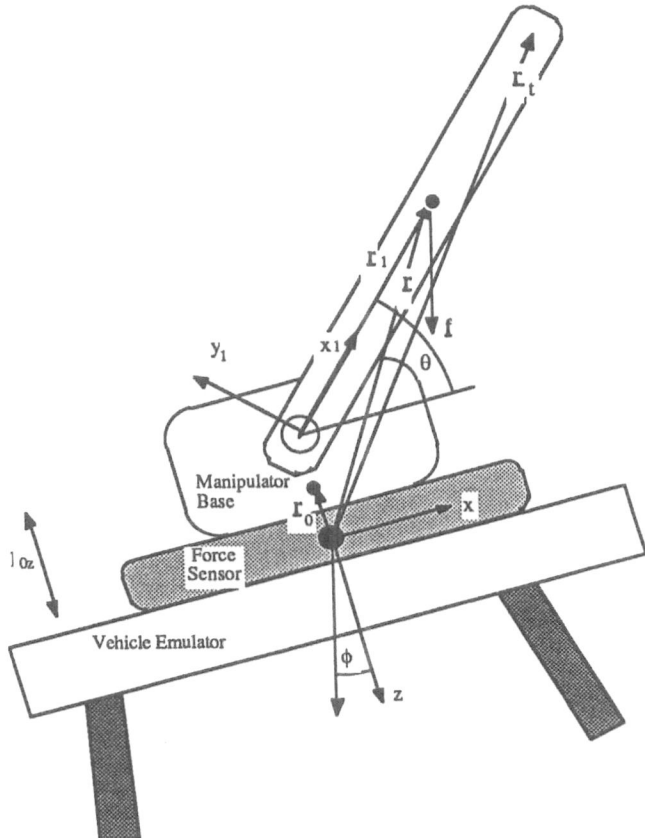

Figure 2. Simple One-Degree-of-Freedom Manipulator
Mounted on a Compliant Base

where $^B\mathbf{r}$ is the position of the center of mass of the moving link in the base reference frame, $\mathbf{f}$ is the gravitational load on the center of mass of the moving link in the base reference frace, and $\mathbf{m}_0$ is the moment due to the weight of the base of the manipulator.

The moment at the base consists of the terms:

$$\mathbf{m} = \begin{bmatrix} m_x \\ m_y \\ m_z \\ 1 \end{bmatrix} \quad (2)$$

and the gravitational load on the link is given by the product of the mass and the acceleration due to gravity:

$$\mathbf{f} = \begin{bmatrix} -M_1 \ g \ \sin\phi \\ 0 \\ M_1 \ g \ \cos\phi \\ 1 \end{bmatrix} \tag{3}$$

where M_1 is the mass of the link and ϕ is the angle through which the base has complied.

$^B\mathbf{r}$, can be described in terms of $\mathbf{r}_1$, the position of the center of mass in the link frame, using the transformation matrices T_0 and T_1:

$$^B\mathbf{r} = T_0 \ T_1 \ \mathbf{r}_1 \tag{4}$$

where T_0 is given by:

$$T_0 = \begin{bmatrix} 1 & 0 & 0 & l_{0x} \\ 0 & 0 & 1 & l_{0y} \\ 0 & -1 & 0 & -l_{0z} \\ 0 & 0 & 0 & 1 \end{bmatrix} \tag{5}$$

and T_1 is given by:

$$T_1 = \begin{bmatrix} C_1 & -S_1 & 0 & 0 \\ S_1 & C_1 & 0 & 0 \\ 0 & 0 & 1 & 0 \\ 0 & 0 & 0 & 1 \end{bmatrix} \tag{6}$$

where C_1 is an abreviation for $\cos(\theta)$ and S_1 for $\sin(\theta)$. The vector $\mathbf{r}_1$ consists of:

$$\mathbf{r}_1 = \begin{bmatrix} r_{1x} \\ r_{1y} \\ r_{1z} \\ 1 \end{bmatrix} \tag{7}$$

Note that that the systems of frames shown in Figure 2 differ from the standard Denavit-Hartenberg notation; frame i is attached at joint i rather than at joint i+1. This notation is used to simplify the analysis.

The resulting moment at the base of the manipulator is:

$$
\begin{bmatrix} m_x \\ m_y \\ m_z \\ 1 \end{bmatrix} = \mathbf{m}_0 +
\begin{bmatrix}
(r_{1z}+l_{0y})\, M_1 g \cos\phi \\
(l_{0z} + S_1 r_{1x} + C_1 r_{1y})\, M_1 g \sin\phi - (l_{0x} + C_1 r_{1x} - S_1 r_{1y})\, M_1 g \cos\phi \\
(l_{0y} + r_{1z})\, M_1 g \sin\phi \\
1
\end{bmatrix}
\tag{8}
$$

where $\mathbf{m}_0$ is the moment of the tilted base and is given by:

$$
\mathbf{m}_0 =
\begin{bmatrix}
r_{0y}\, M_0\, g\, \cos\phi \\
r_{0z}\, M_0\, g\, \sin\phi - r_{0x}\, M_0\, g\, \cos\phi \\
r_{0y}\, M_0\, g\, \sin\phi
\end{bmatrix}
\tag{9}
$$

and $\mathbf{r}_0$ is the location of the center of mass of the base relative to the force sensor frame.

The moments about the y-axis are a function of the position of the manipulator:

$$
m_y = [r_{0z}\, M_0 + (l_{0z} + S_1 r_{1x} + C_1 r_{1y})\, M_1]\, g\, \sin\phi
$$
$$
- [(l_{0x} + C_1 r_{1x} - S_1 r_{1y})\, M_1 + r_{0x}\, M_0]\, g\, \cos\phi
\tag{10}
$$

which can be written in the form

$$
m_y = A\, \sin\phi + B\, \cos\phi
\tag{11}
$$

However, the base deflection angle, ϕ, is a function of the base compliance:

$$
\phi = m_y\, /\, K_y
\tag{12}
$$

where K_y is the base stiffness about the y-axis. The base deflection angle can be solved for iteratively.

310

For relatively stiff bases the angle of deflection will be small and a small angle approximation can be used:

$$K_y \phi = A \phi + B \tag{13}$$

in which case:

$$\phi = \frac{B}{K_y - A} \tag{14}$$

Using this equation, as $A \rightarrow K_y$ then $\phi \rightarrow \infty$, that is to say if the moment due to the out of balance mass increases with ϕ at a faster rate than the restoring moment due to the stiffness of the base, then the system is unstable and may "collapse under its own weight". It can be shown that for $K_y \geq A$ the base will reach a position of stable equilibrium.

The deflected position of the end effector due to base compliance is given by:

$$x_t = R_\phi \, r_t \tag{15}$$

where r_t is the (4×1) vector from the center of compliance to the end effector in the base frame, and R_ϕ is the rotation matrix for an angle ϕ about the y-axis:

$$R_\phi = \begin{bmatrix} C_\phi & 0 & S_\phi & 0 \\ 0 & 1 & 0 & 0 \\ -S_\phi & 0 & C_\phi & 0 \\ 0 & 0 & 0 & 1 \end{bmatrix} \tag{16}$$

An approximation to the appropriate end effector correction for the base compliance is $-\Delta x_t$, however, since ϕ is itself a function of the position of the links of the manipulator, the correction must be made iteratively.

B. Extension to the General Case

The method described in the previous section can be generalized for multi-degree of freedom manipulators mounted on a base with six degrees of freedom of compliance.

The force and moment vector with respect to the base frame is given by:

$$\left[\begin{array}{c} \underline{f} \\ \underline{m} \end{array} \right] = \left[\begin{array}{c} M_t\, \underline{g} \\ M_t\, \underline{r} \times \underline{g} \end{array} \right] = \left[\begin{array}{c} M_t\, g\, \underline{v} \\ M_t\, g\, V\, \underline{r} \end{array} \right] \tag{17}$$

where M_t is the total mass of the manipulator, g is the acceleration of gravity, $\underline{r}$ is the position vector of the manipulator center of mass with respect to the base frame, $\underline{v}$ is a vector whose elements v_i are the components of the unit vector parallel to $\underline{g}$ in the base frame and V is the corresponding matrix whose elements V_i are the components of the unit vector parallel to $\underline{g}$ in the base frame.

$$M_t = M_0 + M_1 + \dots M_n \tag{18}$$

where M_i are the masses of the links of the manipulator.

$$\underline{r} = \left[\begin{array}{c} r_x \\ r_y \\ r_z \end{array} \right] \tag{19}$$

$$\underline{v} = \left[\begin{array}{c} f_x \\ f_y \\ f_z \end{array} \right] \tag{20}$$

and,

$$V = \left[\begin{array}{ccc} 0 & f_z & -f_y \\ -f_z & 0 & f_x \\ f_y & -f_x & 0 \end{array} \right] \tag{21}$$

$M_t g$ represents the weight of the manipulator, and v and V are functions of the orientation (direction) cosines of the base.

The position vector of the center of mass of the whole manipulator, $\underline{r}$, can be calculated as:

$$\underline{r} = \frac{1}{M_t}\, T_L \left(\sum_{i=0}^{n} M_i\, T_i\, \underline{r}_i \right) \tag{22}$$

where r_i is the position of the center of mass of link i with respect to frame i:

$$r_i = \begin{bmatrix} r_{ix} \\ r_{iy} \\ r_{iz} \\ 1 \end{bmatrix} \quad (i = 0,1,...n) \tag{23}$$

T_i is the matrix that transforms a postition vector in frame i to a position vector in the base frame, and:

$$T_L = \begin{bmatrix} 1 & 0 & 0 & 0 \\ 0 & 1 & 0 & 0 \\ 0 & 0 & 1 & 0 \end{bmatrix} \tag{24}$$

T_L is a 3x4 matrix that removes the last element from the 4x1 vectors in the summation.

For a linear compliance, represented by the (6×6) compliance matrix, **C**, the deflection at the base is given by:

$$\Phi = C \begin{bmatrix} f \\ m \end{bmatrix} \tag{25}$$

where:

$$\Phi = \begin{bmatrix} \delta_x \\ \delta_y \\ \delta_z \\ \phi_x \\ \phi_y \\ \phi_z \end{bmatrix} \tag{26}$$

however, since both **f** and **m** are themselves functions of Φ this expression must be solved iteratively.

For small base compliance the deflection at the end effector is given by:

$$\Delta x_t = R_\phi' r_t \tag{27}$$

where Δx_t is the (6×1) vector:

$$\Delta \mathbf{x}_t = \begin{bmatrix} \delta_x \\ \delta_y \\ \delta_z \\ \delta\theta_x \\ \delta\theta_y \\ \delta\theta_z \end{bmatrix} \tag{28}$$

and $\mathbf{R}_\phi{}'$ is now given by:

$$\mathbf{R}_\phi{}' = \begin{bmatrix} 0 & -\phi_z & \phi_y & \delta_x \\ \phi_z & 0 & -\phi_x & \delta_y \\ -\phi_y & \phi_x & 0 & \delta_z \\ 0 & 0 & 0 & \phi_x \\ 0 & 0 & 0 & \phi_y \\ 0 & 0 & 0 & \phi_z \end{bmatrix} \tag{29}$$

Again an approximation to the appropriate end effector correction for the base compliance is -Δ $\mathbf{x}_t$, however, since ϕ is a function of the position of the links of the manipulator the correction must be made iteratively.

3 End Point Control of a Manipulator on a Compliant Base: Experimental Evaluation

By making use of equation (27) it is possible to improve the accuracy of the manipulator by modelling the base compliance and compensating for the deflection at the end effector. However, for most compliant bases that are likely to be encountered in practice the compliance will be difficult to measure, non-linear, and time-varying. Inaccuracies in the estimates of the mass properties of the manipulator or of the load at the end effector will also reduce the usefulness of model-based compensation for obtaining precise positioning at the end effector.

Further improvement in the accuracy of the manipulator can be achieved by directly measuring the position of the end effector relative to the task frame, and controlling the motion of the end effector relative to the task. However, the conditions for stability of a manipulator on a compliant base under endpoint control are not clear. For simplistic models of simple manipulators on a linear compliant base, stability using PD control of the end point can be readily verified [Fasse 1987]. But for real manipulators and real suspensions with non-linear parameters that are difficult both to identify and to model, the conditions for stability are not known. Nonlinear computer simulations can be run using programs such as *Simnon* to verify particular control schemes, but they are only as accurate as the model they simulate and cannot provide general results.

Rather than invest research effort in developing a complex non-linear computer simulation, simple linear models in the literature were investigated to gain insight into the behavior of idealized systems, and the real robot on a complaint base was simulated *experimentally* using the MIT Vehicle Emulation System. By simulating experimentally we could be sure that none of the non-linear terms had been neglected, and that our model was complete. An industrial manipulator was controlled on a linear compliant base using end point feedback, providing an existence proof for this possibility.

The experimental setup is sketched in Figure 3. A six degree of freedom manipulator, PUMA 250, is mounted on the Vehicle Emulator System. The VES admittance model is programmed to give the desired base compliance matrix. The manipulator makes simple one degree of freedom moves at the shoulder resulting in a change of the moment and force at the base. The static mass properties of the manipulator have previously been measured [West, Papadopoulos, Dubowsky and Cheah 1989] and can be used to give an estimate of the base rotation, ϕ, and the resulting deflection at the end effector, Δx_t. Applying this end effector correction brings the manipulator to approximately the desired position. More precise positioning is achieved by endpoint control.

The endpoint position feedback sensor is a photodiode measurement of the relative location of the end effector to a referencing laser beam. The photodiode can provide two axis measurement, but in this implementation only the height of the end effector was controlled. The output of the diode amplifier is a voltage proportional to the height of the center of the light spot. The signal from the photodiode is continuous and linear, better than 0.5% linearity at the center of the diode, so the sampling rate is limited only by the cycle time of the control program. The endpont controller regulated the position of the shoulder axis of the manipulator to maintain the position of the laser spot at the center of the phottdiode. A PID controller was used to eliminate steady state error. The controller modified the set point of the manipulator controller with a cycle time of 30 ms.

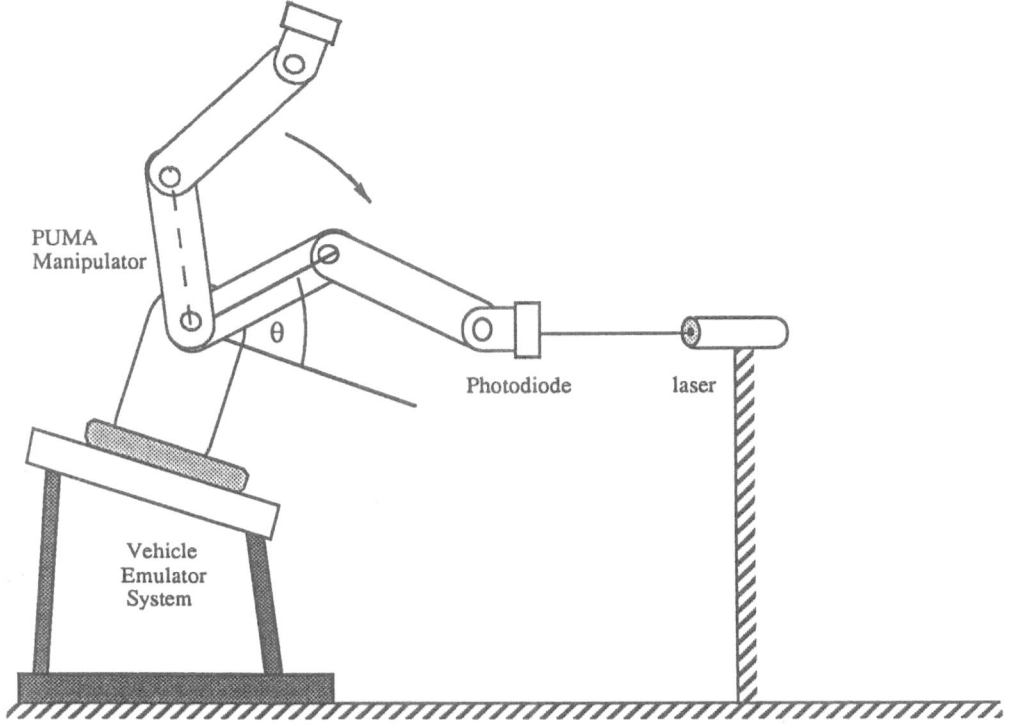

<u>Figure 3. Sketch of Manipulator under Endpoint Feedback Control on a Compliant Base</u>

Experimental Results

Three experiments are documented in the graphs below.

A. VES Stationary
     Endpoint controller: Kp = 0.3, Kv = 0.0, Ki = 0.036
     Initial E. E. position error 9 mm

Under end point control manipulator is able to correct its position error with a rise time of less than 1/30th of a second, and then reduce steady state error down to the sensor noise level within 0.5 seconds with essentially no overshoot, Figure 4.

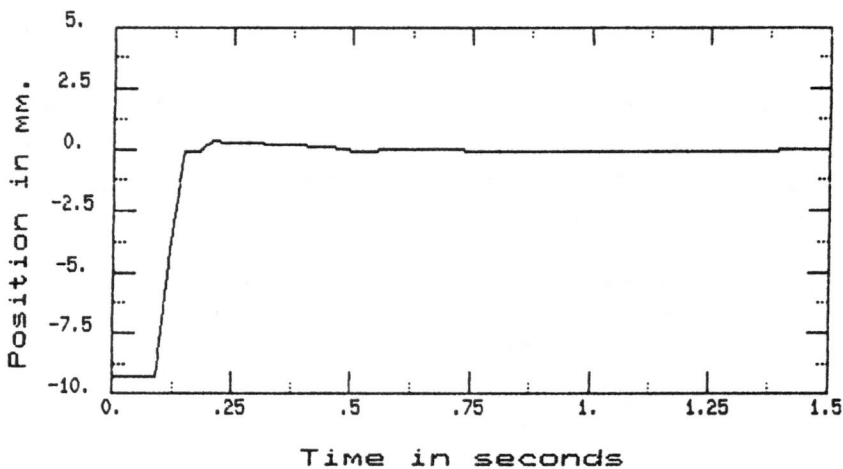

Figure 4. Endpoint Servoing of Manipulator on Rigid Base

B.   VES Compliant: $I = 18$ Kgm^2, $B = 34$ Nm/rad s^{-1}, $K = 215$ Nm/rad,
                            $\zeta = 0.3$, $\omega_n = 0.6$ Hz

Endpoint controller: $Kp = 0.3$, $Kv = 0.0$, $Ki = 0.036$

Initial E. E. position error 17.5 mm

With base compliance the rise time of the endpoint controller is slower and the response more oscillatory, Figure 5. The manipulator is able to correct its position error with a rise time of approximately 1/10th of a second, and then reduce steady state error down to the sensor noise level within 2.0 seconds. The impaired performance may be partly an artifact of the apparatus, for example vibration of the Vehicle Emulator System may be adding a significant amount of noise to the postion and force measurements. Examination of the base torque and angle show that the base compliance contributes to the overshoot of the endpoint controller. The spike on the moment plot clearly indicates the time at which end point control was initiated.

Figure 5. Endpoint Servoing of Manipulator on A Stiff Compliant Base
     (a) Base Torque and Angle as the Manipulator Moves
     (b) Detail of Base Torque and Angle
     (c) End Point Position Error Under End Point Control

317

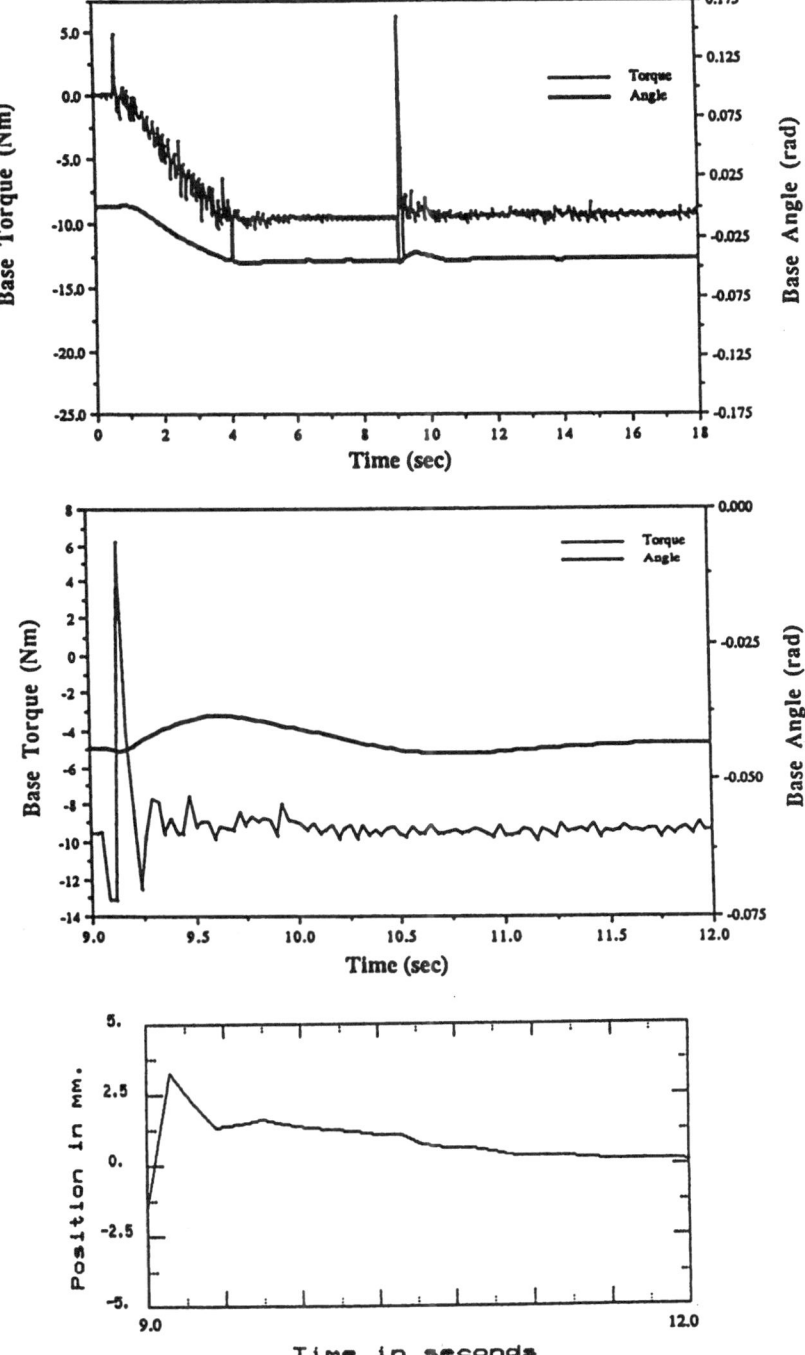

C.   VES Compliant: $I = 10$ Kgm^2, $B = 23$ Nm/rad s^{-1}, $K = 100$ Nm/rad,

$\zeta = 0.36$, $\omega_n = 0.5$ Hz

Endpoint controller: $Kp = 0.3$, $Kv = 0.0$, $Ki = 0.036$

Initial E. E. position error 57mm

Reducing the stiffness of the base makes the interaction between the motion of the manipulator and the motion of the base more apparent, Figure 6. On a soft compliant base the end point feedback sensor is out of range. The manipulator makes an initial correction using a model of its base compliance to bring the end point sensor into range, and then uses end point control as before. The response is slower and more oscillatory, as would be expected.

The initial correction was made using a model of the base compliance and estimates of the mass properties of the manipulator. The mass properties were measured using the method described in [West, Papadopoulos, Dubowsky, and Cheah 1989]. The parameters identified were:

$$p1 = M_1\,r_{1x} \qquad\qquad = +0.518 \quad \text{Kg m}$$
$$p2 = M_1\,r_{1y} \qquad\qquad = +0.309 \quad \text{Kg m}$$
$$p3 = M_0\,r_{0x} + M_1\,l_0 \qquad = -0.006 \quad \text{Kg m}$$
$$p4 = M_0\,r_{0y} + M_1\,(l_{0y} + r_{1z}) \qquad = +0.831 \quad \text{Kg m}$$
$$p5 = M_0\,r_{0z} + M_1\,l_{0z} \qquad = +3.883 \quad \text{Kg m}$$

These values were substituted into equation (10), and then equation (12) was used to iterate for the expected base moment and deflection. For the initial arm position described by $\theta 1 = -90^o$, $\theta 2 = -28^o$, $\theta 3 = -124^o$, the base moment was measured to be 3.1 Nm. After the arm was moved to a new position described by $\theta 1 = -90^o$, $\theta 2 = 62^o$, $\theta 3 = -124^o$, and with a base compliance of 100 Nm/rad, the change in the base moment, m_y, was calculated to be -10.5 ± 0.2 Nm. Adding the initial value, the base moment corresponding to the new arm position was predicted to be -13.6 ± 0.2 Nm. The actual value of the base moment was measured to be -13.5 ± 0.2 Nm, and the corresponding final value of the base angle was measured to be -0.135 radians. Using equation (27) a new arm position was calculated to compensate for the base compliance, bringing the end point sensor into range.

Figure 6. Endpoint Servoing of Manipulator on A Soft Compliant Base

(a) Base Torque and Angle as the Manipulator Moves

(b) Detail of Base Torque and Angle

(c) End Point Position Error Under End Point Control

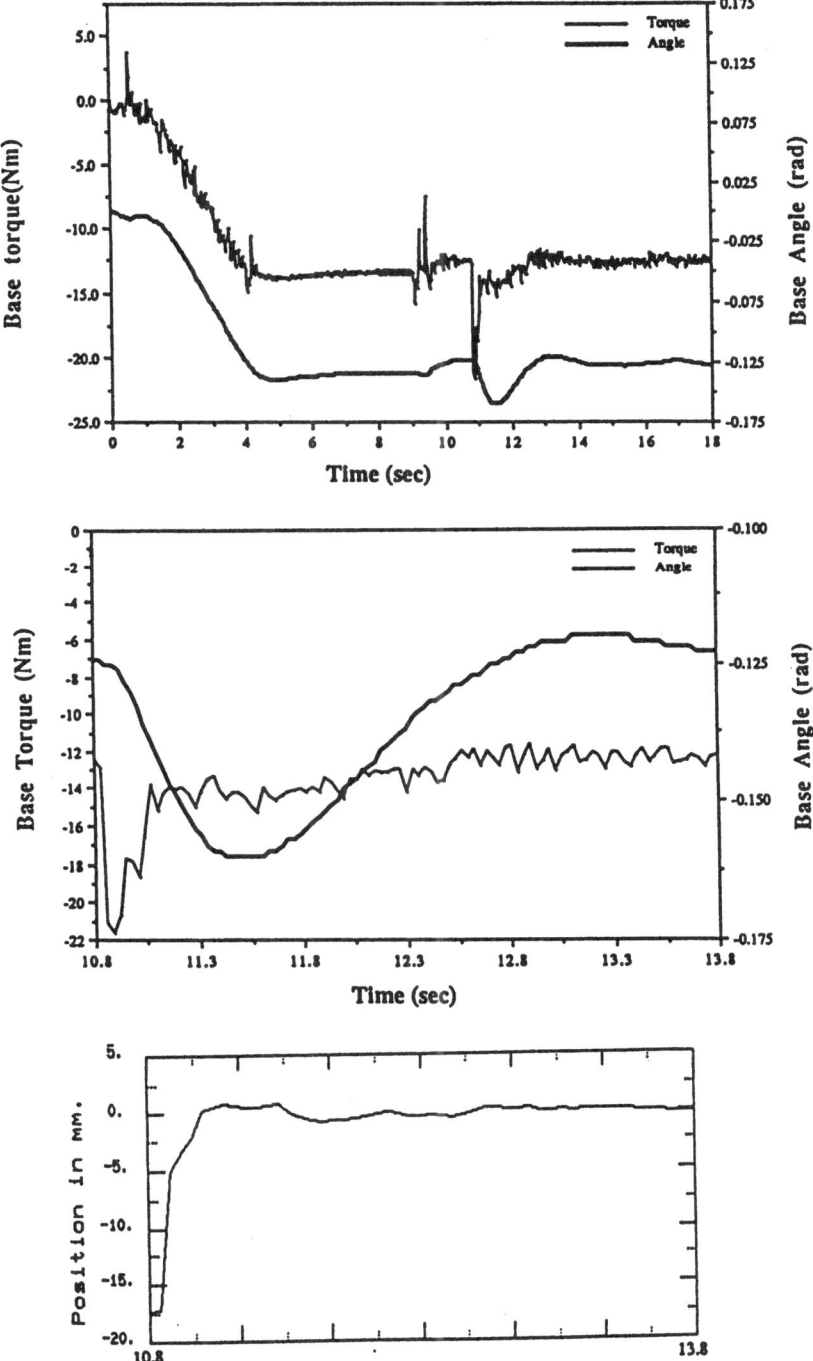

4   Conclusions

In the future, robotic manipulators will be increasingly used in applications in which they are not fixed to the factory floor, but are mobile or are mounted on vehicles. Mobile manipulators will be characterized by compliant rather than rigid bases, and such compliance will impair the performance of the manipulator. Statically the base compliance may lead to error in the position of the end effector, and dynamically the base compliance may interact with the motion of the manipulator and cause instability.

A procedure has been developed for calculating the static deflection at the end effector of a manipulator mounted on a compliant base by modelling the compliance characteristics of the base and the mass properties of the manipulator. Using this model, the position of the end effector can be corrected to compensate for the deflection of the base. More accurate positioning of the manipulator can then be achieved using end point feedback control.

The stability conditions for a manipulator on a compliant base under endpoint control are not clear. Rather than develop a complex non-linear computer simulation to investigate this issue, a real robot on a complaint base was simulated *experimentally* using the MIT Vehicle Emulation System. The robot was found to be stable under end point control for a wide range of linear base compliances.

The Vehicle Emulator System has been built to emulate base compliance. The VES operates under admittance control and can experimentally simulate a wide variety of linear and non-linear six-degree-of-freedom compliances, including terrestrial vehicle compliance and free floating conditions in space. The VES is being used to experimentally evaluate sensors and algorithms for the control of mobile manipulator.

Future work will investigate the stability conditions and performance limits for end point control of manipulators on compliant bases; and will develop algorithms, sensors and other hardware for their control.

Acknowledgements

The support of this work by NASA (Langley Research Center Automation Branch) and the DARPA (US Army Human Engineering Laboratory and the Oak Ridge National Laboratory as agents) is acknowledged. A number of MIT students have made significant contributions to this work; they are Charles Graves, Vijay Krish, Naveed Ismail, and Evangelos Papadopoulos.

References

Dubowsky, S., I. Paul, and H. West, "An Analytical and Experimental Program to Develop Control Algorithms for Mobile Manipulators", Proc. RoManSy '88, Udine, Italy, 1988.

Dubowsky, S. and A. B. Tanner, "A Study of the Dynamics and Control of Mobile Manipulators Subjected to Vehicle Disturbances," Proc IV Int Symp of Robotics Research, Santa Cruz, CA, 1987.

Fasse, E. D., Stability robustness of impedance controlled manipulators coupled to passive environments, S. M. thesis Dept. of Mech. Eng., MIT 1987.

Fresko, M., The Design and Implementation of a Computer Controlled Platform With Variable Admittance, M.S. Thesis, Dept. of Mech. Eng. MIT, Cambridge, MA January, 1986.

Lynch, R., Analysis of the Dynamics and Control of a Two Degree of Freedom Robotic Manipulator Mounted on a Moving Base, S. M. Thesis, Dept. of Mech. Eng., MIT, Cambridge, MA 1985.

National Academy, Application of Robotics to Reduce Risk and Improve Effectiveness: A Study for the United States Army, Published by the National Academy Press, Washington, D.C. 1983

Stelman, N., Design and Control of a Six-Degree-of-Freedom Platform with Variable Admittance, S.M. Thesis, Dept. of Mech. Eng., MIT, Cambridge, MA.

West, H., E. Papadopoulos, S. Dubowsky, H. Cheah, "A Method for Estimating the Mass Properties of a Manipulator by Measuring the Reaction Moments at its Base," IEEE Conference on Robotics and Automation, Scottsdale, AZ, 1989.

Appendix - The Vehicle Emulation System

The Vehicle Emulation System consists of a manipulator mounted on an emulator system [Fresko 1986, Stelman 1988]. The experimental manipulator is a six degree of freedom PUMA 250 with a custom joint controller, and the emulator system comprises a six degree of freedom Stewart mechanism, six degree of freedom force sensor, and computer controller. The experimental manipulator and the emulator system are controlled by individual DEC PDP-11/73's, and can be coordinated using a communication link between the two computers. A sketch of the system is shown in Figure A1.

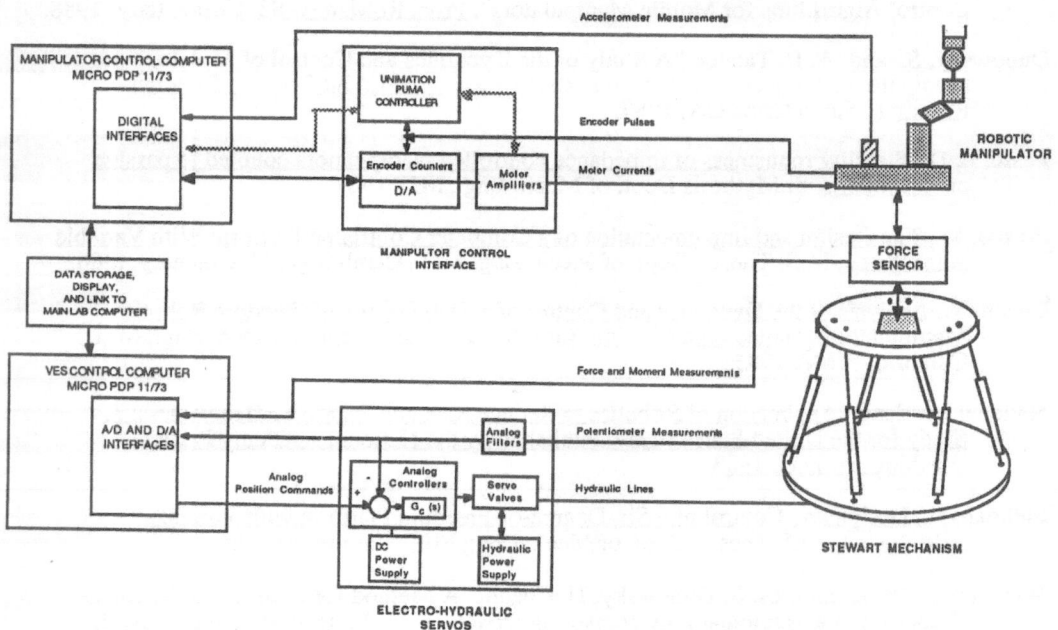

Figure A1. Overview of Space Vehicle Emulation System

The Stewart mechanism and force sensor can, respectively, move and measure forces in any of the three translational directions (x,y,z), and any of the the three rotational directions, $(\theta_x, \theta_y, \theta_z)$ possible in general three dimensional space. The platform stands approximately 3 feet of the ground in its home position. The major hardware elements of the Space Vehicle Emulation System are shown in Figure A2, and a photograph of the System is shown in Figure A3.

Figure A3. Photograph of Vehicle Emulation System

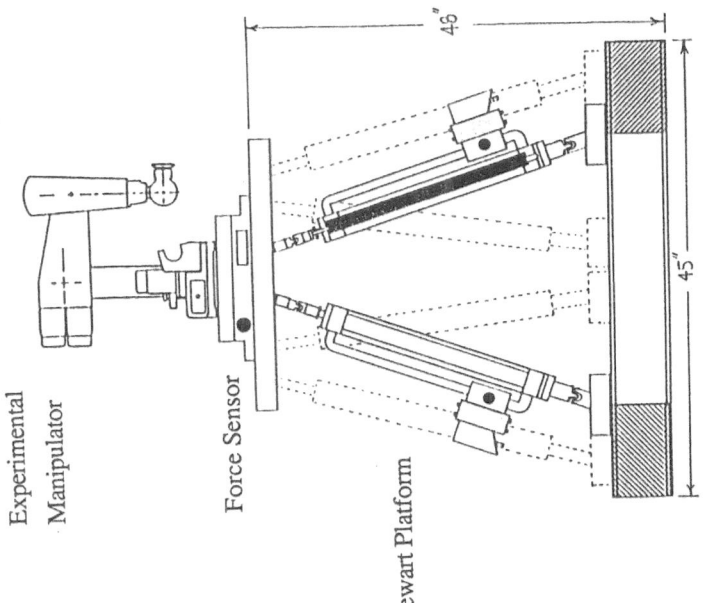

Experimental
Manipulator

Force Sensor

Stewart Platform

46"

45"

Figure A2. Mechanical Hardware Elements
of Vehicle Emulation System

A micro-PDP11/73 is used for trajectory calculation, and also to provide position commands to the analog joint controllers of the Stewart mechanism. The microcomputer is also used to subtract the calculated gravitational load of the manipulator from the force measured by the force sensor for simulating weightless conditions. In addition to trajectory calculation and control, the computer performs a supervisory function checking for approaching violations of the kinematic constraints, and verifies that the joints are following the desired trajectory within allowed error bounds.

The Vehicle Emulator System may be operated in either position driven or admittance control modes. In the position driven mode the VES imposes an arbitrary motion onto the base of the experimental manipulator corresponding, for example, to the motion of a vehicle moving over rough ground. In the admittance control mode, dynamic forces due to manipulator motions are measured by the force sensor and cause the platform to move as if, for example, the manipulator were mounted on a truck suspension or free floating in space.

The admittance control method can be used to simulate vehicles with any dynamic model of the general form:

$$\underline{x}'' = \underline{f}(\underline{x}, \underline{x}', t) \tag{A1}$$

where $\underline{x}$ is the position of the vehicle. The structure of the admittance control mode is shown in Figure A4. The output of the admittance model is the acceleration of the base corresponding to the measured force. The base acceleration is integrated twice to give the desired position trajectory of the compliant base, and then an inverse kinematic model is used to calculate the corresponding leg lengths of the Stewart mechanism. The desired leg lengths are updated each program cycle (every 30 ms for these experiments), and the actual leg lengths are regulated about this series of set points. Under admittance control the Vehicle Emulator System is able to simulate a wide range of different vehicles, for example a robot mounted on a truck with a soft suspension and stabilizing outriggers.

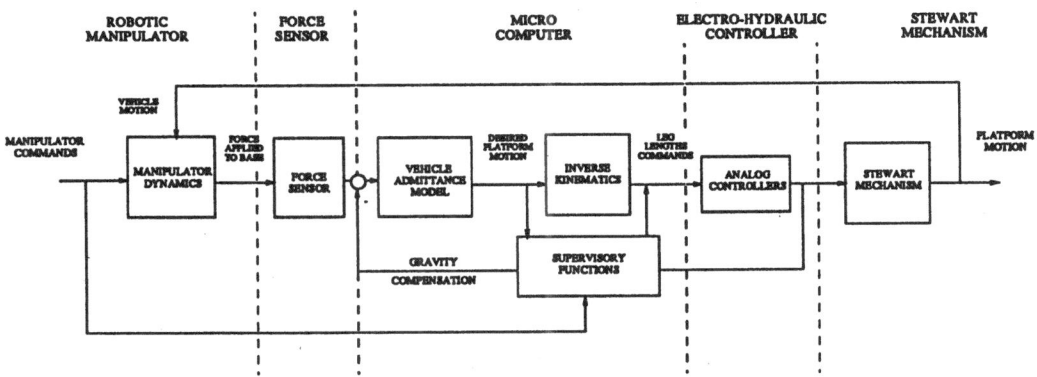

Figure A4. Admittance Control System

The VES is able to simulate vehicle motions of approximately ±150 mm and ±30 degrees in any direction, and currently has a bandwidth of 5 Hz with a 20 kg load. The performance of the vehicle emulator system in its emulation mode is illustrated in Figure A5, which shows the system response to a step change of 45.0 N in vertical load on the platform. The model used by the emulator was for a vehicle whose suspension consisted of a combination of springs and dampers. The model of the vehicle's behavior in the vertical direction is that of a linear second order system, and is shown in the Figure along with the experimentally measured emulator response. The vehicle is modeled for this example as a mass of 175 kg supported by a suspension with a stiffness of 7000 N/m and with a damping constant of 525 N/ms-1, and should respond with a damped oscillation with a natural frequency of 1 Hz and a damping ratio of 0.25. Figure A5 shows that using simple PD control the VES was able to emulate vehicle motion with good accuracy.

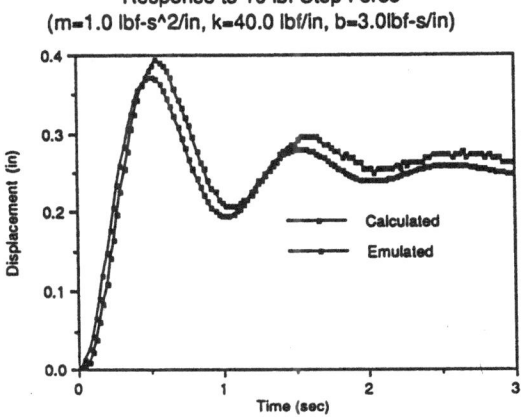

Figure A5. Ideal and Measured VES Response

The force sensor measures the sum of the dynamic reaction forces and the gravitational reaction forces. As the position of the links and the orientation of the base of the manipulator change, the gravitational load also changes. For emulating space conditions this load must be subtracted from the load measured by the force sensor to give the dynamic reaction force at the base. The gravitational load of the manipulator on the force sensor can be calculated from the position of the links of the manipulator and the orientation of the base of the manipulator to the vertical, which are known, and the mass properties of the manipulator [West, Papadopoulos, Dubowsky, Cheah 1989].

An Experimental System for Incremental Environment Modelling by an Autonomous Mobile Robot*

Philippe Moutarlier[†] and Raja Chatila
LAAS-CNRS   7, Ave. du Colonel Roche 31077
Toulouse cedex France

Abstract

Incremental map-making is a necessary function of an autonomous mobile robot. Sensor data are always imprecise, and in the case of a mobile robot, sensor location is itself imprecise and even sometimes false (*e.g.* in case of slippage). We show how sensors data inaccuracies can be processed to produce a consistent environment model and an as precise as possible robot positioning. The experimental system (a mobile robot with a laser range finder and odometry) is presented and the theoretical approach is applied on actual data.

1   Introduction

Incremental map-making involves chaining of important operations among autonomous mobile robots abilities. Environment perception and modelling of the local perception, possibly including feature recognition, local decision about the next goal, and navigation are indeed necessary for this purpose. The whole process is based on perception devices which are always liable to errors. In this paper, we show how sensors inaccuracies cumulate while the perception-modelling process is running. At each stage, earlier errors effects on current computations are analyzed. Especially, we will see how cumulative motion errors add-up with perception uncertainties, how this could lead to inconsistent models, and how to solve this problem.

A rigorous formal approach, related to Kalman filtering, has been developed to deal with stochastic sensory data fusion in its spatial and temporal aspects. We show how an

---

*This work was partially and independantly supported by a contract with SHELL SRSA Grand Couronne, and by ESPRIT project P1560 (SKIDS)

[†]Supported by a CIFRE with SRSA

instability could appear due to the non-linear nature of the problem, and especially if a good sensor noise variance calibration is not available. This is particulary true when the sensor's location is itself imprecise. We propose a solution to this problem within our formalism.

On the other hand, robot localization needs local modelling of raw data for matching with a global model. Indeed, for a laser range finder for example, sensor raw data (depth points) are in general too poor and instable to be directly used as recognizable primitives, and a higher level representation has to be produced (for example segmentation of a laser range-finder depth points, or edge extraction from an image, *etc.*). This segmentation relies on using given primitives (lines, etc.). For a given data set, it may occur that the primitives are not suited for the representation, and an object may be misrepresented (or even not at all). However, the choice of the primitives and their fusion to produce other ones is not the purpose of this paper. We rely of a representation using straight lines (i.e polygonal approximations in 2D).

A real experimentation, using a mobile robot with a horizontally scanning laser range finder and odometry is presented. It includes local perception and modelling, robot repositioning, update of the environment global model (obstacle surfaces).

We present in the following section the basic mathematical tools that we developed to deal with the stochastic data fusion problem. In §3, we derive a model of sensor inaccuracies before applying the tools to build a consistent model of the environment (§4).

2 Mathematical Tools to deal with Sensors Inaccuracies

Incremental map-making uses two kinds of sensors: environment perception sensor(s) (stereovision, ranging devices or a combination of sensors) and position sensor(s) such as odometric systems. Each of them is obviously liable to errors. Furthermore, we have to deal with two important problems when we want to build a global representation of the world from different points of view:

1. motion errors add-up with environment perception inaccuracies in global representation of local data;

2. motion errors are themselves cumulative. Therefore, the noise on the robot (sensor) location state vector, which is not a white noise, induces a drift which can lead to an inconsistent model.

We have proposed and implemented an approach, relying on stochastic methods, to build a consistent representation of the robot's environment while locating it with respect to this environment. In order to deal with spatial (*i.e., between two different features) and temporal (*i.e., between two perceptions of a same feature at different times) correlations, we have developed a formalism related to Kalman filtering. Unfortunately, this method needs explicit variance representations which are not exactly computable in non-linear cases, and are sometimes difficult to estimate, for example in the case of the robot's wheels slippage. This can lead to biases and instability as we will see later. Our formalism enables us to limit this problem.

In this section, we present first our basic choices about environment and robot representations. Then, a minimal variance linear estimator is presented. Basic operations such as integration of a new object, model updating after a new observation, and fusion of two primitives are derived from our estimator. All these results are presented in [10] with more details.

2.1 Basic Choices about Environment Representation and Robot Positioning

Let us consider a mobile robot that moves, perceives objects and models them. An object will be defined as a structured set of geometrical features (points,edges,curves ...) that can be extracted from raw sensor data. In our experiments, a horizontally scanning laser range finder is used, and our objects are polygons. The approach is however not limited by the kind of sensor nor by the type of primitive.

There are two main choices for representing the relations between objects and robot in the environment, and their related uncertainties that we will call "Relational" and "Location " representations.

Relational Representation: An object is related to another one by the uncertain transform between their reference frames. In this approach, we have a network of *relationships* that have to be updated as new observations are made.

   The relation between any two objects i and k is computed by compounding several uncertain relationships $i, i + 1, \ldots, k$. This approach of relational graph is used by several authors [2], [4], [1], [9].

Location Representation: All local object frames are referenced in a unique (arbitrary) frame $\mathcal{R}_0$, called the absolute frame. The mutual object relations are not the basic element in this representation, but rather the relation to the absolute frame, or in other words the *positions* (of the local frames) in the absolute frame. This approach is used by [3] and [11]. If a local relationship between objects i and k is needed, then it is found through transform compounding via the absolute frame $(i, \mathcal{R}_0, k)$. All objects, including the robot, are known in the absolute frame, as well as their associated uncertainties. It is therefore more natural in this approach to use a *single* state vector representing all objects (including the robot).

As it is explained in [10], we chose the second solution for representing the world. We have demonstrated that this representation enables also to compute the relationship between any pair of objects if necessary, since the correlations between their reference frames are known.

Chosing a single reference frame does not mean that we do have one and only one for all the environment (for example for a building with different rooms, etc.), but we have one frame for an entity of this environment within which objects are strongly correlated, for example by robot perception (*e.g.*, one frame per room).

2.2 Notations

Let us consider two objects (*i.e.*, two frames) i, j defined respectively by two random vectors x_0^i, x_0^j. Let r be the frame attached to the robot at the moment of the observation of the two objects.

Let us call:

- ϵ_n^m the error vector on the parameters of frame m in frame n, $\hat{x}_n^m$ its mean value, and $P_n^m = V(x_n^m)$ its variance; ϵ_0^r is then the error on robot position, and P_0^r its variance.

- $C_{nm} = E[\epsilon_0^n \epsilon_0^{mT}]$ the covariance between the error vectors of frames n and m in the absolute frame.

The variance of the relation uncertainty between two frames is then:

$$P_i^j = E[\epsilon_i^j \epsilon_i^{jT}] = J_i P_0^i J_i^T + J_j P_0^j J_j^T + J_j C_{ji} J_i^T + J_i C_{ij} J_j^T \qquad (1)$$

J_i and J_j being the jacobians of the transform T_i between the absolute frame $\mathcal{R}_0$ and the frame i in which frame j is expressed.

This variance is completely determined if the correlations between the uncertainties of each couple of object frames in the absolute frame are known.

We have derived a minimal variance linear estimator and a formalism adapted to manipulating the stochastic aspect of the multisensory data fusion problem, even in the case of noise correlations or colorations. In order to develop our formalism, the same approach as Kalman [6] for minimizing the error variance was used [10]. We only present here the expressions of state vector and estimation variances according to this formalism.

Let us suppose that we have an optimal estimate $\hat{x}_k$ of state x at instant k, and let $\epsilon_k = x - \hat{x}_k$ be the estimation error. A new information is acquired at k+1 by a measurement z_{k+1}. We call δ_{k+1} the prediction error such as :

$$\delta_{k+1} = z_{k+1} - \hat{z}_{k+1}$$

where $\hat{z}_{k+1}$ is the prediction before the actual measurement given by z_{k+1}.

We have:

$$\hat{x}_{k+1} = A_{k+1} \hat{x}_k + B_{k+1} z_{k+1}$$

The zero mean condition:

$$E[\epsilon_{k+1}] = E[\epsilon_k] = E[\delta_{k+1}] = 0$$

leads to :

$$\boxed{\hat{x}_{k+1} = \hat{x}_k + B_{k+1}\delta_{k+1}} \tag{2}$$

Let:

$$\Gamma_{xz} = E[\epsilon_k \delta_{k+1}^T] \qquad \Gamma_{zz} = E[\delta_{k+1}\delta_{k+1}^T]$$

Then we demonstrate [10] that:

$$\boxed{B_{k+1} = \Gamma_{xz}\Gamma_{zz}^{-1}} \tag{3}$$

and this is the unique expression of B_{k+1} even in the case of colored noise.

The variance $E[\epsilon_{k+1}\epsilon_{k+1}^T]$ of the new estimation error is then:

$$\boxed{P_{k+1} = P_k - B_{k+1}\Gamma_{xz}^T} \tag{4}$$

These expressions enable us to compute directly the new state estimate (*i.e.*, objects and robot positions) and its variance from the previous state and the variance of the measurement noise and its covariance with the state noise. These expressions are valid even in case of correlated and colored noises.

3 Modelling Sensors Inaccuracies

3.1 Odometric Drift Evaluation

The robot has two motor-driven wheels and four free wheels. Our odometric system is based on reading angular encoders on the motorized wheels. Position estimation is given by integrating the differential equation :

$$X_{k+1} - X_k = M_k \Delta R_{k+1} \tag{5}$$

where

$$X_k = [x_k y_k \theta_k]^T$$

is the configuration vector of the mobile robot, and

$$\Delta R_{k+1} = [r_l \Delta \psi_{l_{k+1}} r_r \Delta \psi_{r_{k+1}}]^T$$

is the linear displacement of each wheel r_l (respectively r_r) being the radius of the left (respectively right) wheel and $\Delta\psi_{l_{k+1}}, \Delta\psi_{r_{k+1}}$ being their angular rotation.

M_k is called the kinematic matrix and can be expressed as:

$$\begin{bmatrix} \frac{\cos(\theta_k)}{2} & \frac{\cos(\theta_k)}{2} \\ \frac{\sin(\theta_k)}{2} & \frac{\sin(\theta_k)}{2} \\ \frac{-1}{L} & \frac{1}{L} \end{bmatrix}$$

where L is the half-length of the axle-tree.

In order to get an evaluation of the robot position variance, we have to express the variances of wheels radius estimations and half-length of axle-tree. Unfortunately, the non-linear equation (5) is valid only for small displacements, and we could not to date, get a satisfaying calibration of these parameters. Some authors, ([7], [12]) use more complicated expressions to reduce the linearization error but also need small displacements and do not give methods to calibrate intrinsic parameters. A calibration method is yet to be found.

However, if we assume that errors on constant parameters lead to a global error which is proportional to the distance traveled by each wheel, we can give an evaluation by integrating the differential variance evolution derived from (5).

Expressing (5) as:

$$X_{k+1} = \mathcal{F}(X_k, \Delta R_{k+1})$$

we can write, in a first order approximation, the variance as:

$$V(X_{k+1}) \approx J_{X_k} V(X_k) J_{X_k}^T + J_{\Delta R_{k+1}} V(\Delta R_{k+1}) J_{\Delta R_{k+1}}^T \tag{6}$$

where J_{X_k} and $J_{\Delta R_{k+1}}$ are the jacobians of $\mathcal{F}$ about robot location and wheels displacement estimates, and:

$$V(\Delta R_{k+1}) = \alpha^2 \begin{bmatrix} (r_l \Delta\psi_{l_{k+1}})^2 & 0 \\ 0 & (r_r \Delta\psi_{r_{k+1}})^2 \end{bmatrix}$$

Experimentations gave $\alpha \approx 0.03$.

However, wheel slippage cannot be statically identified and induces unpredictable biases. In conclusion, no reliable variance estimation could be found on robot positioning, but, if we assume slippages not correlated with X_k, equation (6) gives the correlation between two successive positions.

$$\text{Cov}(X_k, X_{k+1}) = E[\epsilon_{X_k} \epsilon_{X_{k+1}}^T] = V(X_k) J_{X_k}^T \tag{7}$$

This remark will enable us to maintain consistency in our world representation in an absolute frame. Indeed, if there was a correlation $C_{mX_k} = E[\epsilon_{m_k} \epsilon_{X_k}^T]$ between a world feature m_k and X_k, a correlation with the robot position remains at instant k+n:

$$\text{Cov}(\mathbf{m_k}, \mathbf{X_{k+n}}) = \mathbf{C_{mX_k}} \prod_{i=1}^{n} \mathbf{J^T_{X_{k+i}}} \qquad (8)$$

and implies a correlation with all new objects perceived at k+n. It is important here to notice that this equation can be applied between two successive observation points $\mathbf{X_k}$ and $\mathbf{X_{k+1}}$, but considering the elementary displacements in the right-hand side product. If $\mathbf{T_r}$ is the update period of odometric readings, $\mathbf{T_{obs_{k+1}}}$ the time separating the observations k and $k+1$, we have:

$$\mathbf{T_{obs}} = \mathbf{n_{k+1}} \times \mathbf{T_r}$$

then 8 becomes:

$$\text{Cov}(\mathbf{m_k}, \mathbf{X_{k+T_{obs_{k+1}}}}) = \mathbf{C_{mX_k}} \prod_{i=1}^{n_{k+1}} \mathbf{J^T_{X_{k+iT_r}}} \qquad (9)$$

For commodity reasons, $\mathbf{X_{k+T_{obs_{k+1}}}}$ will be noted $\mathbf{X_{k+1}}$.

3.2 Laser Range Finder Calibration

Our environment perception device is a time-of-flight laser range finder rotating around a vertical axis.

The raw data is a set of 2-D points in polar coordinates $\mathbf{p} = (\rho, \theta)$ ordered by sensor rotation.

Statistical calibration of these two independant parameters has being made [5] and gives us the following variance matrix:

$$\mathbf{Vp} = \begin{bmatrix} 2.6cm^2 & 0 \\ 0 & 0.0225deg^2 \end{bmatrix} \qquad (10)$$

4 Building a Representation of Objects Surfaces

4.1 Local Modelling from Raw Data

In order to enable the robot to recognize some parts of its environment and organize its knowledge for the sake of its navigation, segmentation into aligned groups is made from raw data [5].

This segmentation uses a split-and-merge algorithm producing a set of segments (supporting line and endpoints). A group $\mathbf{sg} = \{\mathbf{p_i}\}$ of n successive points, is a segment if

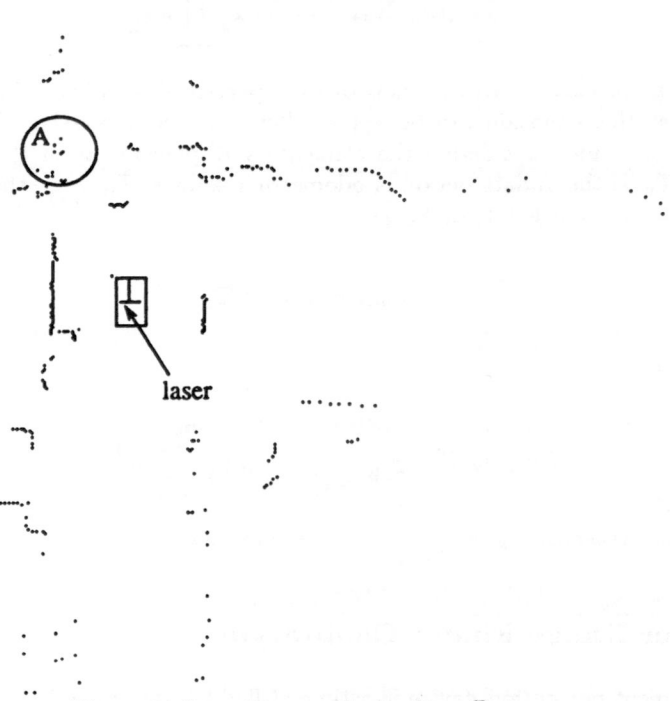

Figure 1: Laser Range Finder Raw Data

$$\frac{1}{n}\sum d_i^2 \leq \epsilon$$

where ϵ is a threshold based on the variance of distance measurement and d_i the orthogonal distance between the point p_i and the segment.

Our extremity points labelling distinguishes edges which are computed by intersecting two supporting lines (labelled "e") from those that are the projection along a line-of-sight (called "p").

Notice that some subsets of raw data cannot be modelled by line segments (see fig(1), area (A), for instance). In order to produce a good representation of free-space, we will have to deal whith this and propose other primitives in this case.

For maintaining a probabilistic representation of the world, we must now compute a stochastic estimation for each segment obtained by the last operation. Using polar coordinates, a supporting line is described by two parameters (d, ϕ) and the line equation is given by:

$$\rho\cos(\theta - \phi) - d = 0 \qquad (11)$$

If p_1, p_n are the endpoints of the segment S, a first estimate is computed such as:

$$[d\phi]^T = \mathcal{F}(p_1, p_n)$$

Expressing the jacobians of $\mathcal{F}$ about p_1, p_n, we get the variance-covariance matrix :

$$V(S) = J_{p_1} VpJ_{p_1}^T + J_{p_n} VpJ_{p_n}^T$$

We can now apply the formalism presented in section 2. For each other point p_j belonging to S, the expression of δ_{p_j} is extracted from (11).

Calling $\mathcal{G}(S, p_j)$,the implicit fonction in (11) and linearizing it, we have:

$$\delta_{p_j} = -\mathcal{G}(\hat{S}, p_j) = J_S \epsilon_S + J_{p_j} \epsilon_{p_j}$$

where ϵ_S and ϵ_{p_j} are the current errors on respectively S and p_j. It is now easy to compute the gain by expression (3), then, to express the new estimate of S (and its variance) by (2) and (4). In this case, a basic Extended Kalman Filtering would lead to exactly the same result.

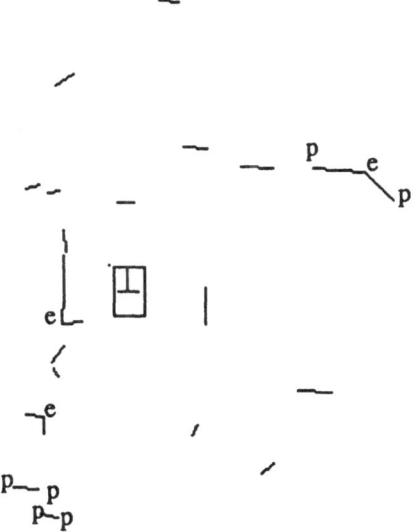

Figure 2: Local Segmentation

4.2   Matching Strategies

As one can see in figure (3), an important drift may appear between two successive robot positions and a correction is needed. Since we do not have a reliable variance on robot location error (see section 3.1), we are not able to use a probabilistic criterion directly for matching segments between the previous environment representation and the new perception. Therefore, in a first step we will use a heuristic method to find matchings with a high degree of confidence. Then, after reducing a possible bias by correcting the robot's position, we detect the correspondance between two primitives by testing the generalized Mahalanobis distance as in [1]. Our approach is thus:

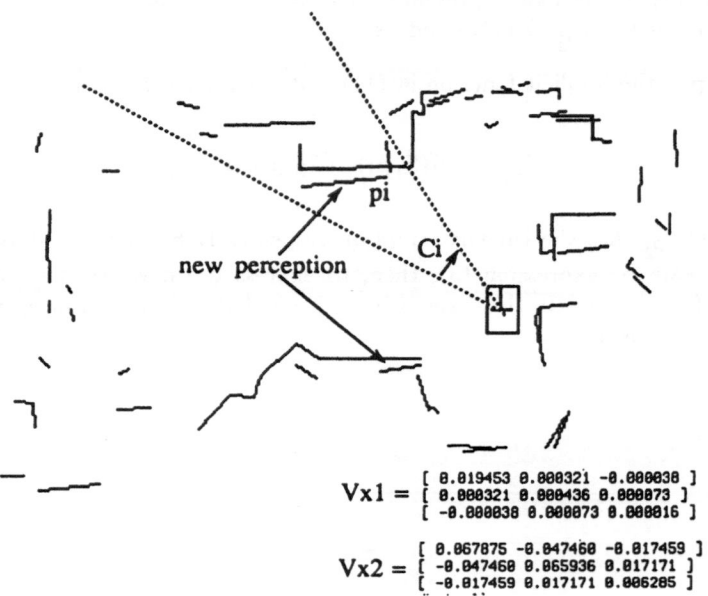

$$Vx1 = \begin{bmatrix} 0.019453 & 0.000321 & -0.000038 \\ 0.000321 & 0.000436 & 0.000073 \\ -0.000038 & 0.000073 & 0.000016 \end{bmatrix}$$

$$Vx2 = \begin{bmatrix} 0.067875 & -0.047460 & -0.017459 \\ -0.047460 & 0.065936 & 0.017171 \\ -0.017459 & 0.017171 & 0.006285 \end{bmatrix}$$

Figure 3: Last Model and Current Perception Superposition

1. After a displacement, and assuming that the global aspect of the scene is stable, a prediction model is extracted from the current model of the world, using the new robot position (fig. 4).

   Then, for the longer segments p_i of the current perception, we build the list L_i of model segments wich intersect the sectors C_i defined by the robot and the segment p_i (fig. 3).

   For each p_i, the angular gaps with all $l_j \in L_i$ are computed. Each one corresponds to a hypothesis on robot rotation error and is weighted by the heuristic:

   $$h(p_i, l_j) = \min(\lg(p_i) \times n(p_i), \lg(l_j) \times n(l_j))$$

   with $\lg(p_i) = $ length of p_i and $n(p_i) = $ number of laser depth points belonging to p_i.

   By grouping similar angle values, we establish a list of rotation hypothesis ordered according to ascending scores.

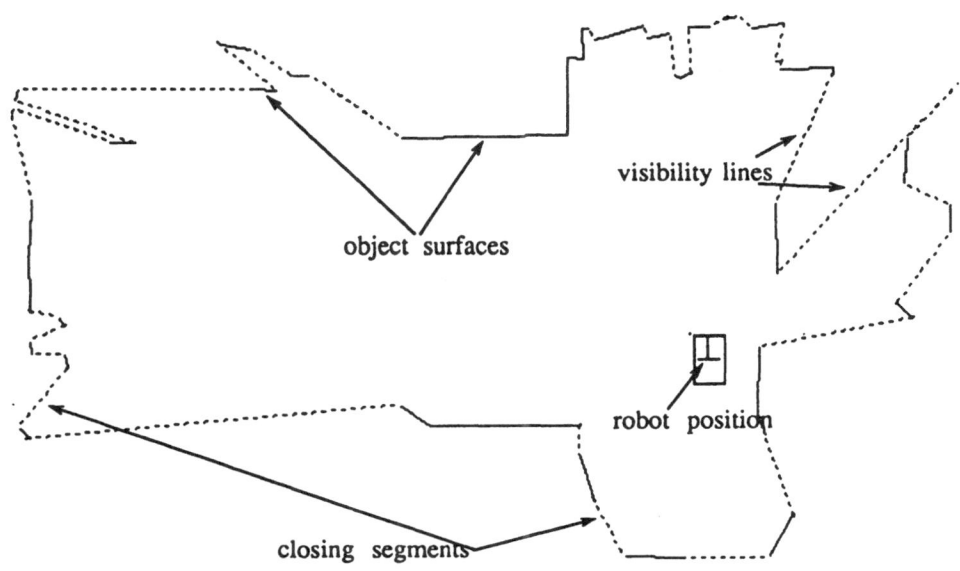

Figure 4: Prediction Model

Taking the set **R** of pairs corresponding to the highest score, we have now to verify the translation compatibility between them. Indeed, as we can see in area C_i , a perception segment may be in correspondence with several model primitives with the same angular gap. The same operation as above is therefore necessary for the translation.

For each non-parallel couple of pairs belonging to **R** a translation, weighted as above, is computed. Finally, the highest scored set **HM** of compatible correspondences is produced (fig 5). Notice that in this last step, every couple of non-parallel pairs gives us a possible translation. In order to verify the consistency, we want the system to produce more than two pairs. If it is not the case, another hypothesis is chosen. If there is no such possibility for the selected rotation hypothesis, another rotation is tried.

2. Assuming now the robot position correction done (as in fig 8), and reliable variances on the objects known (as it will be shown in the next section), we can now use a stochastic method to match perception segments with model features. Let $X_{k+1} = [x_{k+1} y_{k+1} \theta_{k+1}]^T$ be the robot configuration at instant k+1 and $S_{m_k} = [d_{m_k} \phi_{m_k}]^T$ the supporting line of a model segment at instant k. S_{m_k} can be the same actual feature as $S_{p_{k+1}} = [d_{p_{k+1}} \phi_{p_{k+1}}]^T$, a line of the local perception, if and only if :

Figure 5: Heuristic Matchings

(a) the projection $\mathcal{P}(\mathbf{X_{k+1}}, \mathbf{S_{P_{k+1}}}$ of $\mathbf{S_{P_{k+1}}}$ through the robot's position is compatible (probabilistically) with $\mathbf{S_{m_k}}$, *i.e.*, the new measurement of $\mathbf{S_{m_k}}$ by $\mathbf{S_{P_{k+1}}}$ verifies:

$$\mathcal{P}(\mathbf{X_{k+1}}, \mathbf{S_{P_{k+1}}}) = \begin{bmatrix} d_{m_{k+1}} \\ \theta_{m_{k+1}} \end{bmatrix}$$

$$= \begin{bmatrix} d_{p_{k+1}} + x_{k+1}\cos(\theta_{k+1} + \phi_{p_{k+1}}) + y_{k+1}\sin(\theta_{k+1} + \phi_{p_{k+1}}) \\ \theta_{k+1} + \phi_{p_{k+1}} \end{bmatrix} \quad (12)$$

Computing the prediction error on $\mathbf{S_{m_k}}$:

$$\delta_{\mathbf{S_m}} = \mathcal{P}(\hat{\mathbf{X}}_{\mathbf{k+1}}, \hat{\mathbf{S}}_{\mathbf{P_{k+1}}}) - \hat{\mathbf{S}}_{\mathbf{m_k}} = \epsilon_{\mathbf{m_k}} - \mathbf{J}(\mathcal{P})_{\mathbf{X}}\epsilon_{\mathbf{X_{k+1}}} + \mathbf{J}(\mathcal{P})_{\mathbf{P}}\epsilon_{\mathbf{P_{k+1}}} \quad (13)$$

Assuming, at this step, the variance of the model primitive ($\mathbf{C_{mm}}$ as it is defined in section 2.2), of robot location $\mathbf{V}(\mathbf{X_{k+1}})$ and of the local perceived feature $\mathbf{V}(\mathbf{p_{k+1}})$ known , and applying (covar2 m-xk+1) along the trajectory between two successive perception positions $\mathbf{X_k}$ and $\mathbf{X_{k+1}}$, we are able to compute the variance $\mathbf{\Gamma_{zz}}$ of $\delta_{\mathbf{S_m}}$.
Then, relation (12) is valid if:

$$\delta_{\mathbf{S_m}}^{\mathbf{T}} \mathbf{\Gamma_{zz}^{-1}} \delta_{\mathbf{S_m}} < \epsilon$$

where ϵ is a threshold, related in a χ^2 table, corresponding to 95 percent of probabilistic confidence. Notice that this method is theoretically true only for gaussian noises. We assume here that if it is verified, the supporting line matching is true even in a no gaussian case.

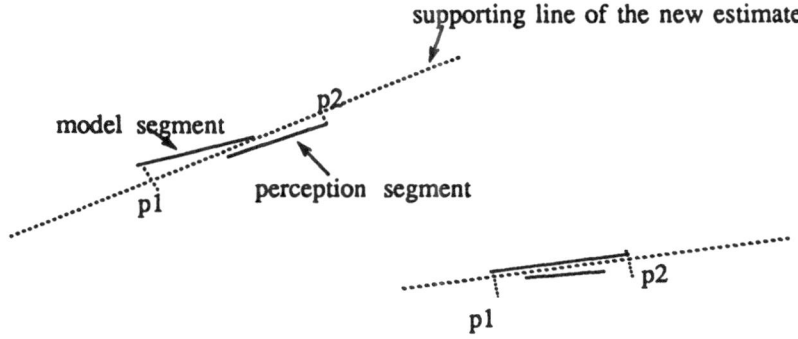

Figure 6: Endpoints Updating

(b) the segments must at least partially overlap (fig 6).

After this algorithm, a set **PM** of probabilistic matchings is produced.

4.3 Fusion between Previous Knowledge and Current Perception

We will now show that a possible bias on the robot's location estimate can be reduced after the first step described in §4.2.

Two different methods were implemented:

1. The first, that we call "Global Approach", is presented in [11]. For each matching belonging to **HM**:

   - the prediction error (13) is computed.
   - the robot's position is updated by (2) with the gain (3) where:

   $$\Gamma_{xz} = E[\epsilon_{x_{k+1}} \delta_{S_m}^T]$$

   and

   $$\Gamma_{zz} = E[\delta_{S_m} \delta_{S_m}^T]$$

   - a new estimate of all segments in the world is computed by the same equations with:

   $$\Gamma_{xz} = E[\epsilon_{m_k} \delta_{S_m}^T]$$

   and Γ_{zz} as above.

Then the set **PM** is produced as in the second step of the previous section and the same processing as above is made on it.

As explained in [10], a bias on position error is propagated to all segments of the model just after processing the first matching. Therefore, when the next matching is used, the model segment is already biased and produces in turn a bias on the robot's location and on all other features, *etc.* This can induce an instability in the non-linear filtering. One can see in fig. 7 the model-perception superposition after the processing of **HM** .

2. Assuming local and model segments non-biased before robot position correction, each matching is able to reduce the robot's position bias by filtering. Inversely, as long as the robot's location is biased, estimation of primitives by projection of the local data may induce a bias on the estimates of the model segments. This remark leads us to another approach:

- For each matching belonging to **HM**:
  the robot's location is updated by (2) with the gain (3) where:

  $$\Gamma_{\mathbf{xz}} = \mathbf{E}[\epsilon_{\mathbf{x}_{k+1}} \delta_{\mathbf{S_m}}^{\mathbf{T}}]$$

  and

  $$\Gamma_{\mathbf{zz}} = \mathbf{E}[\delta_{\mathbf{S_m}} \delta_{\mathbf{S_m}}^{\mathbf{T}}]$$

  no new estimate is computed for model segments
- The set **PM** is produced .
- For each matching belonging to **PM** the robot's location is updated as above.
- For each matching belonging to **HM** and to **PM** a new estimate of all segments in the world is computed by the same way as above , where $\delta_{\mathbf{S_m}}$ is evaluated with the new robot's position the bias of wich is highly reduced.

  Fig.8 shows the results of this approach. Fig. 10 shows an experimental run of this system.

4.4 Updating the Endpoints

Each supporting line being now updated, we have to recompute the segment endpoints. There are three cases (see fig(6):

- a "p" extremity is detected to be close to another one and an intersection point is computed from the two corresponding lines;
- for each other "p" point, there are two subcases:
  - the corresponding segment is not seen in the current perception: currently the projection of the point on the new line is is computed;

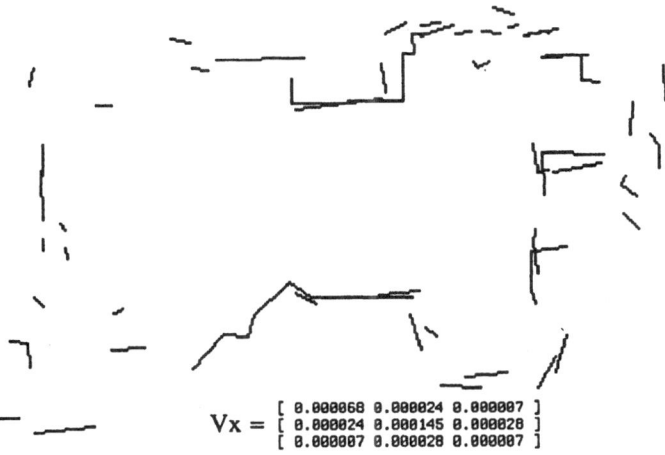

$$V_x = \begin{bmatrix} 0.000068 & 0.000024 & 0.000007 \\ 0.000024 & 0.000145 & 0.000028 \\ 0.000007 & 0.000028 & 0.000007 \end{bmatrix}$$

Figure 7: Global Approach Instability

    — the segment is seen, and we choose to project (between model and perception) the point wich produces the maximum length.

- each "e" endpoint is updated by intersecting the two updated lines passing through it;

5   Conclusion

The system presented in this paper demonstrates an essential capacity for mobile robots: dealing with their perception and motion inaccuracies to build a consistent model of their environment. We investigated two different approaches for doing so: the first one is based on a direct global computation of the new state vector estimate, and the second consists in repositioning the robot first, taking into account its local perception and the known world model, and then fusing the parts of the environment identified as being the same in the perception and the model, before propagating this fusion to all the environment using the correlations between its elements. The second approach gives better results mainly because it enables to reduce position biases due to equation linearization and to possible errors in robot motion . It is necessary for dealing with this problem to have a satisfactory model of sensor inaccuracies, and this is rather difficult for odometry subject to drifts and slippage.

Experimental results showed the possibility of correcting robot position and maintaining an environment model with a known accuracy.

Acknowledgments: The authors wish to thank Pierrick Grandjean and Arnaud Robert de Saint Vincent for their contribution.

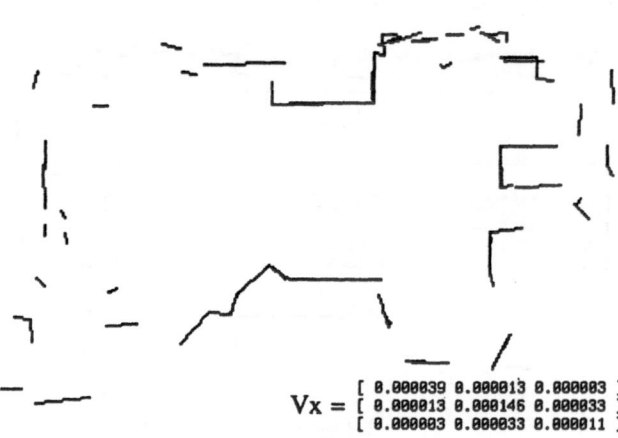

$$Vx = \begin{bmatrix} 0.000039 & 0.000013 & 0.000003 \\ 0.000013 & 0.000146 & 0.000033 \\ 0.000003 & 0.000033 & 0.000011 \end{bmatrix}$$

Figure 8: Robot Repositioning

References

[1] Ayache N. Construction et fusion de représentations visuelles 3D; applications à la robotique mobile. Thèse d'Etat, Université d'Orsay, Mai 1988.

[2] Brooks R. Aspects of mobile robot visual map making. IEEE Int'l Conf. on Robotics and Automation, St-Louis, March 1985.

[3] Chatila R., J-P. Laumond, Position referencing and consistent world modelling for mobile robots, IEEE International Conference on Robotics and Automation, St-Louis, March 1985.

[4] Durrant-Whyte H. Integration, coordination and control of multisensor robot systems. Kluwer Academic Pub. 1988.

[5] P. Grandjean and A. R. de St Vincent. 3D geometric modeling of indoor scenes with planar facets by fusion of noisy range and stereo data. IEEE Conf. on Robotics and Automation, Scottsdale, AZ. May 1989.

[6] Jazwinski A. Stochastic Processes and Filtering Theory. Vol. 64 in Mathematics in Science and Engineering. Academic Press, 1970.

[7] Julliere M., Marce L., Place H. a guidance system for a mobile robot. 13th Int. Symp on Industrial Robots, Chicago 1983.

[8] Matthies L., T. Kanade. The cycle of uncertainty and constraint in robot perception. 4th ISRR, Bolles & Roth Eds., MIT Press, 1988.

[9] Mazon I., Alami R. Representation and propagation of positioning uncertainties through manipulation robot programs. Integration into a task-level programming system. 1989 IEEE Conf. on Robotics and Automation, Phoenix, AZ, 1989.

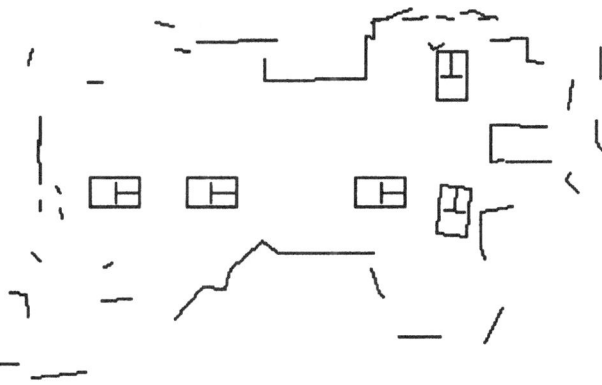

Figure 9: New Model of the World

[10] Moutarlier P., Chatila R. Stochastic Multisensory Data Fusion For Mobile Robot Location And Environment Modelling. 5th ISRR, Tokyo, August 1989.

[11] Smith R.C, Self M., Cheeseman P. Estimating uncertain spatial relationships in robotics. 2nd workshop on uncertainty in Artificial Intelligence (AAAI), Philadelphia, Aug. 1986

[12] Ming Wang C. Location estimation and uncertainty analysis for mobile robots. 1988 IEEE Conf on Robotics and Automation, Railegh, 1988.

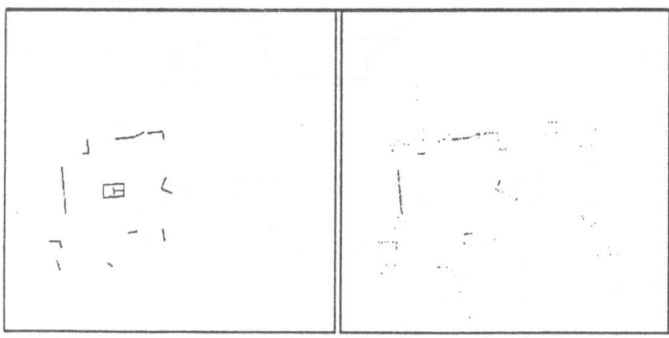

First Perception

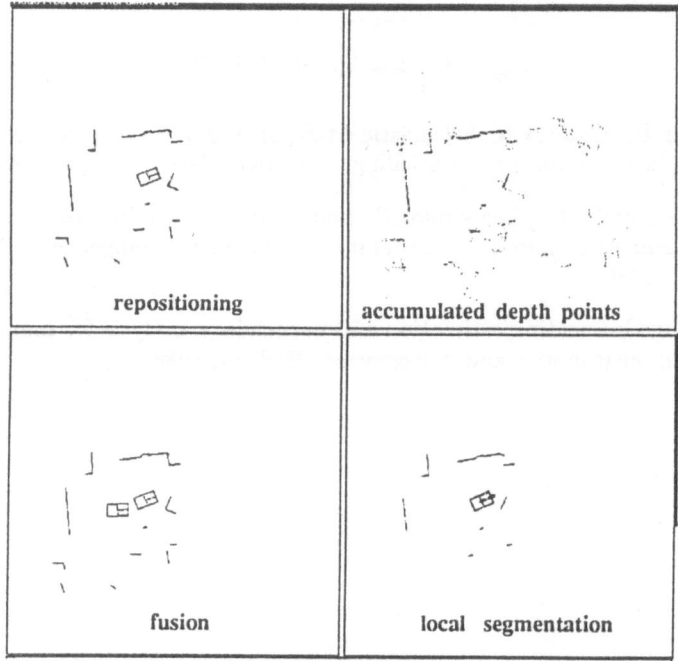

Second perception

Figure 10: Actual experiment of incremental environment modelling

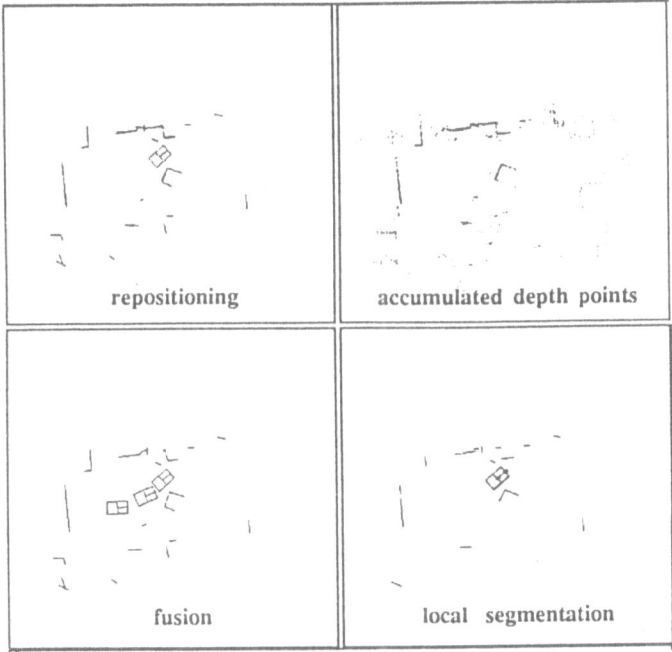

repositioning   accumulated depth points

fusion   local segmentation

Third Perception

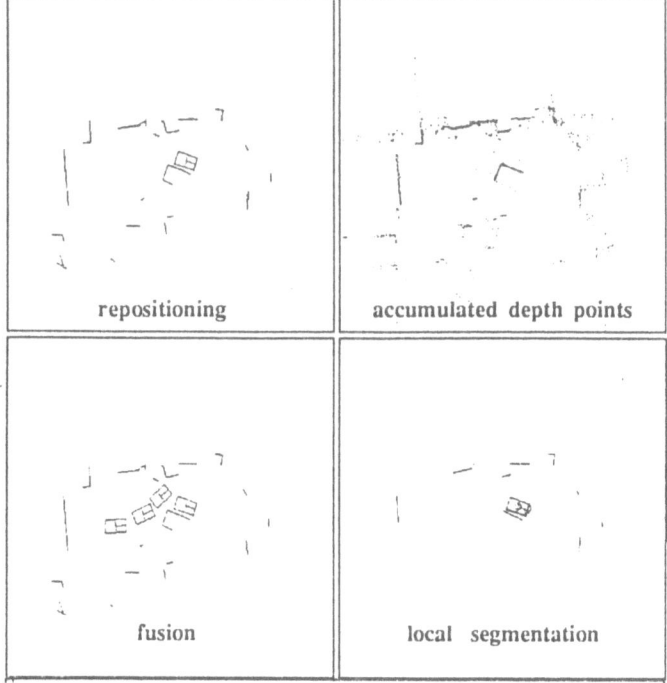

repositioning   accumulated depth points

fusion   local segmentation

Fourth Perception

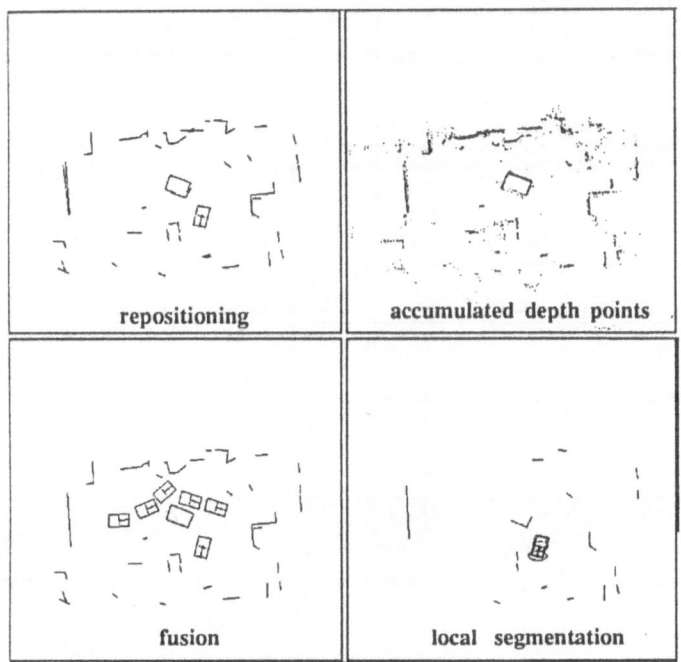

repositioning   accumulated depth points

fusion   local segmentation

Sixth Perception

The modelling system PYRAMIDE
as an interactive help
for the guidance of the
inspection vehicle CENTAURE

Philippe Even,* Lionel Marcé,
IRISA, Centre INRIA de Rennes,
Campus de Beaulieu, F-35042, Rennes, Cédex, France,

–

Joël Morillon and Raymond Fournier,
Commissariat à l'Energie Atomique, UGRA/SETA,
CEN-FAR, BP 6, F-92265 Fontenay-aux-Roses, Cédex, France

Abstract

Teleoperation of vehicles and manipulators can become much easier if a geometric model of the environment and of the robot is provided. However in case of an accident in hostile environment, we do not have always such models. In order to answer this drawback, we conceived a Generalized Information Management System called PYRAMIDE, capable of modelling a 3-d world from camera views, of locating the robot in this environment and of producing synthetic displays for the assistance of the operator while guiding a robot. This article relates an experimental phase where PYRAMIDE was coupled with the intervention vehicle CENTAURE. Some results on the performance are given, and the potentialities offered in case of intervention are discussed.

1   Introduction

Our study occurs after conclusions of advanced teleoperation research carried out into the frame of the French ARA project [1]. The issue is remote control of manipulators and vehicles for intervention in case of an accident in hostile environment. The goal is to provide the operators with synthetic informations from a geometrical data base in order to fully benefit of the potentialities of a Computer Aided Teleoperation system. These synthetic feedbacks are made adequate by using the operator as a decision maker to indicate what kind of information he needs and what precision the tasks requires.

---

*Supported by a grant from the French Nuclear Agency.

The mobile robot CENTAURE-I was used during an experimental phase to evaluate the consistency of the principles we propound. This robot is aimed at *following the natural path of a man in an installation* for achieving an intervention mission [2]. It cumulates the advantages of wheels, tracks and legs, combining an articulated track principle with a movable payload platform which provides active control of stability. In industrial environments, lot of passages are conceived anthropomorphically, and become obstacles for a robot (stairs, narrow places, ...). The nice abilities of CENTAURE to cross most of these obstacles are not always easy to use, due to the loss of geometrical informations between the remote site and the control station. This reason puts CENTAURE as the best candidate to evaluate our system. We indicate first how a geometrical data base can be used to improve a mobile robot teleguidance.

To increase system capacities, we can improve data feedback and exhibit them in such a way that operators feel like working in the remote worksite (telepresence concept [3]). In some circumstances, for instance in bad visibility, the operators are provided with degraded data, which can be partly restored if an environment model of the world in which the robot is running is available. Our philosophy is based upon the concept of Generalized Information Feedback [4].

A geometrical data base in which the robot is located, can be used to build up a realistic synthetic image of the environment, as it would be seen through robot sensors. In this way some sources of image blurring (smoke, obstacles, ...) can be fictitiously eliminated. The synthetic image can also be symbolic. The geometric model allows transformation of data to be displayed to operators in an easily understandable way. For instance range to an obstacle can be displayed with a visual indicator of bar graph type. This geometric model allows operators to record data, for instance a sign posted direction, or guiding marks, in order to avoid their short term memory overloading.

But the geometrical model affords more than to improve operators information; it also permits the programming of tasks. Trajectories and targets can be specified or automatically computed [5], potential areas can be defined, for obstacles avoidance or objects handling [6, 7]. In the same way as in robotics, the geometrical model can be used for characteristics evaluation during system conception. Operators can for instance estimate if cameras position is accurate (occlusion test [8]). This simulation also allows the display of command effects. This aspect is very important during task debugging. Operators can for instance choose a good way to reach a goal (reach test) and also to get one's hand when no error is allowed. The goal can be to limit error risks in case of irreversible commands as to provide operators with anticipation possibilities when simulation is running on-line. As for aircraft simulators in aeronautics, a geometrical model allows operators training and the reduction of the learning stage.

2   The PYRAMIDE geometric modelling system

The goal of PYRAMIDE is to create, to fill up, or to modify a geometrical 3d model of a robot environment. This one is usually unknown. Some objects are possibly normalized.

Sometimes the operator disposes of a priori geometrical data bases; however some objects may be shifted. The operator himself estimates the degree of precision necessary for his work; typically he can fully model work areas and only partially model the obstacles. These features allow to reduce the modelling time and the display time (the less numerous the objects are to visualize, the more possible is animation). They also allow the suppression of useless data which focus and seize attention.

The modelling process is a cooperation between a priori CAD data bases, an embedded multisensor system including on board cameras, operator's knowledge and the computation power of an IRIS graphics workstation. Computers can provide geometrical models of simple volumes, brought together and scaled to form more complex objects. These virtual objects are set up by operators in the real world of the vehicle and their projection is mixed to the projection of real objects. Operators match very quickly the two projections in tuning up parameters of the synthetic model.

The PYRAMIDE geometrical modelling unit architecture is based on a controller that shares messages between four modules. The data base allows the world model updating, the robot model actualization, the selection of a geometrical primitives library and the dynamic management of data. To insure a flexible man-machine interface conception an Object-Oriented approach was chosen. The different kinds of informations are shared between six windows and managed with interactive interface tools (mouse, potentiometers, button box, ...). The geometrical reasoning module is a library of programs that solve some 3d problems. Amongst these are methods to compute the camera attitude parameters, to retrieve the depth of an object in the scene from its image projection, and to compute the model uncertainties. These points are discussed more thoroughly in [9].

3   Integration of PYRAMIDE in the control station

An experimental phase held in december 1988 at the teleoperation site of the UGRA laboratory. The PYRAMIDE system was coupled with the control station of CENTAURE. The scope was to estimate the feasability of the modelling method in realistic conditions and the relevance of the built data base to help an operator when piloting the robot. Amongst the tested fonctions were the modelling process, the resetting of data bases on the environment, the robot location, the training possibilities, and the assistance tools for the pilotage.

A serial bus RS-232 was used to drive data between PYRAMIDE and CENTAURE control station. Images from an embedded CCD camera equipped with a $6.4 \times 4.8mm$ pickup area and a $8.5mm$ format lens were displayed on a TV monitor. An IBM PC AT equipped with a MATROX PIP 1024 acquisition card was used to digitalize the video images. The eight grey levels scaled 512×512 images were then transmitted to PYRAMIDE on ETHERNET with FTP protocol.

The lens was calibrated at IRISA with a high accuracy method [10]. Images of a 48 points rectangular grid were acquired at five distinct heights of the camera. These image points were identified to the known 3-d points. Then a convergent algorithm was used to estimate the internal and external camera parameters with high accuracy (table 1). The optical center is located from the lower left corner of the image. The focal length is defined in pixels by taking into account the pixel size.

Parameters	X axis	Y axis
Optical centre	267 pels	283 pels
Focal length	722 pels	1019 pels
Y / X ratio	1.41	
Distorsion rate	-8.26×10^{-7}	

Tab. 1: *The lens internal parameters.*

The knowledge of the radial distorsion rate makes possible the off-line image correction. So we can estimate the influence of this factor on our method.

4 Geometrical resetting of the data base on the environment

In this section we discuss some techniques to link the frame of the camera with a frame chosen in the 3d scene. Many difficulty levels are faced according to the prior knowledge on camera external parameters and to the required accuracy of the resetting.

4.1 Full resetting

When operators have not any previous knowledge on the position and attitude parameters of the camera, a full resetting must be proceeded. Six parameters (three translations and three rotations) are to be computed. This work can be performed in two steps; first we determine the camera attitude parameters, then the computed frame is translated to its correct location in the 3-d world.

The computation of the attitude parameters can not be easily done manually. In our implemented method, operators must indicate three image segments, parallel to the 3d world frame axes. These segments do not necessary concur. Typically one of them corresponds to a vertical line and the remaining ones lie in horizontal planes and are orthogonal. Most images of structured environments include such features. From the equations of these three segments, two analytical solutions are computed; one provides operators with a convex frame, the other with a concave one. The implemented algorithm generalizes a method that has been developped at LIFIA / IMAG in Grenoble [11]. More details can be found in [12]. The computed frame is displayed at an arbitrary location on the image. Then the operator selects it, translates it at a convenient place, often an obstacle vertex, and specifies the height of the camera in order to make sure the frame lies on the floor.

To evaluate the accuracy of this method, many series were executed by an operator with a good knowledge of the method. The image we chose comprises many segments with distinct features. These segments are labelled on figure 1. In each series, 3 segments were chosen and a full resetting operation was executed around 26 times. The influence of the segment length was tested. We also evaluate the influence of the distorsion. To do it, we worked on the distorted video image and on a corrected version, and we chose segments at different locations (central or peripheral) and orientations (radial or tangent). The table 2 gives the experimental conditions for each series.

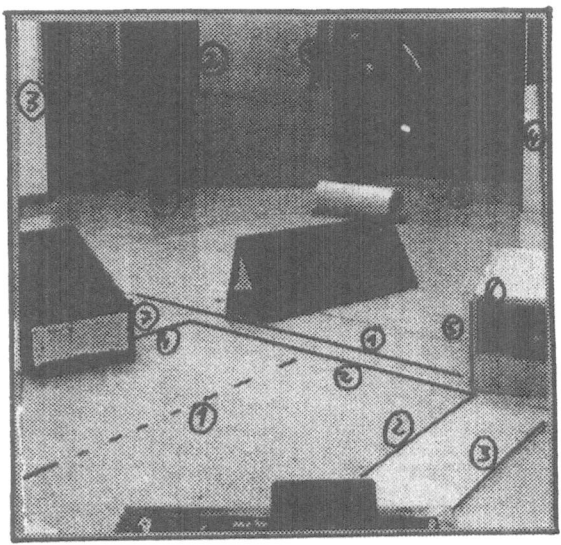

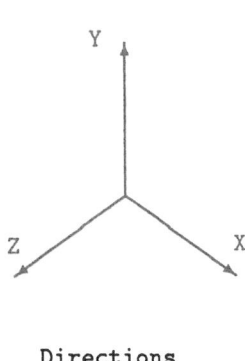

Directions

Fig. 1: *Labelling the segments.*

The figure 2 gives the 5 computed parameters and the standard deviation for each series. The real values are displayed in bold. The accuracy obtained with good segments ($\approx 200 pels$ and with radial orientations), is about $10cm$ for the positional parameters (P_x, P_z), and $0.2°$, $2°$, and $1°$ respectively for the R_x, R_y, and R_z rotational parameters (series 1). Very minutious pointings do not affect considerably the accuracy, The distorsion correction is not very effective with such segments (series 2). On the other hand, important offsets appear when one of these segments lies in the peripheral part with tangent direction (series 3). As shown on series 4, this offset is avoided on the corrected image. If one of the segments is short (under $100pels$), the mean value is still correct but important standard deviations occur (series 5 and 6).

Series	Segments			Chosen image	Execution mean time
	X axis	Y axis	Z axis		
1	2	14	1	video	20 s.
2	2	14	1	corrected	20 s.
3	2	3	1	video	20 s.
4	2	3	1	corrected	20 s.
5	2	1	1	video	20 s.
6	2	1	1	corrected	20 s.

Tab. 2: *Experimental conditions of each series.*

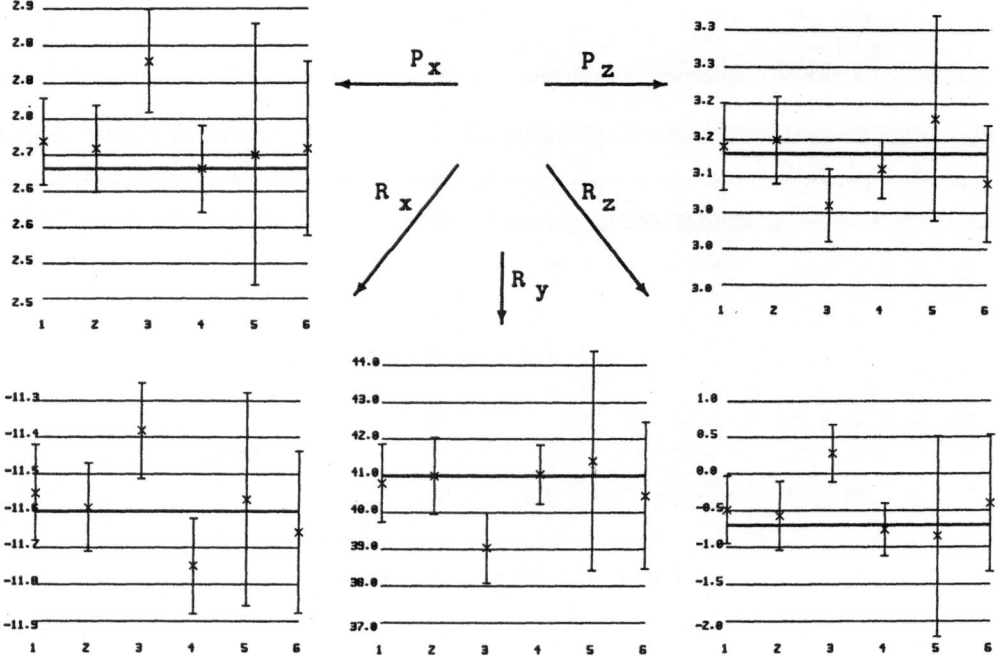

Fig. 2: *The camera position (cm) and attitude (°) for each series.*

To estimate the easiness of use, many unexperienced operators were observed and interviewed. As first result, a poorness of the method is the bad knowledge on the accuracy that can be reached in each experimental situation. The operators try to get the best matching possible. So the execution times increase drastically. Some feedback on the reached accuracy should be produced from the geometrical features of the retained segments. We notice also that a precise pointing on a screen is a visualy very tiresome work. In order to avoid this focus, some automatic segments extraction should be implemented so that the operator must just indicate what segment to be kept.

4.2 Partial resetting

If we have more knowledge on the camera external parameters, a partial resetting can be performed. When modelling an environment, the robot moves on the floor, so the height and the trim of the camera do not change. Only three parameters must be reset, that are two translation parameters from the previous location to the present one, and a rotation parameter corresponding to the azimuthal variation. This can be done manually. First the operator translates the model frame at a location that is visible in the new image. This location is the center of azimuthal rotation of the whole model. Then he selects the whole built model. The wire-frame display is superimposed on the video image. With the mouse and a potentiometer, the operator slides and orientates it until it matches correctly the 2d view.

The robot can be roughly located in a very short time. If a good accuracy is necessary (for example when obstacles are near), more attention is required from the operator to obtain the best matching. In case of a good matching, the robot location is given with an error less than $5cm$. However perfect matchings are seldom. The reason is that during the robot displacement, the trim of the camera changes slightly (a few tenth of degrees). This drawback comes mainly from the floor irregularities, the displacement of the inertial centre and the elasticity of the tracks. In order to reduce this bad effect, buttons have been programmed to adjust the roll and pitch angles. Nevertheless the adjustment of these parameters is difficult. This operation requires from operators some ability to manipulate objects in 3 dimensions. So a few minutes are often necessary to get an already exact matching.

5 Modelling an environment

When modelling a teleoperation environment, the goal is to acquire enough geometrical informations to provide the operator with efficient synthetic feedbacks for an assistance during the pilotage phases. In the frame of intervention scenarios a critical requirement is to build this model as quickly as possible. So the operator manages a compromise between the modelling time and the model accuracy.

5.1 The model acquisition

At the beginning of the scenario, the mobile robot enters a roughly known room, that must be crossed along to reach a valve. Some obstacles are scattered over the floor. The only knowledge the operator has is the height of the camera. So after the first image acquisition he must achieve a full resetting. When this is done, he knows the position and the attitude of the camera related to the the frame that was chosen. So he can now establish geometrical constraints to make the modelling process easier. For example the mouse movements on its plate are linked with objects slidings on the floor, or with vertical displacements. In order to perform the modelling task in best conditions, the resetting operation must be executed as carefully as possible.

To model an obstacle, the operator selects a primitive, translates it from the scene frame to the obstacle, rotates and scales it so that the wired shape fits the obstacle at best. Some geometrical constraints may be used to get a better precision and speed up the task; they are for example identical scalings along both horizontal axes to keep bases square or circular, or contact detection to join objects side by side. If an obstacle shape is not included in the primitive box, the operator can gather many primitives, or take the obstacle in a surrounding volume.

The main visible obstacles are built to obtain a first model. Then the operator guides the robot in this modelled part of the environment and stops it at a location where new features and a known one are visible. From this known feature he achieves a refined partial resetting. To ensure a good modelling some care is required. To make the resetting easier and more precise some surfacing details are often useful. So lines, grids or blobs on the floor may be preferably modelled. When the resetting is over, the operator updates the model.

5.2 Accuracy and execution time

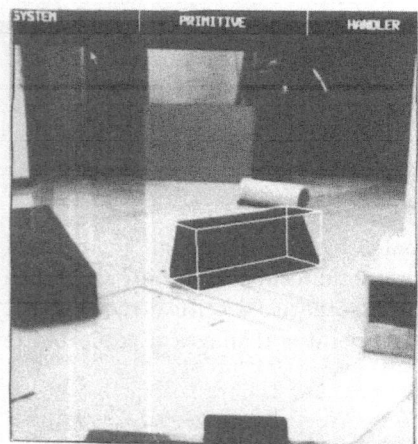 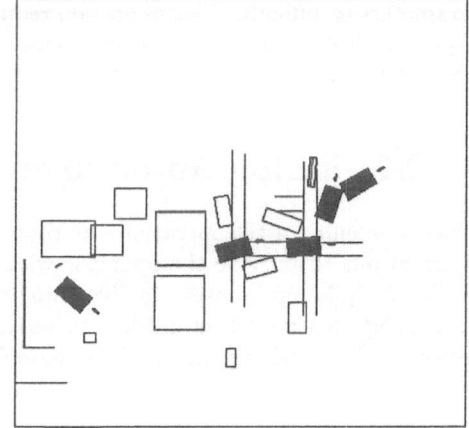

Fig. 3: *An image of the scene and a top view of the model.*

The figure 3 shows the environment modelled during the displacement from the door to the valve and the positions where the robot stopped. Five images (exp1 to exp5) were used to acquire the model. This one was compared with the real world caracteristics. Good accuracy is noticed in the surroundings of the robot path. The dimensions of obstacles near the robot (less than 5 meters) are given with an error less than 5 cm, and their position with an error less than 15 cm. Significant errors are noticeable on more distant obstacles. A very simple model but accurate enough to provide an effective help can be obtained in about fifteen minutes.

To give a better idea of the performance, an experiment was registered. A model was built by an experimented user from five views (exp1 to exp5). In order to obtain an accurate global model, careful partial resettings were performed. The table 3 gives the model acquisition time as the sum of a resetting time, a modelling time and a displacement time, the number of objects and primitives that have been handled on each view, and the number of objects and primitives added to the model. Here for example, 22 objects were built with 32 primitive in forty one minutes. On an image, the number of handled objects may differ from the number of modelled objects; some objects are modified due to the loss of informations on the previous views. The global time seems to be rather long, but it will be largely saved during the future displacements.

Images	Time				Handled		Acquired	
	Resetting	Modelling	Pilotage	Sum	Obj.	Prim.	Obj.	Prim.
exp1	40"	8'20"	2'	11'	13	14	13	14
exp2	40"	2'20"	3'	6'	5	5	3	3
exp3	2'30"	5'	4'	11'30"	7	15	5	13
exp4	1'	0	5'	6'	0	0	0	0
exp5	40"	2'50"	3'	6'30"	2	2	1	2
Global:	5'30"	18'30"	17'	41'	-	-	22	32

Tab. 3: *Acquisition of a model and its acquisition time.*

6 The simulation tools

Some simulation tools have been implemented from the built data base and proposed to many operators with different knowledge levels. The graphics are updated using the mouse or the joystick that usually controls CENTAURE.

Two modes have been tested to update the perspective view. In static mode a cuboid with the dimension of CENTAURE is driven over a video image or a coloured synthetic picture. To obtain a fast display time, only that cuboid is refreshed. This kind of pilotage is rather difficult. Indeed the robot may often go out of the field of view. Moreover it is hard to locate the robot with regard to the obstacles. This is due to the loss of the cuboid hidden parts removal when this one goes behind an obstacle. The only trustable indication is then the position of the cuboid base.

In dynamic mode the graphics is updated in a way to simulate what should be seen from an embedded camera on the fictitious vehicle. This synthetic feedback could be provided when the video information is lost. We meet here the same problem as when we pilot a real robot with the TV feedback, i.e. we do not have any distance or dimension information. This problem is made more drastic on the synthetic display because of the regularity of the displayed surfaces. For instance a correct estimation of the moment when to turn in front of an obstacle requires a good training.

Two animation modes are also proposed to update the orthographic top view. In static mode the environment lies steady in the background and a rectangle is displayed on it for CENTAURE. This tool brings a very noticeable comfort for the robot pilotage. The static mode requires a good knowledge of the elements that are included in the data base. Indeed no height information is visible. So a table may appear as an obstacle, even though the robot can move under it. Moreover when the robot turns back, the command effects are reversed; so many operators turn on the right instead of the left.

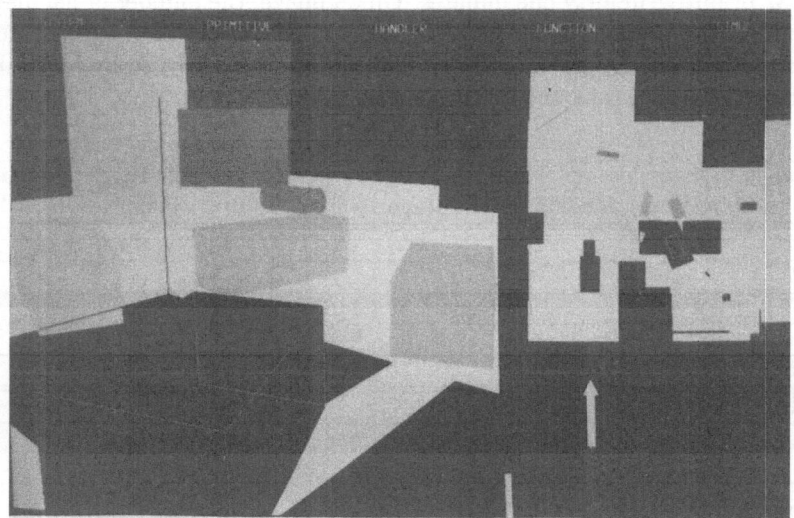

Fig. 4: *A simulation phase.*

In dynamic mode the rectangle lies steady at the center of the view, whereas the environment is updated around it. On this local view, the command effects reversal problem vanishes. So many users like it better. Meanwhile the goal to be reached does not necessarily appear on the view. So some tools should be added to provide the operator with orientation. Moreover this animation mode requires a larger display time.

An on-line collision detection module has been added. The collision test is just graphical and makes use of the IRIS workstation graphics abilities. So its implementation does not increase by far the display time. This tool is useful to know if the robot can pass between two obstacles or under one.

		Top view	
		static	*dynamic*
Perspective	*static*	< 100 ms	200 ms
view	*dynamic*	200 ms	400 ms

Tab. 4: *The refresh time for each simulation mode.*

The table 4 gives the refresh time of these various modes for a scene with an average complexity. For each dynamic mode selected, about 200 facets are to be updated. With a dynamic mode on both view, the image flickering is visible, but as the robot does not go fast, this flickering is still not troublesome for the pilotage. However if a complex object with a lot of facets enters the field of view, some delay arises between the command and the display.

7 The pilotage experiments

These many tools have been integrated into complete simulation experiments. A scenario was conceived in which some enrolled operators tried to pilot CENTAURE from a start point, and to lead it in front of a valve through a path cluttered with many obstacles. Two methods were suggested and observed.

7.1 Pilotage with the synthetic feedbacks

In the first method an execution operator drives the robot using the TV feedback and also the synthetic ones (pilotage phase). The synthetic feedbacks are updated in accordance to the relative positional measurements that are given by an odometry sensor. This one includes two odometry wheels located under the robot centre. The measurements are sent every $x \times 100ms$ to PYRAMIDE, where x is a entire value fixed in relation with the simulation modes the operator chooses (table 4). They are encoded as two translational coordinates X, Y and a rotational coordinate ϕ (figure 5), and transmitted on the RS 232 serial bus.

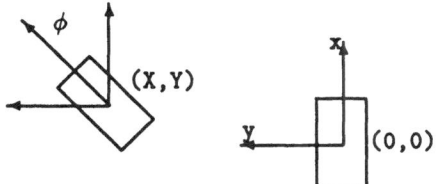

Fig. 5: *The odometry measurements.*

The odometry module is subject to some uncertainties that depend on many parameters (floor irregularities, sliding frictions, ...). As the robot goes on, these uncertainties cumulate. In order to give the operator an idea on the relative location of the vehicule, this one is displayed on the top view with a blue rectangle that increases in accordance to the path covered. The magnification ad hoc formula is given below. T stands for the current size and T_i for the initial one. k_d is the magnification factor in distance. It is set to the initial value $k_i = 4 \times 10^{-5}$. $k_o = 4 \times 10^{-7}$ is the magnification factor in angle. As indicated in this formula, the magnification factor in distance increases after each rotation that has occured.

$$T = T_i + k_d * \sqrt{X^2 * Y^2}, \qquad \text{with} \qquad k_d = k_i + k_o * |\phi|$$

The initial size of the robot is displayed as a red line superimposed on the magnifying rectangle. As the uncertainty on the vehicle location increases, the rectangle overlays the surrounding obstacles. That moment comes at best to proceed to a resetting of the robot position on its environment. So the execution operator stops the vehicle and the supervision operator acquires an image and transmits it to PYRAMIDE. Once the resetting phase is over, the pilotage phase can go again.

This pilotage methodology was submitted to several operators. They found the synthetic display to be a very friendly help. Two ways to use it have been noticed. Many users look at the synthetic top view from time to time, just to check the robot position before steering near a corner. For instance when an obstacle is run along, it disappears from the camera field of view at a certain time. From that time, not any information is accessible to know when to turn without any risk to send the back-sided camera against the obstacle.

Other users better pilot the robot on the top view, and check sometimes the TV image agrees to the fictitious robot environment. The resetting phase is long and arduous. So the operators have a tendency to delay it. This fact is rather troublesome because when they decide to reset, there is not always enough geometrical details in the video image to act easily. Advertised operators do not wait and reset the robot position as plenty of details are still present.

The overall time spent to lead the robot from its start point to the valve varied from about ten minutes up to near an hour. The overall pilotage phase mean time was about four minutes and was negligible compared with the resetting phases durations. Two to six resettings were executed. As these operations were exaggeratedly careful, they were also time consuming.

7.2 Pilotage using the simulation tools

The second method begins with a simulation phase were the execution operator drives the simulated robot on the PYRAMIDE top view. When the simulation is over, the final position is encoded as two entire values, an angle θ and a distance r, and sent to CENTAURE on the serial bus as an order for an automatic displacement phase. This displacement is servoed by the odometry measurements, so resetting phases must be held sometimes.

As main result, the execution operator task is made easier. In counterpart the overall pilotage phase (simulation and automatic displacement) takes more time than in the first method. An other particular interest of that method is a decreasing number of resetting phases. Indeed when a pilot leads his robot with video or synthetic feedbacks, disturbing rotations are introduced while steering. These rotations are also visible during a simulation, but they are not taken into account in the order. So as the odometry uncertainties are especially sensitive to the rotations, the odometry measurements are longer reliable.

8 The fine modelling of a complex object

The last experimented functionality of PYRAMIDE was the fine modelling of a complex object, here a valve. While modelling this valve, many difficulties arised. First plenty of complex shapes are included (pipes, collars, bolts, an elbow, a wheel). Then the object leaned against stairs. So many rotations occured. To benefit of all the potentialities offered by PYRAMIDE, this work was achieved by an experimented user.

First a model was built from a single view. 12 cylinders were assembled in 9 minutes to form the valve. Some results on geometrical features are given in table 5. Other points of view showed a split object with large errors on lot of parameters.

To reach a better accuracy, we acquired an other view of the valve in such a way we could use the interactive stereometry module. The principle of this module is to match a graphical primitive on the chosen detail of the valve on the first view, then to set a modelling constraint to slide the primitive in that first view viewing pyramid, until the matching is reached on the second view. In order to get a good precision, the image acquisitions must be followed by precise full resettings.

Features	Measured value	Single view	Int. stereometry
Wheel diameter	20 cm	21 cm	21 cm
Wheel to pipe axis distance	36 cm	9 cm	36 cm
Pipe diameter	8 cm	8 cm	9 cm
Wheel to pipe base height	60 cm	56.5 cm	61.5 cm
Valve tilt (R_x)	$\approx 10°$	0°	7°
Valve steering (R_y)	$-45°$	$-23.1°$	$-51°$

Tab. 5: *Some modelled features of the valve.*

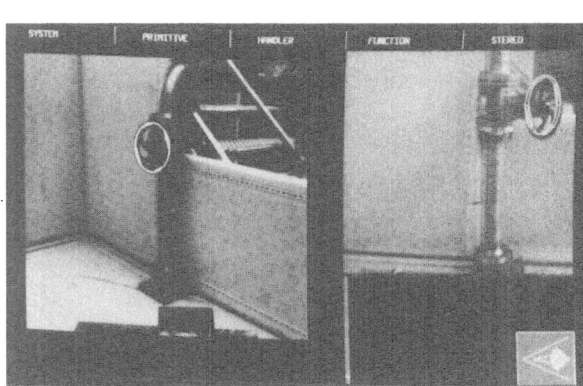
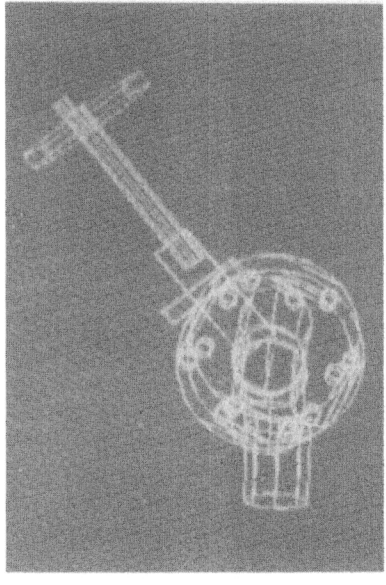

Fig. 6: *The valve and its model.*

We obtained a model of the valve in about fourty minutes. Fifteen minutes were necessary to execute the resettings. The built valve (figure 6) includes 22 cylindrical primitives and 2 cuboidal ones. As shown on table 5, the values of the geometrical features are closer to the real ones. Some a priori knowledge was used to add the bolts. First the operator modeled one bolt with the interactive stereometry principle. When this bolt was built, the perspective and the top views were zoomed, and this first bolt was duplicated, positionned at the same height (on the collar). The eight duplicated bolts were then distributed around the collar on the top view.

In the present version of PYRAMIDE, the fine modelling of fine objects is not enough simple to get quickly a precise model. Some geometrical tools (principally the interactive stereometry module) can be used, but more automaticity must be provided and a data base of standard objects (bolts, collars, ...) should be used.

9 Conclusion

The Generalized Information Management System PYRAMIDE has proved to procure an effective help for the guidance of the mobile robot CENTAURE in case of an accident in a hostile environment. When there is not any a priori model of the environment, a simple model can be built in a very short time. This potentiality can be used to send CENTAURE in the damaged world as a scout to get a model other heavier robots will use. The metric informations provided by the built data base can be used by CENTAURE itself; it is quite possible to know if the robot will pass between two obstacles or to acquire an obstacle height and give that information to an automatic obstacle crossing module [13]. After all the robot can be located accurately enough to enable synthetic feedbacks to help the operator during the guidance phases. In order to produce safer movements, the displacements can be simulated on the synthetic displays and sent as an order for an automatic displacement. Accurate models of complex objects can also be obtained, and used for automatic tasks; however this operation is time consuming.

To remedy the remaining defaults of PYRAMIDE, further works are going on. Topological tools may be estimated to speed up the modelling process while improving the model consistency. Such tools are based on higher level graphics primitives management. Use of telemetric measurements should also provide stronger potentialities.

Acknowledgements

The experiments took place in the UGRA department of CEA (French Nuclear Agency) of which we wish to thank the members, particularly MM. Rouyer, Fraize and Clément for their invaluable help and their friendly collaboration, and MM. Cabirol, Dupont, Mrs. Destolles and the whole team of the UGRA for giving us their sympathetic user suggestions for improvement.
We thank also Mr. Chaumette for providing us with the camera calibration method.

References

[1] B. Espiau, "Advanced teleoperation," *RoManSy 86*, Cracow, September 1986.

[2] G. Clement and E. Villedieu, "Mobile robot for hostile environments.," *Proc. of the Int. Topical Meeting on Remote Systems and Robotics in Hostile Environments*, pp. 270–277, Pasco, Washington, March 29 - April 2, 1987.

[3] S. Tachi and H. Arai, "Study on tele-existence (ii): three-dimensional color display with sensation of presence," *'85 ICAR*, pp. 345–352, Tokyo, September 9-10, 1985.

[4] G. André and A. Fournier, "The generalized information feedback concept in computer aided teleoperation," *RoManSy 86*, Cracow, September 1986.

[5] B. Faverjon, "Obstacle avoidance using an octree in the configuration space of a manipulator," *IEEE Int. Conf. on Robotics*, pp. 504–512, Atlanta, March 13-15, 1984.

[6] O. Khatib, "Real-time obstacle avoidance for manipulators and mobile robots," *IEEE Int. Conf. on Robotics and Automation*, pp. 500–505, St. Louis, March 25-28, 1985.

[7] G. André, "Conception et modélisation de systèmes de perception proximétrique. application à la commande en téléopération," Thèse D.I., IRISA-Rennes I, Octobre 1983.

[8] S. Sakane, M. Ishii, and M. Kakikura, "Hand-eye simulator: a basic tool for off-line programming of visual sensors," *'85 ICAR*, pp. 103–110, Tokyo, September 9-10, 1985.

[9] P. Even and L. Marcé, "Pyramide: an interactive tool for modelling of teleoperation environments.," *IEEE Int. Workshop on Intelligent Robots and Systems*, pp. 725 – 730, Tokyo, Oct. 31 - Nov. 2, 1988.

[10] F. Chaumette and P. Rives, "Réalisation et calibration d'un système expérimental de vision composé d'une caméra mobile embarquée sur un robot-manipulateur.," Publication Interne 454, IRISA - RENNES, Février 1989.

[11] R. Horaud, B. Conio, O. Leboulleux, and B. Lacolle, "An analytic solution for the perspective 4-point problem.," *Computer Vision, Graphics, and Image Processing*, 1989.

[12] P. Even and L. Marcé, "3d modelling of a teleoperation environment with pyramide.," *Third Topical Meeting on Robotics and Remote Systems*, pp. 11,1,1 – 11,1,7, Charleston, South-Carolina, March 13-16, 1989.

[13] R. Fournier, P. Gravez, M. Dupont, and J. Gaillard, "Computer aided teleoperation of the centaure remote controlled mobile robot," *Int. Symp. on Teleoperation and Control*, pp. 97–105, Bristol, July 12-15, 1988.

Experiments and theory with a 0.5 ton mobile robot.

Barry Steer

Robotics Research Group

Department of Engineering Science

Oxford University

OX1 3PJ

Abstract

The long term goal of robotics is to endow artifacts equipped with sensors, computers, and actuators with intelligence. Intelligence [12] is the name we give to the data processing activities of entities/artifacts which enable them to respond to information with behaviour which appears to be intended to be optimal with respect to preset goals. The information includes both that which is immediately sensed and that which is stored. The intelligence of a robot will be a function of the amount, variety, and complexity of its informational input, the arrangement, and connectivity of its storage capacity, the number and complexity of its goals, the degree of optimality achieved in attaining these goals, the amount and complexity of the output instructions involved in the response, and the position on a scale strategic/tactical of the activity involved.

Clearly, an intelligent robot, in the sense outlined above, will need to possess many autonomous "skills". One fundamental skill required by an intelligent robot-vehicle is the means to navigate autonomously. The process of navigation is the establishment ("fixing") of the position and attitude of an object, in our case a terrestrial ground vehicle, relative to some reference system. This contribution describes theory, hardware details, and experimental results obtained between 1980 and 1984 [13] in the U.K. using data processing activities that were implemented on a 0.5 ton wheeled vehicle, equipped with four different types of sensors, and two computer controlled actuation systems that enabled it, amongst other things, to navigate autonomously.

1 Introduction.

A three stage measurement process, involving sensory feedback from the environment and an encoding of the environment, was used to maintain estimates of a robot-vehicle's "global" position and orientation as it moved around. The navigation process relied on "naturally" occuring features in the environment to act as navigation aids. The robot-vehicle was based on the tricycle vehicle architecture. This vehicle was instrumented so as to be able to measure the along track distance moved and the steer angle. These measurements were then input to a kinematic model of the tricycle vehicle whose output represented the vehicle's position and heading. Sensory feedback from the environment was then used to bound the accumulated errors inherent in the dead reckoned estimates. These environmental measurements were obtained from two different types of sensor. Firstly, a "digital" compass to measure the vehicle's heading. Secondly, multiple acoustic sensors distributed around the periphery of the vehicle to measure the vehicle's disposition relative to objects in the environment.

The actual range measurements were correlated with the expected ranges based on a geometric representation of the robot's surroundings. The representation used was designed to take advantage of an encoding of the world in terms of what the sensor could actually observe. The observables were modelled geometrically, and were given the attributes of position, orientation, and semi-length. Sensor fusion experiments performed with the robot-vehicle showed that it was able to keep its estimated position and heading close to their "true" values using this form of indoor "terrain comparison".

The vehicle was also equipped with two computer controlled actuation systems that were able to control its speed and steer angle. A theoretical and experimental study was made of its lateral stability while wall following and techniques for efficient maneouvring at intersections of the type commonly found in industrial interiors [13]

A novel technique that modulated the steering wheel angle was also studied. This technique was able to produce desirable changes in the position and heading of a path curvature limited vehicle, as it moved. Examples of some of the manoeuvres that this technique could achieve were lane changing, parking between two cars, and overtaking other moving vehicles [14].

A number of safety demonstrations were shown over a period of time including "herding" the vehicle around using the sonar sensors to detect a moving person and moving away from that person [6].

1.1   Navigation, and Guidance.

The process of *navigation* must be distinguished from the process of *guidance*. Following standard marine terminology *the process of navigation* is taken to mean the establishment ("fixing") of the position and attitude of a vehicle relative to some reference system. *The guidance process* establishes a control strategy given a difference between the navigated "fix" and a goal state. Navigation has two aspects; the map and the aid. The map is a codified representation of the sensor observables. We make a map by classifying the regularities in the observables and codifying these in a retrieval manner. Making a useful map automatically is a difficult problem. The observables of a prepared map have two basic properties: geometric and topological. Navigation aids encode these two aspects. The aids may be either natural or artifical. The emphasis in this paper is on using naturally occuring sensor observables which can be codified as geometric objects and then correlated with incoming observations.

1.2 Other Work.

An early contribution was by Bauzil et al [3] in which they described an ultra-sonic perception system for navigation with the robot, Hilare. Clemence et al [4] considered the application of acoustic ranging to the automatic control of a ground vehicle. Crowley [5] described a navigation system for a mobile robot equipped with a rotating sensor. The emphasis was on building and maintaining a map. Hilare, originally developed by Giralt and Chatila [10] navigated using, amongst other sensing modalities, ultra-sonics. Brown [2] investigated the techniques of feature extraction for recognising solid objects with ultra-sonic range sensors. Flynn [9] describes a system for combining information from ultra-sonics and infra-red for mobile robot navigation. Drumheller [8] describes a method by which range data from a sonar or other type of rangefinder can be used to determine the 2-D position and orientation inside a room. Brady [1] describes work aimed at providing an AGV with: vision, sonar and direct range sensing. Kuc and Seigal [11] describe a physically based simulation model for acoustic sensor robot navigation. Elfes [7] describes a sonar based mapping and navigation system. The system used range data to build a multi-leveled description of the environment.

The plan of the paper is as follows. In the remainder of section 1 I outline the basic idea. Section 2 presents theory and derives the relevant equations. Section 3 gives details of the experimental hardware. Section 4 presents the results in support of the theory. Section 5 discusses the results obtained. Section 6 concludes the paper.

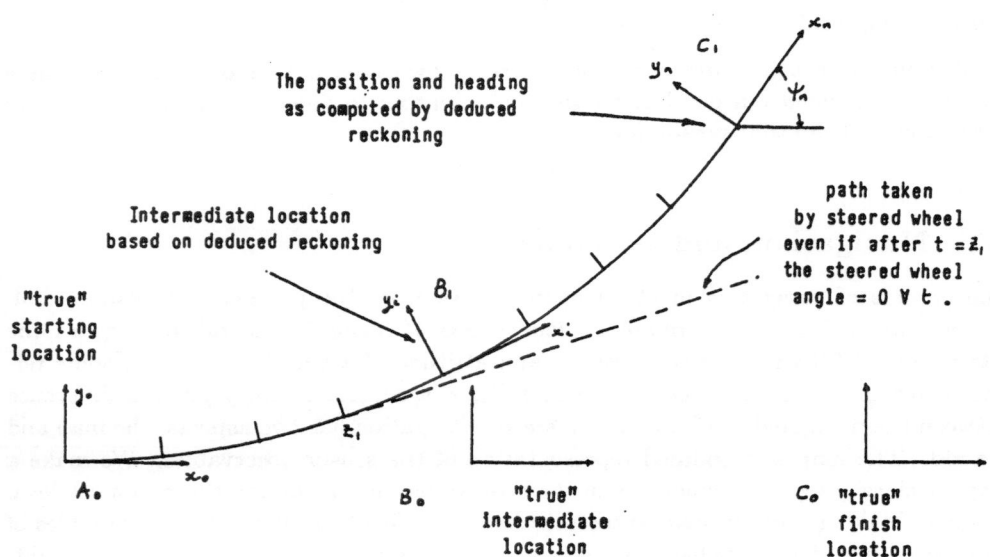

Figure 1: Navigation by deduced reckoning.

1.3 Navigation by deduced reckoning.

The process of deduced or dead reckoning provides an initial estimate of the position (X_e, Y_e) and a heading ψ_e of a robot-vehicle with respect to some global origin. For the vehicle used in this study two measurements were made, the distance moved by the steered wheel in rolling, and the angle the steered wheel pointed to as it rolled. An alternative would have been to measure the distance moved by the two rear wheels. Refering to figure 1 the robot's location is symbolised by the coordinate frame whose zero is positioned at the differential point of the vehicle, and whose orientation is parallel with the X-axis.

Three frames are illustrated at A_0, B_0, and C_0, to indicate the vehicle having actually moved forward with zero curvature and these locations are samples from its "true"path. The eventual divergence of any dead reckoned estimate from the actual track is symbolised by the frame at B_1, and at C_1. The short normal lines along the deduced track illustrate the discrete nature of the technique. The divergence is illustrated for a constant error in the steer angle. Although an error may exist for a "short" duration once present its effect remain. The tangential line is the track which would result even if after location Z_1, there was no further error in the steer angle. Using dead reckoning the estimates (X_e, Y_e) and ψ_e become increasingly uncertain.

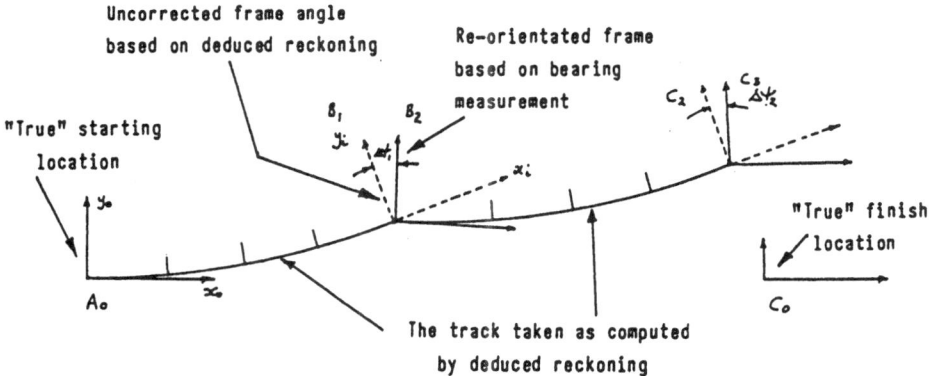

Figure 2: The corrected orientation of the robot's heading based on a direct measurement of the compass bearing.

1.4 The need to correctly orientate the local frame.

To bound this uncertainty two independent measurements can be made. The first bounds the uncertainty in the heading estimate, the second uses this improved heading estimate in conjunction with a map of the surroundings to feedforward expectations of the measured range to be compared with the actual measurements. The difference can then be used to drive the robots actuation system to compensate for these errors.

Refering to figure 2 the vehicle is again shown as having started from location A_0, and actually having moved to location C_0. By dead reckoning the computed position and heading is B_1. At this location a compass bearing measurement becomes available. This information is shown as having been used to modify the computed heading estimate, and this is reflected in the frame shown reorientated by $\Delta\psi_1$. The odometer and steer angle measurements continue then to be used to deduce the track from B_2 to C_2. At C_2 another compass bearing measurement becomes available. The information thus gained is illustrated as having been used once more to reorientate the frame this time by $\Delta\psi_2$.

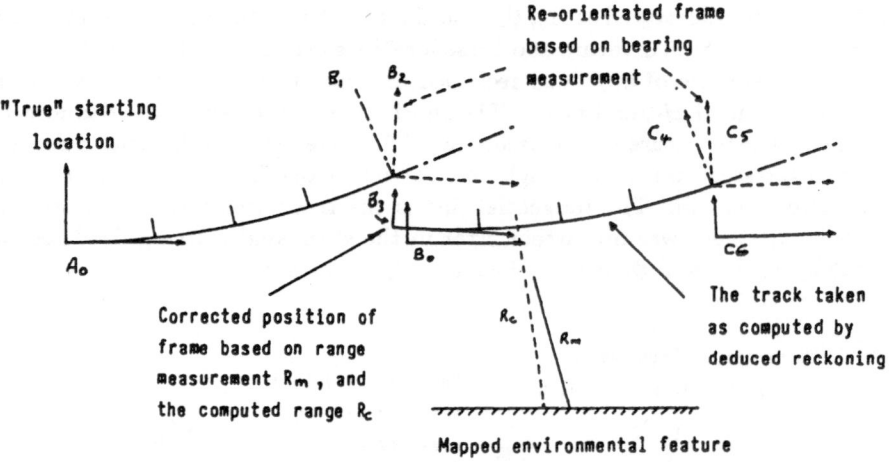

Figure 3: The repositioning of the zero of the local frame of reference based on range measurements.

1.5 Range sensing to reduce the remaining error.

Although the orientation of the local frame of reference has been corrected its position can still be in error. This difference between where the robot-vehicle is actually positioned and where it estimates itself to be can be reduced. To do this we correlate the expected range measurement with actual range measurement, minimise the difference, and update the position of the robot-vehicle accordingly. We model the environment in terms of geometry. Objects are described in terms of geometric objects more specifically as line segments whose mid points are encoded as a global positions and whose orientations are encoded in terms of an outward pointing unit vectors and each segment has a semi-length. Together these geometric objects and the data base they form and their implicit relationships are the map. If the vehicle carries additional sensing instruments to acquire information from the ambient about the geometric disposition of itself relative to these modelled objects then this form of "terrain comparison" can be used to correlate expected ranges with actual ranges. Section 2.3 describes the tech-

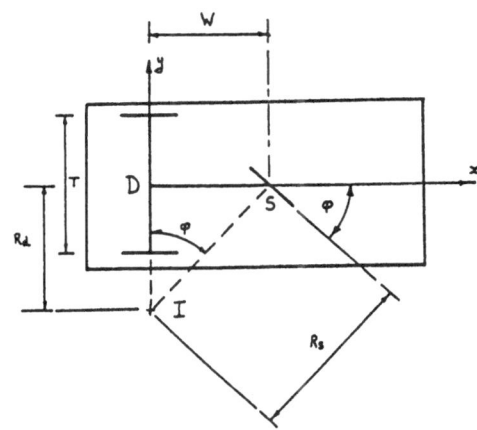

Figure 4: A plan view defining the geometry of the tricycle vehicle. x and y are the longitudinal and lateral axes of the local frame of reference.

nique in more detail. The use of multiple sensors distributed aroud the periphery of the vehicle leads to techniques in which we minimise according to some weighting function the difference between all the computed ranges and the measured ranges. In figure 3 the robot-vehicle is symbolised as having moved forward with zero curvature from A_0 to B_0. The deduced position and corrected heading of the robot is illustrated at B_2. This computed location being based on odometer, steer angle, and compass bearing measurements, as previously described. At B_2 the first sonar range measurement becomes validly available (there are a number of other conditions which need to be met for a range to be *validly* included, these are discussed in section 2.3). This first valid measured range is R_m. It is made from the robot-vehicle's actual physical location. Based on the robot's estimated position, and the known position and orientation of the range sensor of the range sensor a computed range R_c can be calculated. The position and relative angle between the sensor and the heading of the robot-vehicle is known because we placed the sensor there. The difference between these ranges can be used to notionally reposition the zero of the coordinate frame, the differential point, from B_2, to B_3. As the vehicle continues to move the odometer, steer angle and compass bearing measurements are once more used to locate the robot-vehicle. At C_5, the next range measurement becomes available and this can be used to reposition the frame from C_5, to C_6. This location C_6 being "near" where the robot is physically.

2 Theory

The objective of this section is to show how the position and heading of the robot-vehicle can be determined by a three stage measurement process. We derive the equations which describe the relation between what can be measured and what needs to be controlled to initially estimate the position and heading of the vehicle. We then show how compass measurements of the bearing can be used with the odometer to perform a similar, but improved navigation process. Finally, we derive and show how range measurements can be used to augment these estimates using a map to reduce the uncertainty in the estimated position.

2.1 Navigation by deduced reckoning.

We derive the equations of motion governing a tricycle vehicle. A plan view of the model tricycle vehicle is shown in figure 4

The two rear wheels are assumed to be driven via a differential mechanism from a single motor M_T. The rear wheels are separated by a distance T, the track. The distance between the point of contact of the front steered wheel and the differential point is called the wheelbase W. R_d and R_s are the radii of the circles from the instantaneous centre of turning to the differential point and the steered point with the steer angle held constant.

The front steered wheel is not driven, but the angle the steered wheel makes with the longitudinal axis is controlled by a motor M_S. The steered wheel angle, ϕ, can vary between

$$- \phi_{max} \le \phi \le +\phi_{max} \tag{1}$$

where a negative steer angle is taken to mean the vehicle turns to the starboard.

As the robot moves forward (or in reverse), the instantaneous angle the steered wheel makes with the longitudinal axis determines the instantaneous curvature of the path. When the steered wheel is exactly parallel with the x-axis, and if the surface moved over can be treated as plane, the robot would move along a zero curvature path, a straight line. If the steered wheel is rotated through to $\pm\pi/2$ no vertical rotation of the steered wheel is possible. There exists some maximum steer angle, during the application of drive forces from the rear wheels, which is consistent with the front wheel being able to rotate through its vertical axis. If ϕ is held constant, at some non zero angle, and the rear wheels are made to rotate all parts of the robot's the structure will move along a path of constant curvature. The path of the steered wheel as it moves along one constant curvature segment is shown in figure 5

For a constant steer angle ϕ, if the vehicle moves at some constant speed along an arc of length Δs then the steer radius R_S is

$$R_S = W/\sin(\phi) \tag{2}$$

and for an arc with this radius, the angle $\Delta\psi_S$ the steered wheel moves through, is

$$\Delta\psi_S = \Delta s/R_S \tag{3}$$

Using equation 2, we can express the change in heading as a function of the steer angle and the along track distance. These can both be measured.

$$\Delta\psi_S = \frac{\Delta s \, \sin(\phi)}{W} \tag{4}$$

With a local coordinate frame whose origin is at the notional point of contact of the steered wheel with the ground, and whose longitudinal axis is aligned along the axis of

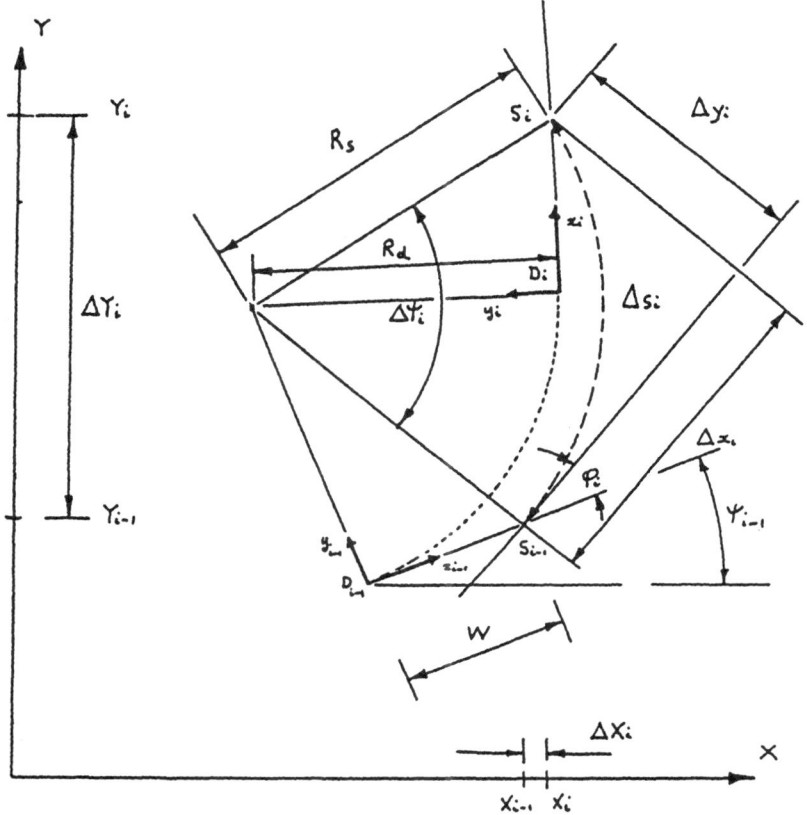

Figure 5: The path of the steered wheel as it traverses the ith segment.

the wheel, the longitudinal distance Δx, moved in this coordinate frame by the steered wheel is

$$\Delta x = R_S \sin(\Delta \psi_S) \tag{5}$$

and the lateral distance Δy, moved by the steered wheel in the same frame is

$$\Delta y = R_S \left(1 - \cos(\Delta \psi_S)\right) \tag{6}$$

Using the previously obtained values for the steer radius, and the incremental change in heading and substituting into equations 5, and 6, we obtain

$$\Delta x = \frac{W}{\sin(\phi)} \left(\sin \left(\frac{\Delta s \, \sin(\phi)}{W} \right) \right) \tag{7}$$

$$\Delta y = \frac{W}{\sin(\phi)} \left(1 - \cos \left(\frac{\Delta s \, \sin(\phi)}{W} \right) \right) \tag{8}$$

In the global frame of reference, we want to determine the incremental changes ΔX, and ΔY, at the steered wheel, these are given by

$$\left[\begin{array}{c} \Delta X \\ \Delta Y \end{array} \right] = \left[\begin{array}{cc} \Delta x & -\Delta y \\ \Delta y & \Delta x \end{array} \right] \left[\begin{array}{c} cos(\psi_S + \phi) \\ sin(\psi_S + \phi) \end{array} \right] \tag{9}$$

The location of the steered point is given by the previous estimate plus the latest update

$$\left[\begin{array}{c} \psi_{S(i)} \\ X_{S(i)} \\ Y_{S(i)} \end{array} \right] = \left[\begin{array}{c} \psi_{S(i-1)} \\ X_{S(i-1)} \\ Y_{S(i-1)} \end{array} \right] + \left[\begin{array}{c} \Delta\psi_{S(i)} \\ \Delta X_{S(i)} \\ \Delta Y_{S(i)} \end{array} \right] \tag{10}$$

The location of the differential point is given by

$$\left[\begin{array}{c} X_{D(i)} \\ Y_{D(i)} \end{array} \right] = \left[\begin{array}{c} X_{S(i)} \\ Y_{S(i)} \end{array} \right] - W \left[\begin{array}{c} cos(\psi_{S(i)}) \\ sin(\psi_{S(i)}) \end{array} \right] \tag{11}$$

where of course the heading angle remains unchanged.

2.2   Navigation with an odometer and a compass.

The section shows how to automatically incorporate measurements made using a compass into the kinematic equations to bound the error in the estimated heading. The compass measurements, and hence the corrections to the heading estimate, should be made prior to performing the range based corrections to the position. The next section discusses why this sequence is necessary.

The incremental change in position and heading in the global frame can now be written as

$$\left[\begin{array}{c} \Delta X_{c,i} \\ \Delta Y_{c,i} \end{array} \right] = \left[\begin{array}{cc} \Delta x_{c,i} & -\Delta y_{c,i} \\ \Delta y_{c,i} & \Delta x_{c,i} \end{array} \right] \left[\begin{array}{c} cos(\psi_{m(i-1)}) \\ sin(\psi_{m(i-1)}) \end{array} \right] \tag{12}$$

with

$$\Delta x_{c,i} = R_{S,i} sin(\Delta\psi_{m,i}) \tag{13}$$

and

$$\Delta y_{c,i} = R_{S,i}(1 - cos(\Delta\psi_{m,i})) \tag{14}$$

with the change in the measured bearing

$$\Delta\psi_{m,i} = \psi_{m,i} - \psi_{m,(i-1)} \tag{15}$$

Assuming a constant change in bearing at each stage, then

$$R_{S,i} = \Delta s_{m,i}/\Delta\psi_{m,i} \tag{16}$$

substituting equation 16 into equation 13 and equation 14 and then substituting equation 13 and equation 14 into equation 12, and rearranging, gives

$$\Delta X_{c,i} = \Delta s_{m,i} sinc(\Delta\theta_{m,i}) cos(\overline{\theta}_{m,i}) \tag{17}$$

$$\Delta Y_{c,i} = \Delta s_{m,i} sinc(\Delta\theta_{m,i}) sin(\overline{\theta}_{m,i}) \tag{18}$$

where

$$\overline{\theta}_{m,i} = \left(\psi_{m,i} + \psi_{m,(i-1)} \right) /2 \tag{19}$$

and

$$\Delta\theta_{m,i} = \left(\psi_{m,i} - \psi_{m,(i-1)} \right) /2 \tag{20}$$

the distance moved at each stage being

$$d_i = \left(\Delta X_{c,i}^2 + \Delta Y_{c,i}^2 \right)^{\frac{1}{2}} \tag{21}$$

or

$$d_i = \Delta s_{m,i} \, sinc(\Delta\theta_{m,i}) \tag{22}$$

2.3 Navigation using Range Measurements to reduce the positional uncertainty.

We now present a theory which shows how range measurements to objects in the environment can be used to reduce and bound the accumulated uncertainty in a mobile robot's estimate of its position.

The environment is modelled, as previously described, as a data structure consisting of acoustic observables. These are modelled as line segments. They possess a midpoint, a semi-length, and an orientation. The robot-vehicle is modelled as possessing length, width, a local origin for its frame of reference, and a-priori information about the positions, and orientations of its range sensors. The orientation of the outward pointing normal from the sensors is calculated with respect to the longitudinal x-axis of the vehicle. It is for this reason that the heading of the vehicle needs to be estimated accurately. Modelling a robot as a point will then be inadequate. The origin of the local coordinate frame of reference is the rear (drive) wheel differential, and the coordinate frame heading is parallel to the vehicle heading. Figure 6 shows the geometrical arrangement of the robot near to a wall in the environment.

The wall was represented in the robot's "map" by a line segment. Four range sensors with their associated orientations are shown in the four quadrants relative to the heading.

The symbols have the following meaning.

$\underline{x}_d$ = The estimated coordinates of the differential point.

ψ_h = The estimated heading of the vehicle.

$\underline{x}_{s,j}$ = The coordinates of the jth sensor position as determined from the differential point.

$\hat{\underline{n}}_{s,j}$ = The outward pointing unit vector of the jth range sensor whose direction cosines are with respect to the central longitudinal axis of the vehicle.

$\underline{x}_{wi}$ = The position coordinates of the ith line segment.

$\hat{\underline{n}}_{w,i}$ = The outward pointing unit vector of the ith line segment whose direction cosines are with respect to the frame of reference.

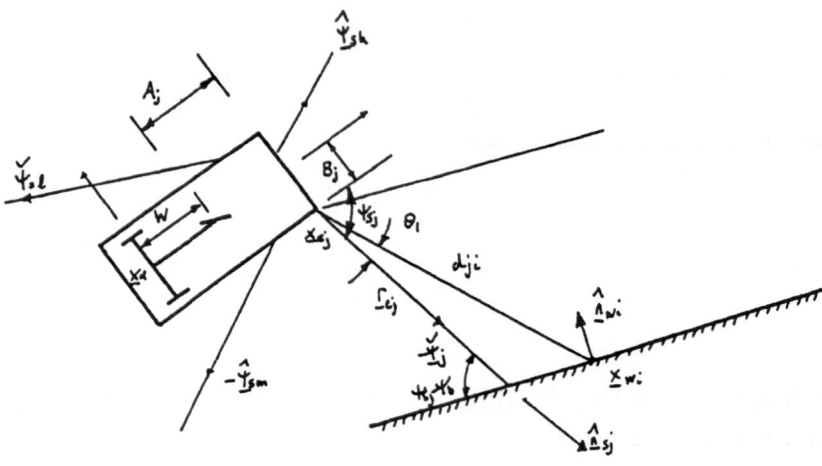

Figure 6: Robot-vehicle near to a wall.

$p_{w,i}$	$=$	The distance from the position of the jth sonar sensor perpendicular to the ith line segment.
$h_{j,i}$	$=$	The projection onto the ith line segment of the distance from the estimated position of the jth sonar head to the central coordinates of the ith line segment.
$r_{m,j,i}$	$=$	The range measurement made by the jth sonar ranger.
$r_{c,j,i}$	$=$	The computed distance from the jth sonar to the ith segment.
θ_b	$=$	The effective beamwidth of the sonar.
$\|d_{c,j,i}\|$	$=$	The computed distance from the jth sonar to the central coordinates of the ith line segment.

It is important to be explicit about the conditions which must prevail before a line segment may be validly included in the correlation process between measured and estimated ranges. The three conditions which need to be satisfied are:-

$r_{m,j,i}$	$\leq$	The maximum range measurable by the jth range sensor.
$h_{j,i}$	$\leq$	The semi-width of the candidate line segment.
$\hat{n}_{s,j} \cdot \hat{n}_{w,i}$	$\leq$	The cosine of the effective beamwidth, where effective means that a range signal will be returned. This will vary with the acoustic reflecting properties of the material.

These conditions impose powerful constraints on whether a modelled line segment need be included into the correlation process. As will be shown in the experimental stage there are circumstances in which no correlations can be made. In these circumstances the uncertainty in the estimated position grows until a valid sensor observable is available for correlation. The angle between the outward pointing normal from a line segment and the outward pointing normal from a range sensor must be less than the effective beamwidth for it to be validly included. The effective beamwidth means the range of angles over which sufficient sound energy returns to the receiver.

The range measurement is assumed to be made from a "true" position which is near to the estimated position. The measured range is assumed to lie along the outward pointing normal from the sonar ranger for orientations with respect to the wall of less than ± 13 deg. This assumption is based on pointing the sonar at various orientations and observing at what angle false range readings were obtained. The positional error is defined to be

$$e_{j,i} = r_{m,j,i} - r_{c,j,i} \tag{23}$$

The distance from the sensor position $\underline{x}_{s,j}$ to the central wall coordinates is computable, and for the ith identified segment

$$\underline{d}_{j,i} = \underline{x}_{w,i} - \underline{x}_{s,j} \tag{24}$$

with

$$\underline{x}_{s,j} = \underline{x}_d + [W + A_j].\hat{\underline{\psi}} \pm B_j.\check{\underline{\psi}} \tag{25}$$

The unit vector in the direction of the central coordinates of the identified segment can be computed by substituting equation 25 into equation 24. This is possible since $\underline{x}_d$, W, (the wheelbase) A_j, (the distance from the steered point to the sensor along the longitudinal axis), B_j, (The lateral offset from the longitudinal axis to the sensor), $\hat{\underline{\psi}}_h$, and $\check{\underline{\psi}}_h$ are either a-priori known or estimated values. The angle φ, between the normal from the sensor and the direction of the central wall coordinates can then be computed.

$$\cos(\varphi) = -\hat{\underline{d}}_{j,i}.\hat{\underline{n}}_{s,j,i} \tag{26}$$

The angle the computed range makes with the line segment is

$$\pi - (\psi_{s,j} - \psi_h) \tag{27}$$

The computed range to the ith segment is then given by

$$r_{c,j,i} = \| d_{j,i} \|. \sin(\psi_{s,j,i} - \psi_h) / \sin(\psi_{s,j,i} - \psi_h - \varphi) \tag{28}$$

The difference between this computed range and the measured range can now be used as the error in whatever algorithm is chosen to reposition the notional coordinate frame so as to reflect more accurately where the robot really is. Two possible methods of correcting the positional error are presented. They illustrate the technique. An extension would be to use a Kalman filter to "blend" the measurements with the expectations in a mathematically "correct" manner.

$$\underline{x}_{d,i} := \underline{x}_{d,i} + C_1.e_{j,i} \tag{29}$$

and the second correction algorithm is

$$\underline{x}_{d,i} := \underline{x}_{d,i} + C_2.e_{j,i} + C_3.e_{j(i-1)} \tag{30}$$

and C_1, C_2, and C_3, are suitably chosen gain constants.

3 Experimental Apparatus

3.1 The robot-vehicle.

The experiments were performed using a pneumatic tyred battery powered, electrically actuated, tricycle vehicle. It weighed about 500 kg. The vehicle was originally an industrial burden carrier. Its movement from place to place was directed by a human driver operating the electric drive motor or braking mechanism, through foot pedals, and by turning the steering wheel, while receiving sensory feedback about the results of these actions in the fashion used to control the speed and direction of a motor car.

To perform a similar function automatically the vehicle was modified. The vehicle was converted so that it could be driven either manually or under computer control. Two computer controlled actuation systems were designed and built to enable digital control over the speed v, and the steer angle ϕ. The power for all the on-board systems was obtained from six 190Ah 6V traction batteries. There was no umbilical. The modifications enabled a digital computer to receive information about the surrounding environment from a compass and nine time-of-flight acoustic sensors, the vehicle's internal state, the odometer, and steer angle, and its actions, the state of relays, for example.

The environment used in the experiments was the interior of a factory. this was built from brick, wood, and steel, variously painted. The surface moved over was concrete, and was smooth and flat enough to be treated as a plane for the purposes of the experiments conducted. The design and conversion from manual to dual computer/manual operation took about two years, involved the senior investigator, two research assistants, and a graduate student, namely the author.

3.2 Actuation systems

The traction interface. The computer was able via this interface to control and receive data on a variety of vehicle functions. Control to the vehicle consisted of; the release and application of a parking brake, the direction of rotation of the traction motor, and hence the direction of motion of the vehicle, the amount of electro-magnetic braking effort, the speed of rotation of the traction motor and thus the speed of the vehicle, clearing the odometer counter, and data from the vehicle about the latest value in the odometer register. The main motor was a 2 h.p. 36 volt series wound d.c. electric traction motor. The output from this motor was connected to the rear wheels via a differential mechanism. The speed was controlled by altering the phase angle of a thyristor controller. If the required bit pattern was not output by the controlling computer every 1200 ms the vehicle was placed in the "park" state with the brake on. *Steering control system.* This was an H-bridge controller. The steered wheel rotated vertically within the front fork assembly. The steered wheel was able to rotate ±60°. The motor was a printed circuit motor controlled by a standard pulse width modulation controller. The angle feedback was obtained from a potentiometer attached to the steering shaft. More details can be found in [14]

3.3   The sensors

An odometer. The resolution of this was 0.04 [m]. *Steer angle measurement.* The angular resolution was 1 degree. *A multi-channel sonar system.* This was a multiplexed system that sampled a range once every 60 ms. Figure 7 shows the coverage of the nine sensors that were used in these experiments. *An opto-magnetic compass.* This was a magnetic compass that was adapted so that it could read the magnetic bearing to ± 5 degrees. All this information was capable of being read by the computer. *Miscellaneous features.* The robot-vehicle was also equipped with a mechanical fender, various low-level status indicators, some low-level controlling functions, and a joystick for moving the vehicle around.

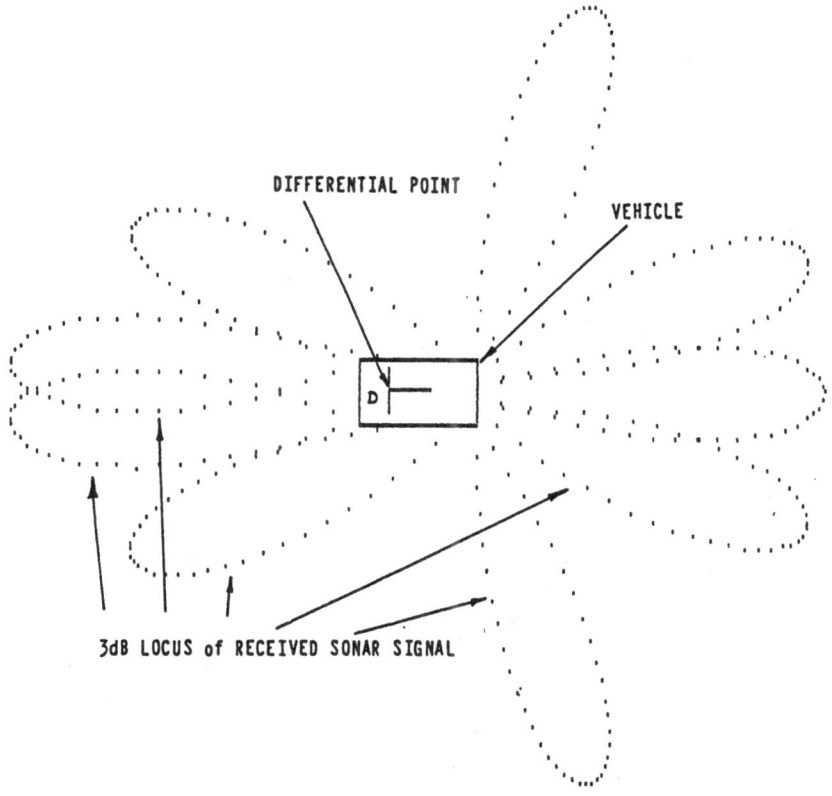

Figure 7: Plan view showing areas "covered" by the 3dB beamwidth of the multi-channel sonar system used in these experiments.

3.4 Calibration

Calibration is needed because the computer outputs digital values to the control systems and humans in general output their commands to robots in metres and degrees. An inherent difficulty with a steer control system is that for the vehicle to truly move along a straight line the steered wheel has to remain pointing in the same direction, and thus there is no inbuilt "on-board" reference for the steered wheel to "know" when it is pointing dead-ahead. It is essential to use a good resolver and to ensure that the feedback element and the steered wheel are accurately aligned. More details of the calibration procedure and results can be found in [13].

4 Experimental Results

The vehicle was driven forward in straight line and the steering was constrained so that the vehicle actually moved in a straight line. At one metre intervals range readings were acquired. At the same time odometer, steer angle, and compass measurements were made. The "true" path of the vehicle is shown in figure 8

4.1 Dead Reckoning.

The data gathered from the experiment were input to the navigation equations and these are shown below in figure 9

4.2 Compass and Odometer.

The data gathered from the experiment were again input to the navigation equations but this time augmented by the compass measurements. The results are shown in figure 10

4.3 Compass, odometer, and range measurements.

The data gathered from the compass, the odometer and the sonar ranger were all input to the full set of navigation equations as implied by equation 29. The results are shown in figure 11. Figures 12, and 13. show the effect of introducing a lateral error in the initial position estimate. The application of equation 29 is shown in figure 12. While the application of equation 30 is shown in figure 13. Figure 13 graphically demonstrates the numerical oscillation that can occur in the estimates of position when both the latest and the previous estimates are used.

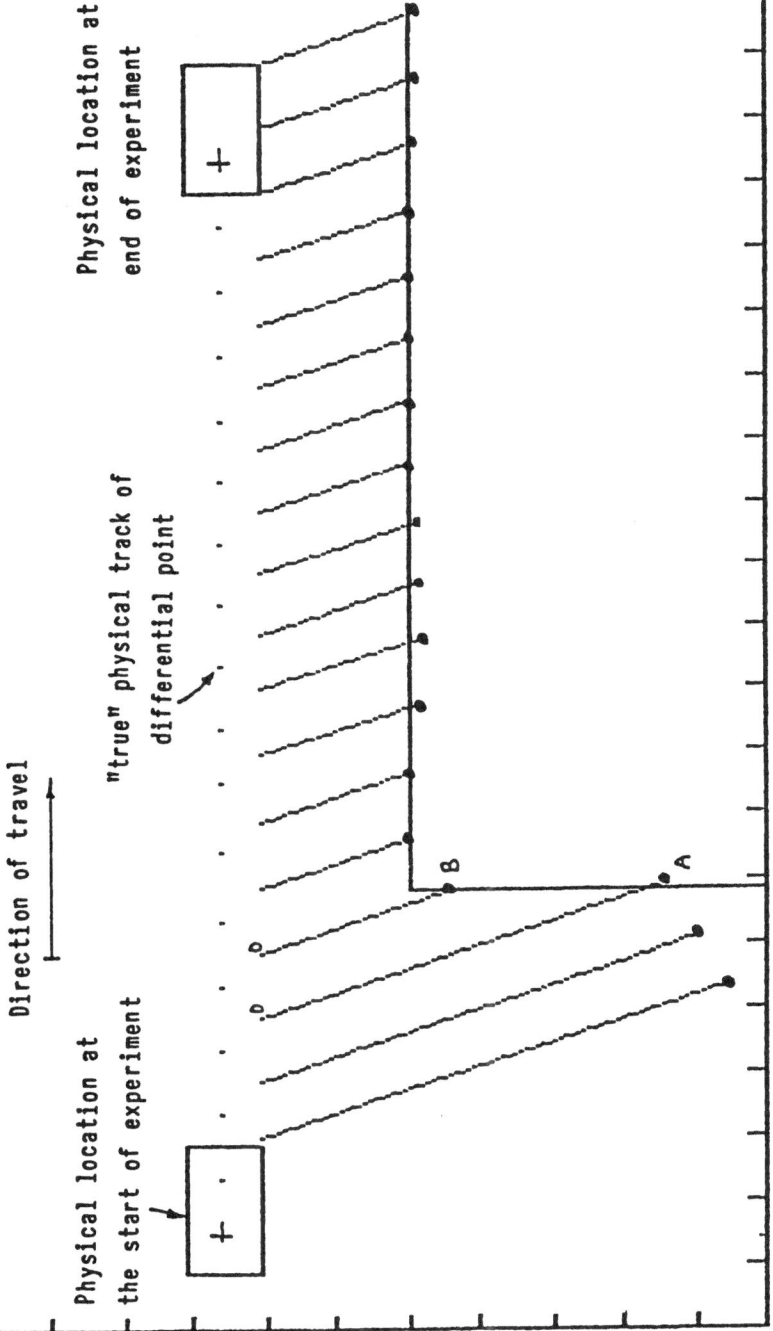

Figure 8: The actual track followed by the robot-vehicle. The scale is marked at one metre intervals. The orientation of the outward pointing normal is represented by the slanting dotted lines that represent the sound pulse travelling away from and back to the transducer after having been reflected from the wall.

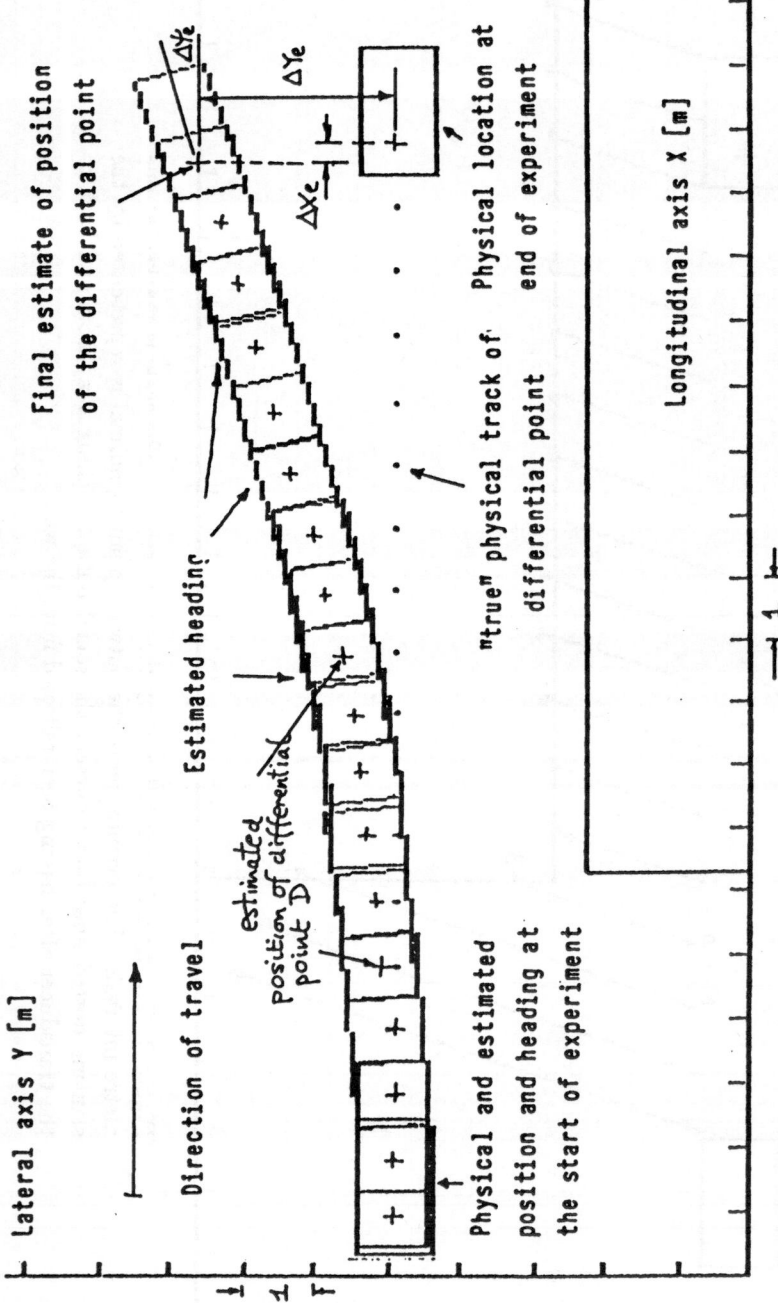

Figure 9: Dead reckoning the position of the differential point. The steer angle and along track distance moved by the steered wheel were used. Also shown is the computed heading of the robot, and its relationship to an environmental feature at the start and finish of the experiment. The scale is marked at one metre intervals. The dotted horizontal from the start position is the path the vehicle actually followed as determined by external measurement.

379

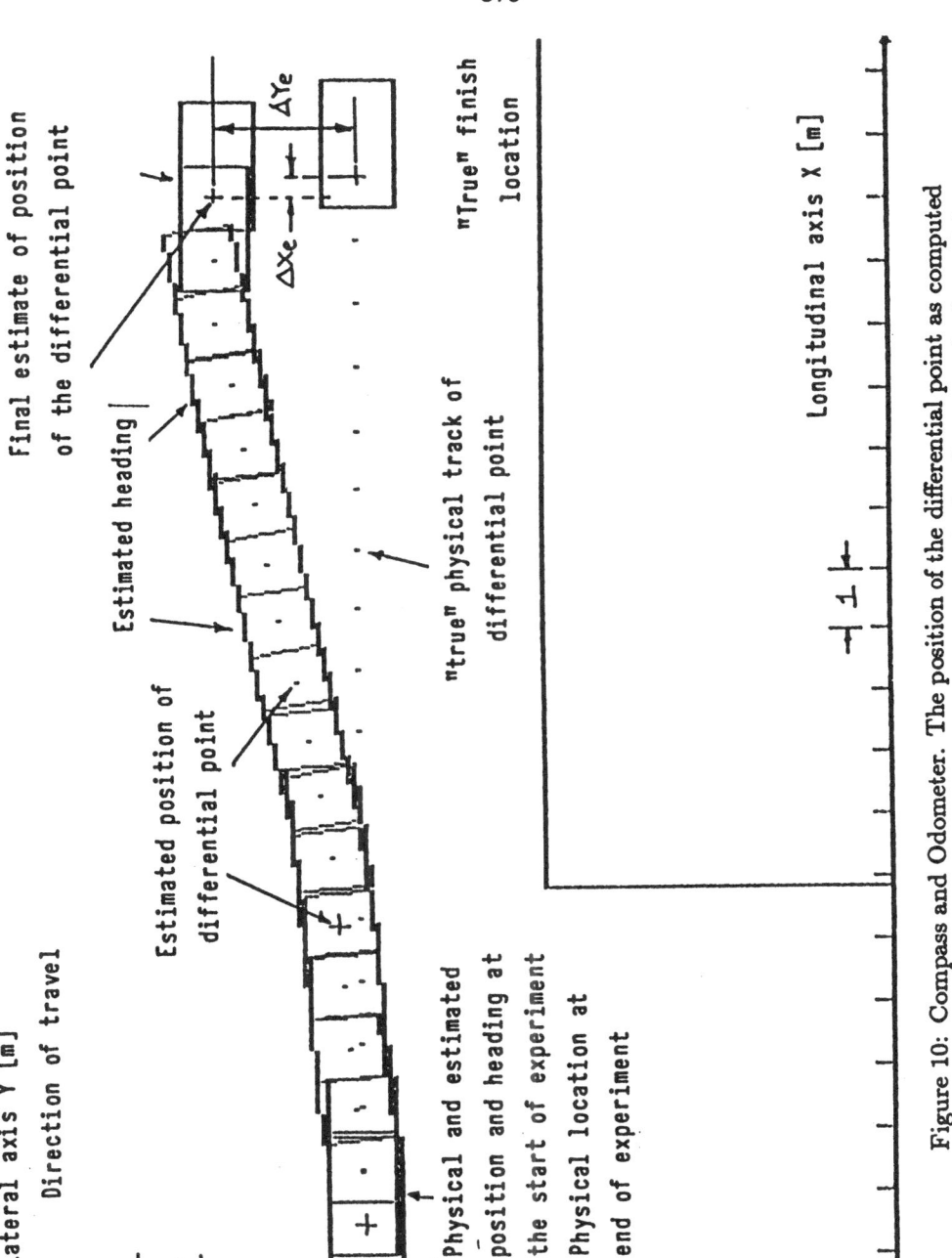

Figure 10: Compass and Odometer. The position of the differential point as computed using a direct measurement of the vehicle's bearing. The heading error in figure 9 has now been constrained by the bearing measurment. See text for further details.

5 Discussion

5.1 Dead Reckoning

Although the vehicle was driven in a straight line the sampled measurements of the steered wheel showed offset from the demanded zero value. As a result the computed heading diverges from the real heading which increases in proportion to the distance travelled, and is given by

$$\Delta \psi_{c,i} = \frac{C\phi_{e,i}}{nW}\left(1 + \frac{\Delta r_f}{r_f}\right) \tag{31}$$

where C, is the rolling circumference of the steered wheel, ϕ is the steer angle, n is the number of segments that C has been digitised into, r_f is the rolling wheel radius, and Δr_f is a small change in this radius, and the measured steer angle has an error, e, associated with it.

$$\phi_{m,i} = \phi_{t,i} + \phi_{e,i} \tag{32}$$

where t = true, c = computed, and m = measured. The most serious form of error is that given by equation 31. This error increases withour bound. Since the $i-1th$ computed heading is used in the ith computed position estimate, this accumulated heading error feeds into and distorts the computed position. This is in addition to the ith measurement errors. Without another means to determine the true heading no correction is possible.

The use of deduced reckoning to compute the position and heading of a vehicle is not new. The next two sections discuss the beneficial effect of using the compass and the ranger to bound and correct these errors.

5.2 Compass and odometer.

Without filtering, and with "low" resolution the compass odometer system was able to experimentally locate the robot more precisely than with the odometer steer angle system. The lateral error was 0.75[m] less than with the steer angle odometer system. The dynamic behaviour of the compass needle as it rotates can be expected to be at least second order. There will be a delay in the measurement of the bearing, and the computation to produce the "best" estimate of the vehicle's heading. Thus the estimated heading will lag behind the "true" heading of the robot. It is evident there will be some adverse influence on the stability. Alternatives would have been to use a solid state compass or a flux gate magnetometer. A detailed comparison between the growth in the error between the odometer steer angle and compass odometer systems show that under certain conditions the odometer steer angle system can result in smaller errors than when using the compass odometer. Even with a compass the lateral position errors remain, and these have to be corrected by other means.

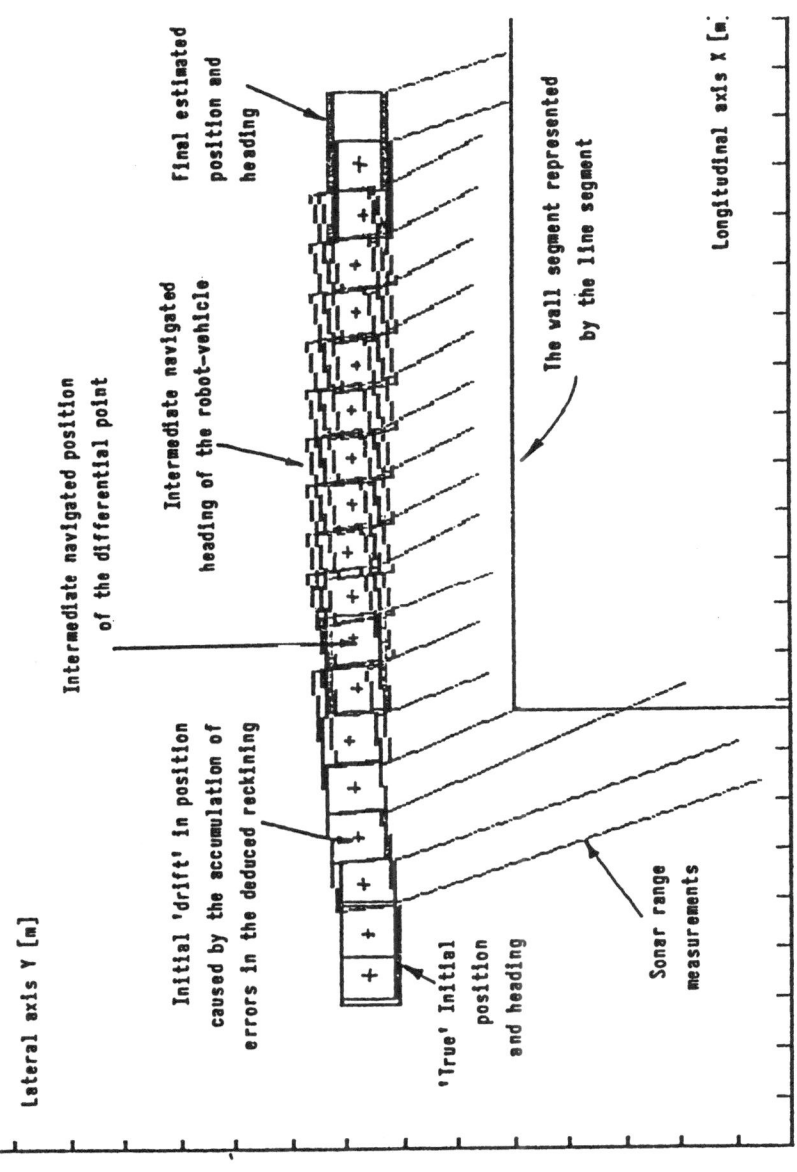

Figure 11: Experimental results showing that using the odometer, the steer angle, the compass, and correlating the sonar range measurements with the expected ranges the robot-vehicle could maintain "good" estimates of its true position and heading. The corrections to the estimated *heading* are based on the *compass* measurements. The corrections to the estimated *position* are based on the sonar *range* measurements. See text for further details.

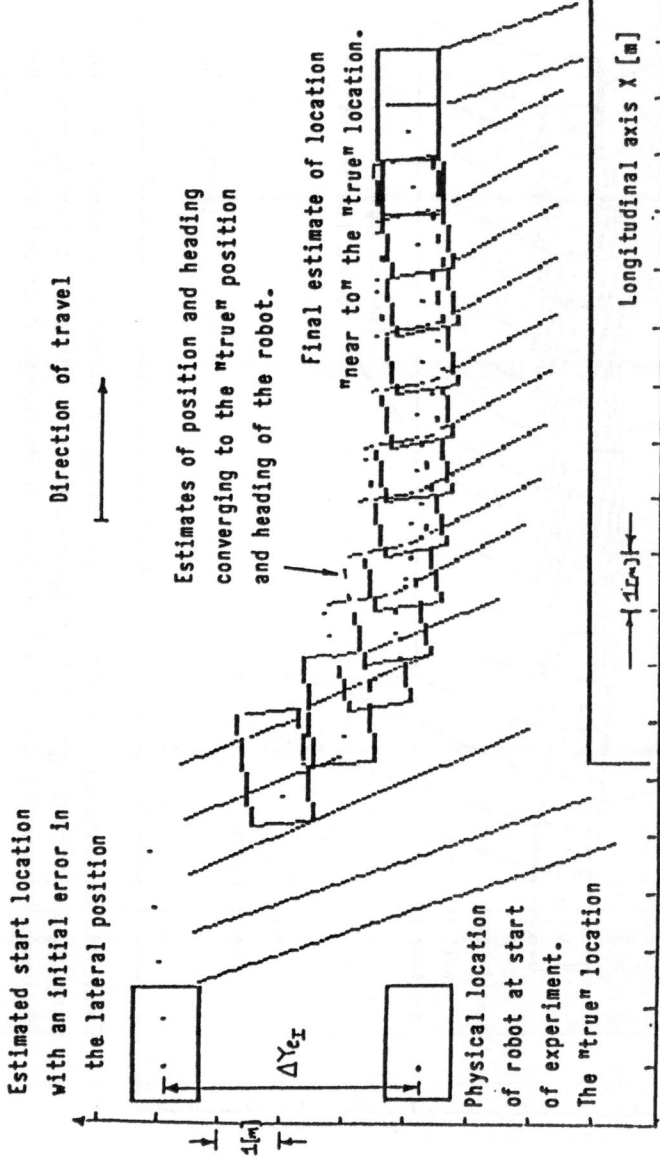

Figure 12: The illustration shows the robot-vehicle starting with an initial error in the estimate of its lateral position. As the vehicle moved forward the range measurements that were obtained experimentally are shown as being able to correct the robot's position estimate. The correction algorithm on which this is based, used the latest expected and measured range. The proportion of the error used, at each stage, to correct the vehicle's estimated position influences the rate at which the position estimate converges to its "true" value.

383

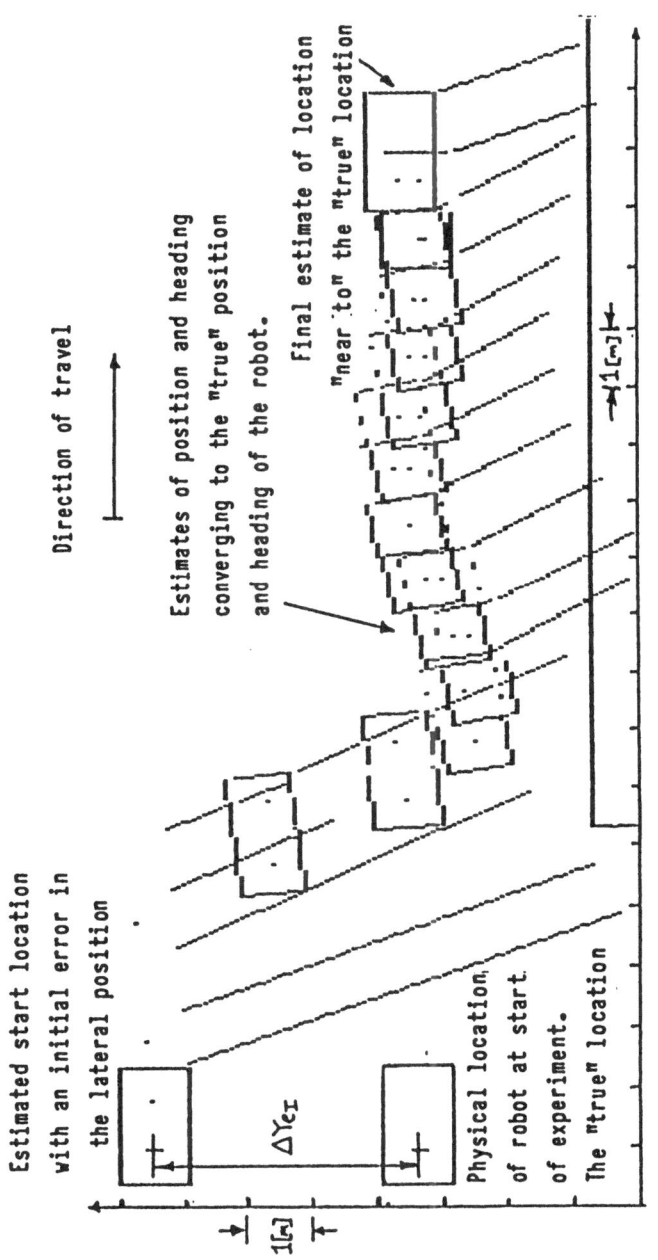

Figure 13: The effect of laterally repositioning the robot-vehicle's computed position based on the range measurements obtained experimentally. The correction is based both on the latest and the previous computed and measured range. The illustration shows the effect of a particular value of gain to correct the position.

5.3   Compass, odometer and sonar range measurements.

There was good agreement between the actual position and heading (as determined by external measurement) and that estimated using the navigation equations presented earlier. The final position and heading at the end of the experiment were close (about 5 cm in position and a few degrees in heading) to their "true" values.

In a correlation process such as we have here there are a number of important sources of error. Firstly, the map may be in error. It may be inaccurate (the attributes of the features may have been recorded with insufficient precision) or incorrect (features have been omitted or added). Secondly, the measurement may be in error, again it may be inaccurate or incorrect. Deciding where the error lies is not trivial. Although not shown here, it was found that when the robot vehicle's sensor horizon was within parts of the world that were not included in the model then range measurements obtained in these regions would cause the navigation process to incorrectly locate the robot-vehicle. This highlights a basic weakness of the method, namely that the world model must be complete. This is particularly annoying because the specular nature of sonar will often cause the measurement to be incorrect as well.

Referring to figure 11 it can be seen that the first six metres of travel of the robot illustrate the effect of the robot-vehicle being in a region where no correlation can take place. There is the expected increase in the uncertainty in the estimated position caused by using dead reckoning alone. Detailed simulation reveals that it is the lateral position errors which accumulate most rapidly. This can been seen in the results obtained using dead reckoning alone. At about six metres into its travel the robot was able to use its observations, the measured ranges, and correlate these with its prestored map. The three conditions neccesary for a valid correlation to take place are repeated below, these are :-

$r_{m,j,i}$	$\leq$	The maximum range measurable by the jth range sensor.
$h_{j,i}$	$\leq$	The semi-width of the candidate line segment.
$\hat{n}_{s,j} \cdot \hat{n}_{w,i}$	$\leq$	The cosine of the effective beamwidth, where effective means that a range signal will be returned. This will vary with the acoustic reflecting properties of the material.

In this experiment only the lateral position could be validly corrected. If the robot had wanted to constrain its position along the longitudinal axis, for example, it would have needed to make a sensor measurement along that axis. When moving down "long" corridors this type of feature is, by definition, not available. More generally, the position of the robot can only be constrained along both degrees of freedom when two orthogonal surfaces are near enough for two orthogonal range measurements to be made.

We have constrained our world model so that it could be modelled and manipulated easily. However our experience with sonar suggests that this simple model is inadequate. We are exploring ways to improve our models.

The sonar ranger and the compass both produce signals. Any signal has an associated noise component. The noise component in the passband of the signal can be expected to introduce additional errors. The specification of the statistics of this unwanted noise in the signal then places a further limit to the reduction of the uncertainty that is possible along each of the coordinate axes. The numeric accuracy of the algorithms may also influence the accuracy of the navigated location.

The rate at which these environmental measurements can be made may also be expected to limit the rate at which the uncertainty along each of the coordinate axes can be reduced, even if there were no measurement error. For example, using a mechanically scanned sonar imposes a limit. The rate of convergence of the algorithms used to process the measurements will also limit the rate at which the uncertainty can be reduced.

Both the lower resolution limit and the rate at which the uncertainty is reduced may all be different along each of the three coordinate axes. The characteristics of the growth and decay of this uncertainty may vary from place to place. It may also vary at different times at the same place.

6   Conclusion

6.1   Dead reckoning.

The technique of dead reckoning can be used to estimate the position and heading of a tricycle vehicle, whose front wheel has been instrumented to measure the steer angle and the along track distance. This can be done for both non-zero constant curvature motion and zero curvature motion.

For zero curvature motion the lateral position error accumulates more rapidly than the longitudinal error.

The heading error with the front wheel instrumented is proportional to the error between the "true" steer angle and the measured steer angle.

For constant curvature motion there is a steer angle which results in a minimum value for the heading error.

If a robot-vehicle is to navigate using dead reckoning, and the needed accuracy in the location estimate, for some travel distance can be specified, then this specification of the accuracy will determine the necessary resolution that the measuring instruments must possess for this specification to be met.

6.2   Compass and odometer.

The direct measurement of the vehicle's heading with a magnetic compass has been used to maintain the positional and heading uncertainty below the level achieved experimentally with the steer angle and odometer measurements alone.

These measurements can be used to re-orientate the local frame of reference and thereby counteract the otherwise irreducible accumulated uncertainty in the heading estimate if dead reckoning alone were used.

6.3 Compass, odometer and sonar range measurements.

The accumulated uncertainty in the dead reckoned position can be reduced by a process of correlating prestored information about the reflecting surfaces in the environment with incoming mesaurements made to those surfaces.

To do this the position and orientation of the environmental segments need to be known a-priori, and an estimate of the vehicle's position and heading are required.

If the direction associated with the range measurement is not that assumed in the calculation then this error will alter the accuracy of the estimated position. This error has two sources. One source is the uncertanty in the orientation of the frame used to calculate the estimated range. The other source of error is from the uncertainty associated with the real direction from along which a returned echo can be received.

6.4 General

Navigation is a fundamental component in the repertoire of an intelligent robot-vehicle. Without a means to establish its position and heading relative to some frame of reference the robot-vehicle is effectively lost.

To demonstrate the principle one line was used experimentally. In an operational system many more would be need to be included. As more segments are added and as mathematically more rigorous methods, for example, Kalman filter techniques are used, we would expect the navigation process described here to provide better estimates of the vehicle's position and heading.

This paper then has presented theory and experimental support for one technique of autonomous navigation for a robot-vehicle using a three stage measurement process. This involved sensory feedback from the environment and an encoding of the environment. The environmental measurements we used consisted of two different types of sensors. Firstly a "digital" compass to measure the vehicle's heading and secondly multiple acoustic sensors distributed around the periphery of the vehicle to measure the vehicle's disposition relative to objects in the environment.

The range measurements were correlated with expectations based on a geometric representation of the robot's surroundings. The representation used was designed to use an encoding of the world in terms of what the sensor could actually observe. The observables were modelled geometrically, and were given the attributes of position, orientation, and semi-length.

Sensor fusion experiments performed with the robot-vehicle showed that it was able to keep its estimated position and heading close to their "true" values using this form of indoor "terrain comparison".

Acknowledgements

This work was carried out with financial support from the Science an Engineering Research Council, Lansing Bagnall, Basingstoke, U.K., and RARDE(Chertsey) UK. I thank Dr M.H.E. Larcombe whose pioneering work in the field of mobile robotics permitted this robot-vehicle to briefly come into existence and who designed the robot.

References

[1] Brady, M. *Progress toward a system that can acquire pallets and clean warehouses.* 4th Int. Symposium Robotics Research. Ed. R. Bolles and B Roth.

[2] Brown, M.K. *Feature extraction techniques for recognising solid objects with an ultra-sonic range sensor.* IEEE Journal of Robotics and Automation, VOL, RA-1, NO. 4. Dec 1985.

[3] Bauzil, G. Briot, M. and Ribes, P. *A Navigation sub-system using ultra-sonic sensors for the mobile robot Hilare.* 1st ROVISEC (Robot Vision and Sensory Control) Stratford-Upon-Avon, UK. Pub. IFS Publications, Bedford, UK. 1981.

[4] Clemence, G.T. and Hurlburt, G.W. *The application of acoustic ranging to the automatic control of a ground vehicle.* IEEE Trans on Vehicular Technology, VOL, VT-32, NO. 3, Aug 1983.

[5] Crowley, J.L. *Navigation for an intelligent Mobile Robot.* IEEE Journal of Robotics and Automation, VOL. RA-1, NO. 1 March 1985.

[6] Presentation by B. Steer to DARPA representatives W. Isler, and C. Thorpe(CMU), and MOD representatives P. Bateman, and M. Larcombe, at the Austin Rover Canley Works, Coventry, UK. Oct. 1985.

[7] Elfes, A. *High resolution maps from wide angle sonar.* IEEE Int. Conf. on Robotics and Automation IEEE Mar 1985.

[8] Drumheller, M. *Mobile Robot Localization Using Sonar.* IEEE Trans. on Pattern Analysis and Machine Intelligence, VOL. PAMI-9, NO. 2, Mar 1987.

[9] Flynn, A.M. *Combining sonar and infra-red sensors for mobile robot navigation.* The International Journal of Robotics Research. VOL. 7 NO. 6. Dec. 1988.

[10] Giralt, G., Sobek, R. and Chatila, R. *A multi-level planning and navigation system for a mobile robot — A first approach to Hilare.* 6th Int. Joint Conf. on Artificial Intelligence, Tokyo, 1979.

[11] Kuk, R. and Seigal, M.W. *Physically based simulation model for acoustic sensor robot navigation.* IEEE Trans. on Pattern Analysis and Machine Intelligence, VOL. PAMI-9, NO. 6 Nov 1987.

[12] Serebriakoff, V. *The future of intelligence.* Pantheon Press, UK. 1988

[13] Steer, B. *Navigation for the Guidance of a Mobile Robot.* Ph.D. Thesis, Department of Computer Science, University of Warwick, Coventry. 1985.

[14] Steer, B. *Trajectory planning for a mobile robot.* Int. Journal Of Robotics Research. Oct. 1989

Planning and executing sensory based grasping operations in a partially known environment

A. Ijel, C. Laugier and J. Troccaz

LIFIA/IMAG, 46 Av. Félix Viallet

38031 Grenoble Cedex, France

Abstract

Making sensing and acting techniques to cooperate in order to achieve a given manipulation task in a partially structured environment, is one of the major issues towards autonomous robotics. In this paper, we describe the way we contributed to solve this problem in the context of automatic grasping. The method we have developed for that purpose, leads to combine commands for moving the robot end-effector towards some selected positions, with several sensing operations aimed at acquiring the missing information. It operates in three phases leading to (1) select a relatively uncluttered region from which it is theoritically possible to see the chosen features of the object to be grasped, (2) construct a model of the local environment of these features using the 3D vision sensor located on the robot end-effector, and (3) determine the grasping parameters and the required robot motions by reasoning on the gripper configuration space. The current implementation of the system and some experimental results are also presented in the paper.

1 Introduction

Making sensing and acting techniques to cooperate in order to achieve a given manipulation task in a partially structured environment, is one of the major issues towards autonomous robotics. In this paper, we describe the way we contributed to solve this problem in the context of automatic grasping. The tackled problem may be stated as follows: *find a goal configuration and a safe trajectory for the robot end-effector, in order to grasp a given object located in a partially known environment; only a partial geometric model of the robot workspace and a description of the object to be grasped are explicitly given to the system.*

It comes from these hypotheses that CAD based models and sensory data have to be combined for planning and controlling the execution of the required robot actions. This means that the system has to cope with both path planning problems (for generating the grasping trajectories and the robot movements which are required for positioning the vision sensor), and sensory data interpretation problems (for dealing with visibility analysis, vision/robot calibration, data interpretation and data fusion issues).

As we will see further, our method for solving these problems operates in three phases leading to (1) select a relatively uncluttered region from which it is theoritically possible to see the chosen features of the object to be grasped, (2) construct a model of the local environment of these features using the 3D vision sensor located on the robot end-effector, and (3) determine the grasping parameters and the required robot motions by reasoning on the gripper configuration space.

Several projects have already addressed the problem of combining computer vision and robot capabilities for automating grasping operations [1] [2] [3] [4]. But, most of these "hand-eye" projects have been initiated at an explicit programming level, avoiding thus most of path planning and accessibility analysis problems. In such approaches, grasping operations have mainly been studied as an interesting field for vision experiments, especially for bin picking. This is why the methods which have been developed in this context focus on the problem of finding specific image features relevant for grasping (for instance: parallel straight lines [1], or predifined sets of features representing stable 2D grasps [2]). We only know one work (the one described in [3]), which explicitly analyzes the accessibility of the selected features, by applying a simple interference checking algorithm on 3D vision data. In the other approaches, it is generally assumed that the "highest" object in the heap can be safely reached by the gripper, and that visible features are accessible.

Other works have also addressed the grasping problem in the context of automatic robot programming (see [5]). Several aspects of the problem have thus been studied, especially those concerning stability analysis and potential grasp checking. But, except at MIT [6] [7] and at LIFIA [8] [9], very few work has been done on accessibility analysis and path planning for grasping.

2 Outline of our approach

Grasping an object in a partially known environment requires to answer two main questions: how to select a good viewpoint for the vision sensor and to construct a model of the local environment of the object to be grasped ? how to choose a grasping strategy satisfying the accessibility constraints imposed by the robot workspace ? Our approach for solving the related problems operates in three phases respectively aimed at selecting a viewpoint, modelling the local environment of the object, and determining the grasping parameters. It also relies on the fact that the grasping problem can be split into two complementary subproblems:

1. *Choosing a stable potential grasp* π (i.e. a set of geometric features of the object to be grasped) which is compatible with the task to perform. The related computing step is performed by reasoning on an accurate geometric model of the object to be grasped. The techniques which have been developed at LIFIA for dealing with this problem are described in [8] [10] and [5].

2. *Determining the dynamic parameters of the grasping operation.* These parameters are the final configuration of the gripper (relatively to the selected object features), and the path allowing the gripper to reach safely this configuration. The related computing step is performed by reasoning on an explicit model of the "empty space".

In this paper, we make the assumption that the first grasping step has been performed off-line (see the example given in section 6), and that the selected grasping features of π are explicitly given to the on-line system. This means that we will focus on the second step, after having constructed an explicit model of the empty space using 3D sensory data. Our three phases approach operates as follows:

The first phase reasons on the available geometric information (for example, a CAD model of the nominal environment). It leads to determine the regions of the workspace from which it is theoritically possible to see the selected features of the object to be grasped. These regions are represented using spherical domains located on a "vision sphere" centered on the object. In the current implementation of the system, it is assumed that any point on this sphere can be reached by the robot gripper without generating a collision, provided that this point is located in the accessible robot workspace. This means that there is no obstacle outside of the sphere, and that the initial positioning operations are executed without moving the robot through the vision-sphere. It is the purpose of further developments to remove this constraint. The viewpoint selection algorithm applied in this phase is described in section 3.

The second phase executes sensing operations for acquiring additional information on the local environment of the object to be grasped. For that purpose it makes use of a CCD camera and of a laser strip, both fixed onto the robot gripper (see figure 1) . In a first step, the system verifies that the model based predictions are consistent with the observed scene (i.e. that the selected features are not hidden by some unknown obstacles in the analyzed 2D image). The next step consists in scanning the region located in the vicinity of the object using the laser strip. It leads to construct a simple 3D model (an ordered set of 2D curves) of the potential obstacles opposed to the gripper motions. The computer vision based techniques applied in this phase are roughly described in section 4.

The last phase constructs an explicit representation of the "empty space", in order to determine the dynamic parameters of the grasping operation. It makes use for that purpose of the 3D model which has been obtained at the end of the second phase. As we will see further, the empty space representation is mapped on a two dimensional space (thanks to the motion constraints imposed by the selected potential grasp) which characterizes the set of safe configurations for the robot gripper. Since we are mainly concerned with this step with the end-effector motions, we made the assumption that the *selected safe trajectories* for the gripper are often collision-free for the robot arm. This means that we will first plan trajectories for an "isolated gripper", before checking for the validity of such trajectories for the rest of the arm using interference checking techniques. This assumption is consistent with the fact that the generated trajectories are made of short straight lines executed in a relatively uncluttered region (it is the purpose of the first planning phase to select such regions). But, it is clear that this approach would probably fail in a too constrained environment.

The method applied in this phase is directly derived from previous work done at LIFIA in the scope of the SHARP project (see [9] and [11]). It is roughly described in section 5.

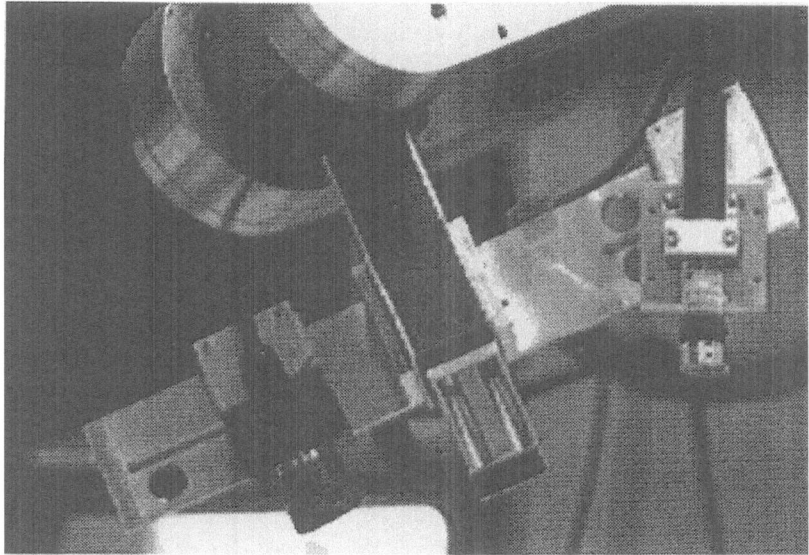

Figure 1: The robot gripper and its sensory equipment

3   Selecting a viewpoint avoiding occlusions

Using a vision sensor for acquiring additional information on the local environment of the object to be grasped, requires to plan an adapted sensing strategy. The purpose of this section is to describe the way we solved this problem, by choosing a location for the sensor within a computed set of valid viewpoints.

Let f_1 and f_2 be the object features which have been selected for performing the grasping operation. Our approach for solving the viewpoint selection problem operates in three steps leading to (1) determine the set of "grasping windows" which can be associated to the pair (f_1, f_2), (2) compute the occlusion-free regions of the workspace for each generated grasping window, and (3) select a particular viewpoint in the resulting occlusion-free space model.

3.1   Constructing the set of "grasping windows"

Since the purpose of the method is to plan a sensing strategy aimed at acquiring visual information on the space swept by the gripper when moving towards the object to be grasped, the associated viewpoint has to be chosen in such a way that it permits to see the set of potential grasping positions. As we will see further, the reference point of the gripper can only move in a plane GP parallel to f_1 anf f_2 during the grasping operation. This means that the set of potential goal positions for the gripper can be seen, from a vision point of view, as the set of points located on the intersection of GP with the convex volume V_{12} generated by the pair (f_1, f_2) of grasping features (figure 2 illustates).

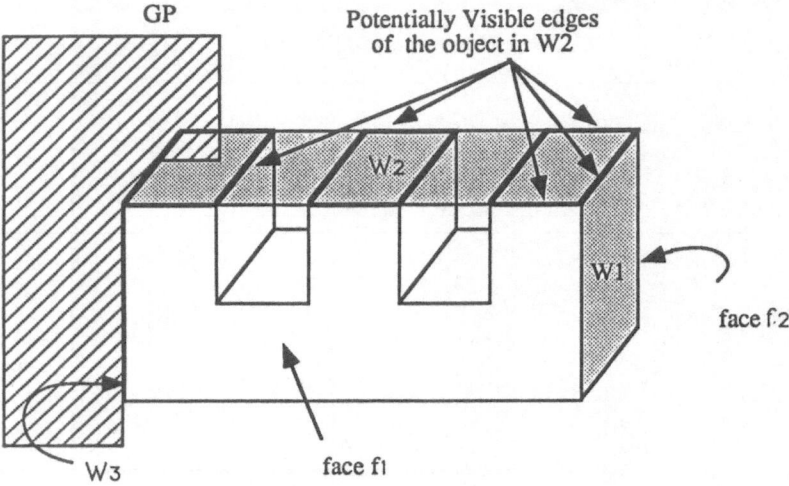

Figure 2: A pair of grasping features and their set of grasping windows.

Since we are only concerned with polyhedral objects, this set is a polygonal line and the associated potentially visible edges of the object are located on the boundaries of the faces which delimit V_{12}. These faces are refered as the "grasping windows" associated to the pair (f_1, f_2) of grasping features.

A practical algorithm for constructing the set S_W of the grasping windows which can be associated to the pair (f_1, f_2) is the following (see figure 2):

1. Project f_1 and f_2 onto GP, and construct the convex hull P of the resulting domain.

2. Conctruct, for each edge e_i of P, a rectangular face W_i characterized by the orthogonal projections of the two extremities of e_i onto the planes containing respectively f_1 and f_2. Remove the faces containing an edge of the object which is in contact with the table.

3.2 Computing the occluding regions

3.2.1 The problem to solve

The problem to solve is to determine, for each window W, the set of points of the workspace from which it is theoritically possible to see W. Let O be the set of potential obstacles, and F_W be the half-space defined by both the plane containing W and its associated external normal vector. A point $v \in F_W$ is an occlusion-free viewpoint for W, iff $vp \cap O = \emptyset$, $\forall p \in W$. Two types of occluding regions in F_W can be defined using this property (see figure 3a):

- Full Occluding Region: $FOR(W) = \{v \in F_W \mid \forall p \in W : vp \cap O \neq \emptyset\}$

- Partial Occluding Region: $POR(W) = \{v \in F_W \mid \exists p \in W : vp \cap O \neq \emptyset\}$

Then, it is clear that $FOR(W)$ is included in $POR(W)$, and that the resulting occlusion-free space $OFS(W)$ is the complement of $POR(W)$ relatively to F_W. In case of failure (i.e. $OFS(W) = \emptyset$), a weaker definition may be applied using $FOR(W)$ instead of $POR(W)$. This property holds when considering a set S_W of grasping windows, but in this case the set $FOR(S_W)$ —resp. $POR(S_W)$— is defined as the union of the sets $FOR(W_i)$ —resp. $POR(W_i)$—. Then, determining the occlusion-free space associated to a set of grasping windows S_W and to a set of obstacles O, can be done by computing separately the occluding regions associated to each pair (W_i, O_j), $\forall W_i \in S_W$ and $\forall O_j \in O$.

3.2.2 The algorithm

For the sake of clarity, we will first describe the method in the case of a pair (W, O), where W and O are polygonal surfaces (in the 3D space) respectively representing a grasping window and an obstacle. Then, we will discuss the modifications which are required for dealing with polyhedral obstacles.

Intuitively, it is clear that the boundaries of the occluding regions are composed of planar surfaces belonging to the planes which are "tangent" to both W and O. Let $\Pi(W, O)$ be the set of planes in which each element π is generated by a pair (v, e), where v is a vertex of either W or O and e is an edge of either O or W. We consider also the subset $\Pi_1(W, O)$ of $\Pi(W, O)$ in which each element is generated by a pair (v, e), where v is a vertex of O and e is an edge of W. Some planes in $\Pi(W, O)$ separates the space in two half-spaces containing respectively W and O (see figure 3b); we will call $\Pi_{sep}(W, O)$ the set of tangent planes verifying this property. Some planes in $\Pi_1(W, O)$ define one empty half-space and another half-space containing both W and O (see figure 3c); we will call $\Pi_{bound}(W, O)$ the set of tangent planes verifying this property. In the sequel, the half-space defined by a plane π and containing a polygonal surface P (P is not included in π) will be refered as the "P-containing half-space of π"; the other half-space will be refered as the "¬P-containing half-space of π".

In the 2D space, a straightforward method for constructing the occluding regions consists in solving a system of inequalities, where each basic inequality defines the "O-containing half-space" of a either a separating or a bounding line, and where the last inequality defines the "¬W-containing half-space" of the line O (see figure 3a). It can be proved that a similar property holds in the 3D space (see [15]):

case 1: The plane π_0 containing O does not intersect W.

$$POR(W) = \left(\bigcap_{\pi \in \Pi_{sep}(W, O)} F_\pi \right) \cap F_0$$
$$FOR(W) = \left(\bigcap_{\pi \in \Pi_{bound}(W, O)} F_\pi \right) \cap F_0$$

where F_π is the "O-containing half-space of π", and F_0 is the "¬W-containing half-space of π_0".

case 2: The plane π_0 containing O intersect W.

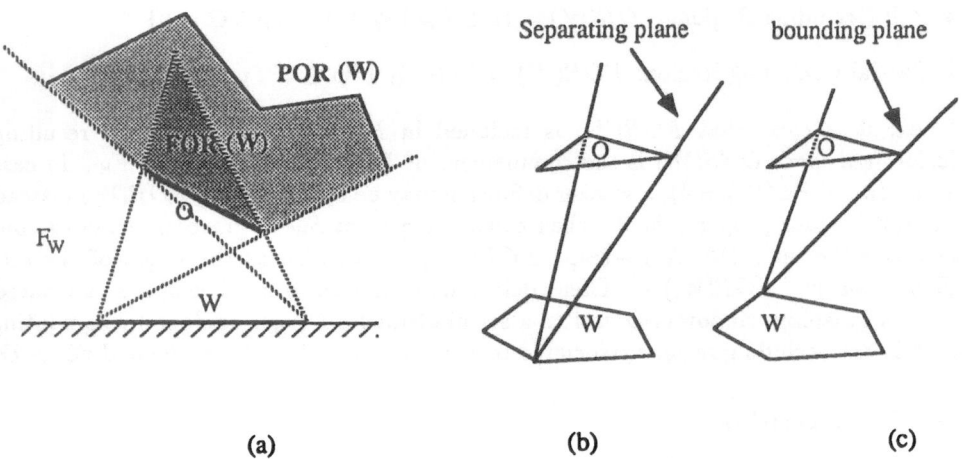

Separating plane   bounding plane

(a)   (b)   (c)

Figure 3: (a) A 2D illustration of the occluding regions. (b) Separating plane for W and O. (c) Bounding plane for W and O.

$$POR(W) = \bigcap_{\pi \in \Pi_{sep}(W,O)} F_\pi$$
$$FOR(W) = \emptyset$$

The same method can be applied when O is a polyhedral obstacle. Nevertheless, in order to improve the efficiency of the algorithm, we can modify the functions which compute the sets $\Pi_{sep}(W,)$ and $\Pi_{bound}(W,O)$ by discarding some of the pairs (v,e). This is due to the fact that some pairs (v,e) of geometric entities generate planes which do not belong to $\Pi(W,O)$ —because they intersect O—.

3.3   Selecting a valid viewpoint

A viewpoint is considered as valid if it is possible, according to the known model, to see (completely or not) one or several grasping windows. Then, selecting a "good" valid viewpoint requires to first construct an explicit representation of the occlusion-free space. In our system, this representation is based on the concept of "visibility sphere" (such a sphere is centered at the barycenter of the considered object features, and its radius is chosen according to the size of the robot workspace). Then, occluding regions can be represented using spherical domains obtained by intersecting the visibility sphere with the previously computed occluding regions. It should be notice that some FOR regions may be completely included in the vision sphere (consequently, such regions generate no occluded domains on the sphere).

In order to reduce the amount of computations, we have chosen to construct an occlusion-free space representation containing two types of regions (see figure 4): the

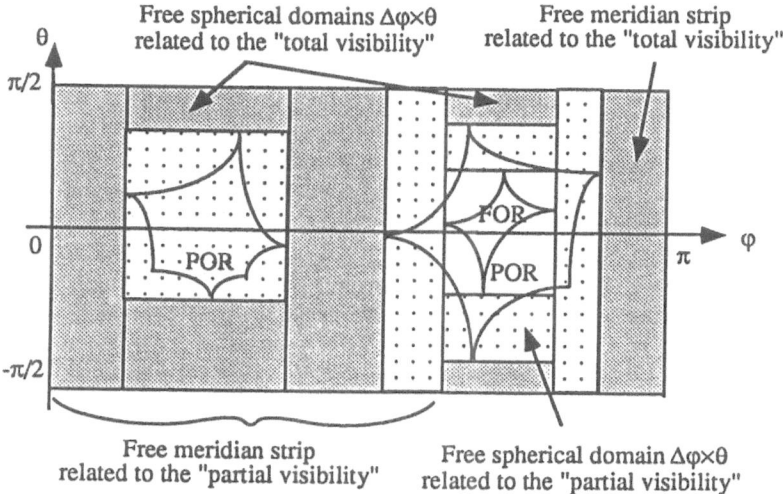

Free spherical domains Δφ×θ related to the "total visibility"

Free meridian strip related to the "total visibility"

Free meridian strip related to the "partial visibility"

Free spherical domain Δφ×θ related to the "partial visibility"

Figure 4: Representing the occlusion-free space.

free meridian strips and the spherical domains of the type $\Delta\varphi \times \vartheta$. In practice, this representation is mapped onto a plane (φ, ϑ), as shown in figure 4. It is also structured into free domains associated respectively to the "total visibility" problem and to the 'partial visibilty" problem.

Finally, the selection of a particular viewpoint in the occlusion-free space is done using an heuristic which attempts to both maximize the size of the surrounding occlusion-free region and the number of the related grasping windows. The selection function is first applied on the basis of the total visibility criterium. For that purpose, free meridian strips and free spherical domains are successively analyzed by the system. In case of failure (the resulting occlusion-free space is empty), the same operation is applied using the partial visibility criterium.

4 Processing vision data

4.1 Calibration

Several calibration operations are required for making 3D perception possible when achieving a vision based manipulation task using a robot and a mobile 3D sensor: camera-scene, camera-laser and camera-robot calibrations.

The camera-scene calibration is needed for determining the perspective transform existing between the real world (the observed scene) and the 2D image. Such a transform is basically characterized by two types of parameters: the *intrinsic parameters* of the camera (vertical and horizontal scales, focal length, and intersection of the optical axis with the image plane), and the *extrinsic parameters* defining the position and the orientation of the camera relatively to a given reference frame. In our system, these parameters are computed using the method of Faugeras & Toscani (see [12] for more details).

The camera-laser calibration is needed for determining the location of the laser plane in the reference frame of the camera. Let R_{cam} and R_{scene} be respectively the reference frames of the camera and of the observed scene. The previous calibration operation (camera-scene) has provided us with the geometric transform T_{cal} existing between R_{cam} and R_{scene}, and with the relations $f(u) = x/z$ (1) and $g(v) = y/z$ (2) holding between the three-dimensional coordinates of a point M in R_{cam} and its associated image pixel (u, v). Let z_s be the Z-coordinate of M in R_{scene}. The relation $z_s = a_1 x + a_2 y + a_3 z + a_4$ (3) can be easily derived from T_{cal} (the terms a_i are those of the third column vector of the matrix representing T_{cal}). Then, a practical way for calibrating the camera-laser device consists in analyzing the image of m points M ($m > 3$) located at the intersection of the laser plane with a known plane $z_s = k$ in R_{scene} (see figure 5a). The coordinates (x_i, y_i, z_i) in R_{cam} of each analyzed point M_i are computed using the relations (1), (2) and (3). The fact that M_i belongs to the laser plane can be expressed by the relation $-z_i = b_1 x_i + b_2 y_i + b_3$, where the b_j ($j = 1 \cdots 3$) are the unknown parameters of the analytic equation of the laser plane in R_{cam}. These terms can be computed by solving the obtained linear system using a least square method.

The camera-robot calibration is needed for determining the relative location of the camera and of the robot gripper. Let X, A and B be respectively the geometric transform representing the unknown relation existing between R_{cam} and R_{robot} (R_{robot} is the reference frame of the gripper), the change in R_{robot} due to an arm movement, and the resulting displacement of R_{cam}. A and B are known, since A can be computed from the encoder values and B can be found by calibrating the camera using the camera-scene calibration procedure (see figure 5b). Then, X can be computed by solving a system of two homogeneous transform equations of the form $AX = XB$ (see [14]). Two arm movements having non parallel rotation axes are required for obtaining a unique solution to the system.

4.2 Modeling local obstacles

Local obstacles to the gripper motions are modelled using the 3D vision sensor located on the robot end-effector. As previously explained, this sensor is composed of a CCD camera and of a laser strip. It provides the system with a depth map of the profiles of the objects crossed by the laser strip [13]. Several profiles may be obtained by moving the sensor using the robot.

Let (O_T, X_T, Y_T, Z_T) be the reference frame associated to the robot gripper T, as shown in figure 6. Since the laser plane is almost parallel to the $X_T O_T Z_T$ plane, the resulting profiles are 2D curves representing the "highest" bi-dimensional obstacles opposed to the gripper motions. Then, scanning an hypothesized path for the gripper (i.e. a spatial corridor ending in the vicinity of the chosen features of the object to be grasped) can be done by combining the n 2D curves obtained by moving the laser plane in the Y_T direction. As we will see further, these curves are projected on a plane parallel to $X_T O_T Z_T$, in order to be combined in a 2D model consistent with our path planning method. This is done

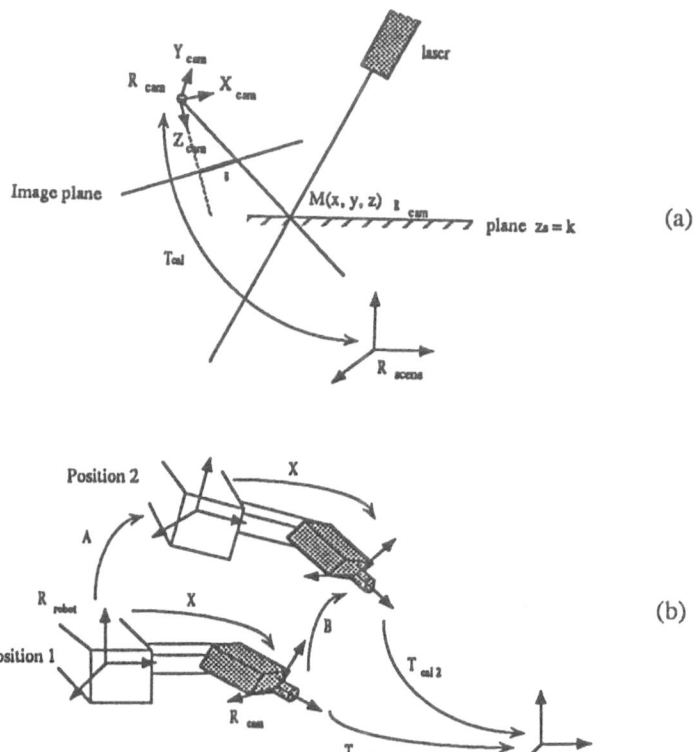

Figure 5: (a) Camera-laser calibration. (b) Camera-robot calibration.

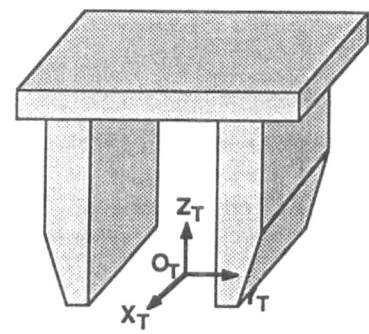

Figure 6: A sample grasping tool

after having discretized each parametric curve of the type $z = f(x)$ according to the required precision (such curves are in fact represented by the polygonal lines obtained by segmenting vision data). This approach leads to represent a profile by a k-dimensional vector $(z_1 z_2 \cdots z_k)$, where k is the chosen precision and z_i is the Z-coordinate of the obstacle associated to the X-coordinate x_i in (O_T, X_T, Y_T, Z_T).

4.3 Verifying hypotheses on the grasping environment

Before starting to plan robot motions, the system makes hypotheses on both the spatial position and the visibility of some features of the object to be grasped. The purpose of this module is to verify (using 2D vision data) that these hypotheses are consistent with the actual world. This is done by analyzing some selected regions of the image, in order to find the visual features corresponding to the chosen grasping features. The applied algorithm operates as follows:

1. Compute the expected image of each edge belonging to the selected grasping features, according to the known world model and to the characteristics of the vision process (point of view and intrinsic parameters of the camera). For the purpose of decreasing the algorithmic complexity of the vision process, the system compute a surrounding rectangular region (called "window") for each obtained straight line segment. The size of these regions is chosen according to both the length of the associated straight line segment and the hypothesized position uncertainty.

2. Determine for each window of the real image (the image provided by the camera located on the robot end-effector), if there exists a contrast line which match with the hypothesized straight line segment. This is done using the Hough method.

In the current version of the system, the verification process is considered as successful when most of the computed contrast lines match with the generated hypotheses. In case of failure due to an unforeseen obstacle which hides several grasping features, the system stops. It is the purpose of further developments to implement strategies for dealing with such failures.

5 Computing grasping parameters

5.1 The problem to solve

This section deals with the accessibility analysis problem related to automatic grasping. More precisely, the problem to solve is to choose a *goal configuration* and a *safe path* for a 3D object (the gripper) moving among 3D obstacles. Thanks to the motion constraints imposed by the selected potential grasp π (see section 5.2), such motions can be studied by reasoning in a two dimensional space. Our method for constructing the required 2D representation consists in discretizing the space swept by the gripper using a set of parallel planes. Then, the accessibility analysis is performed in each of these planes using a configuration space approach, and the resulting local motion constraints are combined in order to compute global constraints for the whole gripper. This is done using data provided by the 3D vision sensor.

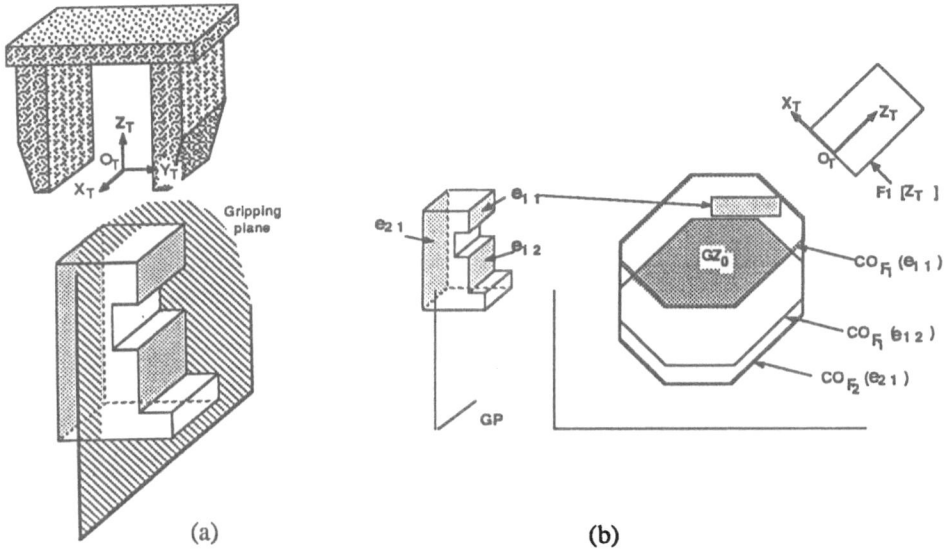

Figure 7: (a) The gripping plane. (b) Computing valid configurations.

Grasping requires a particular path planning. Indeed, compared with the general find path problem, path finding for grasping is different due to the fact that choosing a potential grasp π restrains the set of available configurations of the gripper. Two criteria must be considered for building such a set:

1. A configuration must be *valid*: the gripper must be in contact with each feature of π.

2. A configuration must be *safe*: no collision may occur while grasping between the gripper and any object of the workspace.

So, dealing with accessibility analysis requires the development of two mechanisms aimed at determining respectively valid and safe configurations.

5.2 Computing valid configurations

Let T be the gripper and O_T be its reference point (O_T is the origin of the reference frame of T). In the sequel, we will make the assumption that any potential grasp π generates at least a planar contact with one of the jaws. This means that the orientation of Y_T in the reference frame of the object is completely determine by π (because of the planar contact), and that the gripper can only move between two planes parallel to the contact plane when reaching the grasping position. Let GP be the plane on which O_T will lie; this plane, whose normal vector is Y_T, is called the *gripping plane* (see figure 7a).

A potential grasp π is a structured set of geometric features of P (edges, vertices, faces) that will be in contact with the internal faces of the jaws of T. Determining the valid configurations of T for π is done by computing, for each feature e_i of π, the set of configurations for which the internal face of the involved jaw overlaps e_i. This computation is performed after having hypothesized an orientation for the gripper. This means that the gripper is assumed to keep this constant orientation (characterized by $Z_T{}^1$) during its final approach. Then, the set of valid configurations of O_T can be represented by a particular region of the gripping plane (called the *gripping zone*). This set is computed using the following algorithm (figure 7b illustrates):

1. Project O_T and each feature e_i of π onto the gripping plane GP.

2. Compute the configuration space model associated to the projection of O_T for each projected feature $Proj_{GP}(e_i)$. This is done by growing $Proj_{GP}(e_i)$ inversely to the shape of the internal face of the related jaw.

3. Compute the set of valid configurations of T by intersecting the domains obtained in step 2.

5.3   Searching safe among valid configurations

Searching the safe configurations of T in the previous domains is performed using a method which combines interference checking techniques with a configuration space approach. The basic idea consists in discretizing the volume swept by the gripper using a set $\{P_i\}$ of planes parallel to the gripping plane. This is done using the laser strip (see section 4.2). Then, each obtained plane P_i implicitly contains the local accessibility constraints which are associated to a particular "slice" of the gripper when moving among bi-dimensional obstacles. Converting these local constraints into global constraints to O_T is done by iteraltively pruning the current set of solutions GZ_{i-1} into a new set $GZ_i \subseteq GZ_{i-1}$, according to the accessibility constraints drawn from the plane P_i. This algorithm stops when GZ_i is empty, or when all the planes have been processed.

Let $Obst_i$ and T_i be respectively the part of the obstacle and the slice of the gripper which are associated to the plane P_i. $Obst_i$ is obtained by "cutting" each object of the grasping environment (including the part to be grasped) with the plane P_i. T_i is obtained in a similar way by slicing the gripper; it is referenced by a point O_i, orthogonal projection of O_T onto P_i (see figure 8). Then, safe configurations for O_i can be easilly computed using a bi-dimensional growing, and the resulting motion constraints can be converted into constraints for O_T using an orthogonal projection onto the gripping plane. These constraints are used for pruning the initial gripping zone (i.e. the set of valid configurations of O_T).

As explained in section 4.2, the profiles $Obst_i$ are represented using the k-dimensional vectors computed by the vision process. But, merging the resulting data requires to previously convert them in explicit position constraints to O_T. This is done by first projecting each vector $Obst_i$ (for i=1,n) onto GP, and then by "translating" it according to the distance between O_i (the reference point of T_i) and the lowest point of T_i in the Z_T direction. Since the orientation of the gripper has been previously fixed by the system, the required growing operation can be performed in a straightforward way by "prolongating" the resulting profile in the $-Z_T$ direction (see figure 9).

---

[1]In the sequel, computing any solution X will mean $X[Z_T]$

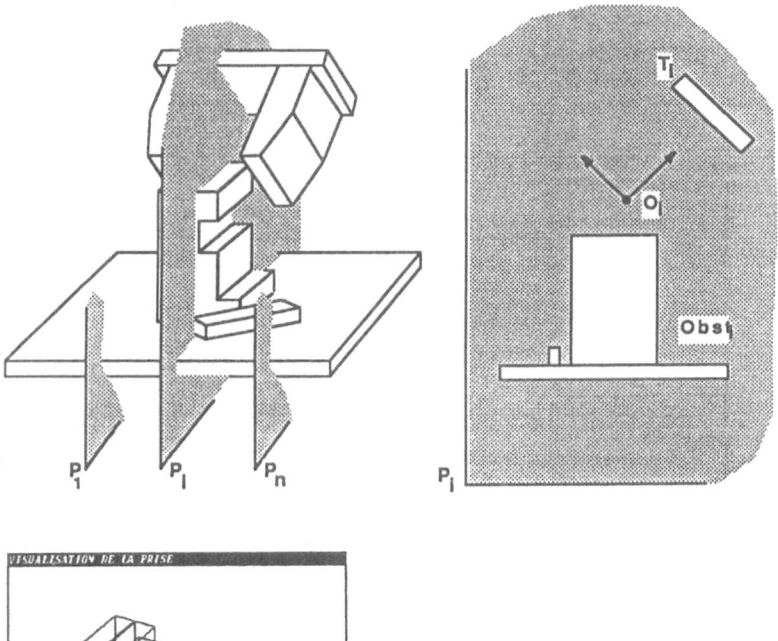

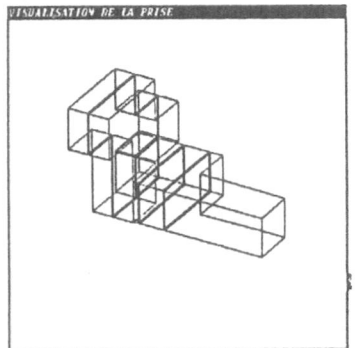

Figure 8: Slicing the space swept by the gripper

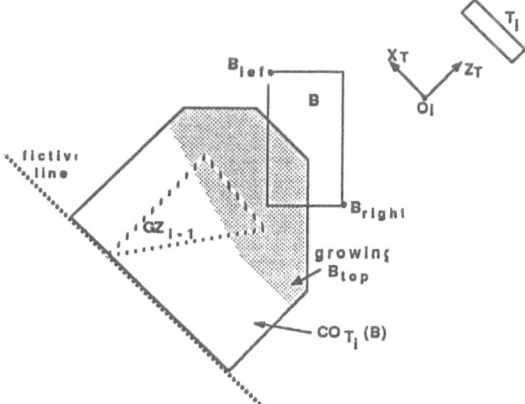

Figure 9: Processing the slices

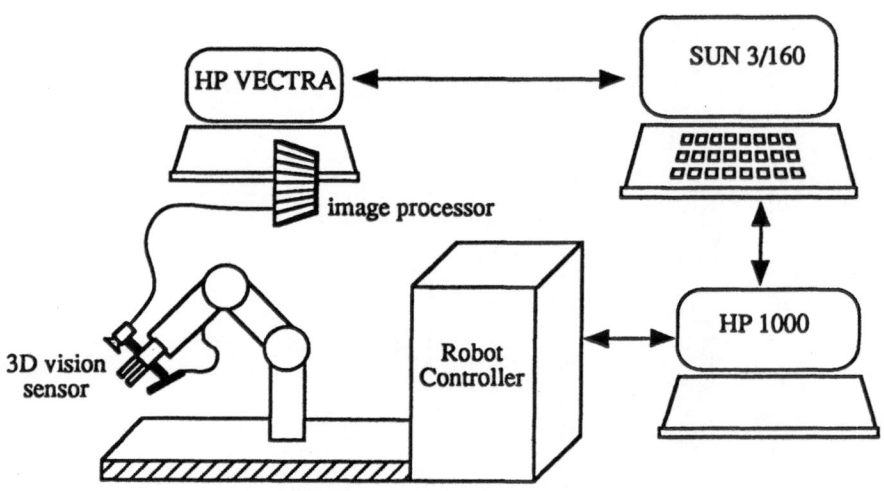

Figure 10: Hardware configuration.

Once the set of possible solutions (i.e. safe and valid configurations of T) has been constructed, a particular solution is chosen in this set using an heuristic trying to maximize the stability of the generated grasp (distance to the center of mass and surface of the involved contacts). A path to this point is generated within computed free space (the gripper will keep its orientation constant along the path). Then, the feasibility of this solution according to the robot arm constraints is verified using simple kinematic based computations and interference checking techniques (for checking collisions for the arm).

6   Implementation and experiments

6.1   Hardware configuration

The current hardware configuration is composed of a six axes SCEMI robot controlled by an ITMI controller, a small CCD camera Pulmix, a laser diode (780nm and 20mW), a HP-VECTRA microcomputer, and an image processor EDGE performing a real time contrast point extraction. The system described in the paper has been implemented in LUCID-LISP on a SUN 3/160 workstation. This workstation has been connected to both the VECTRA microcomputer and the robot controller using two RS-232 serial connections (see figure 10).

As explained in section 4, the camera and the laser diode have been mounted onto the robot gripper. Some mechanical and software modifications of the robot stops have also been executed, in order to guarantee that no collision can occur between the 3D vision sensor and the robot links.

6.2   Experimental results

Figure 11 represents the scene in which the robot will operate (let us remember that the system need not this full representation of the robot workspace thanks to its sensors).

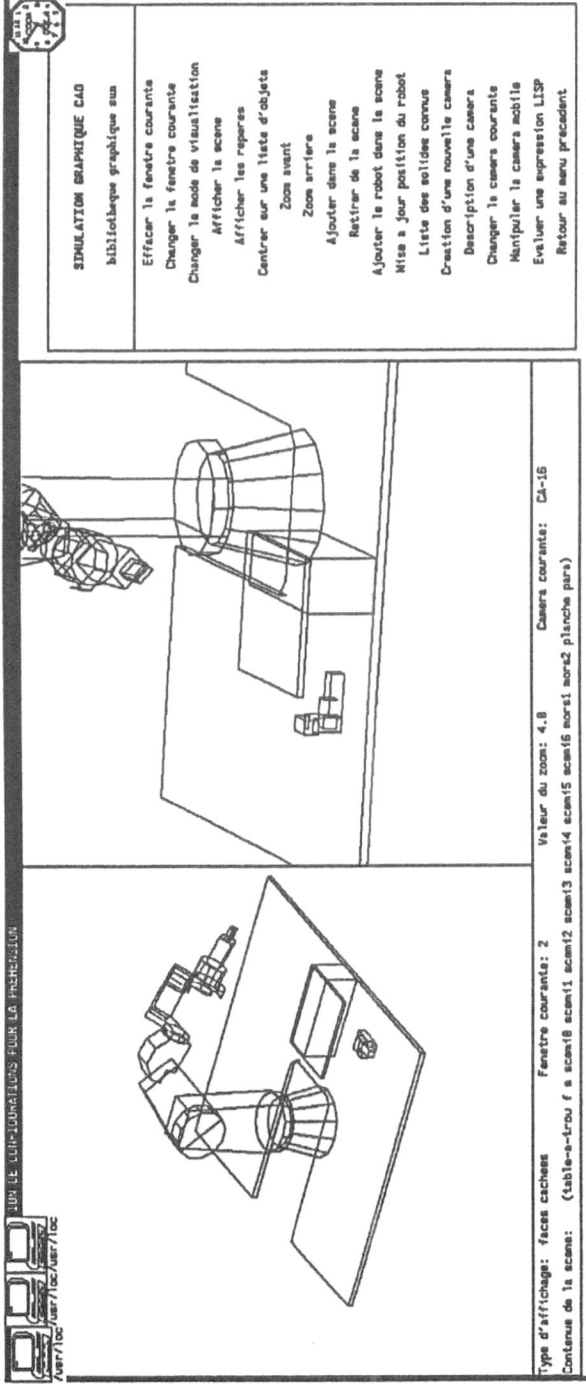

Figure 11: The robot workspace

Figure 12 shows a screendump of some objects of the environment; the right object, named "F", is to be grasped by the robot, while the left one, named "S" is the closest obstacle to "F". In figure 13, we can find the trace of the grasp generation; 9 grasps (couples of geometric features of "F") have been generated and one of them is visualized.

Figure 14 shows the occlusion analysis for the top most face of "F"; in the left window is the planar representation of the occluded regions (shaded regions) on the vision sphere. c is the selected viewpoint in the free domain of the vision sphere. The scene seen from c is simulated in the right window. Figure 15 shows the enviroment with the camera at position c; in the back of the scene, we can see the monitor on which is the image taken by the camera. This image is reproduced in figure 16. This selected viewpoint is used for verifying hypotheses on the grasping environment (see section 4.3)

Figure 17 shows the laser strip when the laser plane is parallel to the gripping plane of the grasp visualized in figure 13. Finally, the gripping position is generated and the object is grasped as shown in figure 18.

The system has been tested in rather simple environments with polyhedral objects. Most of its failures are due to calibration problems. In particular, the calculation of the intrinsic parameters in the camera-scene calibration is not very robust : we think that it comes from both the numerical methods used in this calibration phase and the hardware configuration (calibration device and camera lens). We are currently trying to improve this phase. The camera-robot calibration would also gain by using an iterative method (i.e. using more than 3 arm positions).

7 Conclusion

In this paper we have presented a method for planning and executing sensory based grasping operations in a partially structured environment. This method leads to plan and to control the execution of the required grasping strategies along with their associated vision based sensing operations.

The method has been implemented in LUCID-LISP on a SUN 3/160. It currently operates on a six axes SCEMI robot controlled by an ITMI controler. A 3D vision sensor composed of a small CCD camera and of a laser strip, has been constructed and mounted onto the robot end-effector. An appropriate calibration procedure has also been developped for the purpose of constructing an interface between the vision system and the manipulation system. Some examples have been successfully executed using polyhedral objects located in rather simple environments.

Current work deals with the implementation of the method on a mobile robot (Robosoft) equipped with a six axes arm (AID), in order to make use of three additional degrees of freedom for positioning the end-effector in more cluttured environments (see figure 19). For that purpose, we plan to integrate more complete path planning algorithms allowing an unified control of the robot motions.

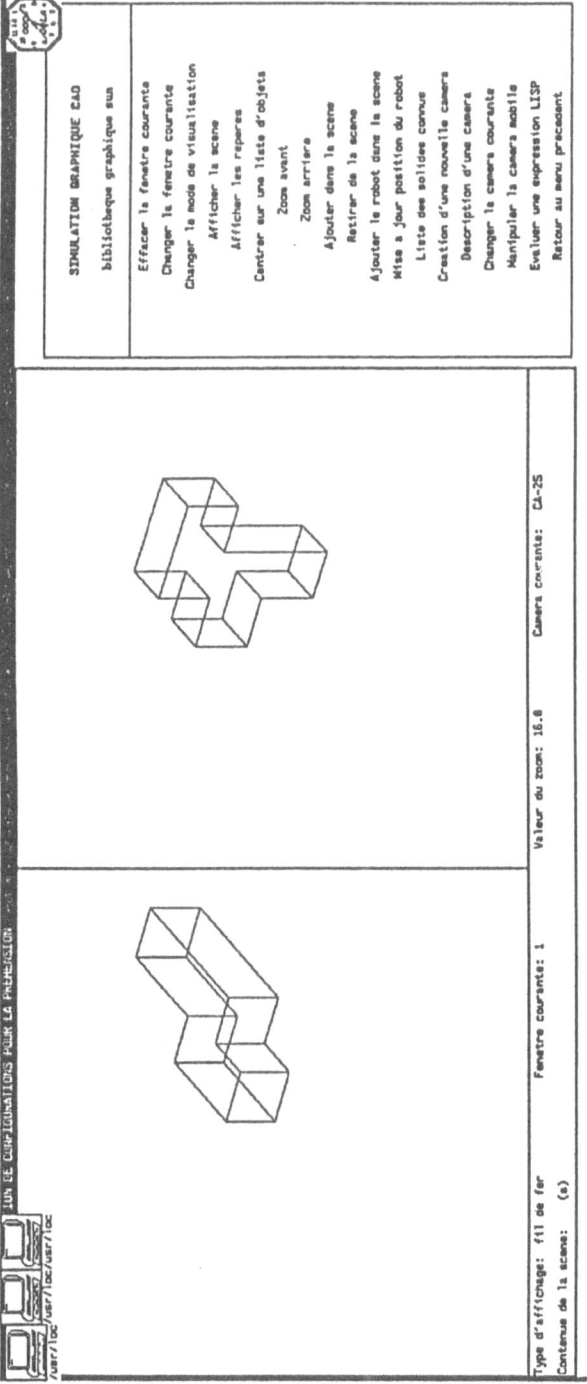

Figure 12: CAD models of the "F" and the "S"

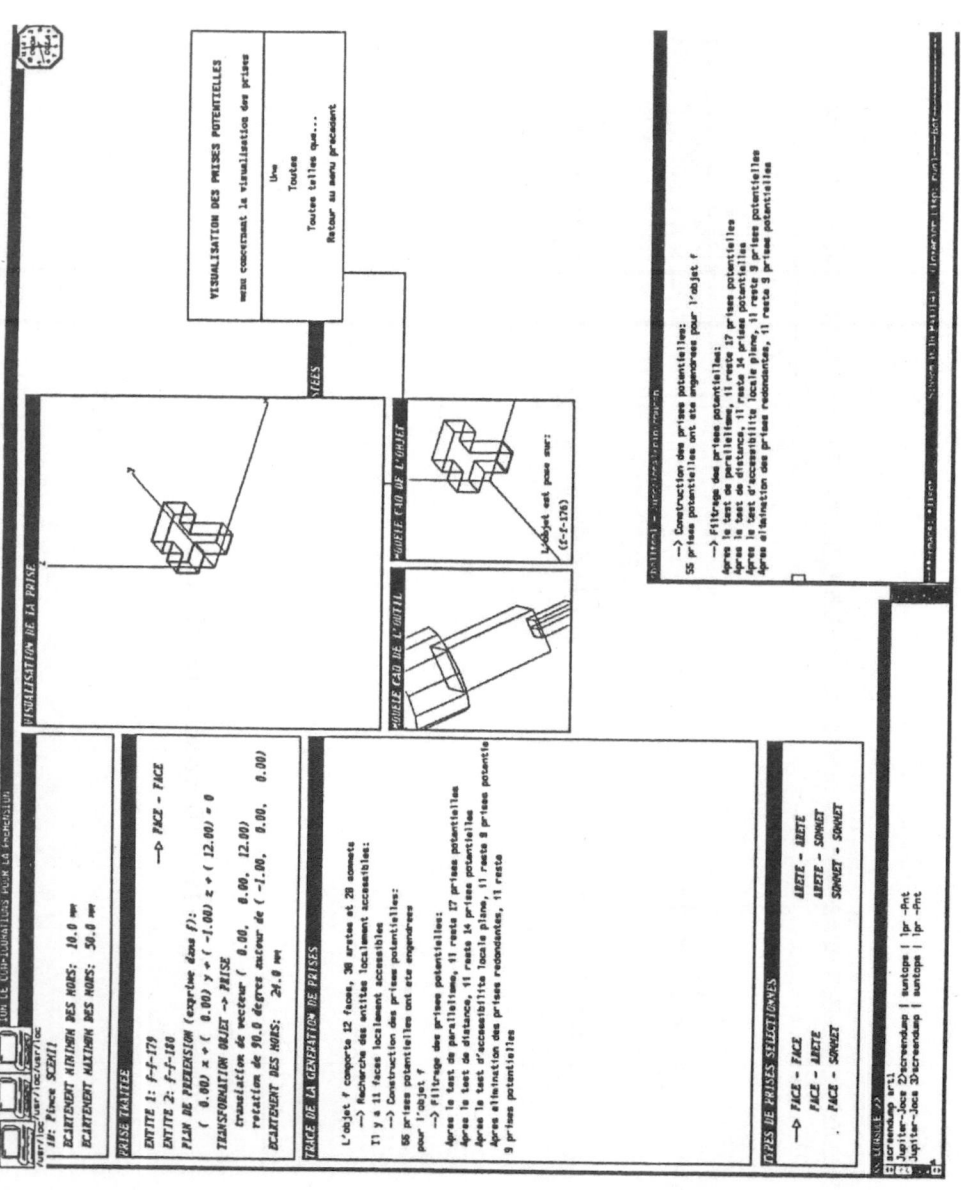

Figure 13: Generating potential grasps for "F"

407

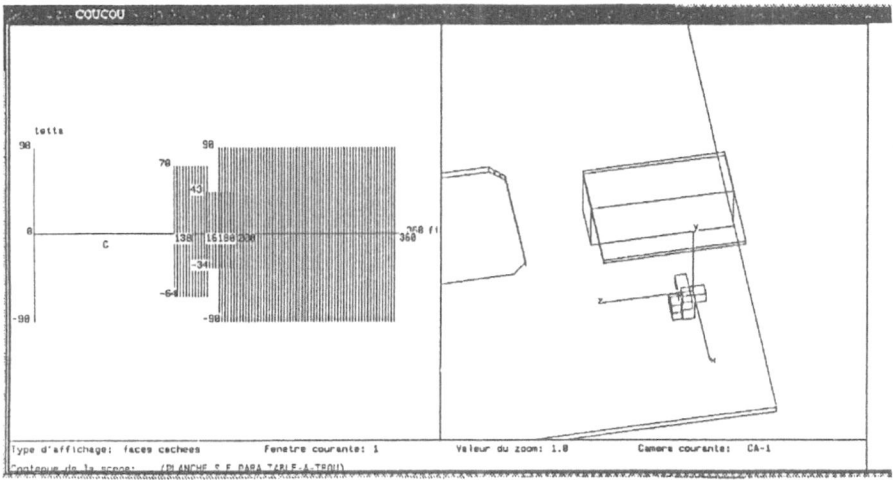

Figure 14: Choosing a viewpoint for the camera

Figure 15: The camera located at the selected viewpoint

Figure 16: Looking from the selected viewpoint

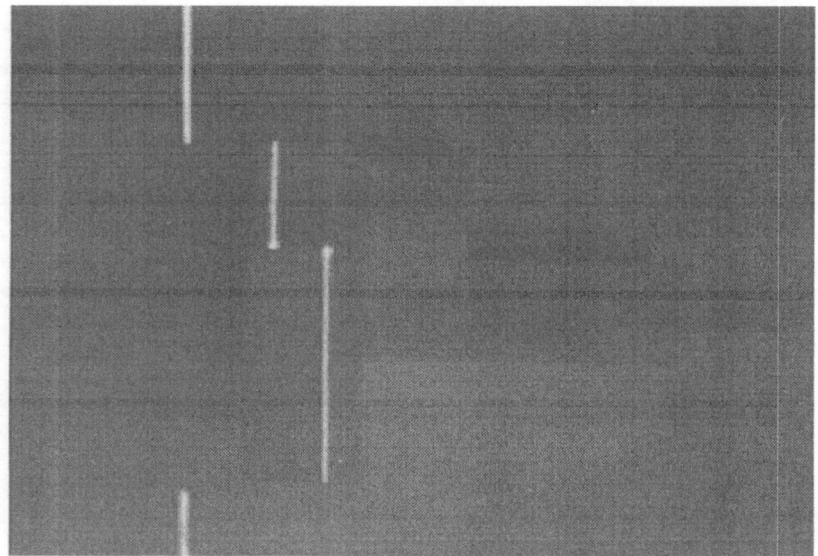

Figure 17: Getting the slices of obstacles

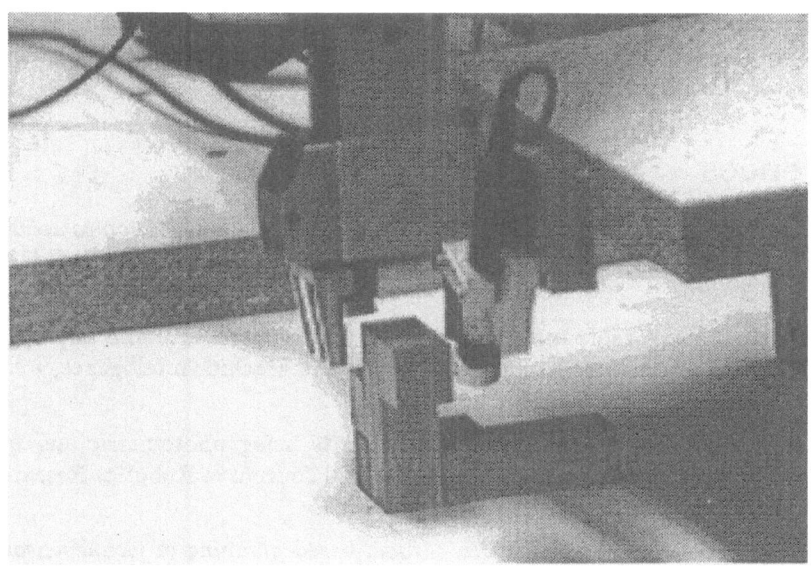

Figure 18: Grasping "F"

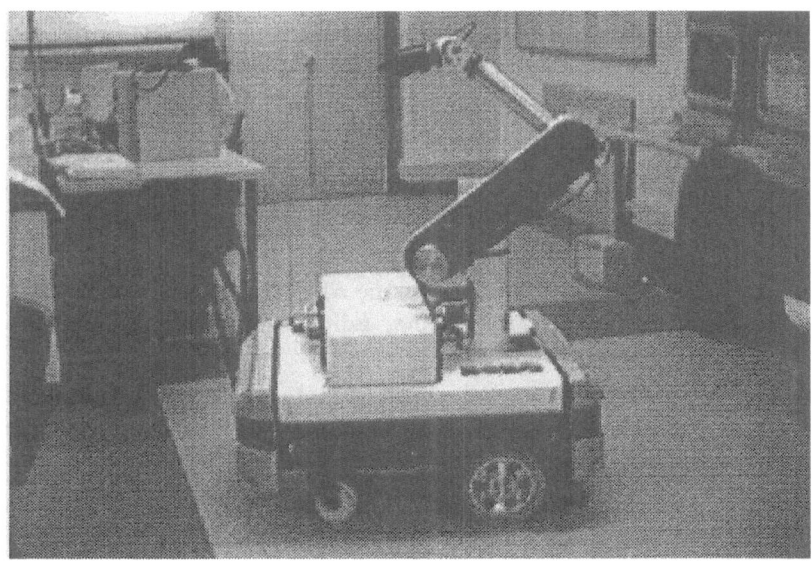

Figure 19: The mobile robot and its arm

Acknowledgements This work has been partly supported by the DRET (Direction des Recherche et Etudes Techniques) agency, the CNRS French National Center of Scientific Research, and the INRIA Institute.

References

[1] R.B.Kelley and al.: *"General methods to enable robots with vision to acquire, orient and transport workpieces"*, NSF final report, University of Rhode Island, December 1984.

[2] J.D.Boissonnat: *"Stable matching between a hand structure and an object silhouette"*, IEEE Transactions on Pattern Analysis and Machine Intelligence, vol. PAMI-4, no. 6, November 1982.

[3] K.Ikeuchi and al.: *"Determining grasp points using photometric stereo and the PRISM binocular stereo system"*, International Journal of Robotics Research, vol. 5, no. 1, 1986.

[4] S.Sakane, T.Sato and M.Kakikura: *"Model-based planning of visual sensors using a hand-eye action simulator system HEAVEN"*, 3rd International Conference on Advanced Robotics, Paris, October 1987.

[5] J.Troccaz: *"Geometric reasoning for grasping: the computational point of view"*, Published in CAD-Based Programming for Sensory Robots, Edited by Bahram Ravani, NATO ASI Serie F, vol. 50, Springer Verlag, 1988.

[6] T.Lozano-Perez: *"Automatic planning of manipulator transfer movements"*, IEEE Transactions on System, Man and Cybernetics, SMC-11,10, 1981.

[7] T.Lozano-Perez and al.: *"Task-level planning of pick-and-place robot motions"*, IEEE Computer Journal, March 1989.

[8] J.Troccaz: *"Modélisation du raisonnement géométrique pour la programmation automatique des robots d'assemblage"*, Thèse de Doctorat, INPG, March 1986 (in french).

[9] C.Laugier: *"Planning robot motions in the SHARP system"*, Published in CAD-Based Programming for Sensory Robots, Edited by Bahram Ravani, NATO ASI Serie F, vol. 50, Springer Verlag, 1988.

[10] C.Laugier: *"Raisonnement géométrique et méthodes de décision en robotique. Application à la programmation automatique des robots"*, Thèse d'Etat, INPG, December 1987 (in french).

[11] J.Troccaz: *"On-line automatic robot programming: a case study in grasping"*, IEEE Conference on Robotics and Automation, Raleigh, April 1987.

[12] O.Faugeras and G.Toscani: *"Camera calibration for 3D computer vision"*, Proc. International Workshop on Machine Vision and Machine Intelligence, Tokyo, February 1987.

[13] P.L.Borianne: *"Contribution à la vision par ordinateur tridimensionnelle"*, Thèse de 3'eme cycle, INPG, April 1984 (in french).

[14] Y.C.Shiu and S.Ahmad: *"Finding the mounting position of a sensor by solving a homogeneous transform equation of the form AX=XB"*, IEEE International Conference on Robotics and Automation, San Francisco, June 1986.

[15] C.Laugier, A.Ijel and J.Troccaz: *"Planification/exécution d'opérations de manipulation en environnement partiellement structué"*, Research Report on DRET 85-198 research convention, November 1988 (in french).

Visual Servoing Based on a Task Function Approach

Patrick Rives*, François Chaumette**, Bernard Espiau***

* INRIA-centre de Sophia Antipolis, 2004 Route des Lucioles,
06565 Valbonne, France
** IRISA-INRIA campus de Beaulieu 35042 Rennes-cedex, France
*** ISIA-ENSMP, rue Claude Daunesse, Sophia Antipolis,
06565 Valbonne, France

Abstract

Recent advances in vision sensors technology and image processing authorize to hope that the use of vision data directly into the control loop of a robot is no more an utopic way. Commonly, the general approach in robot vision is the following: processing vision data into the frame linked to the sensor, converting data into the frame linked to the scene by mean of inverse calibration matrix, computing, with respect to the robot task, the control vector of the robot into the frame linked to the scene, controlling the robot by using the inverse kinematic model. This scheme works in open loop with respect to vision data and cannot take into account inaccuracies and uncertainties occuring during the processing. Such an approach needs to perfectly overcome the constraints of the problem: geometry of the sensor (for example, in a stereovision method), the model of the environment and the model of the robot. In some cases, this approach is the only one possible but, in many cases, an alternative way consists to specify the robot task in terms of control directly into the sensor frame. This approach is often referred as visual servoing [8] , [3] or sensor based control [2]; in this case, a closed loop can be really performed from the vision data and allows to compensate for the perturbations by a robust control scheme. The work described in this paper deals with such an approach using a mobile vision sensor mounted on the end effector of a robot manipulator. It is characterized by two main points:the use of vision sensor as *local* sensor providing relatively poor instantaneous information but at a rate consistent with the bandwith of the robot controller and the exploitation of the vision data into a robust control scheme based on a task function approach [7].

1 The Framework of Sensor-based Control

1.1 Background

We are interested in this paper in the design of control systems which work in closed loop with regard to the environment, with the point of view of automatic control.This means that sensors exclusively devoted to supplying of symbolic information will not be considered, and that only high data rate sensors are used. This excludes, for example, high-level visual functions while this includes force, proximity, local range, and local visual sensing. In all cases , we shall therefore consider a sensor as a device rigidly linked to a mobile body which provides continuously with a p-dimensional vector of data denoted s. These sensor data are assumed to be only dependent of the *interactions* of the sensor and the environment.

Before designing the related control loops, modelling and analysis of these interactions are necessary steps, which will be done in the following subsection. The next one will briefly show how to realize a sensor-based task and will exhibit the needed related properties. All extended developments concerning the described approach may be found in [7].

1.2   Modelling the Interactions Sensor/ Environment

Basic Notations.   A sensor (S) is linked to a rigid body (B), with related frame F_S. (S) interacts with an environment (objects) to which is associated a frame F_T. A fixed reference frame F_0 is also given. F_S and F_T are mobile with respect to F_0. The position (location and attitude) of (B) is an element $\bar{r}$ of the Lie group of the displacements , called SE_3, which is a 6-dimensional differential manifold. Its tangent space is called se_3 and its dual se_3^*. A screw, H, is an element of se_3, which is also defined by its vector and the value of its field in some point O. Shortly, $H = (H(O), u)$. When expressed in a given frame, H will be represented by a vector in $\mathbf{R}^6$.

The velocity screw of a frame F_k with respect to a frame F_l is denoted T_{kl}.

The Concept of Elementary Signal.   Let us consider a one-dimensional component s_j of s. Our basic assumption is that s_j is in a mathematical sense a mapping from SE_3 to $\mathbf{R}$; more precisely, we assume that, given (S) and an observed environment (T), s_j is a function of the relative position of (S) with respect to (T) only. Therefore:

$$s_j = s_j(F_s, F_T) \in \mathbf{R} \tag{1}$$

This function will be assumed to be twice differentiable. Equation (1) defines a so-called *elementary signal*. This is for example the output of an ultrasonic sensor, an optical reflectance sensor, a strain gauge, or in the present case any 'relevant' parameter computed from an image.

Now, if (S) belongs to a mechanical system with several degrees of freedom (rigid manipulator, mobile robot), it may occur that the joint coordinates q constitutes a local chart of SE_3. In another way, the sensed object (T) may move independently with respect to F_0; its motion is then parametrizable by the time variable t. From these two arguments, (1) may also be written:

$$s_j = s_j(q, t) \tag{2}$$

Note that when q is not a chart of SE_3 (case of a plane mobile robot for example), the system evolves inside a submanifold with dimension smaller than 6. Nevertheless, the presented approach may often be applied in this smaller configuration space.

The Concept of Interaction Screw. Owing to its definition s_j admits a derivative represented by an element of se_3. The differential of s_j is a mapping from se_3^* to $\mathbf{R}$ which may shortly be written:

$$\dot{s}_j = \frac{\partial s_j}{\partial \bar{r}} \bullet \frac{d\bar{r}}{dt} = H_j \bullet T_{ST} \tag{3}$$

where $\bullet$ denotes the screw product. Note that H and T_{ST} are both functions of q and t. Let us now suppose that (S) belongs to the last body, (B_6), of a six-jointed robot. Equation (3) may then be expressed, for example in F_0, giving:

$$\dot{s}_j = [u_j^T H_j^T(O_6)](q,t) J_6(q)\dot{q} + [u_j^T H_j^T(T)](q,t) \begin{pmatrix} V_T(t) \\ \Omega_{F_T}(t) \end{pmatrix} \tag{4}$$

where J_6 is the Jacobian matrix associated with the frame F_6 linked to (B_6), with origin O_6, and T a fixed point of (T), all elements being expressed in F_0.

From now, we assume without loss of generality that (T) is motionless in F_0; we then finally define the *interaction screw* between the elementary sensor with output s_j and the environment as the screw H_j such that:

$$\dot{s}_j = H_j \bullet T_{SO} \tag{5}$$

H_j may be understood as a kind of Jacobian of the elementary sensor relative to its displacements with regard to the environment. It should be emphasized that all the information about the interactions between a sensor and its environment is contained in H_j. Unfortunately, the knowledge of H_j is generally partial, because of the existing uncertainties on environment and sensors. Consequently, only models of H_j may be used in practice; such models will be denoted $\hat{H}_j$ in the following.

The Concept of Virtual Linkage. Let us now search for a velocity screw T^* such that $\dot{s}_j = 0$, i.e. *the elementary sensor output is invariant with respect to the motion defined by T^**. T^* is a solution of:

$$H_j \bullet T^* = 0 \tag{6}$$

i.e. is a screw reciprocal to H_j. Considering now the full sensor (S) with *vector* output s shows that the motions T^* leaving s unchanged belong to a subspace reciprocal to the subspace S spanned by the set $H_1,..H_j,..H_p$.

The set (S) ; $H_1,..H_j,..H_p$ constitutes a *virtual linkage*.

This concept is an extension of the classical one used in the description of mechanical contacts between rigid bodies. Its *class*,N (number of "allowed motions"), is the dimension of S. When a frame and a basis are chosen, the virtual linkage is fully defined by the properties of the $6 \times p$ matrix

$$L = \begin{pmatrix} u_1 & . & . & . & u_p \\ H_1(P) & . & . & . & H_p(P) \end{pmatrix} \tag{7}$$

with rank $6 - N$, evaluated at some point P. Recall that $H_j(P) = H_j(P') + \Delta_{PP'}u_j$, where $\Delta_{PP'}$ is the antisymmetric matrix associated with the cross product $\vec{PP'}\times$.

The concept of virtual linkage is an easy way to the user to *specify* the task he wishes to realize through the sensors. It allows to determine the motions which may be sensor-controlled and the ones which remain free. In the case of physical contact, and with some assumptions on the used force sensors, this virtual linkage reduces to an *actual* linkage. An important fact to be emphasized is that a *virtual stiffness* may also be associated with the linkage, through the performed control. All these points are discussed in [7].

1.3   Sensor-based Task Functions

General Background.   Let us consider a rigid robot, the dynamic equation of which is

$$\Gamma = M(q)\ddot{q} + N(q, \dot{q}, t) \; ; \; dim(q) = n \qquad (8)$$

where M is the kinetic energy matrix and N gathers Coriolis, centrifugal, gravity and friction terms. From the point of view of automatic control, (8) is the *state* equation of the system (with natural state vector $(q, \dot{q})$), which fully describes its dynamics and involves only its intrinsic properties. The user's requirements, i.e. the task to be performed, have only to be specified as an *output function* associated with equation (8). The problem will be considered as well-posed if the 'passage' from the output space to the space where the control is actually performed is regular in some sense.

Further considerations ([2], [6]) allows finally to state that, in robotics applications, most of the user's objectives may be defined through a C^2 n-dimensional function depending both of q and t, $e(q, t)$, to be servoed to zero, associated with an initial condition q_0 and a time horizon $[0, T]$. Provided that a solution exist, some other conditions among them the regularity of the task-Jacobian $\frac{\partial e}{\partial q}$ in a certain domain, are also required.

A Major Property Needed for Stability.   Given equation (8), and a task function $e(q, t)$ to be servoed, a control scheme has to be designed. A general form is the approximated decoupling/linearization scheme performed on a transformation of model (8) in the task space ([6]). "Approximated" means that *models* $\hat{M}$, $\hat{N}$, $\frac{\partial \hat{e}}{\partial q}$, $\frac{\partial \hat{e}}{\partial t}$, are used instead of the true expressions often practically unavailable. In [6] and [7],C. Samson has exhibited some sufficient conditions ensuring the closed-loop system to be stable. One of the most important is certainly:

$$\frac{\partial e}{\partial q}(\frac{\partial \hat{e}}{\partial q})^{-1} > 0 \qquad (9)$$

in the sense that a $n \times n$ matrix A is positive if $x^T A x > 0$ for any nonzero $x \in \mathbf{R}^n$.

Clearly, this condition involves the task itself. More precisely, it states the minimal knowledge which is needed about the task for ensuring a good behaviour of the system. This property will be largely used in the following.

Case of Sensor-based Tasks. A common example of task function is the tracking of a trajectory in a given space, for instance $e(q,t) = \begin{pmatrix} x(q) - x_r(t) \\ d(r(q), r_d(t)) \end{pmatrix}$ where x is the location of O_6, x_r the related trajectory to be tracked, and $d(.)$ a 3-dimensional vector representing the error between a desired rotation $r_d(t)$ and the actual one $r(q)$. The use of sensors is also a way of constituting a task function, when the needed assumptions are satisfied. However, due to the fact that the virtual linkage which is desired to be realized is often of class N greater than 0, the related sensor based task may have an intrinsic dimension less that the required n. It is therefore necessary to complete it, provided that the 'added' objective is *independent and compatible* with the sensor-based task. It may be shown ([7],[2]) that an efficient way for doing this is the following: starting from s, a first step is to build a vector function $e_1(s)$, generally linear ($e_1 = Ds - e^*$) with the right dimension, i.e. the one of S, and characterizing in a simple manner the desired virtual linkage (the example of visual-based tasks will be presented later). The second step consists in defining a secondary objective, to be realized under the constraint of achieving $e_1 = 0$. This secondary goal is expressed as the minimization of a coast function h_s. It may then be shown that the resulting (global) task function takes the general form:

$$ e = W^+ W e_1 + \alpha (I - W^+ W) \frac{\partial h_s}{\partial x} \tag{10} $$

where x is an element of any working space such that $\frac{\partial x}{\partial q}$ is nonsingular, α is a positive scalar function and W a matrix function such that, for decoupling requirements, range (W^T) = range $(\frac{\partial e_1}{\partial x})^T$.

A particularly interesting case, used in all the following, is $x = \bar{r}$, with the frame of the sensor system used as working frame. Then, $\frac{\partial e_1}{\partial \bar{r}}$ is directly related to the matrix L of the interaction screws; furthermore, it is sometimes possible to choose W such that $W^+ W$ is diagonal, with entries 0 or 1, then called a selection matrix. In that case, and if $\frac{\partial h_s}{\partial \bar{r}}$ has the meaning of a trajectory tracking in SE_3, the realized task is called an *hybrid task,* by analogy with the wellknown *hybrid control.*

A last thing is that the important positivity condition (9), when applied to equation (10), may be sometimes easily satisfied by a simple choice of the model $\frac{\partial \hat{e}}{\partial q}$: it is shown in [7] that, when W may be chosen such that $\frac{\partial e}{\partial \bar{r}} W^T > 0$, and if α is not too large, then $\frac{\partial \hat{e}}{\partial \bar{r}}$ is *itself* positive. Condition (9) is therefore satisfied with the simple choice $\frac{\partial \hat{e}}{\partial q} = \frac{\partial \bar{r}}{\partial q}$, i.e. the basic Jacobian matrix of the robot. This is the reason why some hybrid control schemes may work even with a bad knowledge of the true interactions.

In the case of visual sensors, it is very difficult to know precisely the interaction screws, because they will be shown later to depend on an unknown parameter, the depth. In opposition to force or range measurements, it is thus not possible to find a priori what are the right *models* to be used, and, more, what is the allowed "amount of uncertainty". From an other point of view, given certain simple choices of models, it is not easy to determine their domain of validity. This is why, in the case of visual servoing, often only an experimental analysis may allow to validate the choices done in the control schemes. This way of applying the previous theory is the topic of the following developments.

2 Application to a visual servoing approach

2.1 Problem statement

Let us consider a vision sensor moving across a three dimensional environment (3D-scene). We assume that the motion of the sensor is fully controlable and can be characterized by its velocity screw. In practice, this vision sensor can be mounted on the end effector of a manipulator or carried by a mobile robot. We use the classical "pinhole" approximation for modelling the perspective geometry of the sensor, and we assume the focal length equal to unity (figure 1). Using the same definition as above for the diverse frames, we can state that, due to the motion of sensor and objects, a point $P_i = (x_i y_i z_i)^T$ linked to an object , moves with a relative velocity with regard to F_S. This velocity can be expressed by the velocity screw T_{ST} by means of:

$$\dot{P}_i = V_T + \Omega_{F_T} \times SP_i \tag{11}$$

At each instant, the point P_i projects onto the image plane as a point p_i with coordinates $(X_i Y_i)^T$

$$X_i = \frac{x_i}{z_i} \; ; \; Y_i = \frac{y_i}{z_i} \tag{12}$$

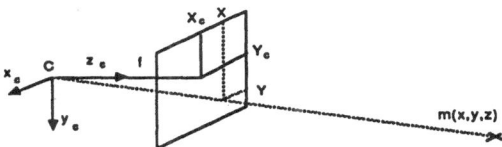

Figure 1: Perspective model

A velocity field projected onto the image plane corresponds to the motions of the sensor and/or objects in the 3D-scene ; this velocity field is often called as the Optical Flow field. By differentiating (12) and using (11) , we can derive the wellknown equation relating optical flow measurement to 3D structure and motion in the scene. When expressed in the F_S frame, we obtain:

$$\begin{pmatrix} \dot{X}_i \\ \dot{Y}_i \end{pmatrix} = \begin{pmatrix} \frac{1}{z_i} & 0 & \frac{X_i}{z_i} & X_i Y_i & -(1 + X_i^2) & Y_i \\ 0 & \frac{1}{z_i} & \frac{Y_i}{z_i} & 1 + Y_i^2 & X_i Y_i & X_i \end{pmatrix} \cdot \begin{pmatrix} V_T \\ \Omega_{F_T} \end{pmatrix} \tag{13}$$

This equation shows the basic structure of the *interaction* between the vision sensor and its environment. As we underlined in the first part, the two interaction screws appearing as the lines of the matrix given in (13) depend on the inverse of the depth z_i expressed in the F_S frame. Generally, without some *a priori* knowledge about the environment, it is not possible to know the true value of the interaction screws and only some rough approximation will be available for the control purpose. As we shall show later, we will be able to compensate for this lack of accuracy by a robust control scheme.

Now, let us consider more complex geometrical primitives than points. In a general way, we assume that a 3D geometrical primitive can be represented as a vectorial function $h(x(t), y(t), z(t), Q(t)) = 0$ which projects onto the image plane under the form $g(X(t), Y(t), R(t)) = 0$ where $Q(t) = Q_i(t); i = 1, m$ and $R(t) = R_i(t); i = 1, n$ are the parameters of the primitives respectively in the 3D scene and in the image plane.

From these assumptions, an important part of the visual servoing problem will be devoted to find, for a given 3D primitive, a wellsuited parametrization of h and g (i.e. without singularity of representation) and to establish the interaction screw $H_i(R, Q)$ such that:

$$\dot{R}_i = H_i(R, Q) \bullet T_{ST} \tag{14}$$

According to virtual linkage defined in section 2.2, we can search for a velocity screw T^* such that $\dot{R}_i = 0$ leaving R unchanged which can be formulated by:

$$(\forall T^* \in S^* \subset S) \; if \; T^* \in Ker(H_i) \Longleftrightarrow \dot{R}_i = 0 \tag{15}$$

The following section is devoted to the definition a set of *elementary visual signals* and their associated virtual linkage from a set of low level geometrical primitives.

2.2   Modelling elementary visual signals

Assuming that g is a C^2 function on a time interval $[0, T]$, let us derive a general solution for computing $H_i(R, Q)$:

$$g(X(t), Y(t), R(t)) = 0; \; \forall t \in [0, T] \Rightarrow \dot{g}(X(t), Y(t), R(t)) = 0; \; \forall t \in [0, T] \tag{16}$$

after developments, we obtain:

$$\sum_{i=1}^{n} \frac{\partial g}{\partial R_i} \dot{R}_i = -\frac{\partial g}{\partial X}\dot{X} - \frac{\partial g}{\partial Y}\dot{Y}; \; \forall (X, Y) \; such \; that \; g(X, Y, R) = 0 \tag{17}$$

Eq.(16) allows us to relate the variation of the parameters R_i characterizing the 2D primitive in the image plane to the optic flow components and thus, to the velocity screw of the camera by means of eq.(13). An *elementary visual signal* will be defined as a function $s_j = f(R_i, \cdots, R_k)$, $i, \cdots, k \in n$ characterizing usual geometrical properties in the image like distance between two points, orientation between two lines, surface, mass centroid and so on.

Case of points

Using one point.   The equation of h and g are trivial, and we find again the classic optical flow equation:

$$h : \begin{cases} x - x_1 = 0 \\ y - y_1 = 0 \\ z - z_1 = 0 \end{cases} \longrightarrow g : \begin{cases} X - X_1 = 0 \\ Y - Y_1 = 0 \end{cases} \tag{18}$$

From the eq.(17), we have:

$$\dot{g}=0 \Rightarrow \begin{cases} \dot{X}_1 &= [\ \frac{1}{z_1}\ \ 0\ \ \frac{X_1}{z_1}\ \ X_1 Y_1\ \ -(1+X_1^2)\ \ Y_1\ ]\ \cdot T_{ST} \\ \dot{Y}_1 &= [\ \ 0\ \ \frac{1}{z_1}\ \ \frac{Y_1}{z_1}\ \ 1+Y_1^2\ \ X_1 Y_1\ \ X_1\ ]\ \cdot T_{ST} \end{cases} \quad (19)$$

Finally, we can define two elementary visual signal $s_1 = X_1$ and $s_2 = Y_1$;the set ($S = [s_1 s_2]^T$) ; H_1, H_2 constitutes a virtual linkage the class of which is $4 = dim(S)$ (S is the subspace spanned by H_1, H_2). A basis of S can be easily found, for example:

$$\mathcal{B}\ :\ \begin{pmatrix} X_1 & 0 & z_1(1+X_1^2+Y_1^2) & 0 \\ Y_1 & 0 & 0 & z_1(1+X_1^2+Y_1^2) \\ 1 & 0 & 0 & 0 \\ 0 & X_1 & -X_1 Y_1 & 1+X_1^2 \\ 0 & Y_1 & -(1+Y_1^2) & X_1 Y_1 \\ 0 & 1 & 0 & 0 \end{pmatrix} \quad (20)$$

Each velocity screw T_{ST}^* which belongs to S leaves unchanged the projection of the 3D point in the image plane.

Using several points. The case of two different points in the image frame leads to the following equation:

$$\begin{pmatrix} \dot{s}_1 \\ \dot{s}_2 \\ \dot{s}_3 \\ \dot{s}_4 \end{pmatrix} = \begin{pmatrix} \dot{X}_1 \\ \dot{Y}_1 \\ \dot{X}_2 \\ \dot{Y}_2 \end{pmatrix} = H \bullet T_{ST} \quad (21)$$

the class of the linkage is 2,ie the dimension of the subspace corresponding to the allowed motions without changing the nature of the image.

The case of three points is more interesting. As it is wellknown, the inverse perspective problem (recovering location and orientation of 3D objects from one 2D image) has not an unique solution. By this fact, several relative attitudes between the camera and the object can lead to the same image. These attitudes correspond to the solving of a 8-degrees analytic equation . In the most cases, a continuous motion does not exist from one solution to another one which leaves unchanged the image . In consequence, for a particular attitude between object and camera, the dimension of the subspace S is 0 (ie $rank(H) = 6$). However, for some particular cases, like, for example, when the three points are colinear in the 3D space , singularities may be occured (i.e. $rank(H) < 6$). For a number of points greater than three (four or more), we are in a case of redundancy: the rank of H is always equal to 6 and the dimension of T_{ST}^* equal to 0. We will see later how we can use these properties to perform a task of target tracking.

Other elementary visual signal based on points In the previous paragraph, we only presented the case where the application $s_j = f(R_i, \cdots, R_k)$, $i, \cdots, k \in n$ is the identity. In many cases, it can be fruitful to consider more relevant signals with regard to the task. For instance, let us consider the problem of positioning an end effector with regard to an object which is characterized by three points. A natural way for specifying this task is to control, on the first hand, the location of a particular point of the object, by example the mass centroid of the three points and on the other hand some geometrical characteristics like the distance between the three points. For doing that, we choose the following signals vector:

$$s = \begin{pmatrix} L_{12} \\ L_{13} \\ L_{23} \\ X_G \\ Y_G \end{pmatrix} \tag{22}$$

with

$$s_i = L_{jk} = (X_j - X_k)^2 + (Y_j - Y_k)^2 \; ; \; i = 1 \cdots 3; \; j = 1 \cdots 3, \; k = j+1 \cdots 3 \tag{23}$$

and

$$s_4 = X_G = \frac{\sum_{i=1}^{3} X_i}{3} \; ; \; s_5 = Y_G = \frac{\sum_{i=1}^{3} Y_i}{3} \tag{24}$$

Then it is possible to compute the interaction screw $H(X_1, X_2, X_3, Y_1, Y_2, Y_3, z_1, z_2, z_3)$ and the associated virtual linkage. Let us remark that this choice of s lets a rotation around the mass centroid to be free (ie $rank H = 5$); The full control of this rotation from the image needs another component characterizing the orientation in the image plane like, for instance, the orientation of the main inertial axis with regard to the image frame has to be add to s.

Case of lines. For many reasons (accuracy, robustness with regard to the noise...) it is often interesting to use in image analysis more structured primitives than simple points. In the case of visual servoing, using lines as primitives seems to be natural. A 3D line will be represented by two planes which intersect:

$$h(x, y, z, Q) : \begin{cases} a_1 x + b_1 y + c_1 z + d_1 = 0 \\ a_2 x + b_2 y + c_2 z + d_2 = 0 \end{cases} , \; with \; d_1, d_2 \neq 0 \tag{25}$$

A 3D line in the scene projects onto the image plane as a 2D line (except on some degenerate cases). Some attention has to be taken concerning the parametrization of the 2D line. As we underline in the first part, an elementary visual signal has to be twice differentiable. Let us consider the classical representation of a 2D line: $Y = aX + b$. This parametrization needs two charts for representing 2D lines into the cartesian space (i.e. the lines: $X = 1$ and $Y = 1$ belongs to two different charts). If we use the parameters a and b as elementary visual signals, the condition of differentiability is not preserved during the passage from one chart to the other. For this reason, we choose a representation ρ, θ and we obtain:

$$\mathcal{D} : g(X, Y, R) = X cos\theta + Y sin\theta - \rho = 0 \; , \; \theta \in \left] -\frac{\pi}{2}, \frac{\pi}{2} \right] \tag{26}$$

deriving this equation,

$$\dot{g} = 0 \Rightarrow \dot{\rho} + (X\sin\theta - Y\cos\theta)\dot{\theta} = \cos\theta\dot{X} + \sin\theta\dot{Y} \ , \ \forall(X,Y) \in \mathcal{D} \qquad (27)$$

finally, using eqs.(13), (25), (26), we obtain:

$$\dot{\theta} = -(\frac{b_1}{d_1}\cos\theta - \frac{a_1}{d_1}\sin\theta)(\cos\theta \ \sin\theta \ -\rho) \cdot (V_T) - (\rho\cos\theta \ \rho\sin\theta \ 1) \cdot (\Omega_{F_T}) \qquad (28)$$

and

$$\dot{\rho} = (\frac{c_1}{d_1} + \rho\frac{a_1}{d_1}\cos\theta + \rho\frac{b_1}{d_1}\sin\theta)(\cos\theta \ \sin\theta \ -\rho) \cdot (V_T) + ((1+\rho^2)\sin\theta \ -(1+\rho^2)\cos\theta \ 0) \cdot (\Omega_{F_T})$$
$$(29)$$

Eqs.(28)(29) define the associated interaction screws. Typically, the vector $s = (\rho, \theta)^T$ will be used to characterize a 2D line, but if necessary, we will be able to control some other characteristics like, for instance, orientation between two lines ($s = |\theta_1 - \theta_2|$) or all other measurement built from θ and ρ.

Case of circles. In a similar manner, let us consider a circle in the 3D scene. It projects onto the image plane as an ellipse:

$$h(x,y,z,Q) : \begin{cases} (x-x_0)^2 + (y-y_0)^2 + (z-z_0)^2 - r^2 = 0 \\ (z-z_0) - \alpha(x-x_0) - \beta(y-y_0) = 0 \end{cases} \qquad (30)$$

$$\Rightarrow \ g(X,Y,R) : \frac{(X - X_C + e(Y - Y_C))^2}{a^2(1+e^2)} + \frac{(Y - Y_C + e(X - X_C))^2}{b^2(1+e^2)} - 1 = 0 \qquad (31)$$

After some tedious developments, we can relate the variation of the parameters of the ellipse to the motion into the 3D scene by means of the interaction screws such that:

$$\begin{pmatrix} \dot{X}_C \\ \dot{Y}_C \\ \dot{e} \\ \dot{a} \\ \dot{b} \end{pmatrix} = H \bullet T_{ST} \qquad (32)$$

2.3 Visual sensing and task function

At this step, we have defined a set of elementary vision signals; in this section, we will investigate some ways of using these signals in a robust control scheme based on a task function approach, as defined in section 3.1. The problem can be stated as follows: is it possible to specify a robotic task in term of reaching a particular configuration of a set of features in the image frame? If so, from a running observation of these features in the image, are we able to perform this task? This means that we have to design a control scheme which allows us to reach this particular configuration (target image). Let us consider an example: a task for a mobile robot consists in going in front of a door at a given distance. This task may be translated in the image frame by the following characteristics:

- matching in the image a set of polygonal lines which can be associated with the model of the door.

- controlling the motion of the robot for bringing these lines to form a rectangle centred on the image frame with two parallel and vertical lines and the surface of which is equal to a given value.

With regard to the general form of sensor based task functions presented in the first part, we define a task vector $e(q(t), t)$, associated to a visual task function, where q is a local chart of SE_3, (obviously, in the case of a camera mounted on an end effector of a manipulator q is a chart of the configuration space) and such that :

$$e(q(t), t) = \begin{pmatrix} s_1(q, t) - s_1^*(t) \\ \vdots \\ s_p(q, t) - s_p^*(t) \end{pmatrix} \tag{33}$$

where $s^*(t)$ can be considered as a reference trajectory to be tracked in the image frame. If the robotic task just consists in controlling a given attitude of the sensor with respect to a motionless target, (i.e. *positioning problem*) then s^* will be time independant.

Considering the control problem as an output regulation problem, we can assume that the concerned task is perfectly achieved during $[0, T]$ if $e(q(t), t) = 0$ for all $t \in [0, T]$. For simplifying the derivation of the control, we shall consider that the motions of the camera and the robot are slow enough to allow the use of a full dynamic model (8) to be avoided. With intend to simplify our purpose, we only consider the case where the matrix $H(R, Q)$ is regular but this approach can be extended to any cases without loss of generality. We therefore focus on the robustness with respect to uncertainties on the interactions by using a gradient-based approach as a particular case of the general approach evoked in section 2.3. In this approach, we assume that a linearly decreasing function $T_d = f(e(q(t), t))$ relates the desired control vector to the task function with a limit condition $\lim_{e \to 0} f = 0$. Under these assumptions, we may choose:

$$T_d = -\mu C \cdot e(q(t), t) \tag{34}$$

where $\mu > 0$ and C is a constant positive matrix. From eqs.(3) and (34), we obtain:

$$\dot{e} = \dot{s} = H \bullet T_d \Rightarrow \dot{e} = -\mu H \cdot Ce \tag{35}$$

An exponential convergency will be ensured under the following sufficient condition:

$$HC > 0 \tag{36}$$

A good and simple way to try to ensure this matrix positivity is to enforce $H(R^*, Q^*) \cdot C = I$ (where I represents the identity matrix) for the equilibrium position $s = s^*$. In this case, the control matrix C takes the following form:

$$C = H(R^*, Q^*)^{-1} \qquad (37)$$

Let us note that the computation of C depends of the parameters R referring to the image which are known or easily measured and of the parameters Q referring to the 3D scene. Without an a priori knowledge on the 3D scene, these last parameters cannot be inferred from the image and, by this fact, the matrix C cannot be exactly computed. Fortunately, in many cases, the interaction screw H can be factorized in a product of two matrices such that:

$$H(R^*, Q^*) = B(R^*) \cdot D(R^*, Q^*) \qquad (38)$$

where $D(R^*, Q^*)$ is a positive diagonal matrix. For instance, when the target is constituted by three 3D points and when the equilibrium position is such that the target is parallel to the image plane , the matrix D has the following form:

$$D(R^*, Q^*) = \begin{pmatrix} \frac{I_3}{z^*} & | & 0 \\ --- & --- & --- \\ 0 & | & I_3 \end{pmatrix} \quad with \ z^* > 0 \qquad (39)$$

In this cases, we can choose a control matrix such that $C = B(R^*)^{-1}$ since the convergency condition $\|B \cdot D \cdot B^{-1}\| > 0$ is always verified.

Obviously, the results presented above ensure the positivity only at the equilibrium position where $s = s^*$. Far from this position, the convergency property is not always insured and the positivity of $\|H(R(t), Q(t)) \cdot C(R^*, Q^*)\|$ should be checked for all $t \in]0, T]$. Fortunately, as we will see in the next section, outside the task singularities, the satisfaction of this condition is shown to be preserved in practice.

3 Results:

The approach developed in this paper has been validated both in simulation [4] and in an experimental cell constituted by a CCD camera mounted on the effector of a 6dof manipulator [1].

3.1 Simulation results

We focused our investigations on the problem of positioning the sensor with respect to a static or a moving target. We investigated the use of different types of primitives as elementary signals (points, lines, circles). For each case, we performed a first serie of experiments corresponding to the following configurations:

- static target, noise-free synthetic image, noise-free control vector.

- static target, the image noise had a Normal distribution with zero mean and a standard deviation of two pixels, the control vector noise had a Normal distribution with zero mean and a standard deviation of 1 centimeter in translation and two degrees in rotation.

In all cases, the chosen control was of form given by eqs. (34), (37). Some results are given in the figure 2 (noise-free) and 3 (with noise) in a positionning task of the camera with respect to a target constituted by four points. The coordinates of every point, X_i and Y_i are taken as elementary signals in the image frame. In these figures, the two top windows presents the relative attitude of the camera (symbolized by a pyramid) and the target viewed by an outside observer. The two bottom windows presents the target as seen by the camera. The windows on the left correspond to the initial position and the windows on the center to the final desired one. The two windows on the right show the behaviour of the error and of the control vector during the visual servoing. The top window corresponds to the mean quadratic error on the set $\epsilon(t) = \sum_{p=1}^{p=8}(s_p(t) - s_p^*)^2$ and the bottom window the behaviour of each component of the control vector (velocity screw of the camera) during the positionning task.

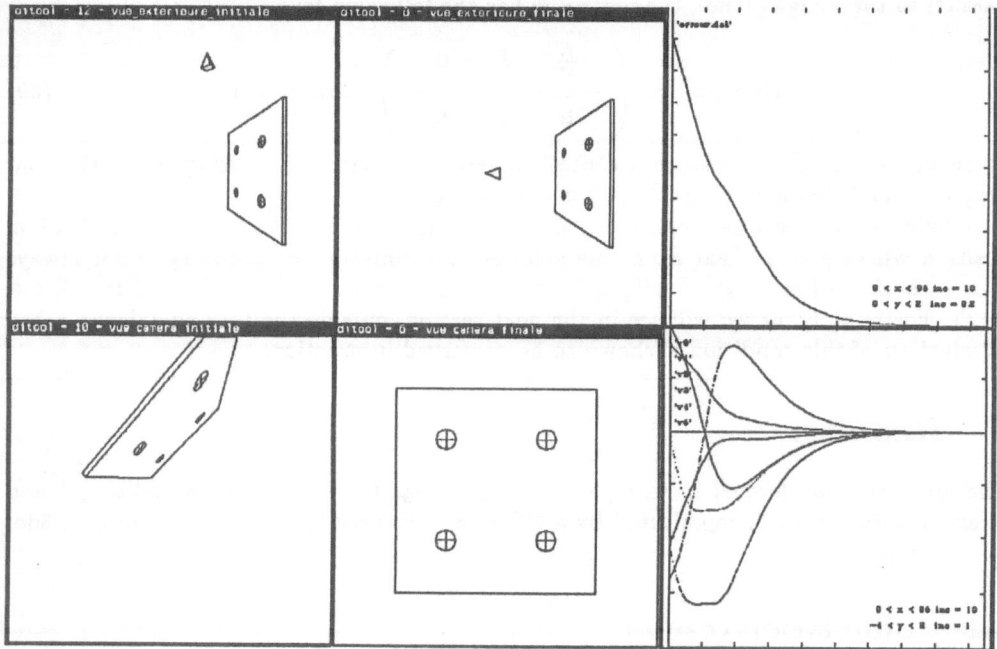

Figure 2: static target, noise-free synthetic image

As shown on the figures, the convergency to the desired configuration of the target in the image is always performed in spite of the distant initial position. Same experiments are successfully performed when the target is moving through the scene with any motion consistent with the sampling of the servocontrol loop. Analog results are obtained with lines and circles primitives.

3.2 Experimental results:

The approach has been also validated in real environments by using a CCD camera mounted on the end effector of a 6dof manipulator (figure 4). The behaviour of the

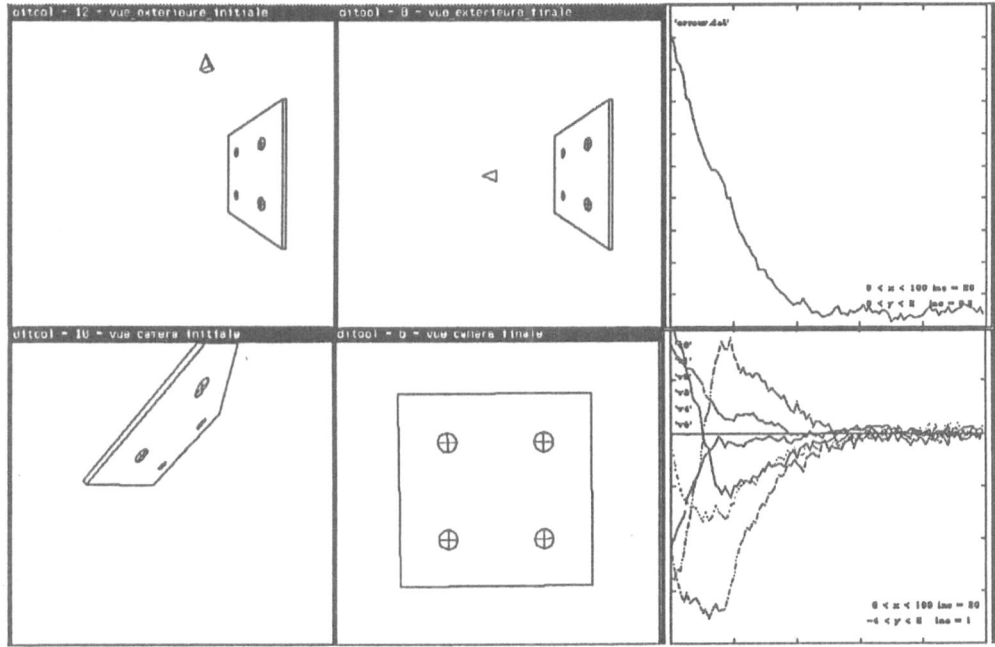

Figure 3: static target, noisy synthetic image

real system is extremly close to the one expected by the simulations. This a posteriori justifies in the studied cases the choice of a control law of the form given by eqs. (34), (37). The choice of $H(Q^*, R^*)$ (i.e. the value of H at the desired equilibrium position) as a control matrix has therefore been experimentally shown to be robust, even for rather large initial position errors.

The four points experiment described in simulation, has been successfully implemented. The figure 5 presents on the first hand, the sequence of images during the positionning task from the initial position to the final one, on the second hand, the behaviour of the mean quadratic error as in the simulation case, and the behaviour of each elementary signal error involved in the control loop.

4 Conclusion

This paper discusses the problem of using vision data directly as an input of the robot control loop. This approach, often referred as visual servoing, is characterized by the use of vision sensor as *local* sensor providing relatively poor instantaneous information but at a rate consistent with the bandwith of the robot controller, and by the exploitation of the vision data into a robust control scheme based on a task function approach. Concerning the first point, vision data are modelled as a set of *elementary signals*. Each elementary signal is associated at a 2D geometric primitive into the image frame (point, line, circle...) corresponding to the projection of a 3D primitive in the scene frame. We show that if

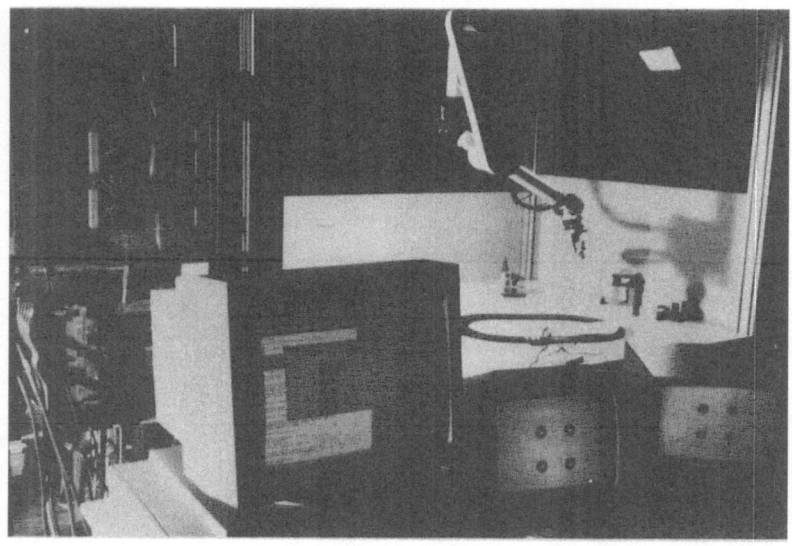

Figure 4: experimental cell

the signal is only dependant of the relative geometry between the sensor and the scene, then the interaction between the sensor and the scene can be described by a screw which relates observed motion in the image frame (optic flow) to relative motion between the camera and the objects in the 3D environment frame.

Concerning the second point, we propose a robust control scheme based on a task function approach. We show that the problem of control can be stated as a problem of regulation directly into the image frame. From the desired image target and the currently image observed by the camera, it is possible to define an error function and to express the problem of regulation as a problem of minimization of this error function. In this case, a class of control based on gradient techniques allows to perform correctly the task with good convergence properties (assuming only the definite positivity of a certain matrix).

This approach has been successfully validated both in simulation and in real experiments. In a next future, we should hope to address the problem of programming complex robotics tasks in terms of a succession of elementary subtasks which can be described by constraints between the frame linked to the effector and the frame linked to the scene (ie colinearity between axis, following of surface at a constant range...).

5   Acknowledgements

This work is supported by the French National Council of Scientific Research (CNRS) under the national project ORASIS-PRC Communication H/M.

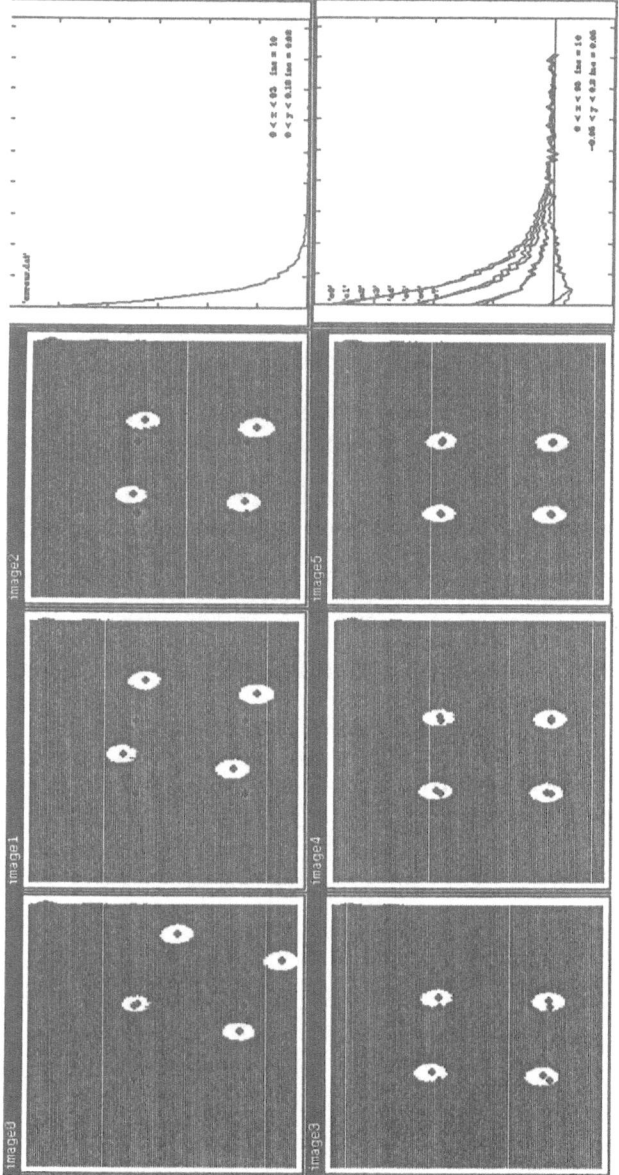

Figure 5: Real experiment

References

[1] F. Chaumette, P. Rives, *Réalisation et calibration d'un système expérimental de vision composé d'une caméra mobile embarquée sur un robot-manipulateur* Rapport de Recherche INRIA n.994, Mars 1989

[2] B. Espiau, *Sensory-based Control: Robustness Issues and Modelling Techniques. Application to Proximity Sensing*NATO Workshop on Kinematic and Dynamic Issues in Sensor Based Control, Italy, Oct.1987

[3] J. T. Feddema, C. S. G. Lee and O. R. Mitchell, *Automatic selection of image features for visual servoing of a robot manipulator* Conf. IEEE Robotics and Automation, Scottsdale, Arizona, USA, May 14-19, 1989

[4] P. Rives, G. Hegron, *Design of a simulation tool for robots using vision sensors* NATO Workshop on Languages for Sensor Based Control in Robotics, Italy, Sept. 1986

[5] P. Rives, *Dynamic vision: theoretical capabilities and practical problems.* NATO Workshop on Kinematic and Dynamic Issues in Sensor Based Control, Italy, Oct.1987

[6] C. Samson, *Une approche pour la synthèse et l'analyse de la commande des robots manipulateurs rigides* Rapport de Recherche INRIA n.669, Mars 1989

[7] C. Samson, M. Leborgne, B. Espiau, *Robot control: the task function approach* to appear, Ed. Oxford University Press 1989

[8] L. E. Weiss, *Dynamic Visual Servo Control of Robots. An Adaptive Image based Approach* Technical Report, CMU-RI-TR-84-16; Carnegie Mellon

VIDEO-RATE VISUAL SERVOING FOR ROBOTS

Peter I. Corke

CSIRO Division of Manufacturing Technology,

Preston, Victoria, 3072. Australia.

Richard P. Paul

Department of Computer and Information Science,

University of Pennsylvania, Philadelphia, PA, 19104. USA.

Abstract

This paper presents some recent experimental results in robotic visual servoing. A general-purpose computer in conjunction with special purpose video processing hardware, in particular a newly available hardware region-growing and moment-generation unit, has been used to visually close the robot position loop at video field rates, 60Hz.

The paper reviews prior work in the area of visual-servoing, and the related topic of real-time image segmentation. The architecture of an experimental system is discussed, and its' dynamics are investigated both experimentally and analytically.

1 Introduction

This paper presents some recent experimental results in robotic visual servoing, utilizing an off-the-shelf hardware region-growing and moment-generation unit. This unit processes a binary scene, containing many regions, at video data rates and yields fundamental geometric parameters about each region. Such a device opens up many applications that were not previously possible due to the computational complexity involved.

The enormous amount of data produced by vision sensors, and technological limitations in processing that data rapidly has meant that vision is not exploited as a sensing technology when high measurement rates are required. A vision sensor's output data rate, for example, is several orders of magnitude greater than that of a force sensor for the same sample rate. Existing systems tend to use a simple "look then move" approach to vision controlled motion. This paper uses the term "visual servoing" to describe an approach which treats the camera-processing subsystem as a sensor with its own inherent dynamics, which the closed-loop position control must take into account.

Researchers in robotic force control[38][10] have discovered that problems occur as the bandwidth of the sensor increases, allowing dynamic effects in the sensor, robot and environment to dominate the closed loop response. This paper discusses some of the dynamic effects evident in the use of a high bandwidth vision sensor to close the robot position loop.

These experiments are based on rapid analysis of binary scenes, since the technology presently exists to process these at video rates. Processing and interpretation of grey scale or color images is the goal of much research, and when those techniques are implemented at video rates then the results that follow will be applicable.

This paper presents a vision system capable of providing endpoint relative positioning information. This could be exploited for robotic pickup of parts swinging on overhead transfer lines, endpoint control of very flexible manipulators or gantry crane cables, or control of underwater robots using visual reference points.

The paper comprises four major parts. Firstly the literature on dynamic scene analysis, segmentation and visual servoing is reviewed. Secondly a system architecture capable of high bandwidth visual servoing is described. The third section analyzes the dynamic performance analytically and experimentally, and discusses some appropriate control laws. The fourth section provides a discussion of further work and conclusion. An extensive bibliography of prior work, and relevant image processing techniques is provided.

2 Prior work

2.1 Dynamic scene analysis and segmentation

There are many approaches to segmenting a complex scene so as to locate a possibly moving object. A general discussion of motion detection schemes is presented by Martin[24].

Perhaps the simplest approach is image differencing[19] in which the difference between consecutive frames eliminates all detail from the scene except the edges of moving objects. This approach will fail if the velocity of the object is too low to differentiate motion from sensor noise, or if the camera is moving, thus necessitating some hybrid strategy to initially locate a static object.

A more sophisticated approach is optical flow analysis[17] which yields a dense velocity vector field from spatio-temporal changes in pixel intensities. This analysis is computationally expensive, particularly the smoothing operation[1] and requires special processing hardware to implement in real time. Other problems include interpretation of the velocity field for general camera motion with rotations[36] and sensitivity to camera noise[26].

The derived velocity field must still be segmented to separate background from foreground. A related technique is based on feature matching, where features are generally corner points[32][27], edges[11], or regions. The change in position of various feature points between scenes can be used to estimate the velocity of those points. However features such as corners or lines are a sparse representation of the scene, and it is difficult to relate those features to the objects of interest in the environment without complex world models.

Sanderson[31] discusses visual servoing of robots and proposes a control scheme based on features extracted from the scene such as face areas of polyhedra, but gives no detail of the vision processing required to obtain those features. Kabuka describes two segmentationless approaches. One uses an adaptive control based on parameters derived from the entire scene[21], while another is based on Fourier phase difference between consecutive frames[20].

Image segmentation is only the first step towards visual servoing, the next problem is to identify the position of regions within the scene, and must be done whether the input image represents intensity, velocity or range data. One approach to object location is to use a multispectral spatial, or pyramid, decomposition[2] so as to reduce the complexity of the search problem. Others have used moment generation hardware to compute the centroid of an object within the scene at video data rates[5][35][12].

This paper discusses a region feature approach: a hardware unit performs segmentation of the intensity image to identify region features. The regions are then further investigated in software to identify the target, and tracking commands are then generated from the measured position of the identified target region. This approach is very general and works regardless of the state of motion of camera and target. The task of robustly segmenting a complex scene in real-time is not trivial, but is nevertheless the basis of this work.

2.2 Visual servoing

There is relatively little literature on visual servo implementations for robot position control and many papers present only simulation results. Approaches to date could be classified as either stop-start (not dealt with here) or continuous.

Hill and Park[16] describe a closed-loop position controller for a Unimate robot. They use binary image processing for speed, but also propose the use of structured lighting to reduce computational burden and to provide depth determination. Coulon and Nougaret[7] describe a digital video processing system for determining the location of one object within a processing window, and use this information for closed loop position control of an XY mechanism. They report a settling time of around 0.2s to a step demand. Kabuka[21] describes a two axis camera platform and image processor controlled using an IBM-PC/XT, and reports a minimum time of 30s to center on an object. Makhlin[23] discusses aspects of accuracy, tracking speed, and stability for a Unimate based visual servo system. Prajoux[28] demonstrated visual servoing of a 2DOF mechanism for following a swinging hook. The system used a predictor to estimate the future position of the hook, and achieved settling times of the order of 1s.

A number of related areas also require real-time visual tracking. Gilbert[13] discusses automatic object tracking cameras used for rocket tracking. Andersson[4] describes a ping-pong playing robot that uses a real-time vision system to estimate ball trajectory for

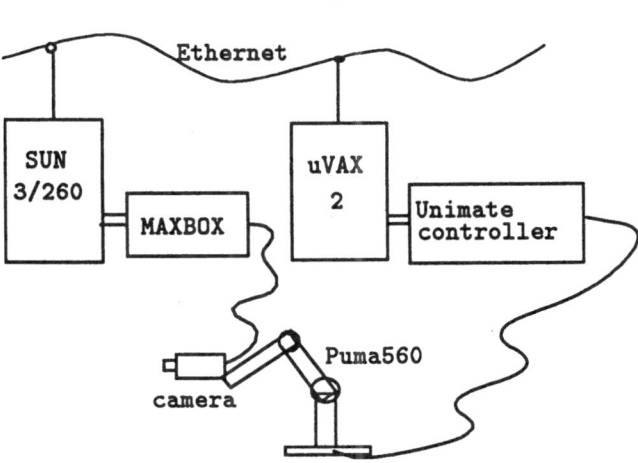

Figure 1: First experimental setup.

a subsequent paddle trajectory planning algorithm. While that system is not a closed-loop position controller many of the principles and problems are the same.

3   The experimental setup

The first version of this system was created at University of Pennsylvania in 1988[6], and is shown schematically in Figure 1. It comprised two computers; a MicroVAX-2 for robot control using the RCI interface of RCCL[22], and a Sun-3/260 for vision coordination with attached Datacube[9] image processing hardware. The two machines ran Unix and communicated via Ethernet using UDP/XDR protocol. Three special device drivers were involved to give Unix the appropriate "real-time" response. Since RS170 format video was used, measurements from 60 video fields per second were obtained.

The current version is shown schematically in Figure 2. It uses a single processor in a VME bus chassis, with a VME bus/Unimate interface and Datacube image processing hardware. The processor runs the VxWorks[39] real-time operating system, and processes communicate via shared memory. Robot control is performed by ARCL (for Advanced Robot Control Language). That software embodies many of the concepts of RCCL[15], is written in C, is highly modular, and allows for real-time, sensor-based trajectory modification. Since CCIR format video is used measurements from 50 video fields per second are obtained.

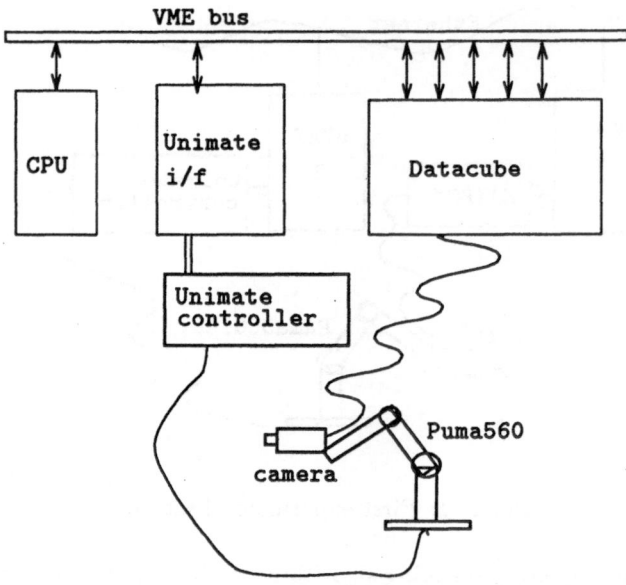

Figure 2: Current experimental setup.

3.1 Robot control

The robot control task in this application implements a cartesian rate server: that is, it accepts cartesian velocity commands (in the tool reference frame). The robot servo interval is either 14 or 28ms.

The server will accept velocity commands from a number of different clients such as the vision task or a graphical teach-pendant emulator.

3.2 Image processing

3.2.1 Segmentation

Robust segmentation of a scene remains one of the great problems of machine vision. Haralick[14] provides a survey of techniques applicable to static images. Unfortunately many of the algorithms proposed for segmentation are iterative and thus not suitable for real-time applications. This being the case, most real-time implementations use contrived situations with dark backgrounds and simply threshold the video data. Much work has been done on automated approaches to threshold selection[37][29].

Many suggest that incorporating edge information into the segmentation process increases robustness. 2D histograms of edge and intensity information have been used to extract targets from noisy FLIR (Forward looking infrared) imagery in real time[30][25][8].

An adaptive approach to segmentation is currently being investigated, using the capability of this hardware region-grower to perform many trial segmentations per second. One of the principle problems is how to rate the success of a segmentation so that an automatic system may modify or adapt its parameters[3]. However, at this stage the system described, like many others[16][21][4], uses simple thresholding.

3.3 Architecture for preprocessing

A functional representation of the image processing subsystem is shown in Figure 3, and is based on Datacube pipeline processing modules.

The Datacube family of video processing modules are VME bus boards that perform various operations on digital video data. The inter-module video data paths are patch cables installed by the user. The boards are controlled by a host computer via the VME bus. The video data paths run at 10Mpixels/s and are known as MAXBUS. Horizontal and vertical timing is established by a separate timing bus linking all boards. All images are 512 x 512 pixels, and each pixel is represented as an 8 bit 2's complement quantity.

The incoming video stream is digitized, optionally deinterlaced by the FRAMESTORE module, offset by an intensity threshold *thresh*, and translated by a lookup table which maps pixels to one of two grey levels, corresponding to the binary values *black* or *white*. Binary median filtering on a 3x3 neighbourhood is used to eliminate one or two pixel noise regions which may overwhelm the host processor.

The APA-512[34] (for Area Parameter Accelerator) is a hardware unit designed to accelerate the computation of area parameters of objects in a scene. The APA binarizes incoming video data and performs a single pass connectivity (simple-linkage region growing[14]) analysis, and subsequent computation of moments upto second order,

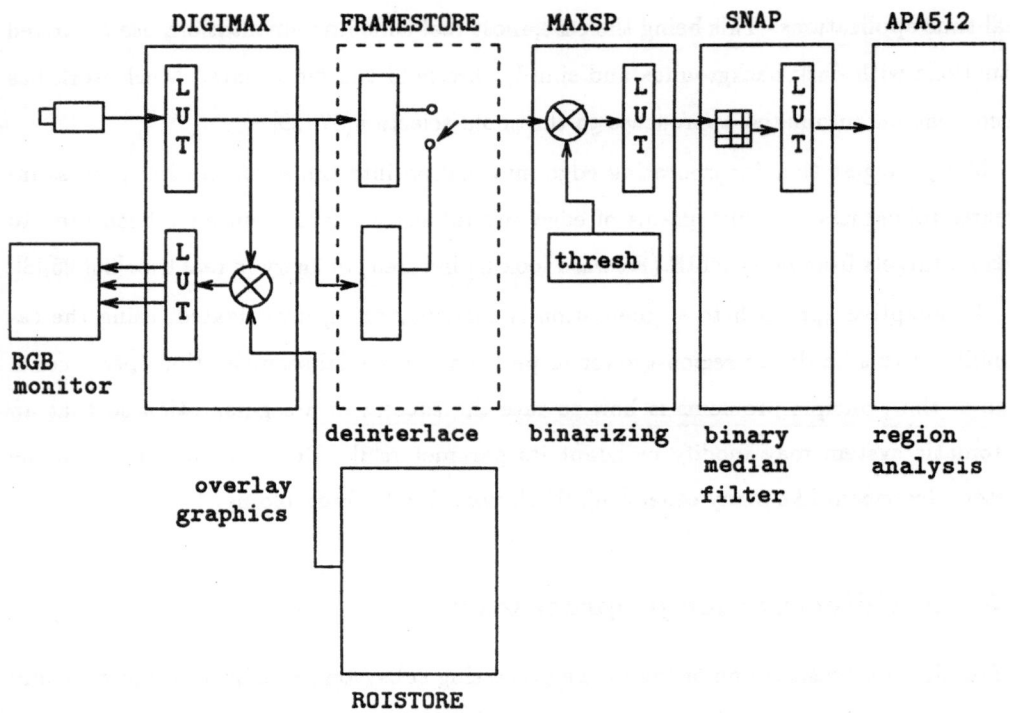

Figure 3: Image processing architecture.

perimeter and bounding box. The APA performs very effective data reduction, reducing a 10Mpixel/s stream of grey-scale video data via a MAXBUS interface, to a stream of tokens representing objects in the scene, available via onboard shared memory. The host processor screens the tokens according to their parameters, and thus finds the objects of interest. Appendix A provides additional details of APA operation.

The ROISTORE is used by the "statistics" task to display real-time overlay graphics and performance data.

Field rate processing is not discussed in the APA documentation[34], but can be achieved by treating the interlaced data as two consecutive frames of half vertical resolution. Frame-rate processing requires that the interlaced video data be deinterlaced using double-buffered framestores in the FRAMESTORE module, which introduces an additional one frame-time delay. Field-rate processing approach eliminates this delay, at the expense of having to process twice as many regions per second.

3.4 Coordination

Much of the processing involves sending commands to the various hardware image processing modules and this must be synchronized with the video vertical blanking interval,
which is only 1.2ms wide for RS-170 format video data. This necessitates highly responsive
host software, either a device driver under Unix or a high-priority task under VxWorks.

The main computational burden is involved in post-processing regions detected by the
APA. Small regions due to noise are screened out quickly and the region labels are returned
to the APA. Larger regions are then screened according to object feature descriptions,
which contain specifications for color (0 or 1), and permissible ranges for size and shape.
If a matching object is found then the manipulator's velocity in the 2D plane orthogonal
to the camera axis is set as a function of the displacement of the object's centroid from
the center of the image.

4 Experimental results and modelling

4.1 Measurements

The step response of the system was measured by a test rig with two spatially separated
LEDs, of which only one was illuminated at a time, and controlled by a square wave
signal source. Since the two LEDs are arranged horizontally, most of the robot motion
during visual servoing is done by the waist joint. The encoder signals from this axis were
processed by an encoder to voltage converter to yield the step response data, shown in
Figure 4 for a proportional gain of 0.010.

An FFT analyzer was then used to measure the closed-loop transfer function, see
Figure 5, given the square wave demand to the LEDs and the robot waist position. The
effects of a low frequency pole at approximately 0.3Hz can be seen in the magnitude
plot, while the phase plot shows evidence of substantial time delay. Autocorrelation
measurements proved unsatisfactory for estimating delay since it is influenced by the slow
rise time of the plant output signal. Instead, delay was estimated directly from the phase

438

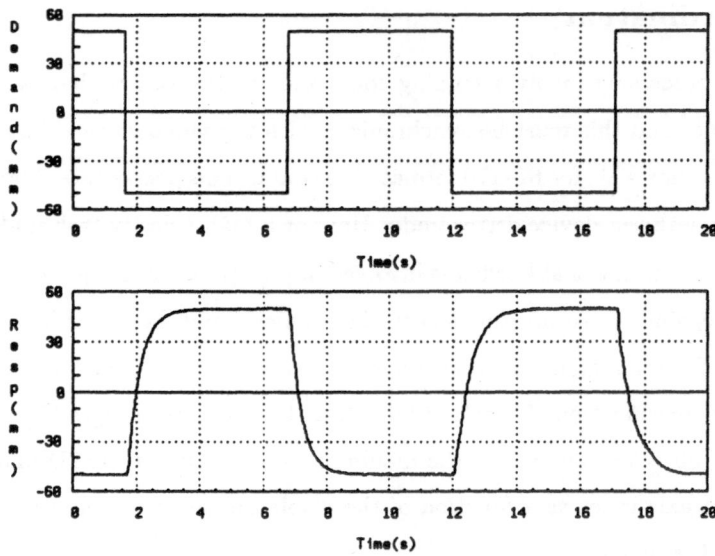

Figure 4: Step response (proportional gain of 0.01)

data. The phase data is plotted on a linear frequency scale in Figure 6 along with a curve representing a pure delay of 60ms, or 3 CCIR video field times. The experiments reveal that delay is the dominant dynamic effect in this system, as has been found by others[16][23][31].

4.2 Modelling

A block diagram of the control loop is shown in Figure 7, and where y is the robot position and u is the target position. The error $e = y - u$ is measured in pixel space by the camera, since it sees the object with respect to the robot end-effector.

The vision system is modelled as a gain and a delay. The gain in mm/pixel is a function of camera object distance, lens focal length, sensing chip geometry and the video digitizer. A camera calibration is required to relate real world dimensions to those sensed by the camera in pixel space. Due to the aspect ratio of standard TV screens an image of a circle will appear as a vertically elongated ellipse in pixel space, thus two calibration constants K_X and K_Y are determined. These are defined such that at a distance u one pixel represents a rectangle of width $K_X u$ and height $K_Y u$. The consequence of this is

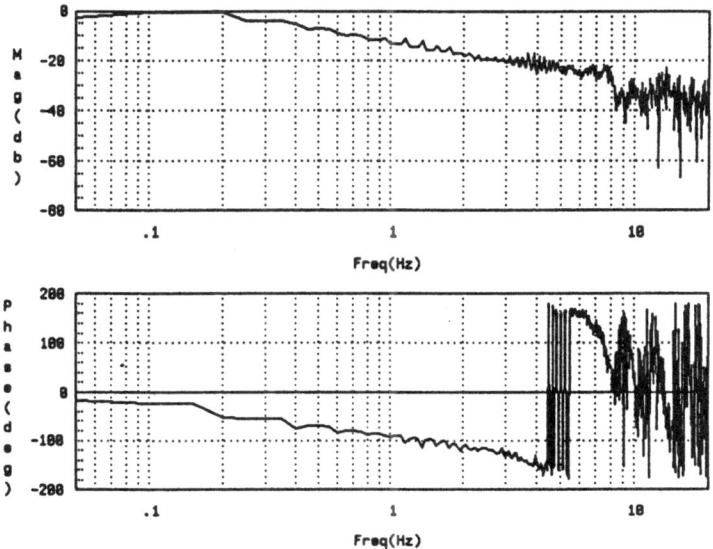

Figure 5: Closed loop frequency response (measured)

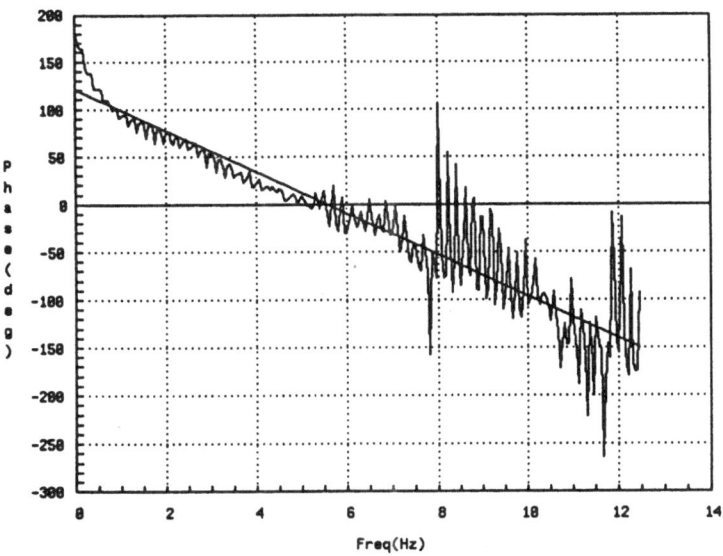

Figure 6: Phase response in detail

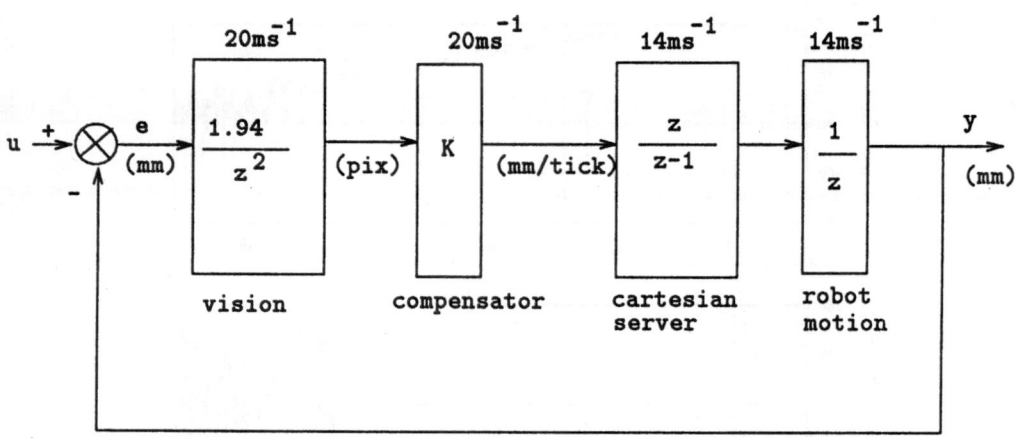

Figure 7: System block diagram

that loop gain is different for the X and Y directions, as well as being dependant upon object distance.

The field rate system exhibits 60ms, or 3 field times, of delay as determined above. By comparison a frame rate system exhibits 120ms, or 3 frame times, of delay. These delays can be accounted for by

- camera latency. The CCD camera used integrates each field for one field time, then transfers it to a shift register for output. Thus a one field time delay is introduced by the camera.

- raster scan latencies,

- processing time,

- deinterlacing for the frame rate version, and

- robot finite motion time.

This delay is partitioned into a two field time delay in the vision subsystem, incorporating camera and scan latency, as well as computational delay. The robot control

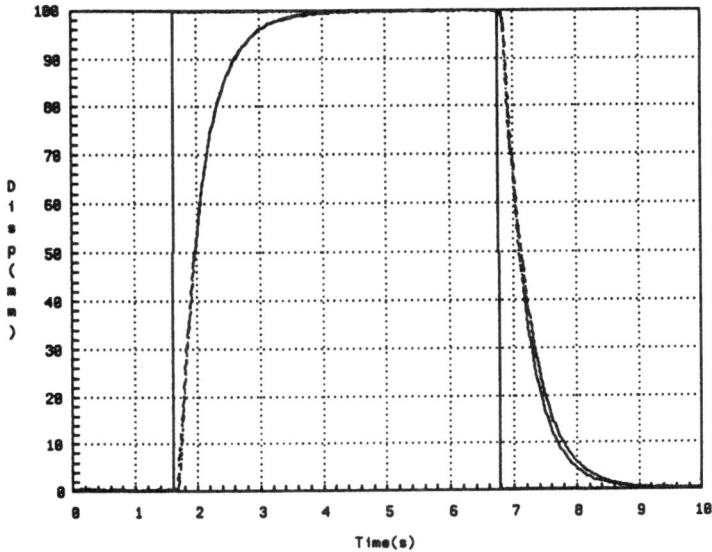

Figure 8: Comparison of measured demand, response and simulated response

subsystem is modelled as an integrator in the cartesian velocity server and a one time-step delay due to finite motion time. Mechanical and servo dynamics of the robot are ignored.

Unfortunately the model of Figure 7 is multi-rate and difficult to analyze using classical techniques. A full multi-rate simulation was created using $MATRIX_X$[18], and the model and measured response are compared in Figure 8 for the same demand. This shows close agreement between the model and measured responses.

An approximate single-rate open-loop model

$$\frac{1.94}{z^3(z-1)} \tag{1}$$

has been used for some analysis. It exhibits a 3 time-step delay as does the measured system. A root-locus diagram is shown in Figure 9; the closed loop poles corresponding to a proportional gain of 0.015 are marked by boxes. For this low gain the dominant pole is at $z = 0.968$, which corresponds to a real pole at $s = -0.26$Hz. This agrees well with the measured closed loop transfer function. Root locus analysis also indicates that the poles leave the unit circle at a gain of 0.23, the multi-rate simulation becomes unstable at a gain of 0.20.

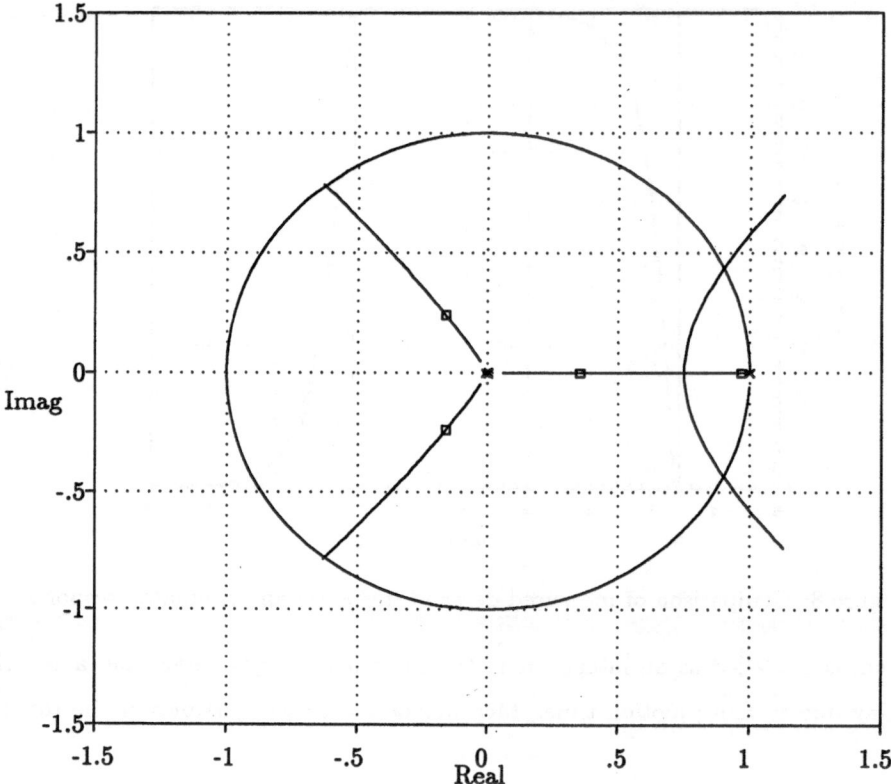

Figure 9: Root locus of approximate system.

4.3 Control design

Equation 1 is of type 1 and will thus exhibit steady-state error for a constant velocity target. A compensator of the form

$$D(z) = K\frac{z - a}{z - 1}$$

introduces another pole at $z = 1$ to produce a type 2 system, while the zero is used to position the closed-loop poles. In practice however, to cause the closed loop poles to move into the unit circle, the zero must be very close to z=1, which causes unacceptable overshoot.

As the proportional gain is raised to reduce rise time the control becomes very rough. The run-time statistics indicate that this is due to the system losing "sight" of the object,

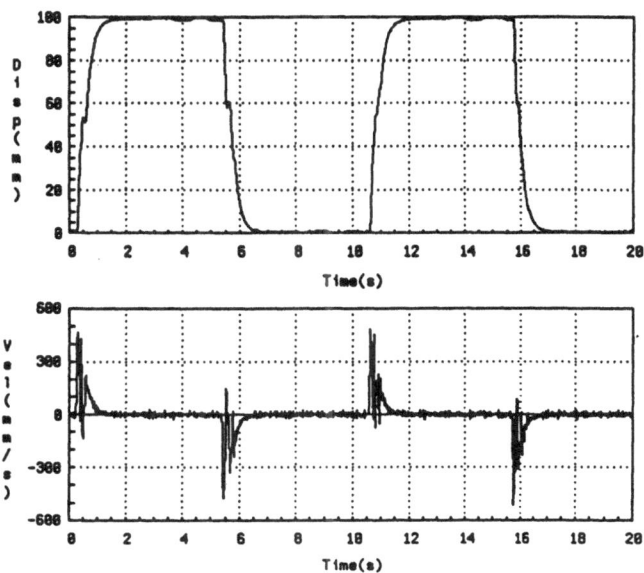

Figure 10: Robot position/velocity for large proportional gain

due to the high camera velocity. Figure 10 shows such a time response, and glitches in displacement at t=0.5s and t=5.5s are evident. Figure 10 also shows that the velocity peaks are over 300mm/s, or 11pixels/field. Since the width of the objects being tracked (the LED) is only 10 pixels the image will be considerably blurred and diminished in intensity since the LED energy is spread over more pixels.

Increasing the proportional gain to reduce rise-time is not productive. What is needed is a better trajectory, one that maintains a high velocity below the blur threshold for as much of the move as possible. This is achieved in the current system by limiting the velocity command output by the coordination task. Time responses for the velocity limited system are shown in Figure 11. These are for a proportional gain of 0.05 and a velocity limit of 260mm/sec. Use of a higher gain is of no advantage since the trajectory is mostly at constant velocity.

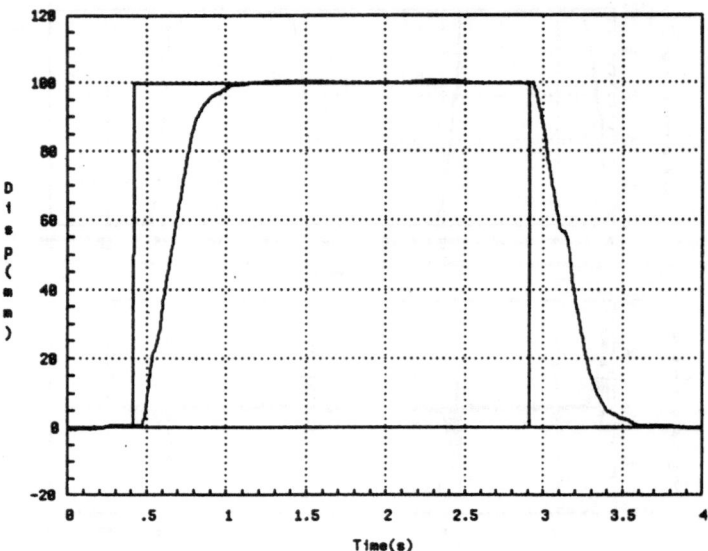

Figure 11: Response with velocity limiter

5 Further work and discussion

The system works well on a scene containing many regions but good contrast between target and background is still required.

Experiments have shown that interlaced frames are substantially distorted due to the delay between exposure times of the component fields, but this problem is eliminated when working at field rate. High speed camera motion causes image blur which is currently held to an acceptable level by limiting the velocity of the camera. An additional problem with high speed motion is that the target intensity appears diminished, and may fall below the intensity threshold. Motion blur effects have been investigated in detail by Anderson[4].

This work, and others[4] show that video standards and conventional cameras are now a limiting factor in high performance video sensing applications. Video technology is currently tied to the needs of the television industry where conflicting requirements such as small bandwidth and flicker have to lead to such devices as interlaced scanning. However when the end user is a machine vision system the requirements are very different. Fortunately, a new generation of cameras with both high resolution and high speed non-interlaced data formats are becoming available.

Areas for further work include:

1. Investigate adaptive segmentation based on feedback of segmentation results from the APA so as to modify scene thresholds.

2. Work is commencing with a CIDTEC high-resolution non-interlaced "automation" camera, with electronic shutter for blur control.

3. Extend the techniques to determine depth and orientation.

4. Investigation of more sophisticated control algorithms to improve the position response time of the closed-loop system.

6   Conclusion

This paper has described a high performance image processing system capable of providing robot positioning commands at rates of up to 60Hz, and demonstrated closed-loop position control. The image processing subsection is comprised entirely of off-the-shelf components, notably a hardware region growing and moment generation unit. The dynamic performance of the system has been investigated analytically and experimentally. Delay has been shown to be the dominant dynamic effect.

The region oriented approach used here enables the system to identify the target object whether or not there is relative motion between object and camera.

7   Acknowledgments

This research was conducted at the General Robotics and Sensory Perception (GRASP) Laboratory, University of Pennsylvania.

The authors are grateful to Vision Systems Ltd, Adelaide, Australia, for providing the APA-512 unit upon which this project is based. Professor Kwangyoen Wohn of the University of Pennsylvania provided advice and comments. The first author's visit to University of Pennsylvania was possible due to a CSIRO Overseas Fellowship.

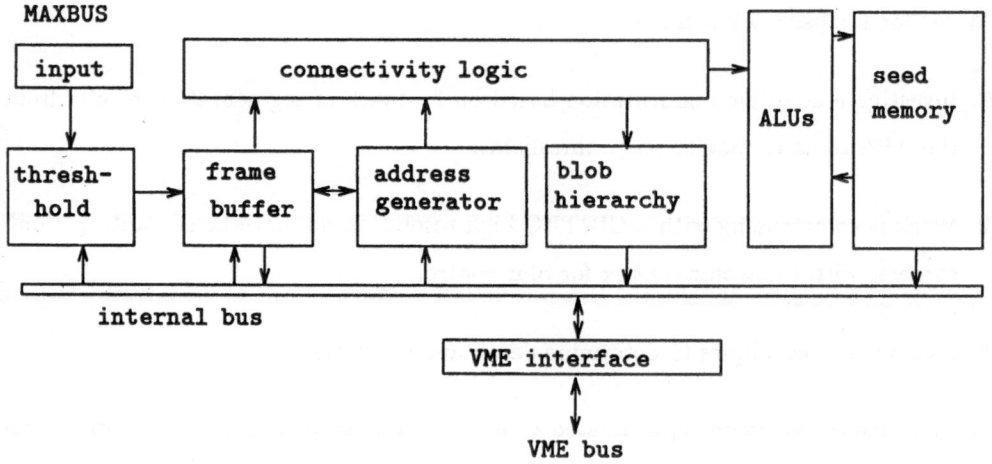

Figure 12: APA-512 block diagram.

A   APA-512

The APA-512[34] is a hardware unit designed to accelerate the computation of area parameters of objects in a scene. It was conceived and prototyped by the CSIRO Division
of Manufacturing Technology, Melbourne, Australia, in 1982-4, and is now manufactured
by Vision Systems Ltd. of Adelaide, Australia. The first author was a member of the
design team.

The unit has some similarities to other hardware implementations of binary image
processing systems[35][33][12].

Andersson's unit[5] computes moments of grey-scale data so as to improve accuracy
when dealing with a quickly moving object. However it has no capability to detect and
generate moments for multiple regions, and multiple regions will, if present, be merged
into one moment set.

The APA binarizes incoming video data and performs a single pass connectivity
(simple-linkage region growing) analysis. The connectivity unit commands a bank of
8 ALUs which update the region parameters (referred to as *seeds*) stored in seed memory.
The ALUs are implemented by custom gate arrays. The seed memory is dual ported to
the host VMEbus so that seed parameters of completed regions may be read.

The APA performs very effective data reduction, reducing a 10Mpixel/s stream of grey-scale video data via a MAXBUS interface, to a stream of tokens representing objects in the scene. The host processor screens the tokens according to their parameters, and thus finds the objects of interest.

For each region the following parameters are computed:

- Σi, number of pixels (zeroth moment)

- Σx, Σy (first moments)

- Σx^2, Σy^2, Σxy (second moments)

- minimum and maximum x and y values for the region

- perimeter length

- a perimeter point

- region color (0 or 1)

- window edge contact

From these fundamental parameters, a number of commonly used area parameters such as

- area

- centroid location

- circularity

- major and minor equivalent ellipse axis lengths

- object orientation (angle between major axis and horizontal)

may be calculated by the host processor. The perimeter point is the coordinate of one pixel on the region's perimeter, and is used for those subsequent operations that require traversal of the perimeter. The edge contact flag, when set, indicates that the region touches the edge of the processing window and may be partially out of the image, in this case the parameters would not represent the complete object.

Perimeter is computed by a sophisticated scheme that examines a 3x3 window around each perimeter point and produces an appropriate perimeter length contribution depending upon the slope of the perimeter at that point. Experiments reveal a perimeter error of less than 2% with this scheme.

The APA-512 computes these parameters for each of up to 256 *current* regions within the scene. Processing of the data is done in raster scan fashion, and as the end of a region is detected the region label is placed in a queue and the host is notified by an interrupt or a pollable status flag. The host may read the region parameters and then return the region label to the APA for reuse later in the frame, thus allowing processing of more than 256 objects within one frame. This feature is essential for processing non-trivial scenes which can contain several hundred regions of which only a few are of interest. Maximum processing time is one video frame time.

An additional feature of the APA is its ability to return region hierarchy information. When a region is complete the APA may be polled to recover the labels of already completed regions which were topologically contained within that region. This makes it possible to count the number of holes within an object, and compute the area of enclosed holes or internal perimeter.

References

[1] P. Anandan. Measuring visual motion from image sequences. Technical Report COINS 86-16, University of Massachusetts, Mar 87.

[2] C.H. Anderson, P.J. Burt, and G.S. van der Wal. Change detection and tracking using pyramid transform techniques. *Proceeding of SPIE*, 579:72–78, Sept 1985.

[3] Helen L. Anderson, Ruzena Bajcsy, and Max Mintz. Adaptive image segmentation. Technical Report MS-CIS-88-26, University of Pennsylvania, April 1988.

[4] R.L. Andersonn. *A Robot Ping-Pong Player*. MIT Press, 1988.

[5] R.L. Andersonn. Real-time gray-scale video processing using a moment-generating chip. *IEEE Journal of Robotics and Automation*, RA-1(2):79–85, June 1985.

[6] Peter I. Corke and Richard P. Paul. Video-rate visual servoing for robots. Technical Report MS-CIS-89-18, GRASP Lab, University of Pennsylvania, Feb 1989.

[7] P.Y. Coulon and M. Nougaret. *Use of a TV camera system in closed-loop position control of mechnisms.*, pages 117–127. IFS (Publications) Ltd, U.K, 1983.

[8] S. Cussons. A real-time operator for the segmentation of blobs in imaging sensors. *Proc. IEE Electronic Image Processing Conference.*, 214:51–57, 1982.

[9] Datacube Inc. *Installation and Software*, 1988.

[10] S.D. Eppinger and W. P. Seering. Introduction to dynamic models for robot force control. *IEEE Control Systems Magazine*, pages 48–52, 1987.

[11] C.L. Fennema and W.B. Thompson. Velocity determination in scenes containing several moving objects. *Computer Graphics and Image Processing*, 9:301–315, 1979.

[12] J.P. Froith, C. Eisenbarth, E. Enderle, H. Geisselmann, H. Ringschauser, and G. Zimmermann. *Real-time processing of binary images for industrial applications*. Springer-Verlag, Germany, 1981.

[13] A.L. Gilbert. Video data conversion and real-time tracking. *Computer*, 14(8):50–56, Aug 1981.

[14] R.M. Haralick and L.G. Shapiro. Survey: Image segmentation techniques. *Computer Vision, Graphics, and Image Processing*, 29:100–132, 1985.

[15] V. Hayward and R.P. Paul. Robot manipulator control under Unix, RCCL. *International Journal of Robotics Research*, 5(4):94–111, 1986.

[16] J. Hill and W.T. Park. Real time control of a robot with a mobile camera. *9th International Symposium on Industrial Robots*, Mar 13-15, 1979.

[17] B.K.P. Horn and B.G. Schunck. Determining optical flow. *Artificial Intelligence*, 17:185–203, 1981.

[18] Integrated Systems Inc. *MATRIXx User's Guide*, 1986.

[19] R. Jain. Dynamic scene analysis using pixel-based processes. *Computer*, 14(8):12–18, Aug 1981.

[20] M. Kabuka, J. Desoto, and J. Miranda. Robot vision tracking system. *IEEE Transactions on Industrial Electronics*, 35(1):40–51, Feb 88.

[21] M. Kabuka, E. McVey, and P. Shironoshita. An adaptive approach to video tracking. *IEEE Journal of Robotics and Automation*, 4(2):228–236, April 88.

[22] John Lloyd. Implementation of a robot control development environment. Master's thesis, McGill University, December 1985.

[23] A.G. Makhlin. Stability and sensitivity of servo vision systems. *Proc 5th Int Conf on Robot Vision and Sensory Controls - RoViSeC 5*, pages 79–89, 29-31 Oct, 1985.

[24] W.N. Martin and J.K. Aggarwal. Survey. dynamic scene analysis, 1978.

[25] O.R. Mitchell and S.M. Lutton. Segmentation and classification of targets in flir imagery. *SPIE*, 155:83–90, 1978.

[26] D.W. Murray and B.F. Buxton. Reconstructing the optic flow field from edge motion: An examination of two different approaches. *First Conference on Artificial Intelligence Applications*, pages 382–388, Dec 84.

[27] H. Nagel. Represenation of moving rigid objects based on visual observations. *Computer*, 14(8):29–39, Aug 1981.

[28] R.E. Prajoux. *Visual Tracking*. Machine Intelligence Research Applied to Industrial Automation. SRI International, August 1979.

[29] P.K. Sahoo, S. Soltani, and A.K.C. Wong. A survey of thresholding techniques. *Computer Vision, Graphics, and Image Processing*, 41:233–260, 1988.

[30] C.J. Samwell and G.A. Cain. The BAe (BRACKNELL) automatic detection, tracking and classification system. *2nd Int.Conf. on Image Processing and Applications*, IEE-265:164–170, JUne 1986.

[31] A.C. Sanderson, L.E. Weiss, and C.P. Neuman. Dynamic sensor-based control of robots with visual feedback. *IEEE Journal of Robotics and Automation*, RA-3(5):404–417, Oct 1987.

[32] W.B. Thompson and S.T. Barnard. Lower-level estimation and interpretation of visual motion. *Computer*, 14(8):20–28, Aug 1981.

[33] H. Tropf, H. Geisselmann, and J.P. Foith. Some applications of the fast vision system s.a.m. *Workshop on Industrial Applications of Machine Vision Conf. Record*, pages 73–79, May 3-5, 1982.

[34] Vision Systems Limited. *APA-512MX Area Parameter Accelerator User Manual*, Oct 87.

[35] P. Vuylsteke, P. Defraeye, A. Oosterlinck, and H. Van den Berghe. Video rate recognition of plane objects. *Sensor Review*, pages 132–135, July 1981.

[36] A.M. Waxman and K. Wohn. Contour evolution, neighborhood deformation, and global image flow: Planar surfaces in motion. *International Journal of Robotics Research*, 4(3):95–108, Fall 85.

[37] Joan S. Weszka. A survey of threshold selection techniques. *Computer Graphics and Image Processing*, 7:259–265, 1978.

[38] D. E. Whitney. Historical perspective and state of the art in robot force control. *Proc. 1985 IEEE Int. Conf. on Robotics and Automation*, pages 262–8, 1985.

[39] Wind River Systems Inc. *VxWorks Reference Manual*, 1988.

Interpretation of mechanical properties of soft tissues from tactile measurements

A. M. Sabatini, P. Dario, M. Bergamasco
Scuola Superiore S. Anna
Via Carducci, 40
56100 Pisa, Italy
and
Centro "E. Piaggio"
University of Pisa

Abstract

Robotic applications involving manipulation of soft materials are receiving increasing attention. A particular case of this problem is the manipulation of soft body tissues for robotic applications in medicine. The ability to sense and characterize the mechanical properties of soft tissues is a prerequisite for a robot to carry out any real intervention for diagnostic and therapeutic purposes. Identifying regions of different mechanical properties by touch is in fact not only potentially useful per se, but critical for a robot in order to execute, for example, surgical operations. The purpose of this paper is to investigate a method for discriminating mechanical inhomogeneities into a soft tissue, in the simplified case of spherical inclusions. The proposed model has been validated through an experimental approach, also described in the paper, based on the use of a finger-like palpation device equipped with a fingertip piezoelectric polymer film tactile sensor.

1 Introduction

Present industrial robots are usually unable to manage situations in which a contact occurs between the robot and the external environment. This fact poses severe limitations to the use of robots not only in the field of industrial automation, such as in assembly, but also, and even more, in those applications in which the environment is poorly or totally unstructured.

In addition to the difficulty to actively control contact, most robot systems, being conceived and used as precision machines for precision tasks, are unable to deal with operations requiring the manipulation of soft and highly deformable objects. The ability of robot systems to execute these types of tasks, which occur frequently in some economically important industrial fields, for example in the clothing industry, or in other fields, such as agriculture, is critical for a wider diffusion of industrial automation, and of robotics in general, and is therefore receiving increasing research attention [1].

The ability to actively control forces and movements is a fundamental requirement for most robots intended for operations outside the factory floor. Some of those robots, for example the domestic robot, must interact frequently with humans; other robots, such as medical robots, should indeed even not be conceived without the ability to touch and manipulate human patients. This ability could be obtained, in part, through the use of safety devices: this approach is appropriate, for instance, in the case of rehabilitative robots for the assistance to the disabled [2]. A more general solution to the problem of gently manipulating body parts through active procedures opens up, however, new and extremely attractive potential applications of robotics technology to the execution of diagnostic and therapeutic procedures. The ability to sense and characterize the mechanical properties of soft tissues would thus become a prerequisite for a robot system to carry out any real intervention for diagnostic and therapeutic procedures.

An interesting area of application of soft tissue characterization is automated palpation, a procedure useful for detecting, for example, breast cancer. Two different approaches have been proposed so far for the development of palpation systems. The first, described by Lang et al. [3], does not use any robotic system, but rather a thin film array of piezoelectric polymer sensors (polyvinylidene fluoride, or PVDF) pressed manually with a uniform stress against the tissue containing embedded hardened regions. The signal detected by each PVDF sensor is processed and displayed, normally in the form of a color map, so that the physician (or even the same patient) can assess the presence of embedded regions.

The second approach has been proposed by Kato et al. [4], who developed an automatic breast-cancer palpation robot (WAPRO-4R). The robot has four fingers, each having four pressure sensors. The system has been used in clinical tests, demonstrating the ability to detect tumors of small size (about 6 mm x 6 mm).

In addition to this practical application, the ability to identify regions of different mechanical properties by touch is also critical for a robot in order to execute surgical procedures, especially in the field of microsurgery. In fact, there is a growing interest towards the use of robots as autonomous or teleoperated precision surgery tools [5].

The purpose of this paper is to investigate a method for discriminating mechanical inhomogeneities into a soft tissue, in the simplified case of spherical inclusions. A continuum mechanics model has been investigated and validated through an experimental approach based on the use of a finger-like palpation device equipped with a skin-like PVDF tactile sensor.

2 Theoretical model

The general problem we have elected to address is the following: given a medium in which regions of different Young's modulus are embedded, is it possible to detect size, shape, depth and elastic properties of the inclusions by exerting a uniform compressive stress on the medium? It is conceivable that only one sensing modality could hardly be successful in providing the whole information necessary to completely characterize the mechanical and geometrical properties of the inclusions. While, for instance, ultrasonics imaging is effective in giving details about the geometrical properties of the inclusions, namely size and depth, it is likely to be of little assistance in evaluating the elastic properties in terms of Young's modulus (techniques for material characterization by ultrasound are still in their infancy). Palpation is an attractive technique for sensing mechanical properties of soft materials.

Palpation has been known for centuries as a tool used by physicians for diagnostic purposes. The aim of the present work has been to understand first the theoretical basis of palpation, and then to design and build an experimental robotic system capable of replicating the procedure adopted by a physician during palpation. Palpation is based on the concept that if the skin which forms the boundary of a soft tissue environment is pressed with a uniform stress, the surface will be deformed in such a way as to conform to possible underlying regions of larger Young's modulus. In more technical terms, regions of different viscoelastic properties change the stress-strain relations at a boundary (the skin), and information on the underlying tissues can be obtained by measuring these stress-strain relations. The critical aspects of a procedure for replicating palpation by means of a robotic system are: 1) to provide an accurate description of the viscoelastic behavior of the soft tissue with hardened regions through a continuum mechanics model; 2) to execute appropriate sensory-motor procedures for probing and sensing soft tissue characteristics. The interpretation of the data obtained during the palpation procedure, based on the model of the soft tissue, would allow to extract the desired information on the possible inclusions. The ultimate goal of this procedure is to obtain a system capable of discriminating automatically even tiny local dishomogeneities.

The first step towards the implementation of the robotic palpation procedure is the development of a model for describing the interaction between the palpation device and the viscoelastic environment.

The simplified geometry considered in our model is illustrated in Fig. 1.

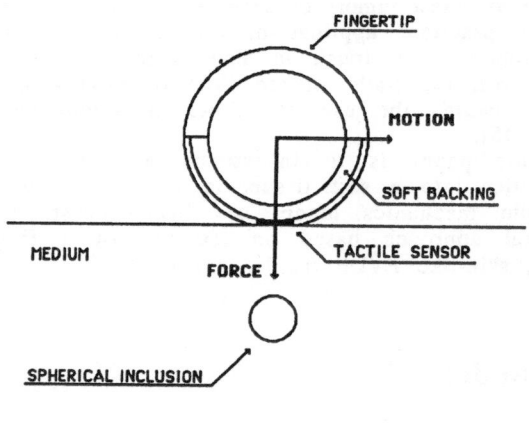

Fig. 1: Model of the behavior of a finger-like palpation device pressed with uniform stress on a soft tissue with hardened inclusions.

A compliant curved sensor located at the fingertip of a robot finger is pressed against a soft infinite medium. The sensor comprises a large number of individual sensing sites.

The finger makes contact with the medium on a surface whose extension after deformation is assumed to be much larger than the size of each sensing site of the tactile sensor. Furthermore, the contact area is considered as planar, at first approximation. The proposed model considers the finger-like palpation device contacting an elastic, homogeneous, isotropic medium in a stress field which can be considered as uniform and infinite in extent, as far as the sensor behavior is concerned.

If the hypothesis of small deformation holds, the theoretical analysis proposed by Goodier [6] can be used to evaluate the strain distribution induced by spherical inclusions of different size and elastic properties, located at a certain depth into the medium.

According to this analysis, the elastic displacements in polar coordinates are complex algebraic functions of a number of parameters, including magnitudes of the Young's moduli E_1, E_2 and Poisson's ratios ν_1, ν_2, depth z and diameter ϕ of inclusions, and value T of the compressive stress. In the following, the subscripts 1 and 2 refer to as the medium and inclusion phases, respectively.

The horizontal and vertical strains in cartesian coordinates can be derived from the strains in polar coordinates by means of simple transformation formulas.

The model assumes that the sensor is thin and compliant enough to "copy" the underlying boundary, without significantly affecting its mechanical behavior. We also assume that the sensor is made out of a thin piezoelectric polymer (PVDF) film.

The constitutive equations of the piezoelectric polymer material can be used to predict sensor response to combined horizontal and vertical strains. The charge Q generated by the sensor can be written in terms of the strains of the sensor S_1, S_2, S_3 as follows:

$$Q = \int_A (e_{31}S_1 + e_{32}S_2 + e_{33}S_3) \, dA \qquad 1)$$

1, 2, 3 are the reference axes of the piezoelectric polymer film (3 being the axis normal to the film surface), A is the sensing site area and e_{31}, e_{32} and e_{33} are the piezoelectric constants along different axes. Values of the piezoelectric constants are reported for the case of biaxially stretched PVDF in [7].

The main features of the function to be integrated, defined as σ, are evaluated by means of numerical calculations. σ takes the following form:

$$\sigma = \sigma_0 \, (T/E_1) \, f(E_2/E_1) \, g(z/\phi, d) \qquad 2)$$

where the geometrical variable d represents distance of the sensing site from the center of the coordinates   The vertical through the center of the inclusion intersects the mock in the origin of the reference system.

The quantity T/E_1 is approximately the vertical strain induced far from the inclusion: an increase of the compressive stress T results in a proportional increase of the charge, at least as long as the limit of small deformation is not exceeded.

The piezoelectric response is of the differential type, namely the difference between the actual response and the response far from the inclusion is considered in all of the calculations.

A peak P_{max} is located at x=0 and two peaks of opposite sign are simmetrically placed around the origin. We refer to the distance between these two peaks as the spatial window B_s.

As for the dependence on E_2/E_1, a non linear relationship holds approaching the asymptotic value relative to the case of perfectly rigid inclusions, as illustrated in Fig. 2 left.

P_{max} depends inversely also on the ratio z/ϕ, as shown in Fig. 2 right.

The theoretical curves σ can be normalized to P_{max}. An example of the normalized curves σ_n is reported in Fig. 3.

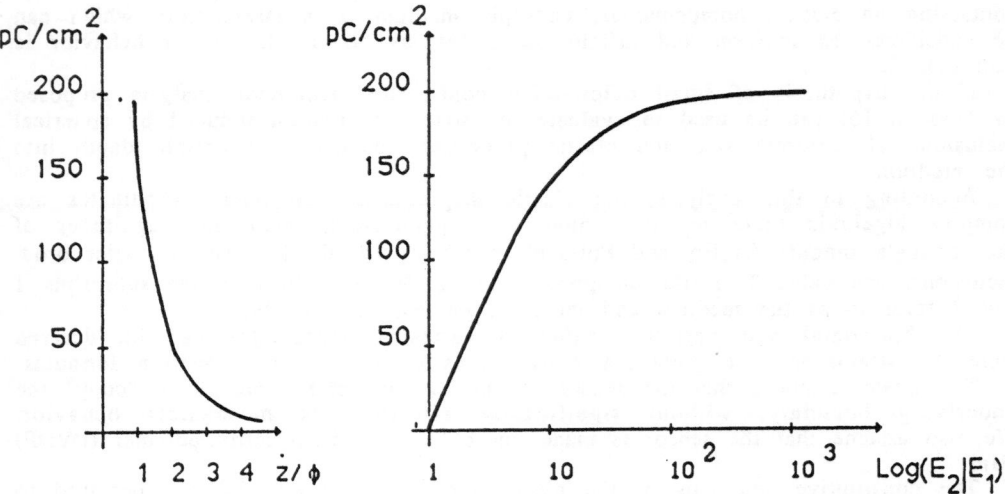

Fig. 2: Graphic displaying the behavior of P_{max} as a function of E_2/E_1 (on the left) and as a function of the ratio z/ϕ (on the right).

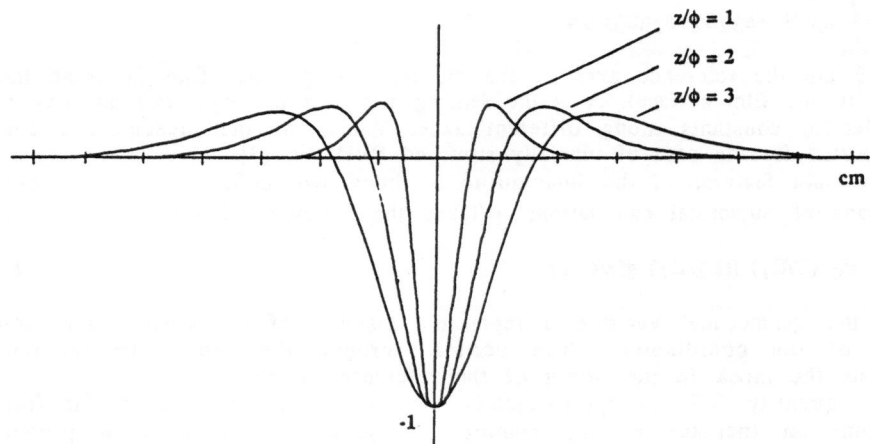

Fig. 3: An example of normalized curves σ_n

The functions:

$$f_m(\alpha y) = [(\alpha y)^2 - 1]/[(\alpha y)^2 + 1]^m \qquad 3)$$

are useful in finding bounds for $\sigma_n(z/\phi, y)$:

$$f_3(\alpha y) < \sigma_n < f_2(\alpha y) \qquad 4)$$

The scale factor α accounts for the spreading of the curves around the origin, due to the dependence of B_s on z/ϕ. Numerical calculations lead to the conclusion that B_S depends just on z, whatever the inclusion size is. Approximately, we can write:

$$B_S = \beta z \qquad \qquad 5)$$

where β is a constant ($\beta=1,8$ when z is expressed in cm).

A fair estimation of the spatial bandwidth of the strain profiles is easily obtained by taking the Fourier transform of $f_3(\alpha y)$. As for sampling requirements, the critical sampling spatial frequency can be related to the depth of the inclusion (see Appendix). A good rule of thumb poses that the number n_S of samples per unit length should be:

$$n_S = 10/z \qquad \qquad 6)$$

For an inclusion located 10 mm below the surface, the exploratory strategy should then involve 1 mm step movements of the finger.

3 Hardware

The finger-like palpation device we have designed in our laboratory is composed of four rigid links individually actuated by DC servomotors via cables. Incremental encoders are used to detect the joint angular displacements; tendon tension sensors consisting of cantilevers instrumented with foil strain gauges and located at the outlet of the conduits guiding the drive cables, are used to measure the joint torques and ultimately to control contact forces during palpation.

The fingertip incorporates a skin-like tactile sensor whose role is not to control contact, but rather to sense specific object features, i.e. physical properties, during tactile exploration [8]. The skin-like sensor is based on the technology of ferroelectric polymers: two 40 μm thick PVDF films are layered in a "bilaminate" configuration on which a "fovea" of 7 circular sensing sites (diameter 1,5 mm, center to center spacing 2,5 mm) are obtained. The tactile sensor is made particularly sensitive to membrane strain by allowing it to operate like a flexible membrane: the contact surface is provided with some compliance by a 5 mm thick silicone rubber layer backing. Thus the tactile sensor can be considered as a membrane sensor responding only to the combination of normal and tangential strains acting at its surface.

Every exploratory strategy involves positioning first the tactile fovea on the soft tissue to examine; secondly, the array sensor is scanned and the desired feature is detected through the sensing element located directly on that area. In the context of the present paper, only one element of the tactile fovea is polled for perceptual tasks: this sensing element is connected to the input of a charge amplifier.

At present, the robot system consists of two blocks. At the bottom level of the control hierarchy, servo loops control finger joint positions and torques through feedback from internal sensors, namely encoders and tendon tension sensors. An upper level has the capability of controlling the execution of appropriate sensory-motor procedures aimed at acquiring useful exteroceptive data from the tactile sensor at the fingertip. A DEC microPDP-11/73 is sufficient to implement the control loops necessary to manage in real time simple tactile exploratory procedures.

A hybrid (position-force) control of the joints is provided by the control computer so that the fingertip sensor can move along the surface of an object while simultaneously exerting a controlled contact force in order to elicit tactile stimuli from the sensor. In such a way the robot finger can carry out quite complex exploratory procedures by finely adjusting position, orientation and contact force. In this preliminary investigation, however, palpation procedures are carried out with a very simple sensory-motor sequence.

Finger operation is controlled through a suitable force law: contact with the mock is recognized when the joint torque sensor read-out at the second distal phalanx exceeds a threshold, adjusted at the upper level; the finger is then moved against the surface as long as a second torque threshold is exceeded. At this point the finger movements are momentarily terminated, and the voltage at the output of the charge amplifier is acquired via A/D conversion and sent to the upper level of the hierarchical architecture for further analysis and processing. The presence of offset levels at the output of the charge amplifier, often not stable in time and resulting in a large, variable drift is removed immediately before the acquisition from the tactile sensor by remotely resetting the charge amplifier; the finger is then commanded to return to the initial configuration. After each exploratory step, the finger moves laterally in order to progressively scan the entire surface of the explored object. In Fig. 4, the overall architecture of the tactile sensing system is presented.

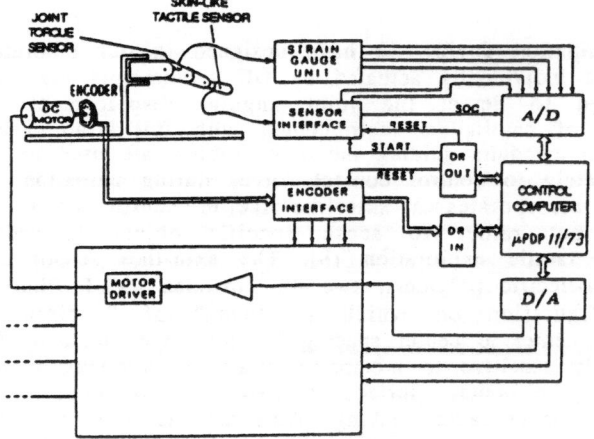

Fig. 4: A block diagram of the architecture of the tactile sensing system.

4 Experimentation

A mock breast, obtained by casting silicone rubber gel into a mold in which spheres of different diameter (from 10 mm to 5 mm) were suspended at a certain height, was used for testing the robotic system. Since it is relatively difficult to fabricate spheres with Young's moduli slightly different from the Young's modulus of the silicone gel, say up to an order of magnitude greater (the range in which tactile sensor response would be significantly affected, see Fig. 2 left), we decided to consider only the limit case of perfectly rigid inclusions: thus, bearing balls made out of stainless steel were located in the rubbery matrix. After curing, the mold was removed; the inclusions were located 1 cm below the surface of the mock. The mock surface was protected with a thin Mylar film to avoid adhesive contact between the finger and the silicone gel.

In order to validate the continuum mechanics model previously outlined, a simple experiment was devised consisting of scanning the mock breast along a linear trajectory passing exactly over the inclusions. As shown in Fig. 5, the experimental curves are in reasonable agreement with the theoretical curves (see Fig. 3): in particular, such salient features described by the model as the presence of the central located peak P_{max} and of the spatial window B_s are nicely reproduced by experimental data.

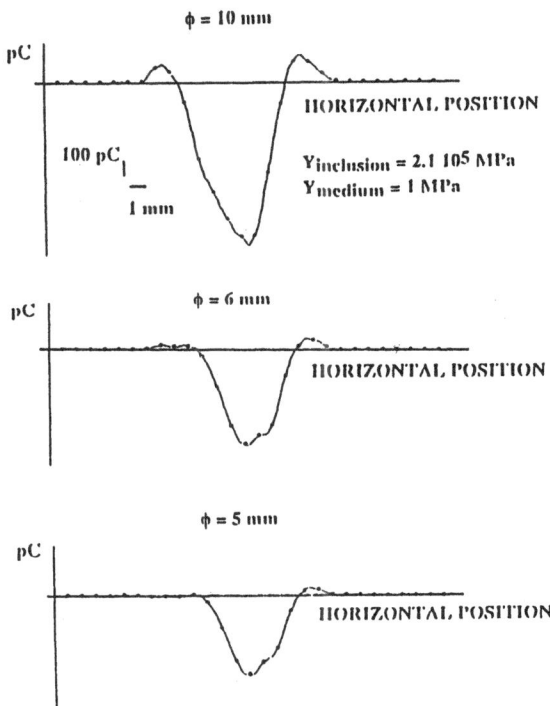

Fig. 5: Experimental curves obtained by scanning a linear trajectory over the inclusions.

It is rather difficult to directly compare charge densities generated by the sensor in Fig. 3 with the charge levels measured by the charge amplifier in Fig. 5. In fact, only at first approximation the strain can be considered as constant over the surface area of each sensing site of the tactile array; furthermore, it is difficult to estimate the real contact area between the finger and the mock, as is necessary to evaluate the stress exerted by the finger, the total force imparted being inferred from the joint torque sensors.

From the viewpoint of an automated palpation procedure, the only significant experimental data are the spatial window B_s (from which the depth of the spherical inclusion can theoretically be calculated), and the central located peak P_{max}.

As for the depth of the inclusion, the agreement between the theoretical predictions and the actual measurements is satisfactory: interestingly enough, the bandwidth of the strain profiles can be quite accurately evaluated just from the spatial window, without any Fourier transform on the data. In a real situation, however, neither the ratio z/ϕ nor the ratio E_2/E_1 are known: some additional information is thus necessary in order to fully exploit the data relative to the central located peak P_{max}.

As a final comment, the reconstruction of the strain profiles from the sensor readouts requires a relatively high sampling spatial frequency: this results in a too much time-consuming procedure to scan even a few square centimeters of the explored object. It is reasonable then to conceive an exploratory strategy involving a preliminary coarse scan of the surface: this is necessary to calibrate the system by evaluating the baseline, namely the readout far from the inclusions. On the other hand, this preliminary scan should also detect "at risk" spots on the surface (detection of the inclusion), that is areas where the signal generated by the sensor becomes negative. The finer scan can thus be planned only in those areas, where it is necessary to measure the spatial window and to detect the point directly over the inclusion in order to locate the inclusion.

5 Conclusion

The investigation described in this paper has provided a number of encouraging indications, along with some observations on aspects to improve and to work on. For example, the continuum mechanics model proposed in this paper to describe the alterations of the stress-strain relations at the surface of a homogeneous and isotropic medium in which spherical inclusions of different size are embedded at a certain depth, is not able to predict exactly those stress-strain relations in a real case: a much more complex behavior than the one predicted by the present simplified model could be expected due to the viscoelastic, hysteretic and non-linear properties of living biological tissues and to the actual shape of the inclusions, which can be different from the particular case of spherical inclusions considered in our experiments. Nevertheless, considering the good agreement between the predictions of the model and the experimental data, the results obtained can be regarded as satisfactory for our purposes.

Looking at the more general problem of gently manipulating soft materials by means of an articulated end-effector, which is expected to become a major issue in future applications of advanced robotics, we consider rather appropriate our approach. In fact, the concept of incorporating into the end-effectors both sensors specifically intended for the "hybrid" control of the robotic manipulator at the low level of the hierarchical control structure, and sensors of the "exteroceptive" type, specifically intended to acquire data useful for perceptual tasks during appropriate motor acts, seems particularly promising. Particularly powerful seems the definition of tactile subroutines, commanded at the upper control level, and managed at the intermediate control level. These subroutines would enable the system to cope with much more demanding experimental scenarios than the present one, in which the sensory-motor sequence is actually quite simple and does not require very fine adjustements of the finger position on the explored object, as would be necessary in such common cases as when the manipulated object would have a complex shape. Unfortunately, our robotic finger is, in the present configuration, unable to move with the precision requested for carrying out such tasks, owing to its mechanical design. A novel design of the finger is under way to overcome the actual limits of the system in terms of friction and mechanical inaccuracies.

A final comment relates to the tactile sensor we employed at the fingertip. Sensor performance in terms of sensitivity was appropriate for detecting inclusions down to 5 mm diameter at 1 cm below the surface, a value that is already considered as clinically significant for the early detection of breast cancer (4). Above all, however, and despite its limitations in terms of spatial resolution, a unique feature of the sensor, and in fact the one that renders the present experiment feasible and future development possible, is its skin-like behavior, namely its capability to sense strains acting at the surface of the mock.

Acknowledgments

This work has been supported in part by the Italian Ministry of Education (MPI 40%) and by National Research Council (Finalized Project on Robotics).

Appendix

From simple properties of Fourier transforms, it is straigthforward to derive:

$$f(x) = 1/(1+x^2) \quad \twoheadrightarrow \quad F(f) = \pi\exp(-|2\pi f|) \tag{A1}$$
$$f_2(x) = (x^2-1)/(1+x^2)^2 \quad \twoheadrightarrow \quad F_2(f) = -\pi j 2\pi f\exp(-|2\pi f|) \tag{A2}$$

For sampling requirements, aliasing phenomena are assumed to be negligible if the value at the tail of the frequency response is down a factor of 1000 from the maximum. .According to this assumption, the bandwidths B and B_3 of $f(x)$ and $f_3(x)$, respectively, result:

$$2\pi B = 3\ln 10 \tag{A3}$$
$$2\pi B_3 = 10,2 \tag{A4}$$

Since

$$f_3(x) = (x^2-1)/(1+x^2)^3 \tag{A5}$$

can be written as $f_3(x)=f(x)f_2(x)$, its Fourier transform is the convolution of $F(f)$ with $F_2(x)$. Therefore, its bandwidth is $B_3=B+B_2$. Including the scale factor α into the calculations, it follows:

$$2\pi B_3 = (3\ln 10 + 10,2)\alpha \tag{A6}$$

where:

$$\alpha = 2\sqrt{2}/B_s \tag{A7}$$

Since $B_s = 1,8z$,

$$B_3 = 4.2/z \tag{A8}$$

According to Nyquist's theorem, an appropriate sampling frequency should be greater than $2B_3$:

$$f_s > 8.4/z \qquad \qquad A9)$$

References

(1)   P.M. Taylor Ed., *Sensory robotics for the handling of limp materials*, NATO ASI Series, Springer-Verlag 1989.

(2)   L. Leifer, *Rehabilitation Research and Development Center: 1988 Progress Report*, Veterans Administration Medical Center, Palo Alto, California, 1988.

(3)   S. B. Lang, B. D. Sollish, M. Moshitzky, E. H. Frei, *Model of a PVDF piezoelectric transducer for use in biomedical studies*, Ferroelectrics, Vol. 24, pp. 289-292, 1980.

(4)   I. Kato, K. Koganezawa, H. Fujimoto, M. Hirata, *The automatic breast-cancer palpation robot: WAPRO-4R*, Proc. of the Int. Workshop on Intelligent Robots and Systems, pp. 73-78, Tokyo, Japan, Nov. 1988.

(5)   Proc. of the 1th Int. Workshop on Robotic Applications in Medical and Health Care, June 23-24 1988, Ottawa, Canada.

(6)   J.N. Goodier, *Concentration of stress around spherical and cylindrical inclusions and flaws*, J. of Appl. Mech., Vol. 55, pp. 39-44, 1933.

(7)   H. Ohigashi, *Electromechanical properties of polarized PVDF films as studied by the piezoelectric resonance method*, J. of Appl. Phys., Vol. 47, pp.949- , 1976.

(8)   P. Dario, G. Buttazzo, *An anthropomorphic robot finger for investigating artificial tactile perception*, Int. J. of Robotics Res., Vol. 6, N. 3, pp. 25-48, Fall 1987.

Determination of Manipulator Contact Information
from
Joint Torque Measurements

Brian S. Eberman

J. Kenneth Salisbury

MIT Artificial Intelligence Laboratory
Cambridge MA 02139

Abstract

Measurements of the net forces acting on a structure can provide information necessary to determine the location of contacts through which the external forces act, under appropriate assumptions. One practical application of this idea resulted in the contact-resolving force-sensing fingertips developed by Salisbury and Brock. These fingertip sensors employ a 6-axis force sensor under a fingertip shell to measure the net applied force. It has been demonstrated that this type of sensor supplies enough information to determine the location and force components of contacts which transmit pure forces. Bicchi recently extended this analysis to permit the same sensor to also measure a moment exerted about the contact normal.

In a similar manner, we can sense contacts with the last link of a robot arm by measurement of the joint torques alone. Because the arm acts as a force sensor with varying geometry, the conditioning of the solution can vary enormously, resulting in configuration dependent accuracy. This paper describes an approach to joint-torque-based contact sensing for single contacts on the last link of a manipulator and discusses factors affecting the accuracy of the solution. Experimental results with the approach are reported for a four degree-degree-of-freedom arm (the MIT-WAM arm).

1 Introduction

By definition, manipulation requires contact between the robot and the object being manipulated. Much of current robotic practice relies on precise prepositioning of objects to manipulated, and uses only joint position information to guide the robot to the object position. As we move toward less structured manipulation, robots become more reliant on sensory (and ultimately perceptual) feedback. In many non-contact situations vision can be used to advantage. However, when contact is occurring (such as in grasping and pushing operations) more precise and intrinsically mechanical information is required about the contact. Traditionally, tactile sensors have been considered for this type of contact sensing. Such devices employ surface mounted arrays of force sensitive elements which can be used to reveal contact locations, shapes and normal force components [Siegel 86].

An alternative approach, which we consider here, relies on the use of force (or torque) information to reveal the contact location and force components. Previous work in this type of contact sensing dealt with static force sensing structures which employed strain gauges to measure the net force acting on a finger or link [Salisbury 84, Brock 87, Bicchi 89]. Here, we will address the use of joint torque information as a means for determining where contacts occur on a manipulator's surface. This is similar to our earlier strain-gauge-based approaches except that now the structure of the "sensor" may vary as the manipulator's joint positions vary. [Gordon and Townsend, 89] addressed the problem of locating contacts on the last link of a planar 2-link mechanism and compared the accuracy possible in this force based approach with that of a tactile sensor based approach.

The advantage of this joint-torque-based approach for determining contacts is that it may be employed in any manipulator (or hand) that has the ability to measure joint torques. It does not rely on complex tactile sensor arrays to determine contact locations and it gives more complete information about the actual forces transmitted through a contact. There are limitations to using a force-based approach to contact sensing. The actual shape and pressure distribution of the contact cannot be determined and multiple contacts can not be distinguished.

2 Elementary Considerations

If the forces which act upon a manipulation mechanism sum together so as to yield no net acceleration of the mechanism, it is said to be in a state of static equilibrium. These forces may arise from actuators, body forces (i.e. weight of links) and contacts with the environment.

Although the net forces on the bodies of a system in static equilibrium will sum to zero, there will be internal forces which are necessarily non-zero (i.e. structural stresses, contact forces etc.) If measurements of some of these forces can be made (i.e. by force sensing elements in the structure or actuators) then they can be checked for consistency with the forces expected for a particular configuration of the mechanism and used to deduce information about contact.

As a trivial example, consider a single link supported by an actuator from which torque and angle measurements are available. Assuming we know the link's mass, at any position we may predict the torque necessary to support the link against gravity in the absence of contact. If the joint is servoed to a particular position and the actuator torque deviates appreciably from our model's prediction, then we can deduce that contact of some sort is occurring on the link. Even in this simple case, the contact induced torque may arise from a wide variety of contact situations and without further sensing or contact constraint, details of the exact nature of the contact cannot be inferred. Furthermore, the contact forces would be unobservable if they acted only along an axis parallel to the joint axis and the contact would not be detected.

This simple example points out two important aspects of force-based contact sensing: the number of independent force measurements must be equal to or greater than the number of variables which affect the contact force and the space which the force sensors monitor must fully span the space of forces which can be generated by the contacts. In other words, to solve for the attributes of contact from force measurements we must have enough sensors and they must be placed correctly. The number of sensors required will depend upon the number of unknowns necessary to characterize the problem. In the following analysis we assume that only single contacts occur and are to be sensed. While multiple contacts may be perceived with a sufficient number of sensors, our analysis will address the more fundamental case of single contacts.

3 Contact Characterization and Constraint

A single contact may be characterized by two attributes: the force and torque transmitted through the contact and the location and shape of the contact area. In this paper we shall restrict our selves to the important class of contacts which may be approximated by a single point. With this restriction we can characterize a contact occurring on the link of a manipulator by the *contact wrench*, $\mathbf{W}$, and the contact position, $\mathbf{r}$.

$\mathbf{W}$ is a 6-vector of the three force and three moments which are transmitted through the contact. In the contact frame of reference we represent this wrench as $^C\mathbf{W} = (\mathbf{f}, \mathbf{m})$, with $\mathbf{f}$ and $\mathbf{m}$ equal to the force and moment contact applied through the contact, respectively. $\mathbf{r}$ is the three element vector giving the contact's position. There will be, in general, nine unknowns which we must solve for in order to characterize a contact. Depending upon the type of contact, we must find an appropriate combination of constraints and measurements to apply in order to solve for these nine quantities.

The shape of the link upon which the contact is to be detected establishes the first set of constraints. If the contact occurs on the surface of a link of whose surface equation is given by

$$S(\mathbf{r}) = 0, \tag{1}$$

then the number of constraints imposed will depend upon the type of surface. If S is a general surface then one constraint is imposed upon $\mathbf{r}$. If S is line, then two constraints are imposed upon $\mathbf{r}$. If S is a point then $\mathbf{r}$ is completely determined by the three constraints imposed upon it.

The second set of constraints arises from restrictions on the applied forces due to friction and contact geometry. For the purpose of discussion, we assume that a measurement of the net force and torque acting on the link, expressed in a base coordinate frame O, is given by $^0\mathbf{W} = (\mathbf{F}, \mathbf{M})$ where $\mathbf{F}$ and $\mathbf{M}$ are actual measured quantities. We assume $\mathbf{r}$ is also expressed in frame 0.

If the contact is a *point contact with friction*, the contact wrench must be a pure force (i.e. a zero-pitch wrench) implying that

$$\mathbf{m} \cdot \mathbf{f} = 0. \tag{2}$$

Furthermore, this contact permits no moment to be exerted in the contact frame of reference, implying that

$$\mathbf{M} = \mathbf{r} \times \mathbf{F}. \tag{3}$$

If the contact is a *frictionless point contact* then, in addition to the constraints imposed by eqn(2) and eqn(3) above, the contact force must be normal to the surface at the point of contact:

$$\mathbf{f} \times \nabla S(\mathbf{r}) = 0. \tag{4}$$

Finally, a *soft finger contact* may exert an arbitrary force through the contact and a moment normal to the contact plane such that:

$$\mathbf{m} \times \nabla S(\mathbf{r}) = 0. \tag{5}$$

Eqn(2) imposes one constraint on the values of $\mathbf{f}$ and $\mathbf{m}$. Eqn(3) imposes two constraints on the value of $\mathbf{r}$ as can be see by solving it for $\mathbf{r} = \frac{\mathbf{F} \times \mathbf{M}}{\mathbf{F} \cdot \mathbf{F}} + \lambda \frac{\mathbf{F}}{\|\mathbf{F}\|}$ leaving the free variable (λ). Eqn(4) and eqn(5) both each indicate that a vector is normal to a surface. The constraints imposed by these conditions will depend upon the type of the surface. If $S(\mathbf{r})$ is a general surface then eqn(4) and eqn(5) each impose two constraints. If $S(\mathbf{r})$ is a line, then the equations only impose one constraint each, since the surface normal defined only to lie in a plane normal to the line's direction. If $S(\mathbf{r})$ is a point then eqns(4) and (5) impose no additional constraint, since the surface normal at a point is indeterminant.

If we subtract the number of constraints imposed by link shape and contact type from the original number of variables (nine) we may determine the minimum number of force sensors which must be employed to characterize a particular contact. For example if a point contact with friction occurs on a general body, the constraints are as follows: contact on general body - one constraint, zero pitch wrench - one constraint, no moment at contact - two constraints. The minimum number of sensors is $9 - 1 - 1 - 2 = 5$. Although this is theoretically the minimum, only in special cases can 5 sensors be arranged so that they always span the forces exertable from all contacts on a surface.

Later in this paper will consider the case of a point contact with friction occurring on a link which is slender enough to be considered to be a line. The constraints will be: contact on a line - two constraints, zero-pitch wrench - one constraint, no moment at contact - two constraints. Thus the minimum number of sensors is $9 - 2 - 1 - 2 = 4$. In this case, our sensors will be the joint torque measurements taken from the four joints through which forces applied to the link must pass.

4 Link Contact Problem

The basic problem is illustrated by considering a frictionless point contact on a line occurring in a plane. In this case there are seven constraints, six from the frictionless contact on a line and an additional one from the planar constraint. We consider in Figure 1, the basic problem of determining contact locations on a link from joint torque information where a force f is applied to the last link at a distance L from the last joint. If we assume that the joint angles are known then the vector of joint torques τ may be written as:

$$\tau = (l + eL)f \tag{6}$$

where l is a matrix composed of elements dependent upon the known link lengths and joint angles and e is a matrix composed of elements dependent only on the joint angles. For the simple planar case shown in Figure 1, if we assume the contact force f to be transmitted through a frictionless contact, then f is the scalar magnitude of the contact force and $l = (l_1 \cos(\theta_2) \; 0)^T$ and $e = (1 \; 1)^T$. Assuming f and L are to be determined from a measurement of $\tau = (\tau_1 \; \tau_2)^T$ we can solve eqn (6) to find

$$f = \frac{\tau_1 - \tau_2}{l_1 \cos(\theta_2)} \tag{7}$$

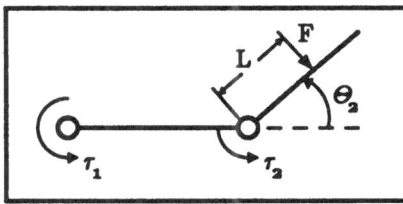

Figure 1: Planar 2 DOF Mechanism contact determination.

and

$$L = \frac{\tau_2 l_1 cos(\theta_2)}{\tau_1 - \tau_2} \tag{8}$$

where θ_1 is assumed to be 0 without loss of generality. An important problem inherent in this approach occurs when $\theta_2 = \pm \pi/2$. In this case the two equations implied by eqn 6 cease to be independent and no solution is possible. Even in the neighborhood of this singularity the equations become ill-conditioned and sensitive to error in torque measurement. This illustrates a problem inherent in solely using joint torque measurement for determining a contact's location and force components. The configuration dependent conditioning renders parts of the workspace inappropriate for force based contact sensing.

One possible remedy to this situation is to utilize a tactile sensor to determine the contact's location and use the joint torque data to determine the magnitude of the contact force(s). With L known, eqn (6) reduces to a pair of linear equations which are over-determined. While either the expression for τ_1 or τ_2 alone could be used to find f, a more sensible approach, assuming equally distributed Gaussian zero mean noise on the torques, would be to use a least squares solution to eqn (6) so that

$$f = (\mathcal{L}_T \mathcal{L})^{-1} \mathcal{L} \tau \tag{9}$$

where $\mathcal{L} = (l + eL)$. This approach would permit us to take advantage of the tactile sensor, which often has poor force resolution, to find the contact location while the joint torque data is used to determine the contact force.

5 Four degree-of-freedom Arm Solution

For the MIT-WAM arm, we will consider single point contacts with friction occurring along the last link and, as a simplifying assumption, treat the last link as a line. This assumption will introduce errors to the extent that the contact force's line of action does not pass through the last link's central axis.

For a point contact with friction, five constraints are imposed on the nine parameters. We can write six independent equations for the components of the contact wrench, four from the magnitudes of the torques measured at each joint, the fifth from the constraint that the line of action of the force must intersect the last link, and the last constraint from the requirement that the wrench be the resultant of a pure force.

Because the arm can assume a wide variety of configurations, it is possible to find arm postures and wrenches where the equations are not independent and solving for

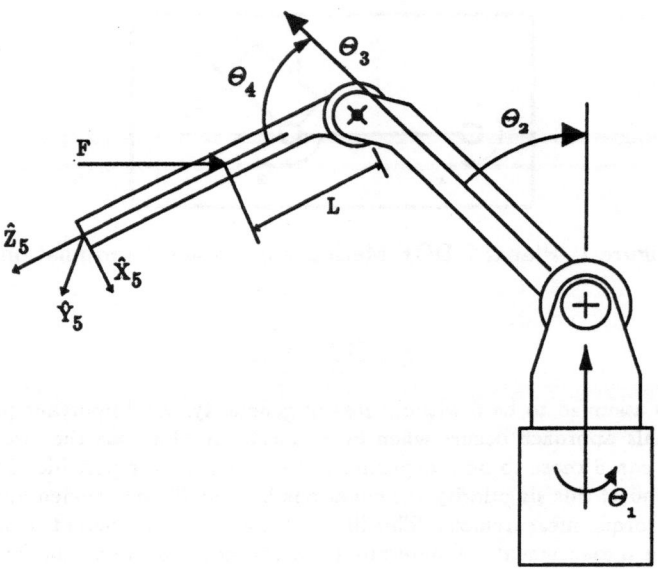

Figure 2: MIT-WAM Kinematics

accurate contact information is impossible. We detail these conditions and define a measure of local conditioning for the problem. Using the conditioning result we will present experimental results for a well conditioned configuration.

5.1 Kinematics and Plücker Coordinates

The kinematics of the MIT-WAM are shown in Figure 2. In this system, $^{i}z_{j}$ is the vector of the jth joint axes expressed in the ith coordinate system. The base system is the 0 frame. The Denavit-Hartenberg parameters for the arm, with this set of frames, are listed in the following table:

coordinate frame number i	link twist (radians) α_{i-1}	link length a_{i-1}	link offset d_i	joint angle θ
1	0	0	0	θ_1
2	$\pi/2$	0	0	θ_2
3	$-\pi/2$	0	L_3	θ_3
4	$\pi/2$	A_3	0	θ_4
5	$-\pi/2$	$-A_4$	0	0

L_3 is the inner link length of 22 inches. The outer link length is 17.5 inches. A_3 and A_4 are the offsets in the manipulator, 1.6 inches and 1.1 inches.

A line is designated by $L = (\hat{L}, \tilde{L})$ where $\hat{L}$ is the normalized direction of the line, and $\tilde{L}$ is the moment of the normalized direction about the origin of the coordinate system.

The virtual product of two lines, $\otimes$, (or a line and a wrench) is

$$L_1 \otimes L_2 = \hat{L}_1 \cdot \tilde{L}_2 + \tilde{L}_1 \cdot \hat{L}_2$$

The virtual product of two lines goes to zero when two lines intersect (including intersections at infinity). The line product of a line and a wrench gives the magnitude of the torque about the line.

5.2 Contact Solution

We consider the problem of determining the line of action and magnitude of a pure force applied along the line through the center of the last link of the MIT-WAM. This force can be written as a general wrench ^0W in the base coordinate system. Four equations of the wrench can be found by taking the virtual product of the wrench with the the lines along each of the four joint axes (^0L_i: $i \epsilon \, 1, 2, 3, 4$) which is equal to the joint torque.

The constraint that the force act along the line of the last link produces the constraint that the virtual product of the wrench with the line through the last link L_5 must equal 0. The final sixth constraint comes from the constraint that the dot product of $\hat{W}$ with $\tilde{W}$ is zero because the wrench is a pure force. The six equations for the six unknown terms of the wrench are:

$$\begin{aligned}
\tau_1 &= L_1 \otimes W \\
\tau_2 &= L_2 \otimes W \\
\tau_3 &= L_3 \otimes W \\
\tau_4 &= L_4 \otimes W \\
0 &= L_5 \otimes W \\
0 &= \hat{W} \cdot \tilde{W}.
\end{aligned} \tag{10}$$

This set of equations can be simplified by expanding L_i in terms of its components in the base frame ($^0\hat{z}_i, {}^0\tilde{z}_i$). In this frame, $\tilde{z}$ is zero for the first three lines. In other words the first three joint torque measure the moment of the wrench which is the moment of the applied force about the base. Performing the expansion results in:

$$\begin{aligned}
\tau_1 &= {}^0\hat{z}_1 \cdot {}^0\tilde{W} \\
\tau_2 &= {}^0\hat{z}_2 \cdot {}^0\tilde{W} \\
\tau_3 &= {}^0\hat{z}_3 \cdot {}^0\tilde{W} \\
\tau_4 &= {}^4\hat{z}_4 \cdot {}^4\tilde{W} + {}^4\tilde{z}_4 \cdot {}^4\hat{W} \\
0 &= {}^4\hat{z}_5 \cdot {}^4\tilde{W} + {}^4\tilde{z}_5 \cdot {}^4\hat{W} \\
0 &= {}^4\hat{W} \cdot {}^4\tilde{W}.
\end{aligned} \tag{11}$$

The solution for W is then found by a two step procedure. First, the first three equations are solved for $^0\tilde{W}$. Then, for increased speed of computation, a rotation is applied to $^0\tilde{W}$ to take it to $^4\tilde{W}$. Finally, given the value of $^4\tilde{W}$, the second set of equations can be solved for $^4\hat{W}$. In the implementation of this procedure, we have symbolically inverted the first three equations.

To determine the location of contact, the plane through the origin of the base frame normal to $^4\hat{W}$ is intersected with the line $^4L_5$. Because the wrench is a pure force $\tilde{W} = r \times \hat{W}$, where r is the vector to the point of application of the force, the point of contact must lie in the plane normal to $\tilde{W}$. The distance to the contact measured along the line from the origin of the fifth frame is then:

$$d = -(^4\hat{W} \cdot {}^4O_5)/(^4\hat{W} \cdot {}^4\hat{z}_5) \tag{12}$$

where $^4O_5$ is the vector to the origin of the fifth frame in the fourth coordinate system.

6 Singularities and Conditioning

Because of the kinematics of the manipulator, the general nonlinear form of the contact problem simplifies to a two step linear problem. We now examine the conditions under which this procedure becomes singular.

The first step, solving for $\tilde{W}$, involves the inversion of a 3×3 matrix consisting of the unit directions of the first three joints. This matrix becomes singular when $\theta_2 = 0$. The conditioning for this matrix is determined by the ratio of the maximum to minimum singular values which is found to be:

$$\frac{1 + |cos(\theta_2)|}{1 - |cos(\theta_2)|} \tag{13}$$

For good performance of the calculation, θ_2 must be near $\pi/2$.

The second step, solving for $\hat{W}$, involves the inversion of a 3×3 matrix involving $\hat{z}_4$ and $\hat{z}_5$ in addition to $\tilde{W}$. To determine singularity we examine when these vectors become linearly dependent. First, by taking the dot product of $\hat{z}_4$ with $\hat{z}_5$ we find that they are always perpendicular and so form a basis against which $\hat{W}$ can be evaluated.

The non-zero length links of the robot always form a plane P. $\hat{z}_4$ is always perpendicular to this plane and $\hat{z}_5$, and all other vectors to points on the robot, lie within P. The applied force, $\hat{W}$, can be decomposed into a vector in the plane of the robot ($\hat{W}_P$) and a force perpendicular to the plane $\hat{W}_N$. Because $\tilde{W}_P$ must be perpendicular to $\hat{W}_P$ and the vector from the origin to the point of contact, which lies in the plane of the robot, $\tilde{W}_P$ must be perpendicular to P or in the direction of $\hat{z}_5$. Therefore, $\tilde{W}_P$ is always linearly dependent on $\hat{z}_5$.

Similarly, $\tilde{W}_N$ can be shown to be perpendicular to $\hat{z}_5$ and so can only depend on $\hat{z}_4$. In order for $\tilde{W}_N$ to be linearly dependent on $\hat{z}_4$, the contact location must be in the direction $^0O_4$. By intersecting this line with the last link, we find the unique contact location which will cause linear dependence to be:

$$L = \frac{A_3A_4S_4 - A_4L_3C_4}{L_3S_4 + A_3C_4}. \tag{14}$$

In summary, in order for there to be a unique solution the force must have a component perpendicular to the plane of the robot, the force must be applied at a

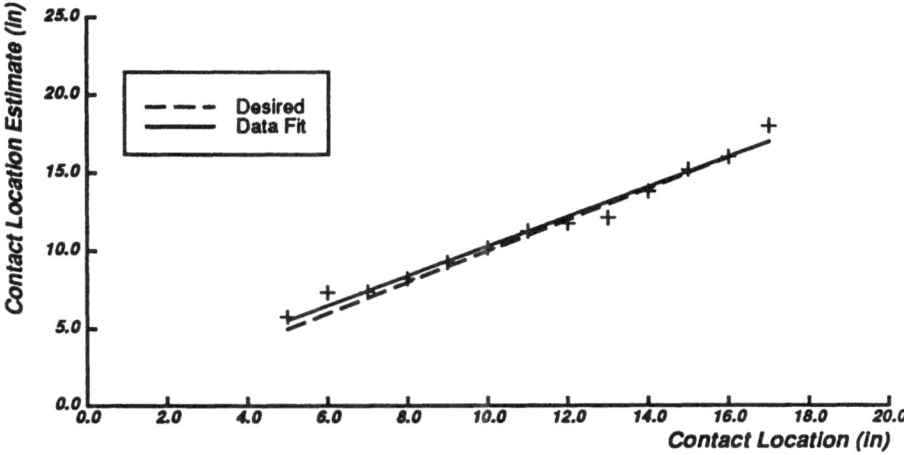

Figure 3: Variation in contact location estimate with actual location

location other than the location given by eqn (14), and θ_2 cannot equal 0. Away from these conditions, a unique solution exists. However, the conditioning of the problem may still be poor. To determine the conditioning, and select experimental locations, we examined the condition number of the differential relationship between the applied wrench and the torques. For a given wrench, the differential relationship is a 6×6 matrix of constants and so a condition number can be determined.

7 Experimental Results

We present results for this method at a well conditioned location. All experiments were done in a static condition.

The joint torques are inferred from the errors in positioning. We assume that the torques acting on the robot consist of the gravity forces, motor ripple, friction, and applied forces. The controller applies a feedforward correction term for both the gravity forces and the ripple. A switching controller is used to decrease the effects of friction. The remaining torque should be due only to the applied forces. Unfortunately, we have found that there remains a significant variation in the torques due to ripple which depends upon the magnitude of the total torque. Since, the gravity torque is greatest at locations with θ_2 near $\pi/2$, causing an increase in the ripple magnitude, this effects seriously degrades the performance of the calculation at the well conditioned locations.

For the well conditioned experiment, we selected joint angles $(0, -1.57, -1.57, -1.57)$. At this location we experimented with varying the contact location along the second link with a constant force of 2 lb in the $^5\hat{y}$ direction (see figure 2). At this location we found numerically that the condition number increased by 0.9 for every inch that the contact location was increased from a starting value of 52.8 at 5 inches. Figure 3 shows the variation in contact location estimate with actual contact location and Figure 4 shows the variation in contact force estimate with contact location.

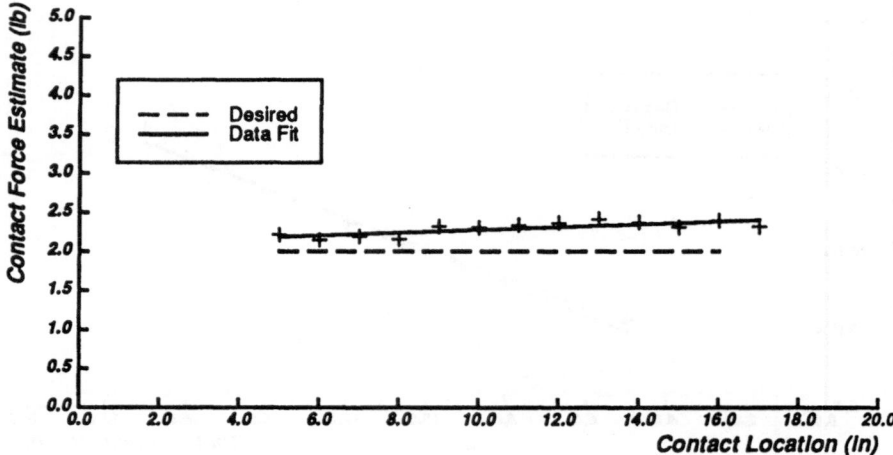

Figure 4: Variation in contact force estimate with contact location

To perform the experiment, we hung a 2lb weight on the last link at the experimental locations. The robot was servoed back to the original location using stiffness and integral control. The torque computed from the errors in positioning and the integral of the position error were used to determine the joint torque. Errors in positioning and the increase in motor torque ripple magnitude with the mean motor torque contribute to the errors.

8 Discussion and Future Improvements

Manipulation requires the sensing and perception of contact information between the robot and the workspace. An internal force sensor can determine the location and magnitude of a contact for certain geometries, providing a robust and simple technique for determining this important information. For the special case of the manipulator joint torques, we have a time varying geometry and thus a time varying conditioning to the problem.

The changes in conditioning with position, and the singularity of the problem to forces in the plane of the robot, imposes restrictions on the utility of this approach for the MIT-WAM with the current configuration of torque sensors. By adding additional torque sensing to the manipulator so as to minimize the condition number over the workspace of positions and forces, we believe the performance of this idea can be greatly improved.

Our procedure also suffers from the inference of motor torques from position errors. Motor ripple and errors, deadband in the motor controllers, errors in calibration of the torque-voltage command curves, and gravity torque calibration errors all contribute to errors in the contact torque measurement. In addition, for higher speed motions, the acceleration of the robot links must be subtracted from the observed torques in order to determine the contact torques creating additional sources of error. Direct sensing and control of motor torque would reduce these problems.

Array tactile sensing suffers from the number of signals that must be analyzed and from an often poor resolution of force. However, a low density tactile array (with only a few elements per square inch) would provide sufficient resolution of location for many tasks. Given a value for the contact location, from an array tactile sensor, determination of the the contact forces from the joint torques becomes a linear problem. This approach remains to be investigated.

9 Acknowledgments

The authors would like to gratefully acknowledge the financial assistance of the Office of Naval Research for support in this research under ONR contracts N00014-86-K-0685 and N00015-85-K-0124.

References

[1] Bicchi, A. "Strumenti e Metodi per il Controllo di Mani per Robot," Phd. Thesis, University of Bologna, 1989.

[2] Brock, D.L., "Enhancing the Dexterity of a Robot Hand Using Controlled Slip," Master's Thesis, Department of Mechanical Engineering, MIT, June, 1987. MIT AI-TR 992.

[3] Eberman, B.S., "Whole Arm Manipulation: Kinematics and Control," Master's Thesis, Dept. of Mechanical Engineering, MIT, Cambridge, MA, January 1989.

[4] Hunt, K.H.,*Kinematic Geometry of Mechanisms*, Oxford University Press, London, England, 1978.

[5] Salisbury, J.K., Townsend, W.T., Eberman, B.S., DiPietro, D.M., "Preliminary Design of a Whole-Arm Manipulation System (WAMS)," Proc. of 1988 IEEE International Conf. on Robotics and Automation, Philadelphia, PA, April 1988.

[6] Salisbury, J.K. "Whole-Arm Manipulation," Proc. of the 4th International Symposium on Robotics Research, Santa Cruz, CA August 1987. Published by the MIT Press, Cambridge MA.

[7] Salisbury, J.K., "Interpretation of Contact Geometries from Force Measurements," *proc. 1st International Symposium on Robotics Research*, Bretton Woods, NH, September 1984, published by the MIT Press, Cambridge MA.

[8] Shimano, B.E., "The Kinematic Design and Force Control of Computer Controlled Manipulators," Ph.D. thesis, Department of Computer Science, Stanford University, March 1978.

[9] Siegel, D.M., "Contact sensors for dexterous robotic hands," Technical Report 900, MIT Artificial Intelligence Laboratory, Cambridge, MA, August 1986.

[10] Townsend, W.T., "The Effect of Transmission Design on Force-Controlled Manipulator Performance," Ph.D. thesis, Department of Mechanical Engineering, Massachusetts Institute of Technology, April 1988.

Using Tactile Data for Real-Time Feedback

Alan D. Berger[1] and Pradeep K. Khosla

Department of Electrical and Computer Engineering
The Robotics Institute
Carnegie-Mellon University
Pittsburgh, PA 15213

Abstract

Object recognition through the use of input from multiple sensors is an important aspect of an autonomous manipulation system. In tactile object recognition, it is necessary to determine the location and orientation of object edges and surfaces. In this paper, we describe a controller that utilizes a Lord LTS 210 tactile sensor in the feedback loop of a manipulator to track edges in real-time. In our control system the data from the tactile sensor is processed in two stages to determine the location of edges. The parameters of these edges are then used to generate a control signal to drive the manipulator. The edge tracker has been implemented on the CMU Direct Drive Arm II system. We describe both theory and experimental implementation of tactile edge detection and an edge tracking controller.

1. Introduction

Object recognition is an important problem in robotics [11, 12, 32], particularly for autonomous manipulation systems. In the most general form, it is the problem of determining the environment from sensory data. The long-term goal of our research is to address the issue of object recognition using tactile data through the process of exploring the environment by moving the sensor. We call this approach dynamic object exploration.

Dynamic object exploration involves scheduling moves of the manipulator based on previously acquired data in order to create a more complete description of the object that is being explored. Thus, there is an interaction between manipulation and sensing. In dynamic exploration, the scheduled move affects the data obtained from the sensor, which in turn affects the next move of the manipulator. The two main steps in dynamic object exploration are: first to create strategies for scheduling manipulator moves; and second, to develop processing algorithms that will extract features of interest from the currently available data.

Researchers have actively addressed issues in both of the above mentioned components of dynamic object exploration and especially so in the context of using tactile data for exploration. Early work in edge and surface tracking is presented in [3, 29]. In these works, the use of a tactile sensor to move about an object to detect features is discussed. Work in object recognition has been discussed in [2, 13, 17, 19, 24, 33, 34, 35]. Some of these groups [13, 24, 36] take the approach of creating tactile subroutines to find particular features of an object. In this approach, a feature is extracted by calling a specific subroutine that moves and takes the appropriate measurements with the sensor.

---

[1]Currently at Xerox, Rochester, NY.

Exploratory procedures were proposed by Klatzky and Lederman in an attempt to model the human haptic system. Recently, Klatzky et al [24] assessed this model's applicability to robotic touch sensing. The model is attractive for robotics since the object recognition task is broken into small, basic tasks. These can be classified as surface properties, structural properties, and functional properties. Each exploratory procedure is a specific operation for the hand that examines one single property. In addition to providing a good set of features for object recognition, they also provide a convenient method for dividing a hierarchical controller into different levels.

Dario, et al [14] have created four tactile subroutines to control a finger through the use of data from their tactile sensor: *Approach, Shape, Texture, Hardness*, and *Thermal*. Subroutine *Approach* manages the finger until it contacts an object. While executing *Approach*, motion of the finger continues until an object is touched with the desired force, or excessive heat or a sharp object is detected. The *Shape* subroutine attempts to discover the shape of an object. It controls the finger in a compliance frame, with no pre-determined positional goal. Instead, the finger is stepped along the tangent plane to the object, maintaining a constant normal force. The *Texture* primitive attempts to determine the texture of the object by moving the sensor and processing the resulting tactile signal. Subroutine *Hardness* measures surface hardness. In this primitive, the dermal sensors are used in a feedback loop to apply a force to the surface. Then, the force on the epidermal sensors is measured. The combination of the force measured by the dermal and epidermal sensors determines the surface hardness. The *thermal* subroutine measures heat flow from the sensor to the object. Dario et al propose to use sequences of these subroutines to perform object recognition.

Stansfield [34] has also implemented a set of exploratory procedures. Primitives for compliance, elasticity, surface normal and contact have been defined. These are simple algorithmic procedures as opposed to the control-theory oriented procedures described above. The *compliance* routine pushes against an object until the force hits a threshold. The distance moved while pushing is a measure of the compliance. The *elasticity* routine pushes against an object, then backs off, and then moves towards the object again until it touches it. This checks to see if and how much the object can be permanently deformed. The *surface normal* routine places the sensor on a surface such that the moments exerted on it are zero. Two points on the sensor are then used to calculate the normal. Texture is determined by measuring the mean and standard deviation of the sensor values. Subroutine *contact* places the sensor in contact with the surface. This is accomplished by moving the sensor until the moments exerted on the sensor are zero. Combinations of these four primitives are combined to detect higher level features such as edges, corners, contours, surfaces, and holes. Recently, recognition of simple objects has been performed by combining tactile exploration with stereo vision [35]. In this multi-sensor system, vision is used to discover edges and taction is used to find surface patches.

Other researchers have taken a completely different approach to object recognition [17, 19, 33]. They have devised algorithms that determine the best path to approach a planar polygonal object such that it can be identified in a small number of discrete moves of the sensor. With this method, the sensor is simply used to determine if an edge of the object exists. In contrast, tactile subroutines use the sensor to track and extract many different properties of an object.

The problem of tactile image processing has received less attention object exploration. Muthukrishnan, et al [30] used a vision-like algorithm to detect edges in a tactile image. It is a three step process. The first is a median filter which replaces the center taxel in a 3 by 3 window by the median of the taxels in that window. The window is sequentially moved over the image such that it is centered over each taxel.

This filtering operation is done iteratively to the image until the image does not change anymore. The purpose of this step is to eliminate noise without destroying the edges. In addition, it tends to remove, or reduce force gradients across the image. The next step is a thresholding operation. A threshold equal to the average taxel (tactile element) value in the image is set. Taxels above this value are set to a constant, and taxels below this value are set to zero. Finally, an edge operator is applied to the image. After this step, edges may be extracted by using a contour following algorithm, or a straight line extracting algorithm such as that proposed by Burns [10].

Fearing and Binford [18] process the signals from their tactile sensor to measure the curvature of an object in two axis. They convolve spatial impulse response of the rubber covering with the strain data from the sensor to recover the surface force distribution. Experiments with objects aligned along the axis of the sensor show curvature estimate errors of less than 10 percent. However, the impulse response is only valid for contact along the axis of the sensor and curvature estimates are not possible if the object is not aligned with the sensor axis.

At Carnegie Mellon, our research group is addressing multi-sensor based manipulation. The goal of our research is to incorporate position, velocity, force, vision, and tactile sensors in the real-time feedback loop to create an autonomous manipulator system. The focus of this paper is to describe the use of a tactile sensor in the real-time feedback loop for edge tracking. We call this system a dynamic edge extractor. Our methodology utilizes a tactile sensor mounted on the end-effector of a manipulator to obtain data about objects. This system consists of both signal processing and control aspects. The role of the signal processing module is to find edges in the data from the tactile sensor, while the control module generates signals to servo the center of the tactile sensor along the edge.

Previous tactile edge detection algorithms do not meet the demanding requirements of high execution speed and robustness in low signal to noise environments. The signal processing algorithm [5] and the Modified Hough Transform [6] that we have developed have been implemented and shown to work experimentally on the CMU Direct-Drive Arm II system. We present the results of an experimental verification of the dynamic edge extractor using the CMU Direct Drive Arm II and a Lord LTS-210 Tactile Array Sensor.

This paper is organized as follows: Section 2 describes the signal processing of tactile images for edge detection and also presents the results of applying our edge detector to tactile data. The edge tracking controller is discussed in Section 3. Sections 4 describes the experimental apparatus used in our edge tracking experiments. Results of these experiments are shown in Section 5. Finally, the paper is summarized in Section 6.

2. Signal Processing

The goal of this section is to describe the signal processing module that is at the heart of incorporating a tactile sensor in the real-time loop for edge and surface exploration [5]. The role of this module is to find edges in the data from the tactile sensor. Our approach views a tactile sensor as a different device compared to a camera, inspite of the similarity in output data format. This is in contrast to earlier approaches to this problem [30]. Further, as our algorithm is coupled to a real-time controller, speed of execution is an important consideration in our methodology and we address the computational requirements of our scheme.

2.1. A Comparsion Between Touch and Vision

We take the approach that a tactile sensor is a different device than a video camera, and thus requires different processing techniques than those used for computer vision. These processing techniques should be based on the physical properties of the tactile sensor and the measured data. In the following paragraphs, we discuss these properties and contrast them with those of a video camera. This comparison should make it clear that an algorithm that works for vision may not always work with a tactile sensor.

Typical tactile sensors are small arrays of force sensing sites. It is this array configuration that makes tactile images appear to be similar to visual images. Tactile sensors have been built using varied technologies such as capacitive, magnetic, piezoresistive, optical, and conductive elastometer. A detailed survey of these technologies and their advantages and disadvantages is beyond the scope of this paper and may be found in [7, 1, 21, 20]. In our experiments, we have used a Lord LTS 210 tactile sensor. In this sensor, the tactile array consists of 10 rows of 16 elements (taxels) each, for a total of 160 sites. These elements are on approximately 1.8mm centers. To protect the sensing elements, and to provide compliance, the sensor is covered with rubber. While the rubber protects the sensor, it also mechanically couples the sensing sites. These characteristics of the Lord LTS 210 tactile sensor create differences between taction and vision. Specifically, these differences lie in the resolution of the images, existence of cross-talk, the noise present, and the structure of the background pixels.

Vision images have substantially higher resolution than most tactile images. One consequence of this is that the discrete approximations used in creating local neighborhood type operators for vision data do not hold very well for tactile images. This is because a 3 by 3 window on a tactile image covers a large physical area, and could correspond to a substantial change in the image. Consequently, the result of applying these operators to tactile images is a significant spreading of edges. Our experiments also show that edge operators that should effectively high pass filter the image sometimes leave a residual low frequency bias in the image.

The mechanically compliant surface of the sensor creates cross-talk in the tactile image. When a force is applied directly to the rubber surface on top of a taxel, it is transmitted through the rubber to neighboring taxels. Although this is somewhat like blurring in vision, it cannot be handled in the same ways. For the Lord LTS-210 (see Section 4 for a description of the sensor) , we found that only the four-connected neighbors are close enough for the cross-talk force to be measurable. This makes the cross-talk effect difficult to remove using traditional techniques. At the same time, since one taxel represents a significant area, it is important to remove this effect.

Background area, or the area of an image where the object of interest is *not* located, is very simple to distinguish in a tactile image. Since background is where the object isn't, it is also where the actual force is zero, assuming no mechanical cross-talk. Remember, though, that the actual force is somewhat different from the measured force due to the presence of mechanical cross-talk. In vision, the background is more difficult to differentiate since it may include other objects at different distances. These background objects will create intensity variations, thus causing edges to appear in the background areas of the image [4].

In summary, the tactile sensor has low resolution and mechanical cross-talk. Also, the background taxels have non-zero force readings due to mechanical cross-talk. Taking these assumptions into account, we devised an edge detecting algorithm that consists of two steps. The first step is an adaptive thresholder to remove cross-talk, and the second consists of an edge detector and the Modified Adaptive

Hough Transform Algorithm (MAHT) that is used to determine the parameters of the edge. For the sake of brevity, we have ommited the details of MAHT and these may be found in [6].

2.2. Adaptive Thresholder

The purpose of this filtering stage in our algorithm is to remove the effects of cross-talk from the tactile image. This operation simplifies the process of detecting edges because with no cross-talk, edges of planar surfaces are the locations where the force goes from a non-zero value to zero. As will be discussed in the following section, the edge detector does not utilize the magnitudes of the taxels. It only uses the state of each taxel, whether it is zero or non-zero. Thus the filter may distort magnitude without negative side effects. In the ensuing discussion of the thresholding algorithm, we show how this property is utilized.

While tactile images have a lot of cross-talk, the cross-talk of interest exists only at the edges of the object's image. In particular, the cross-talk causes taxels that should read a force of zero to have a non-zero value. These taxels always have values that are less then their neighbors that are directly beneath the object. Hence, a thresholder that can choose the appropriate threshold at each taxel may be used to remove the cross-talk. The threshold value is determined by the neighbors of the current taxel, thus making the thresholding an adaptive procedure. The proposed algorithm consists of three basic steps:

1. At each pixel, the force value at each of the four-connected neighbors is checked.

2. If any of these neighbors are large enough to have caused the current pixel to be cross-talk (greater than threshold), the current pixel is set to 0 (no force).

3. Otherwise the pixel is set to a constant.

The threshold for a given taxel value is the minimum value that a neighbor must have in order for the original taxel to be cross-talk. Thus, if all neighbors of a taxel are below threshold, the taxel is considered to be part of the signal. Thresholds obtained with our sensor are summarized in Table 2-1. In this table, the first column is the cross-talk value, and the second column is the smallest value that will cause that cross-talk value. The depicted threshold values are determined through an experimental procedure which is described in [5].

Cross Talk	Minimum Neighbor
2	4
4	10
6	20

Table 2-1: Filter Threshold Values

2.3. Edge Detector

Edge detection in the thresholded tactile image is accomplished very efficiently. We assume that the sensor is placed flush against a surface and that physical edges correspond to a force discontinuity. Thus, the measured force goes to zero on one side of an edge, and is some non-zero value on the other side of the edge. Since the thresholding step filters out the taxels that have non-zero readings purely due to cross-talk, all that remains for the edge detector to do is to find those taxels that are neighbors of taxels with zero values.

Neighbors can be four-connected or eight-connected. We use eight-connected neighbors to ensure that narrow objects at an angle to the sensor are not thinned into a single line. Our edge detection algorithm consists of the following steps:

1. For each taxel, the eight-connected neighbors are checked.

2. If at least one of these neighbors is 0, the current taxel is copied to the edge image.

3. Otherwise the corresponding taxel in the edge image is set to 0.

This algorithm is very fast and minimally distorts the size, shape and position of the object. What does not come out of the algorithm is an estimate of the slope of the edges. Vision researchers have recognized that slope provides a considerable amount of information about the edge [10, 30]. However, since tactile images are small and simple in structure, simply finding the position of edges appears to be sufficient for higher-level processing. In addition, standard vision edge operators that do provide the slope information have a number of undesirable characteristics, as described earlier, when used on tactile images. The slope of an object edge is easily obtained by combining the tactile and position information as the sensor moves along the edge of an object.

2.4. Computational Complexity

Since the proposed processing algorithm is an integral part of a real-time controller, it is essential that it be fast. So, the computational complexity of the algorithm is of interest. A detailed analysis of both our and Muthukrishnan et al's algorithms is presented in [5]. We summarize the results in Table 2-2. In this table K is the number of iterations of the median filter. For a quantitative comparison of the computational requirements, we choose the Lord LTS-210 tactile sensor as an example which has a tactile array of 10 by 16. The numbers depicted in Table 2-2 are for 3 × 3 windows and it shows the relative complexities of the two algorithms utilizing these parameters.

Algorithm	Computations
Berger and Khosla (worst case)	1920
Berger and Khosla (average case)	960
Muthukrishnan et al (best case)	$9129K + 3200$

Table 2-2: Computational Complexities

From our experiments, we found that a conservative estimate of K for most images is about 7. Using this in the formula for Muthukrishnan et al's algorithm, we obtain an estimate of 67,103 calculations in the best case. In the worst case, this is almost 35 times more calculations than our algorithm. On the average, it is 70 times slower.

2.5. Experimental Results

The result of applying our algorithm on two test images obtained with the Lord sensor is shown below. Further examples are shown in [5]. Before we discuss the results, we introduce the original tactile images. They are plotted such that the height in the 3-D picture corresponds to the force at that point on the sensor. The first image, shown in Figure 2-1, is a straight, flat surface placed at an angle to the tactile sensor. The second image, in Figure 2-2, is a small socket for a socket wrench. Note that there is a force gradient across the image from front to back.

480

In the following discussion, the image of the flat surface at an angle will be referred to as Image 1, and the image of the socket as Image 2.

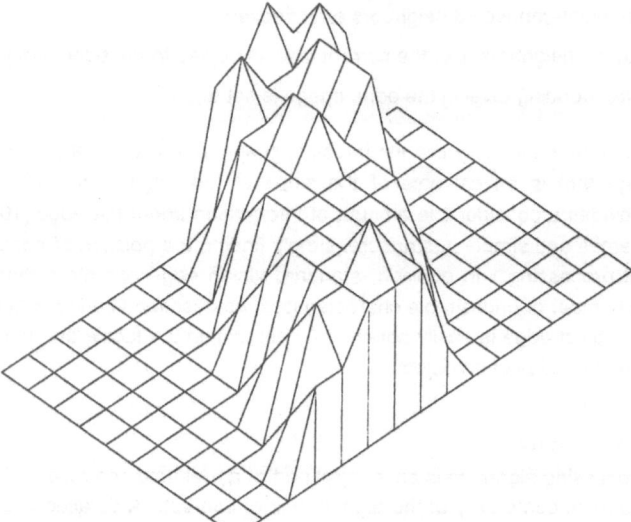

Figure 2-1: Flat Surface (Test Image 1)

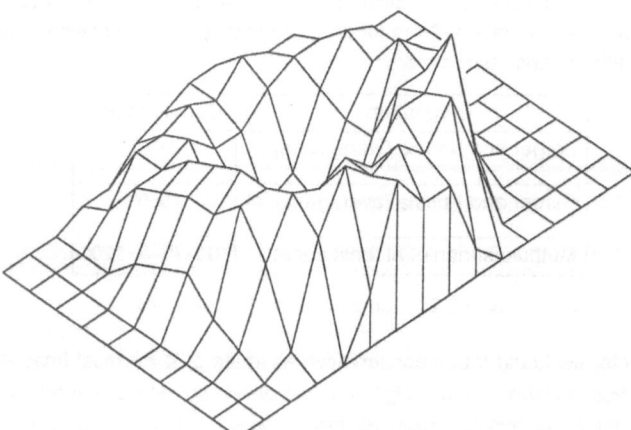

Figure 2-2: Socket (Test Image 2)

2.5.1. Adaptive Thresholder Results

For the first image, the filter performs as desired. The original image (Fig 2-1) has cross-talk at the edges of the ruler. In the filtered image (Fig 2-3) the cross-talk effects are removed, leaving an object four taxels wide and at an angle to the sensor. The result of applying the thresholder to the second test image is shown in Figure 2-4. In this image most, but not all, of the cross-talk from the original image (Fig 2-2) is removed. Further, the shape and size of the object is now clear compared to the original image. These test images show that the adaptive thresholder removes most or all of the cross-talk in an image in addition to maintaining the shape and size of the original object.

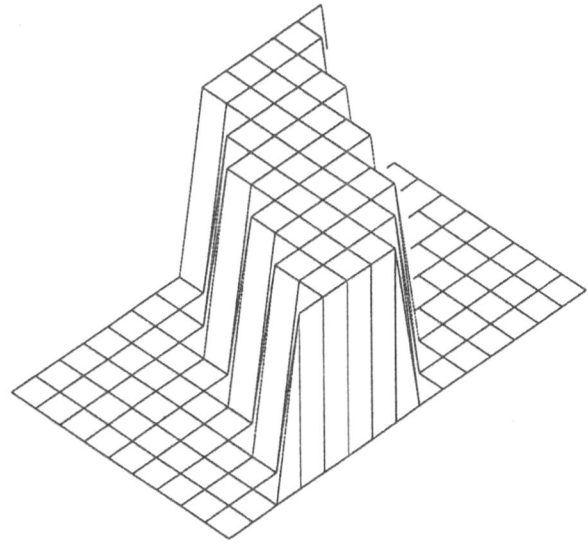

Figure 2-3: Test Image 1 After Filtering (Flat Surface)

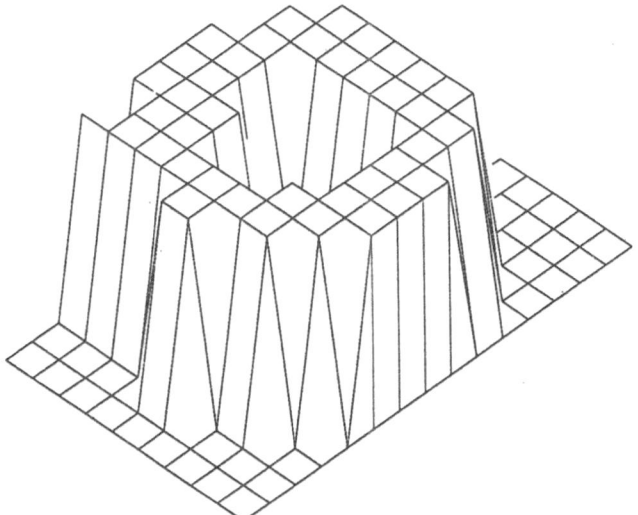

Figure 2-4: Test Image 2 After Filtering (Socket)

2.6. Edge Detection Results

The result of detecting the edges in the thresholded Test Image 1 (Fig 2-3) is displayed in Figure 2-5. This edge image is exactly the desired output of the edge detector; the entire length of both edges of the object are visible, and no cross-talk is present. Figure 2-6 shows the thresholded second image (Fig 2-4) after edge detection. In this figure, all of the object's edges are visible. From these tests, we see that the edge detector produces the correct edges, with minimal cross-talk.

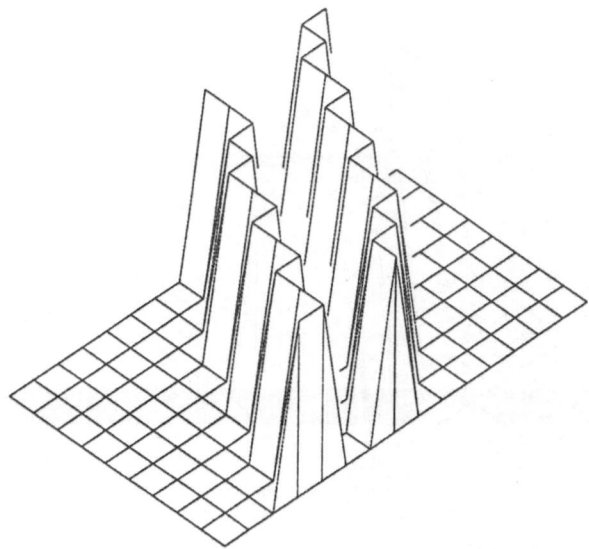

Figure 2-5: Test Image 1 After Edge Detection (Flat Surface)

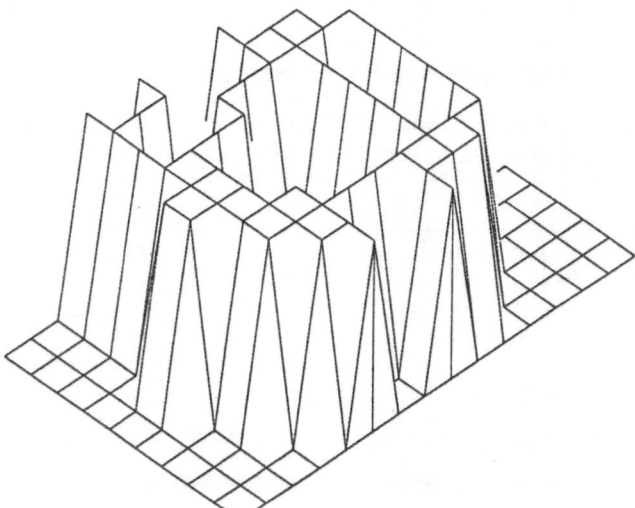

Figure 2-6: Test Image 2 After Edge Detection (Socket)

3. Control

In this section, we discuss the control aspects of dynamic edge extraction [8, 9]. The edge tracker starts on an edge and uses the extracted parameters of the edge to generate control signals to move along that edge. The control scheme is hierarchical, with the tactile controller wrapped around a cartesian space hybrid controller. In the ensuing paragraphs, we describe both the hybrid controller used in our scheme and the tactile controller.

3.1. Hybrid Controller

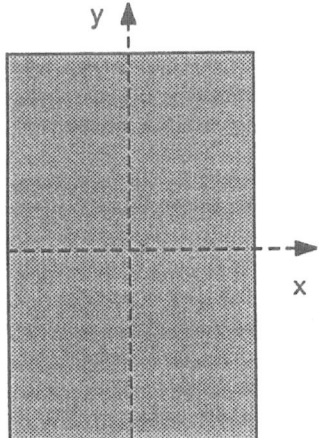

Figure 3-1: Sensor Coordinate Frame

Hybrid force and position control provides the ability to control both forces at the end effector and position of the sensor [27, 31]. Figure 3-1 depicts the sensor frame coordinate axis. The shaded box shows the face of the sensor. The x and y axis lie in the plane of the sensor, and the z axis (not shown) points out of the page. We control the normal force, and torques about the x and y axis of the sensor. Position is controlled in the x–y plane, and about the z axis of the sensor. As discussed in Section 2, normal force control is necessary to ensure that the tactile data is within the middle of the operating range. High forces change the sensor cross-talk characteristics, and low forces result in a very low signal to noise ratio. Controlling torques about the x and y axis of the sensor allows tracking of surfaces that are not flat. Specifically, the desired torques are set to zero in order to place the sensor as flush as possible against the surface. Position control in the plane of the sensor is used because the processed sensor data provides information about the surface in the x–y plane of the sensor. Thus, it is in this plane that we generate position control signals. Further, we control rotation about the z axis of the sensor. The ability to change this angle is not used in the current tracking algorithm. In summary, the hybrid controller commands position in three degrees of freedom, and commands force in the other three. The x and y positions, and the rotation about the z axis of the end effector are controlled. Torques about the x and y axis, and force along the z axis are controlled.

Figure 3-2 is a detailed block diagram of this controller. In this diagram, J is the Jacobian of the manipulator in the end effector frame and the position X_d and velocity $\dot{X}_d$ reference signals are in the world frame. The force reference, F_d is in the end-effector frame. The measured values of position, velocity, and force are subtracted from these signals to produce the error vectors. The error vectors are rotated to the end-effector frame and multiplied by the control gains. This creates the desired effect of controlling motion relative to the end-effector. Subsequently, the control vectors are summed and multiplied by J^T to generate the control torques.

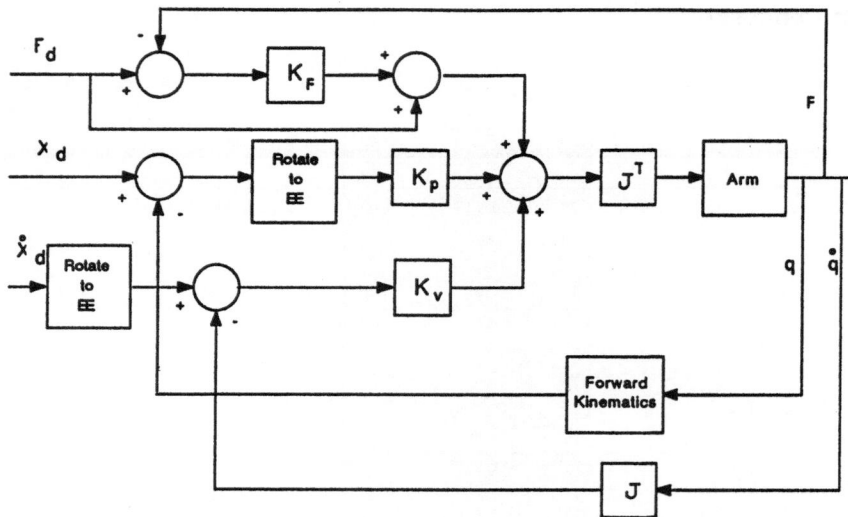

Figure 3-2: Hybrid Controller

3.2. Edge Tracking Controller

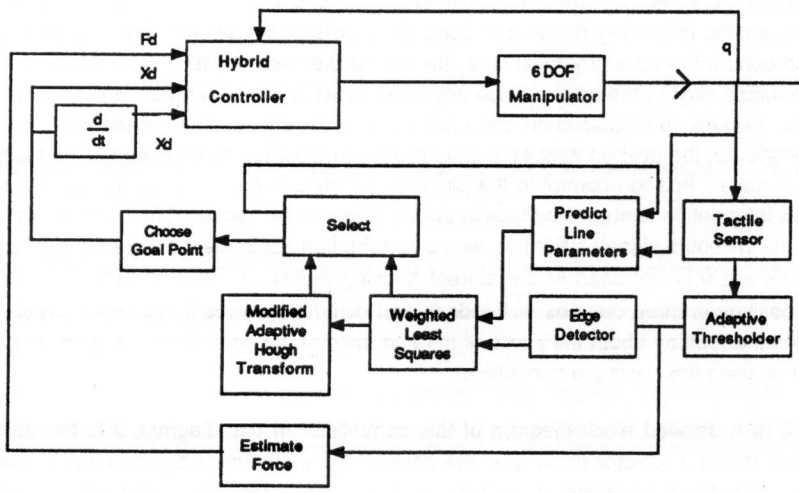

Figure 3-3: Block Diagram of Edge Tracking Controller

The edge tracking controller utilizes the edges extracted from tactile images to generate new reference signals for the hybrid arm controller. Edge tracking is initiated by positioning the tactile sensor on an edge. Through the edge detection technique discussed in the preceding section and the Modified Adaptive Hough Transform, our implementation of the Hough Transform [16, 22], the tracker finds the parameters of the edge. The tracker queries a higher level process to determine which direction to travel, and begins to move the end effector towards that point. After this startup, the edge tracker functions independently of higher level input, utilizing a weighted least squares line fit to the data to determine the current parameters of the line. The Hough Transform is also performed every cycle to determine if any new edges have become visible. Each time through the loop, the robot's reference position is set to be the end point of the line segment on the sensor. Thus, if the edge extends past the end of the sensor, the point where the line intersects the edge of the sensor is selected as the goal point. As the end of a edge becomes visible to the sensor, the reference position is set to the actual end of the edge. In addition, a reference velocity is set such that the end effector should arrive at the reference position at the same time that a new reference position is generated.

Now we consider the controller in detail. Figure 3-3 is a block diagram of the edge tracker. Starting at the upper right corner of the diagram, the tactile sensor is mounted at the end effector of the manipulator. The touch image is first thresholded, with the adaptive thresholder discussed in Section 2. The thresholded image is then input to both the edge detector and the force estimator.

The *Estimate Force* box computes a reference force such that the taxels operate in the middle of their range. Specifically, it takes the thresholded image and counts the number of taxels that are non-zero. The number of non-zero taxels multiplied by the area of each taxel is an estimate of the area of the sensor that is covered by objects. A desired normal force to the sensor may then be generated by dividing the full scale force by the area in contact with the surface. Full scale force is the total force to drive all taxels to mid-range when the entire sensor is on a flat surface.

The thresholded image is passed through the edge detector (discussed in Section 2) and the result is sent to a weighted least squares line parameter estimator, described in detail in [7]. This algorithm is used to estimate the slope and intercept of the edge at the next cycle based on the slope and intercept computed in the previous cycle. All data points in the image are weighted with a gaussian function. The weighting function is oriented such that data points located on the predicted location of the line have the highest weight. As the perpendicular distance of a point to the predicted line increases, the weight of that point decreases. Thus, a line is fit to the data points in the image that are most near the predicted location of the line. An example of the weighting function is shown in Figure 3-4. The coordinate axis are in the plane of the sensor, and the straight line is the a' priori estimate of the line parameters. The gaussians set with their mean along the line show the weights that points at a given position in the tactile image receive. A standard deviation of 0.75 for the gaussians was determined from our experimental work to be the best compromise for both accurate line fitting and adapting of line parameters.

Use of this weighting function allows us to pass all of the data points to the line fitting algorithm without pre-processing to remove points that don't appear to be on the line. After the slope and intercept parameters for the edge are determined, the data points in the image corresponding to that line are removed. Also, the end points of the line are determined at this stage. These computations are the same as those performed by the MAHT, the details of which are discussed in [6]. The point removal and end point computation are part of the *Weighted Least Squares* box in the block diagram.

The weighted least squares computation requires an estimate of the parameters of the previous line segment in the current frame. The *Predict Line Parameters* box in the diagram performs this operation. The end effector will have translated and possibly rotated since the previous set of line parameters were

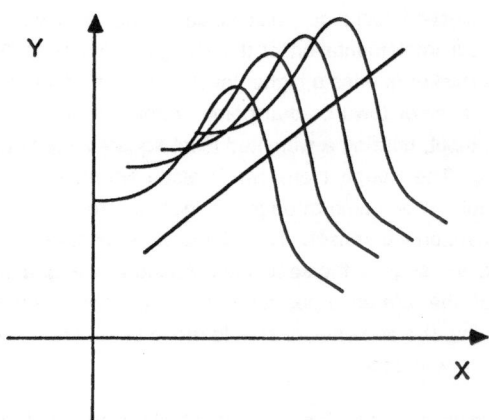

Figure 3-4: Gaussian Weighting Function

determined. Thus the slope and intercept stored from the previous cycle are updated to reflect this change. The predictor calculates the parameters of the current line based on the parameters of the previous line, the position of the end effector in the previous cycle, and the current position.

The remaining image is passed on to the Modified Adaptive Hough Transform (MAHT). The MAHT extracts multiple lines of arbitrary slope from input data that has low signal to noise ratio. Further details on this algorithm are presented in [5]. Any line segments other than the one being currently tracked will be detected by this algorithm. If there are no edges remaining in the image, the transform will exit, and the parameters and end points determined by weighted least squares are passed through the *Selector*. If there are new line segments, the higher level process will be informed. At this point a new line segment may be selected for tracking. When a new segment is selected, the *Selector* passes the parameters determined by MAHT to the predictor, and the end points determined by MAHT to the *Choose Goal Point* process.

Finally, *Choose Goal Point* determines which of the two end points of the segment should be set as the new reference position for the robot. The choice is made such that the robot continues to move in the same direction that it has been moving. The reference velocity is set to the distance to the new goal position divided by the edge tracking sampling period.

3.3. Discussion

The design of the edge tracking controller has several desirable properties. Specifically, it handles the ends of segments smoothly, it can track curves in addition to straight lines, and the design is tolerant of any size sensor and data rate. In the following paragraphs, we discuss each of these points in some detail.

As the tactile sensor approaches the end of a line segment, the controller slows the arm down. When the center of the sensor reaches the end point, the arm stops. This action is a natural consequence of the way that new reference points for the hybrid controller are generated. In each cycle, the visible end of the line segment is chosen as the new reference point. Hence, before the end of the line is under the sensor, the point where the line leaves the sensor is the reference point. However, as the end point becomes visible, the controller chooses that point as the goal. This new goal point is closer to the center of the sensor than the edge of the sensor, and as a result, the velocity of the arm decreases. As the center of the sensor gets closer to the end of the segment, the arm continues to slow down, until it stops when the segment end is below the center of the sensor. This allows the arm to accurately position itself at the end of the segment, and provides an easy way to detect the end of a line segment.

Gradual curves appear as piecewise straight lines to the tactile sensor, allowing it to track them. During each cycle, new line parameters are fit to the segment of the curve that is under the sensor by the weighted least squares method. The parameters that control the weighting are the line parameters from the previous cycle. The old parameters will not be correct, as both the slope and intercept of the new section of the curve may be different. However, the old values are close enough to the correct ones that the weighting function will still be in approximately the correct location, and weighted least squares will extract the correct new parameters. Thus, the procedure of adapting the line parameters each cycle allows the system to track curves in addition to straight lines.

The sampling rate of the sensor only affects the maximum tracking velocity. As discussed above, the reference point for the hybrid controller is set to the intersection of the line with the edge of the sensor. Further, the reference velocity is set to the length of the new reference trajectory divided by the cycle time of the controller, T. As the sampling rate of the sensor decreases, T increases. Thus, desired velocities are reduced, and the reference points are placed closer together. In this scheme, there is no danger of the manipulator traveling faster than the arrival of new data.

4. Experimental Apparatus

In this section, we describe the hardware and software architecture used in our laboratory to implement the tactile edge follower. The hardware consists of the CMU DD Arm II, control computers, and a Lord LTS 210 Tactile Array Sensor. The tactile control software is run on a Sun 3 computer.

4.1. Control Computers

The hardware of the DD Arm II control system consists of four integral components: the Sun workstation, the Motorola M68000 microcomputer, the Marinco processors [26] and the TMS-320 microprocessor-based individual joint controllers. All of the computers, with the exception of the Sun are connected through a common Multibus backplane, as shown in Figure 4-1. The Eurocard Sun 3 is connected to the backplane through a serial line and interface card, operating at 4800 Baud. A simple packet based communications scheme [15] between the M68000 Coordinating Processor and the Sun operates over this serial connection.

Previous control work included the development of the customized Newton-Euler equations for the CMU DD Arm II which achieved a computation time of 1 *ms* on the Marinco processor. The details of the customized algorithm, hardware configuration and the numerical values of the dynamics parameters are presented in [23]. For tactile sensing, we execute a cartesian position controller on one of the Marinco boards, while gravity compensation torques are computed on the other Marinco. The edge tracking controller runs on the Sun. At each cycle, new reference positions are sent from the Sun to the 68000, and the current position is transmitted from the 68000 to the Sun.

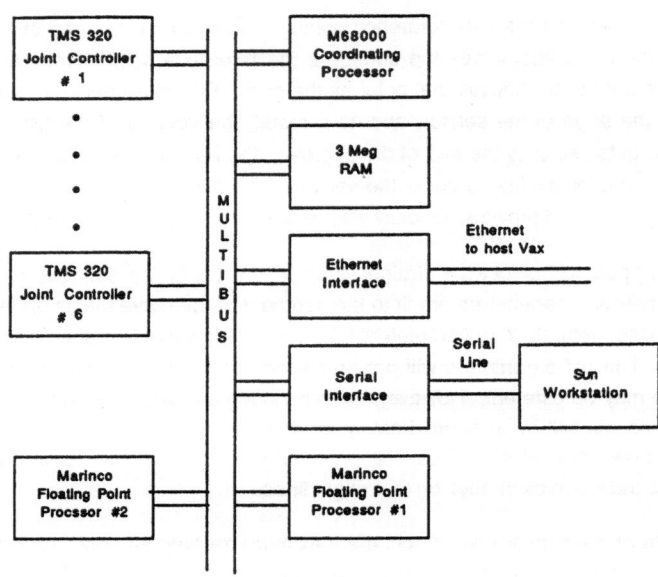

Figure 4-1: System Computational Architecture

4.2. Lord LTS 210 Tactile Array Sensor

To perform our taction experiments, we added a Lord LTS-210 tactile array sensor to the DD Arm II system. This sensor is mounted at the end-effector of the robot. The sensor is an array of 10×16 elements spaced on $1.8mm$ centers [25]. Each sensing site is a small plunger mounted such that as it is depressed, it blocks the light path between a LED and a photodiode [28]. Sixteen different increments in deflection may be read for each site in the sensor. A sheet of rubber protects the top surface of the sensor, but also mechanically couples the sensing sites. The sensor is interfaced to the Sun 3 through a 9600 Baud serial line.

4.3. Sun Software Architecture

The tactile controller software must provide the 68000 with new reference positions, request and obtain the current position of the manipulator, request and obtain the tactile data, and process the tactile data. Careful interleaving of these operations minimizes the time for a single cycle through the tactile control loop. In order to allow for variations in the execution time of the control calculations, a separate check process that ensures that a new reference position is sent to the 68000 at least once every T seconds. T is the nominal sampling rate of the system. Thus if the control calculations take longer than what is considered normal (for example, when the MAHT finds multiple line segments), this check process will send the previous reference position, with a zero velocity vector to the robot. This ensures that the robot does not wander past the visible horizon that its previous goal position was calculated for.

Figure 4-2 shows the relationship between the various operations. Note that all times presented in the following discussion are approximate due both the methods used to time the system and the nature of the Unix kernel. The entities displayed are the system clock process (running at 33.3Hz), transmission of the

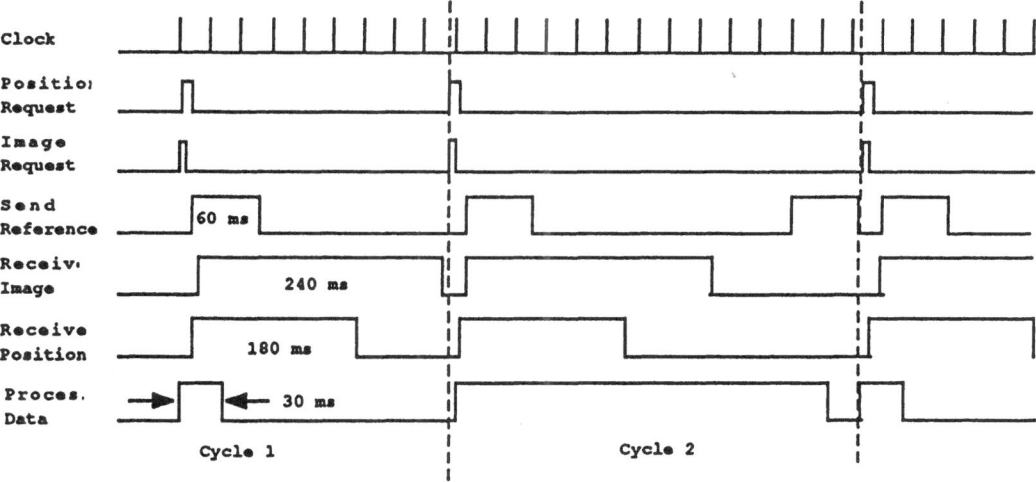

Figure 4-2: Software Timing Diagram

position request packet to the 68000, the transmission of the image request packet to the tactile sensor, transmission of the new reference position and velocity to the 68000, transmission of the tactile image from the array sensor, transmission of the robot's current position to the Sun, and the control calculations process. In the figure, Cycle 1 represents a "normal" cycle. This is a cycle where the control calculations don't run over the sampling period. Cycle 2 is a cycle where the control calculations do run over the allotted time. Below, we examine both scenarios in detail.

We begin with Cycle 1, which represents what occurs during each cycle of the edge tracker while the system is actually moving along an edge. Starting at the top of the diagram, the position request and image request packets are sent at the beginning of the cycle. These requests are for data that will be processed in the next cycle. The control calculations also run starting at the beginning of the cycle. After the position request packet is transmitted, the reference position calculated on the previous cycle is also transmitted. Meanwhile, the new end effector position and the tactile image are received by the Sun. The tactile image is the bottleneck in the system, with delay time from request to complete transmission of 240*ms*. This is longer than the time expected for transmition of 160 bytes at 9600 baud, so there must be some delay built into the tactile sensor. Since the tactile image delay time dominates the system, the minimum sampling period of the system, T is 240 *ms*, for a sampling rate of 4.2*Hz*.

Cycle 2 begins the same as Cycle 1, with position and image requests going out, and tactile images and new end effector positions being received. However, the image processing takes approximately 370 *ms* this time. After 300*ms*, the check process wakes up and forces a send of the previous reference position with a zero velocity vector. During transmission, the control calculations are completed. Thus at the end of data transmission, a new cycle begins.

5. Experimental Results with the CMU Direct Drive Arm II

In the ensuing paragraphs, we present the results of two different edge tracking experiments along with some observations about the use of a tactile sensor for edge tracking. First, we discuss a change in the thresholds used by the adaptive thresholder, and our strategy for orienting the tactile sensor for edge tracking. Then, we show the trajectory followed by the manipulator while tracking both straight and curved edges. The straight edge experiment allows us to view the accuracy of the tracking system, while the curved edge experiment shows the line parameter adaptation capability.

5.1. Observations

Our experiments to determine the threshold values for the adaptive thresholder show that taxel values of 2 are noise if there is a four connected neighbor of value 4 or greater [5]. During early edge tracking experiments, however, we found that after the sensor is moved over a surface for a distance of a few centimeters random 2's appear in the image. Thus, motion of the sensor against a surface makes force values of 2 unreliable. To compensate for this phenomena, the adaptive thresholder parameters were adjusted to always filter out twos regardless of the force on neighbors. No side effects in system capability are produced by the elimination of 2 as a usable force value. As discussed in Section 3, forces on the sensor are maintained above 2 for best utilization of the sensor.

We track edges with the sensor oriented such that it only contacts the edge, and not the surfaces of the object. Although the algorithms presented in the previous sections are general and may be used to track edges with the sensor in contact with the surface, we found that the friction between the object and the sensor is very high when the system is used in this mode. With our approach, two effects combine to reduce the friction. First, less area is in contact with the surface since the sensor is only contacting a line, instead of a plane. Second, lower normal force is required. The normal force necessary to operate the sensor in the mid-region is proportional to the area of the sensor in contact with the surface. Each taxel in contact with the surface must experience a force large enough to keep it in operating range. Thus the normal force that must be exerted by the manipulator is approximately the product of the force each taxel requires and the number of active taxels. Lower forces on the sensor not only help to reduce the requirements placed on the manipulator, but also reduce wear on the sensor.

5.2. Edge Tracking

Figure 5-1 shows the result of tracking a straight edge on a metal box. In each cycle, the position of the end effector was recorded. Dots in the graph correspond to these end effector positions. Thus, the graph shows the distance between samples in addition to the robot's trajectory. The dashed line in the figure is an approximation of the location of the actual edge and is included for reference. In this experiment, the tactile sensor was oriented such that the long dimension (the 16 rows) was parallel to the direction of travel. The end effector traced a path starting at (0.47, 0.1) and ending at (0.72, 0.26), with an average speed of 5 mm/sec.

The plot (Figure 5-1) shows the typical characteristics of our edge tracking system. First, we note that its accuracy is acceptable and the errors are within the width of the lines in this plot. Figure 5-2 is a magnified view of the region with the largest errors, from $x = 0.525$ to $x = 0.565$ $meters$. Here, we observe that the position errors are approximately 1mm. Remember that the tactile sensor resolution is 1.8mm, and the reference line is only an approximation to the actual edge. Thus, we conclude that the position error is well within expectations for the system.

491

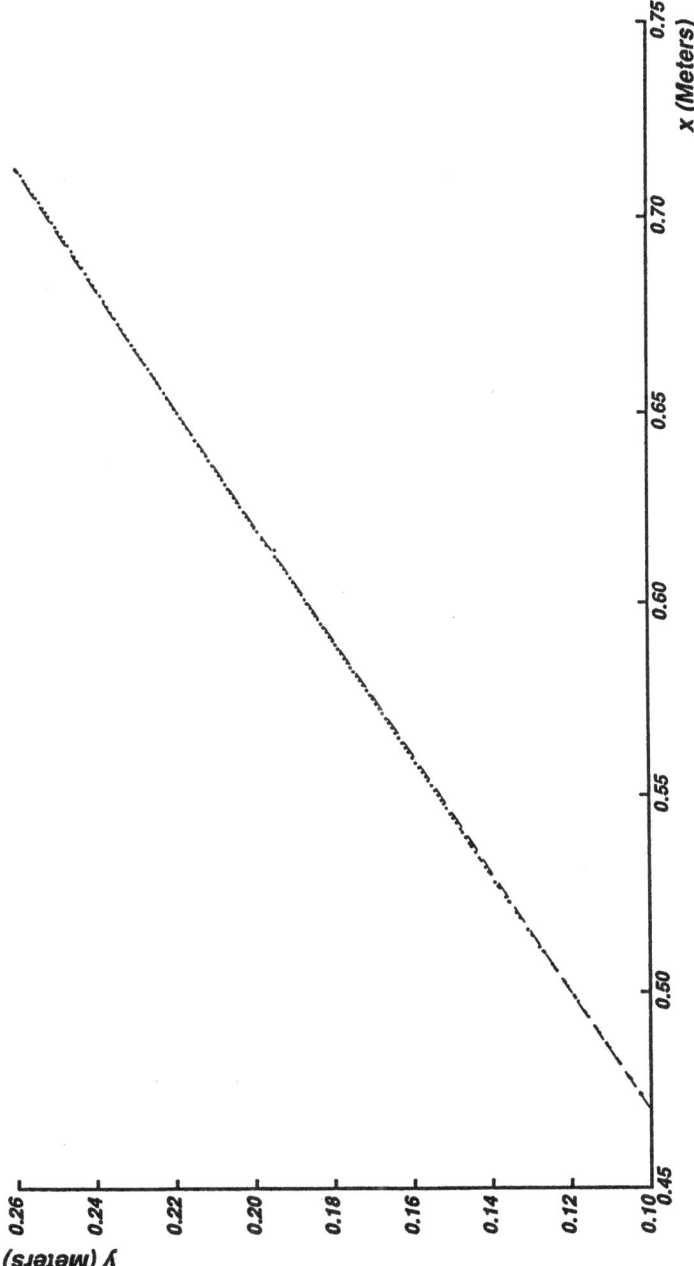

Figure 5-1: Straight Edge Tracking

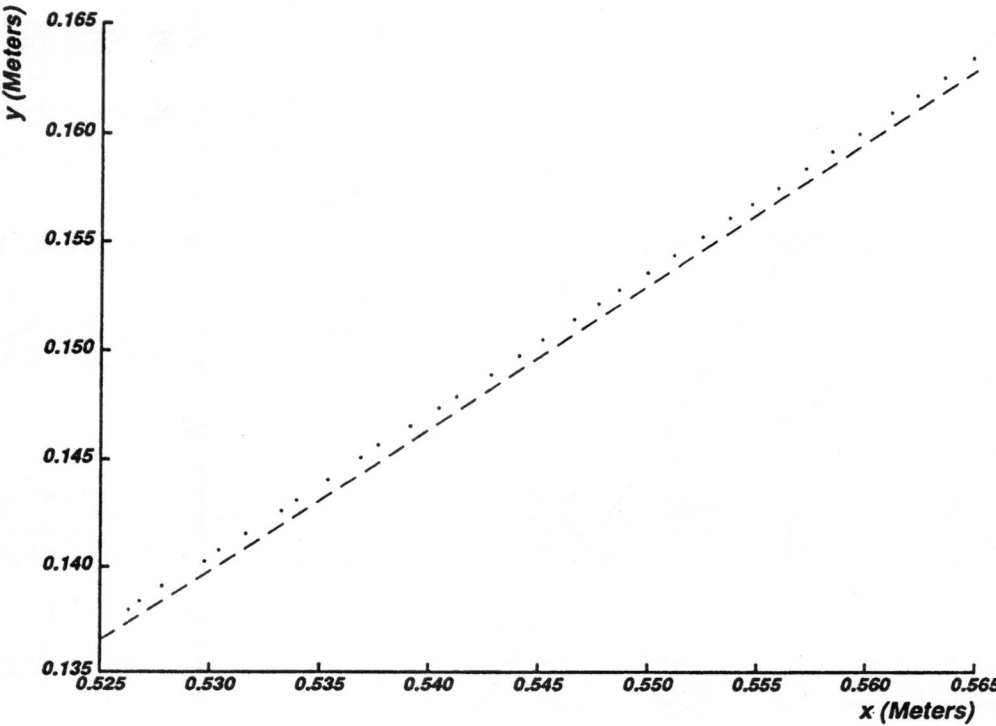

Figure 5-2: Straight Edge Tracking - Exploded View

We return to the picture of the entire trajectory (Figure 5-1) to discuss the start and end points. At the start (0.47, 0.1), the velocity does not appear to be as consistent as the during the remainder of the trajectory. This is to be expected as the end effector moves to place the center of the tactile array on the line, and the estimated line parameters adapt to the edge. Further, at the beginning of the line the manipulator is at rest. Thus, the first move request is a step input to the cartesian controller. Our current controller is somewhat under-damped and requires time to reach steady motion. On this particular run, the motion of the sensor smoothed out after 4 or 5 *cm*. At the very end of the trajectory, the dots become close together, indicating that the end effector slowed down. This is precisely the action designed into the system. The visible end of the line segment is always chosen as the new goal point. Thus, as the end of an edge comes into view, the commanded trajectory length, and end effector velocity decreases.

The next experiment involved tracking a *S* shaped object. Figure 5-3 shows the results when the sensor is started with the long dimension approximately oriented at a positive 45 degree angle to the *x* axis. Tracking follows a smooth arc beginning at (0.45, -0.14) and ending at (0.93, 0.21). The primary result from this experiment is the verification of the line parameter adaptation. The edge tracker always attempts to follow a straight line. Curves are taken to be piecewise linear, with line parameters changing slightly each cycle. The motion shown in Figure 5-3 clearly shows that line parameters are adapting properly. As with the straight line, we note a small amount of oscillation at the beginning of the trajectory, and a decrease in velocity at the end.

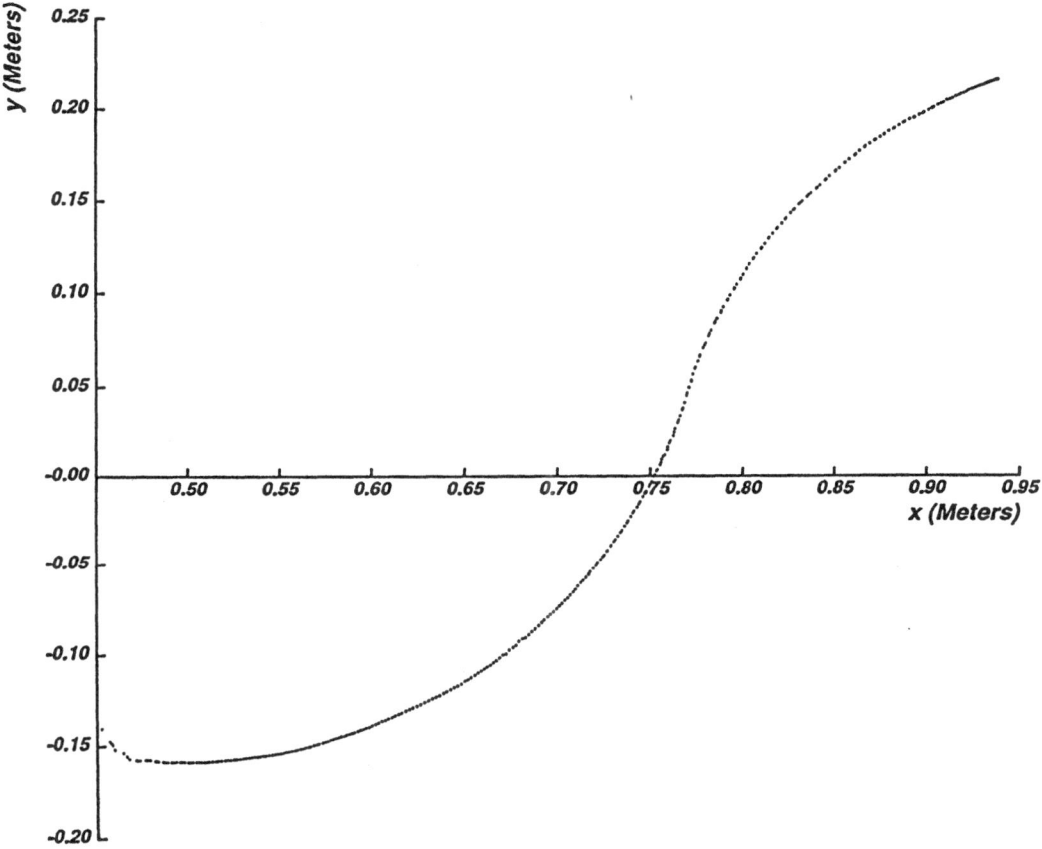

Figure 5-3: *S* Curve Tracking

6. Summary

In this paper we have presented the implementation of a strategy to use a tactile sensor in the real-time feedback loop of a robot controller. There are two main components to our dynamic edge tracker: tactile signal processing and control. We base our tactile signal processing algorithms on the physical properties of the sensor. Thus, we accomplish edge detection by a two step process that first filters mechanical cross-talk noise and second finds edges by looking for transitions from non-zero to zero force. The controller uses detected line segments to generate reference signals for a manipulator. During each cycle of the edge tracker, the estimated parameters of the line are transformed to the current frame. These parameters are used to position a weighting function for a weighted least squares estimate of the new line. Performing this procedure every time through the control loop allows the line parameters to continuously adapt. Continuous adaptation of the parameters, in turn, allows the system to track curved objects in addition to straight objects.

References

[1]   R. Agrawal.
        An Overview of Tactile Sensing.
        University of Michigan, Ann Arbor, 1986.

[2]   P. Allen.
        Surface Descriptions from Vision and Touch.
        Proc IEEE Conf on Robotics and Automation :394 - 397, 1984.

[3]   R. Bajcsy.
        What can we learn from one finger experiments?
        In M. Brady and R. Paul (editor), *The First International Symposium on Robotics Research.* MIT
            Press, Cambridge, MA, 1984.

[4]   D.H. Ballard, C.M. Brown.
        Computer Vision.
        Prentice Hall, Englewood Cliffs, NJ, 1982.

[5]   A.D. Berger, and P.K. Khosla.
        Edge Detection for Tactile Sensing.
        *Proc. SPIE's Cambridge Symposium on Optical and Optoelectronic Engineering: Advances in
            Intelligent Robotics Systems* , November, 1988.

[6]   A.D. Berger, and P.K. Khosla.
        The Modified Adaptive Hough Transform (MAHT).
        Technical Report, Dept. of Electrical and Computer Engineering, Carnegie Mellon University,
            1988.

[7]   Alan D. Berger.
        On Using a Tactile Sensor for Real-Time Feature Extraction.
        Master's thesis, Carnegie-Mellon University, December, 1988.

[8]   A.D. Berger and P.K. Khosla.
        Real-Time Edge Tracking Using a Tactile Sensor.
        Proc. NASA Conference on Space Telerobotics , February, 1989.

[9]   A.D. Berger and P.K. Khosla.
        Real-Time Edge Tracking Using a Tactile Sensor.
        International Journal of Robotics Research, MIT Press , (in press), .

[10]   J.B. Burns, A.R. Hanson, and E.M. Riseman.
        Extracting Straight Lines.
        IEEE Trans. on Pattern Analysis and Machine Intelligence 8(4):425 - 455, July, 1986.

[11]   H. Chang, K. Ikeuchi, and T. Kanade.
        Model-Based Vision System by Object-Oriented Programming.
        Technical Report CMU-RI-TR-88-3, Carnegie Mellon University Robotics Institute, February,
            1988.

[12]   R.T. Chin, C.R. Dyer.
        Model-Based Recognition in Robot Vision.
        ACM Computing Surveys 18(1):67 - 90, March, 1986.

[13]   P. Dario, M. Bergamasco, D. Femi, A. Fiorillo, A. Vaccarelli.
        Tactile Perception in Unstructured Environments: A Case Study for Rehabilitative Robotics
            Applications.
        Proc IEEE Conf on Robotics and Automation 3:2047 - 2054, 1987.

[14]   P. Dario, and G. Buttazzo.
        An Anthropomorphic Robot Finger for Investigating Artificial Tactile Perception.
        International Journal of Robotics Research 6(3):25 - 48, Fall, 1987.

[15]   D.R. Doll.
       Data Communications: Facilities, Networks and Systems Design.
       J. Wiley and Sons, New York, 1978.

[16]   Duda and Hart.
       Pattern Classification and Scene Analysis.
       John Wiley & Sons, New York, 1973.

[17]   R.E. Ellis.
       Acquiring Tactile Data for the Recognition of Planar Objects.
       Proc IEEE Conf on robotics and Automation 3:1799 - 1805, 1987.

[18]   R.S. Fearing, and T.O. Binford.
       Using a Cylindrical Tactile Sensor for Determining Curvature.
       Proc IEEE Conf on Robotics and Automation :765 - 771, 1988.

[19]   W.E.L. Grimson, and T. Lozano-Perez.
       Model-Based Recognition and Localization from Tactile Data.
       Proc IEEE Conf on Robotics and Automation :248 - 255, 1984.

[20]   L.D. Harmon.
       Automated Tactile Sensing.
       IJRR 1(2):3 - 30, Summer, 1982.

[21]   L.D. Harmon.
       Automated Touch Sensing: A Brief Perspective and Several New Approaches.
       Proc. IEEE Int. Conf. on Robotics :326-331, 1984.

[22]   J. Illingworth, and J. Kittler.
       The Adaptive Hough Transform.
       IEEE PAMI 9(5):690 - 698, September, 1987.

[23]   P.K. Khosla and T. Kanade.
       Experimental Evaluation of the Feedforward Compensation and Computed-Torque Control
           Schemes.
       In Stear, E. B. (editor), *Proceedings of the 1986 ACC.* AAAC, Seattle, WA, June 18-20, 1986.

[24]   R.L. Klatzky, R. Bajcsy, and S.J. Lederman.
       Object Exploration in One and Two Fingered Robots.
       Proc IEEE Conf on Robotics and Automation 3:1806 - 1809, 1987.

[25]   Lord LTS-210 Tactile Sensor.
       Installation and Operations Manual.
       Lord Corp., Cary, North Carolina, 1987.

[26]   Marinco APB-3024M Array Processor Board.
       Reference Manual.
       Marinco, Inc., 3878-A Ruffin Road, San Diego, CA 92123, 1983.

[27]   M.T. Mason.
       Compliance and Force Control for Computer Controlled Manipulators.
       IEEE Trans. on Systems, Man and Cybernetics 11(6), 1981.

[28]   G.A. McAlpine.
       Tactile Sensing.
       Sensors :7 - 16, April, 1986.

[29]   Montana, D. J.
       Tactile Sensing and Kinematics of Contact.
       PhD thesis, Division of Applied Sciences, Harvard University, August, 1986.

[30]   C. Muthukrishnan, D. Smith, D. Myers, J. Rebman, and A. Koivo.
       Edge Detection In Tactile Images.
       Proc. IEEE Conf. on Robotics and Automation 3:1500 - 1505, April, 1987.

[31]   M.H. Raibert and J.J. Craig.
        Hybrid Position/Force Control of Manipulators.
        J. Dynamic Systems, Measurement, Control , 1981.

[32]   A.A.G. Requicha.
        Representations for Rigid Solids: Theory, Methods, and Systems.
        ACM Computing Surveys 12(4):437 - 464, December, 1980.

[33]   J.L. Schneiter.
        An Objective Tactile Sensing Strategy for Object Recognition and Localization.
        Proc. IEEE Int'l Conf. on Robotics and Automation :1262-1267, April, 1986.

[34]   S.A. Stansfield.
        Primitives, Features, and Exploratory Procedures: Building a Robot Tactile Perception System.
        Proc IEEE Conf on Robotics and Automation 2:1274 - 1279, 1986.

[35]   S. A. Stansfield.
        Visually Aided Tactile Exploration.
        Proc. IEEE Int'l Conf. on Robotics and Automation :1487-1492, April, 1987.

[36]   S.A. Stansfield.
        Representing Generic Objects for Exploration and Recognition.
        Proc IEEE Conf on Robotics and Automation 2:1090 - 1095, 1988.

Approximate Calculation of Robot Inverse Kinematics Applied to Arc Welding

Jadran Lenarčič and Andreja Košutnik
Jožef Stefan Institute, University of Edvard Kardelj
Jamova 39, 61111 Ljubljana, Yugoslavia

Abstract

The paper presents a new approach to the inverse kinematics calculation which consists in the approximation of Cartesian orientations of the robot end effector. The approach can be well used in variuos practical applications, such as arc welding, and is extremely effective when applied to robot manipulators which do not have a closed–form solution to the inverse kinematics problem. Several numerical experiments are described that show the validity of the proposed approach.

1 Introduction

In many industrial applications of robot manipulators the end effector is programmed to move along the prescribed Cartesian paths. To obtain the joint trajectories which cause such movements, the inverse kinematics problem must be solved. As is well known [1], the solution to the inverse kinematics problem, which includes the transformation of Cartesian coordinates into joint coordinates, can in general be obtained by numerical procedures, since the closed–form solution exists only for special robot mechanisms.

Several authors have studied and reported various numerical methods for the calculation of inverse kinematics. The first of the commonly used methods is based on Newton's integration formula. The main principle of the method is to solve the system of kinematic equations expressed in a differential form. In this method the computational accuracy cannot be controlled and it fails close to the Jacobian singularities. Furusho [2] presented an overview of various modifications which decrease the number of arithmetical operations and increase the accuracy in comparison with the original method. In the second group there is a wide variety of iterative numerical methods. They use iterative procedures in order to decrease the error but at the same time, by repeating the calculation in several steps, they increase the number of arithmetical operations. A typical representative of these methods, based on the Newton–Raphson iterative procedure, was introduced by Goldenberg [3]. Another method that uses different gradient procedures was proposed by Lenarčič [4]. The inverse kinematics can also be calculated by constructing a simple closed–loop stable dynamic system, whose input is the desired Cartesian trajectory and whose outputs are the joint trajectories. The class of first–order and second–order schemes based on the closed–loop formulation of the inverse kinematics problem was surveyed by Siciliano [5]. In recent years, the inverse kinematics calculation was intensively investigated and various new papers describe more and more effective methods or modifications [6,7]. However, it is a common property of all the mentioned numerical methods, that they still cannot be used in real–time control of industrial robots, since they need too much numerical computation or they cannot handle problems of multiple solutions, singularities, and similar drawbacks.

Engineers have managed this problem by designing robot manipulators that have a closed–form solution to the inverse kinematics. For instance, very usually they design robot manipulators with a spherical wrist which have a closed–form solution without regard to the kinematic structure of the major linkage (first three links) [1]. We believe, however, that this fact represents a significant limitation in the design of robot manipulators and their applications.

Let us recall a rather different approach to the inverse kinematics calculation introduced by Milenković [8] and Lumelsky [9]. They reported a method whose main principle is to divide the robot linkage into such parts (major and minor linkage) that, first, a closed–form solution for each of the parts would be attainable, and second, the procedure will converge rapidly. This method guarantees the exact solution for the end effector position, only the error in the end effector orientation has to be minimized by the iterative procedure.

The aim of our paper is to repropose the concept of approximate calculation of robot inverse kinematics and to present an approach that can be used in applications, such as arc welding, where the end effector orientation may vary within a small interval. The concept is based on the idea of approximate calculation that has been introduced by Lenarčič, Oblak and Stanič [10] and uses the mathematical principles of Milenković [8] and Lumelsky [9]. The approach includes very few mathematical operations and, thus can be effectively used for real time control. The paper elaborates different properties of the proposed approach mostly by studying a number of numerical experiments. It starts by introducing the mathematical notation and by representing the example of a robot manipulator which will serve for numerical experimentation in the following sections. At the end, based on the evaluation of experimental results, some general recommendations from the viewpoint of application in arc welding processes are added.

2 Theory

The joint coordinates of a robot manipulator are expressed in terms of the vector $q = (q_1, q_2, ..., q_N)$, where N is the number of degrees of freedom. We will consider in this paper only robot manipulators with $N = 6$ which can be separated into the major and minor linkage. The major linkage is composed of the first three links $q_r = (q_1, q_2, q_3)$ and serves to position the robot end effector, the minor linkage, that is composed of the last three links $q_f = (q_4, q_5, q_6)$ is used to orient it. The end effector position is given by the Cartesian vector $r = (r_x, r_y, r_z)$ and the end effector orientation by the three orientation angles included in the vector $f = (f_x, f_y, f_z)$. The vector **r** represents the position of the reference point of the end effector relative to the reference coordinate frame. The orientation angles **f** are measured between the axes of the end effector coordinate frame and the reference coordinate frame.

As is well known, the end effector position and orientation can be specified in terms of a set of independent trigonometric equations as functions of joint coordinates **q** [1]. Thus,

$$\mathbf{r} = \mathbf{r}_{0,6}(\mathbf{q}_r,\mathbf{q}_f), \tag{1}$$
$$\mathbf{T} = \mathbf{T}_{0,6}(\mathbf{q}_r,\mathbf{q}_f). \tag{2}$$

Here $\mathbf{r}_{0,6}$ connects the origin of the reference coordinate frame and the reference point of the end effector, and $\mathbf{T}_{0,6}$ is the 3x3 transformation matrix between the end effector coordinate frame and the reference coordinate frame. Note that the orientation angles **f** are calculated through the elements of the transformation matrix **T**. It follows

$$\mathbf{f} = \mathbf{f}_{0,6}(\mathbf{q}_r,\mathbf{q}_f) . \tag{3}$$

The problem of inverse kinematics thus consists of calculating the joint coordinates $\mathbf{q}_r$ and $\mathbf{q}_f$ that correspond to the given Cartesian coordinates **r** and **f**. If the robot manipulator possesses revolute joints, the joint coordinates appear in equations (1–3) as arguments of trigonometric functions. In general, we cannot explicitly express joint coordinates as functions of Cartesian coordinates. For some values of Cartesian coordinates there is no real solution for joint coordinates, and besides, if a real solution exists, it can be multivalued.

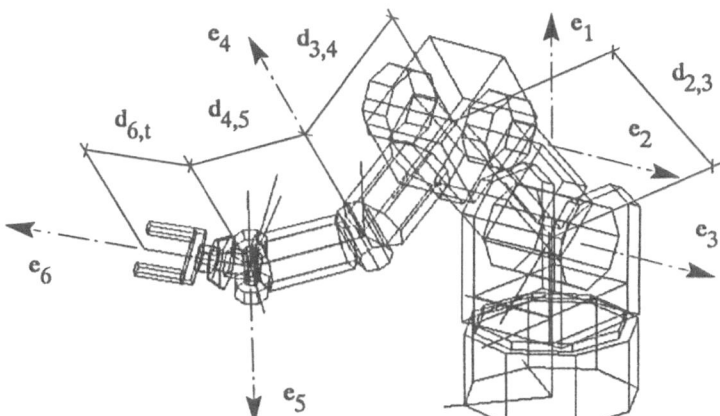

Figure 1:   *A robot manipulator with six revolute joints*

In Figure 1 an example of a six–degrees–of–freedom robot manipulator is given. It has only revolute joints. They are arranged in a way that a closed–form solution of the inverse kinematics is not attainable. Basically, this is because the fourth joint axis does not intersect with the axes of the fifth and the sixth joint. This robot manipulator will serve to illustrate and evaluate the

500

proposed approach and its modifications. According to the notation established by [1], the robot manipulator possesses the following geometry: $e_1 = e_4 = (0,0,1)$, $e_2 = e_3 = e_5 = (1,0,0)$, $e_6 = (0,1,0)$, $d_{0,1} = d_{1,2} = d_{5,6} = (0,0,0)$, $d_{2,3} = (0,0,0.78)$, $d_{3,4} = (0,0.55,0)$, $d_{4,5} = (0,0.40,0)$, $d_{6,t} = (0,0.376,0)$.

3 Experiment 1a

Consider the given robot manipulator (Figure 1) that is required to move along a straight line in Cartesian coordinates starting at the initial point $p(0) = (r(0),f(0)) = (0.6,0.45,1.2,-45,45,75)$ and ending at the end point $p(K) = (r(K),f(K)) = (-0.1,1.5,0.2,45,-45,45)$ as shown in Figure 2. The intermmediate Cartesian coordinates are given as a linear combination of the initial and the end point $(k = 0,1,...,K)$

$$p(k) = p(k-1) + dp(k), \quad dp(k) = (p(K)-p(0))/K. \tag{4}$$

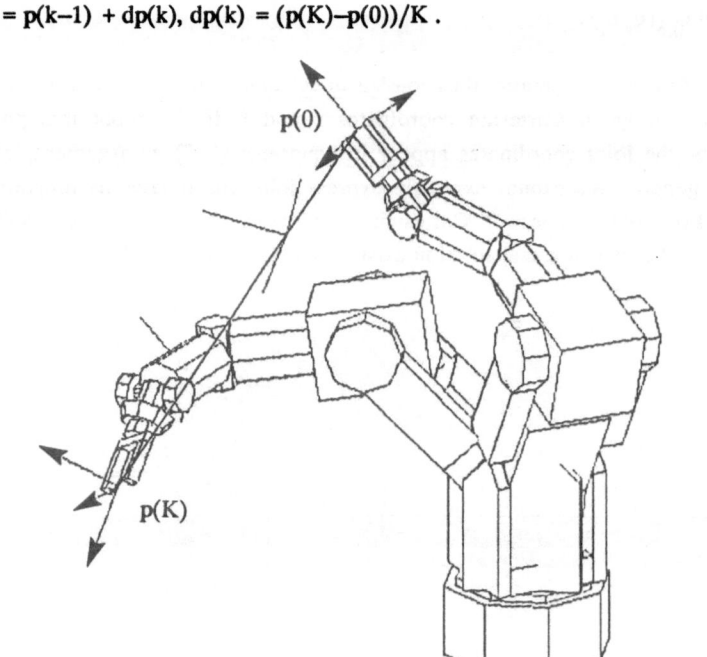

Figure 2:   *A Cartesian trajectory given as a straight line*

Suppose that the joint coordinates q(0) which are related to p(0) and q(K) which are related to p(K) are known. This is the case in many industrial applications where robot programming consists of moving the robot manually from one characteristic point p(0) (named a knot) to another p(K), while the related joint coordinates q(0) and q(K) are spontaneously measured by the proprioceptive sensors of the robot. The most simple approximation of joint trajectories that correspond to the straight line in Cartesian coordinates is given by

$$q(k) = q(k-1) + dq(k), \; dq(k) = (q(K)-q(0))/K, \tag{5}$$

$k = 0,1,...,K$. It is clear that this approximation is far from the exact solution. However, if $p(K)$ is close to $p(0)$, such an approximation could be used. In fact, many industrial robots which are applied in practice, take advantage of this simple approximation to obtain small sections of joint trajectories in order to fit a given Cartesian path. Figure 3 shows the obtained Cartesian trajectories when we calulate the joint trajectories by using equation (5).

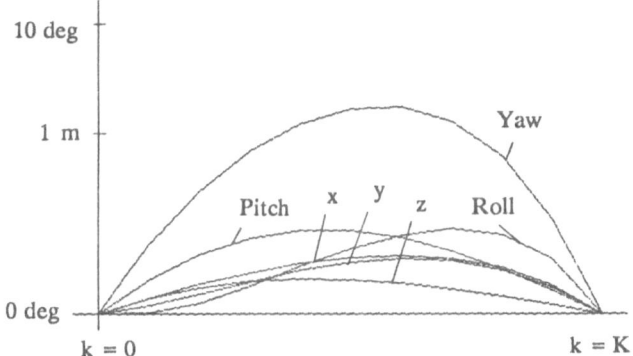

Figure 3: *The error between the required and the obtained Cartesian position and orientation angles if approximation (5) is used*

4 Experiment 1b

Let us now calculate the intermediate joint coordinates of the robot minor linkage by linear interpolation as follows

$$q_f(k) = q_f(k-1) + dq_f(k), \; dq_f(k) = (q_f(K)-q_f(0))/K, \tag{6}$$

and then use equation (1) for each $k = 1,2,...,K-1$ to obtain $q_r(k)$ for the given $r(k)$ and for the calculated $q_f(k)$. Note that q_r can be expressed explicitly as a function of q_f and r without regard to the kinematic structure of the robot manipulator.

By using equation (6) instead of equations (1–3) to determine the joint coordinates q_f, an approximation to the inverse kinematics calculation is introduced. This approximation does not perturb the required end effector position, since it is disenabled by calculating the appropriate joint coordinates q_r, but the error appears in the end effector orientation. Figure 4 shows the error between the required Cartesian orientation angles and the resulting orientation angles of the robot end effector. In contrasst to what we could expect, the orientation error is relatively small. In the given example, although the initial and the end point on the trajectory are very distant, it is keptwithin 7.75 degrees.

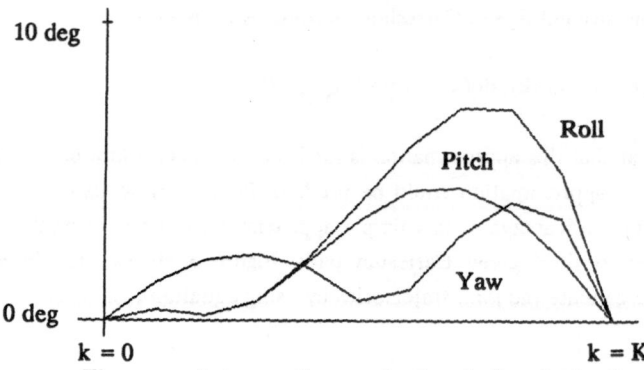

Figure 4: *The error between the required and the obtained orientation angles if approximation (6) is used*

5 Experiment 2a

In the second experiment we will present another simple procedure for the calculation of inverse kinematics . It is based on the idea of approximating the end effector orientations. In comparison with the previously described procedure, it includes more arithmetical operations, but, since it does not require knowledge of the joint coordinates at the end point of the trajectory, it is more general from the viewpoint of applicability. Besides, the previously described procedure is effective only if the task that is being performed is composed of linear Cartesian sections, while for this procedure it is not necessary.

According to the properties of the robot kinematic equations we can rewrite the relationship (1,2) as follows

$$r = r_{0,3}(q_r) + T_{0,3}(q_r)r_{3,6}(q_f) \,, \tag{7}$$
$$T = T_{0,3}(q_r)T_{3,6}(q_f) \,, \tag{8}$$

where $r_{0,3}$ connects the origin of the reference coordinate frame and the top of the third link, $r_{3,6}$ connects the origin of the fourth link and the reference point of the robot end effector, $T_{0,3}$ transforms the local coordinate frame which is attached to the third link into the reference coordinate frame, and $T_{3,6}$ is the transformation matrix between the end effector coordinate frame and the local coordinate frame of the third link.

Assume that the robot end effector is required to fit a given Cartesian trajectory $p(k)$, $k = 0,1,...,K$ which is not necessarily defined as a straight line. In order to control the robot along the given Cartesian trajectory, it should be transformed into joint trajectory $q(k)$, $k = 0,1,...,K$. Assume that joint coordinates $q(0)$ at the initial point of the trajectory are known. In the following procedure for the inverse kinematics transformation, we first estimate the joint coordinates of the major linkage by $q_{r_+}(k)$ and then obtain $q_f(k)$ of the minor linkage by solving equation (8) for the given end effector orientation which is determined by the transformation matrix $T(k)$ and for the estimated $q_{r+}(k)$ as follows

$$T_{3,6}(q_f(k)) = T_{0,3}(q_{r+}(k))^T T(k) , \qquad (9)$$

$k = 1,2,...,K$. Equation (9) has a closed–form solution for the joint coordinates q_f. Once having established $q_f(k)$, we use equation (7) to obtain $q_r(k)$ as follows

$$T_{0,3}(q_r(k))^T (r(k) - r_{0,3}(q_r(k))) = r_{3,6}(q_f(k)) , \qquad (10)$$

$k = 1,2,...,K$. Equation (10) has a closed–form solution for the joint coordinates $q_r(k)$. The resulting end effector positions strictly fit the required–ones, while the end effector orientations are approximated. The orientation error in each time step depends on the value of the estimated joint coordinates $q_{r+}(k)$. To obtain accurate results, it is extremely important that $q_{r+}(k)$ be close to the exact value.

Let us first try the following

$$q_{r+}(k) = q_r(k-1) , \qquad (11)$$

$k = 1,2,...,K$. Note that this corresponds to one iteration of the method introduced by Lumelsky [9]. Figure 5 shows the error between the required and the resulting orientation angles for the linear Cartesian trajectory given in Figure 2. The error can be reduced by increasing the number of time steps K. If K = 10, the error is kept within 10.90 degrees, if K = 100, it is smaller than 1.04 degree.

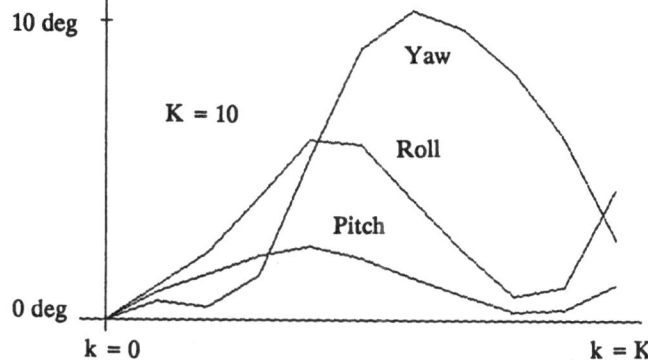

Figure 5:   *The error between the required and the obtained orientation angles if approximation (9) and estimation (11) are used*

6 Experiment 2b

The procedure becomes more accurate by introducing the estimation

$$q_{r+}(k) = 2q_r(k-1) - q_r(k-2) , \tag{12}$$

$k = 2,3,...,K$, $q_{r+}(1) = q_r(0)$. Figure 6 shows the error between the required and the resulting orientation angles. As is evident, the error is much smaller. If $K = 10$, it is kept within 4.72 degrees, if $K = 100$ (which would be a normal resolution for control purposes in the given case), it is less than 0.111 degree.

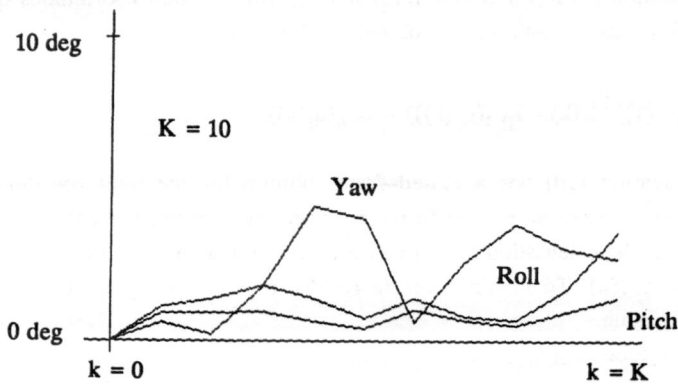

<u>Figure 6:</u> *The error between the required and the obtained orientation angles if approximation (9) and estimation (12) are used*

This experiment assumes constant joint velocities to estimate the joint coordinates of the major linkage in equation (9). Note that a more sophisticated assumption, such as constant joint accelerations, may not necessarily increase the accuracy. But certainly it increases the number of arithmetical operations.

7 Numerical behaviour

The most fascinating property of the approximations introduced for the inverse kinematics calculation is that they need a very small number of arithmetical operations to obtain the solution and that they can be applied with the same effect to robot manipulators possessing a closed–form solution as well as to those without it.

The first procedure presented in experiment 1a is used in practice to approximate the inverse kinematics of short linear sections of a Cartesian path. The approximation error appears in the end effector position and in the end effector orientation. Procedure 1b is similar but gives the exact solution to the end effector position. If the error obtained is neglected, the main disadvantage of both procedures is that one must know not only the joint coordinates of the initial point on the trajectory but also the joint coordinates of the end point. It is not very clear whether the resulting error between the obtained and the required Cartesian path is within the prescribed limits, although it can be decreased by imposing more characteristic points on the trajectory for which the corresponding joint coordinates are known. For this reason, the applicability of procedures 1a and 1b is very limited and, in general, they should be used in combination with one of the more accurate numerical methods to calculate the inverse kinematics in the characteristic points or they should be applied to a robot manipulator that possesses programming capabilities which permit its manual movement from one characteristic point on the trajectory to another and spontaneous measurement of the corresponding joint coordinates.

In contrast, the procedures 2A and 2B are more general from the viewpoint of applicability. This is, first, because the joint coordinates of the end point do not need to be known, and second, the resulting approximation error, which is related only to the end effector orientation, is much smaller. In addition, the required Cartesian path is not necessarily a short linear section, moreover, the Cartesian path can be arbitrarily long and curved.

As said, once having applied one of the approaches 2A and 2B to the inverse kinematics calculation, one must be aware a of possible orientation error of the robot end effector. This error cannot be spontaneously controlled, although it can be kept relatively small, for instance, less than 0.2 degree. It could be decreased by decreasing the distance between two intermmediate points $p(k)$ and $p(k+1)$ on the given Cartesian path (increasing K). Figure 7 shows the value of the Euclidian norm of the difference between the required f and the obtained orientation $f_{0,6}$ as a function of the number K for the procedures 2A and 2B related to the example in Figure 2.

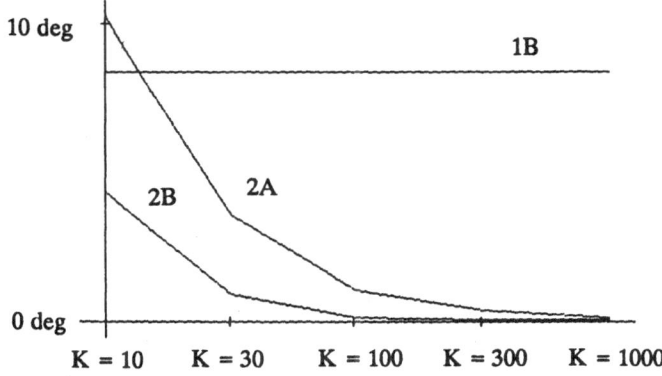

<u>Figure 7:</u>   *Euclidian norm of the orientation error as a function of K*

Evidently, the error in the end effector orientation has a cumulative nature, but an experimental analysis of this property has shown a good numerical behaviour. For example, assume a robot manipulator moving between three points in Cartesian space p_1 = (0.6,0.45,1.2,–45,45,75), p_2 = (–0.1,1.5,0.2,45,–45,45) and p_3 = (0.3,1.6,1.0,0.0,0.0,0.0) making a triangle as shown in Figure 8 which is repeated several times. If the number of intermmediate points in one linear section is 100, the maximum orientation error would be less than 0.74 degree (procedure 2B). It is interesting that the error increases when making the first triangle, but later on it stabilizes at a smaller value.

8 Conclusions

The validity of the introduced procedures is emphasized by the very small number of arithmetical operations required to obtain the solution, which is, very roughly, equal to the number of arithmetical operations included in the analogous closed–form solution. The numerical behaviour of these procedures is as good as of a closed–form solution. The only difference is the resulting error of the end effector orientation, while the end effector position related to the calculated joint coordinates is precisely the requiredone. We can conclude, however, that the accuracy of procedures 2A and 2B is good enough for several applications of industrial robots, especially those where small variations in the end effector orientation do not effect the quality of the job.

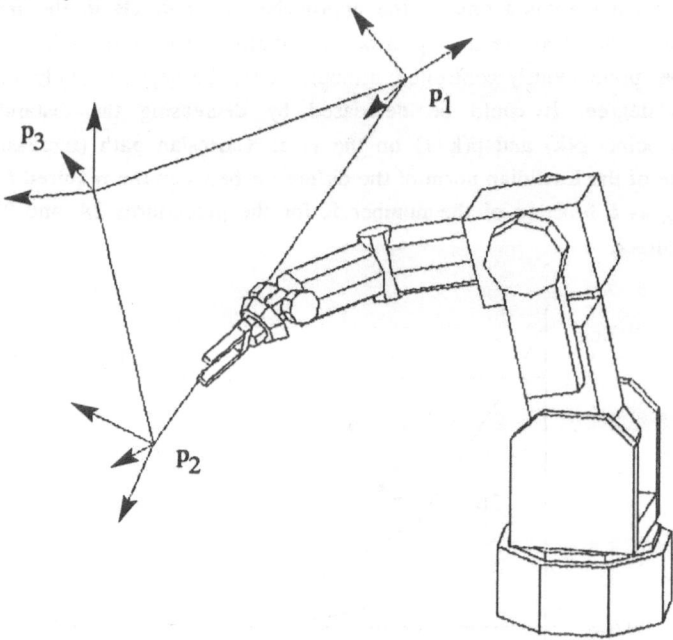

Figure 8: *A Cartesian path composed of three linear sections*

An example for such an application is the arc welding process. Advanced robotic systems for arc welding and similar applications should have good dynamic performance, high repeatibility, spontaneous control of the six degrees of freedom of the robot and additionaly three to six degrees of freedom of the positioners. They should enable Cartesian path control along straight lines and circles, on–line synchronization with the conveyor, adaptive seam treacking automatic change of the execution velocity etc. One of the most important requirements of arc welding technology is the accuracy of robot position in Cartesian space. All these mechanical and programming features call for an effective, fast and accurate inverse kinematics calculation. The application of general numerical methods may introduce unsolvable problems for real–time computation, while closed–form solutions exist only for special robot structures which, on the other hand, may represent a significant limitation in the design of the mechanism.

In this paper we have presented an approach to the inverse kinematics calculation which satisfies the mentioned requirements. The resulting orientation error, which is less than 1.0 degree, does not represent a problem for arc welding, moreover, there are a lot of robot applications, such as adhesive application, water jet and laser cutting, and painting, where small variations of orientation are not crucial from the technological aspect. Different experiments were studied in order to analyse the properties of the approach, mostly based on computer graphics simulation. Experiments on a commercial robot manipulator were also made. But, since the robot kinematic structure enabled the derivation of the closed–form solution, the main advantages of the proposed approach over general numerical methods could not be demonstrated.

9 References

[1]   J. Lenarčič, Kinematics, *International Encyclopedia of Robotics*, John Wiley and Sons, New York, 1988

[2]   J. Furusho, S. Onishi, An Efficient Approach for Solving the Inverse Kinematics of Manipulators, *Proc. of the 15th ISIR*, Tokyo, Japan, 1985

[3]   A.A. Goldenberg, B. Benhabib, R.G. Fenton, A Complete Generalized Solution to the Inverse Kinematics of Robots, *IEEE J. of Robotics and Automation*, RA–1, No. 1, 1985

[4]   J. Lenarčič, An Efficient Numerical Approach for Calculating the Inverse Kinematics for Robot Manipulators, *Robotica*, Vol. 3, No. 1, 1985

[5]   B. Siciliano, Closed–Loop Computational Schemes of Robot Inverse Kinematics, *Proc. of the 1st Int. Meeting Advances in Robot Kinematics*, Ljubljana, Yugoslavia, 1988

[6]   J. Angeles, Iterative Kinematic Inversion of General Five–Axis Robot Manipulators, Robotics Research, Vol.4, No. 4, 1986

[7]   R. Maunseur, K.L. Doty, A Fast Algorithm for Inverse Kinematic Analysis of Robot Manipulators, *Robotics Research*, Vol. 7, No. 3, 1988

[8]   V. Milenković, B. Huang, Kinematics of Major Robot Linkage, *Proc. of the 13th ISIR*, Chicago, USA, 1983

[9]   V.J. Lumelsky, Iterative Coordinate Transformation Procedure for One Class of Robots, *IEEE Trans. on Systems, Man and Cybern.*, SMC–14, No. 3, 1984

[10]   J. Lenarčič, P. Oblak, U. Stanič, Approximate Calculation of Robot Joint Trajectories for Control Along Cartesian Paths, *Proc. of the 15th ISIR*, Tokyo, Japan, 1985

Modelling and Control of a Modular, Redundant Robot Manipulator

Sylvie CHARENTUS and Marc RENAUD

Laboratoire d'Automatique et d'Analyse des Systèmes (LAAS)

7, avenue du Colonel Roche 31077 TOULOUSE Cédex (FRANCE)

Abstract

A new type of robot manipulator which possesses rigidity capabilities and a large workspace is presented. This robot is composed of several modules which consist of elementary manipulators with the parallel structure of the "Stewart Platform". However, the structure of our new robot manipulator raises modelling difficulties. First, one parallel module is studied and a new method of computation of the direct kinematic model is proposed. Then, the computation of the complete direct and inverse kinematic model of our robot manipulator is presented.

1 Introduction

Robot manipulators with an open kinematic chain structure [1] offer the advantage of usually sweeping a wide workspace but are intrinsically inadequate for positioning and orientating accurately their end effector; indeed, position and orientation errors at the different joints increase from the base outward to the end effector. Also, they suffer a relative lack of rigidity which leads to a small useful mass ratio (weight carried by the end effector over robot mass in the order of 1/35 for electric robots [2]). To overcome these difficulties, various robot manipulators using a closed kinematic chain structure [1] have been proposed or are being developed [3] [4]. Among these are the parallel manipulators that employ segments to connect two links, one of which is fixed and serves as a base and the other is mobile and used as end effector. These robots are instrinsically adequate for positioning and orientating the mobile link relative to the fixed one because the different position and orientation errors at the level of the different joints are not cumulative. In addition, they are sufficiently rigid, leading to a useful mass ratio that sometimes exceed 1 (90 in the case of INRIA's prototype [2]). On the other hand, these robots only sweep a limited workspace.

This paper presents a new, globally-closed, kinematic chain structure. It combines the ability to sweep a wide workspace of an open chain structure with the accuracy and rigidity capabilities of parallel structures. In particular, this structure associates in series n identical parallel structures. However, this new structure raises difficulties in modelling, the latter being an essential step in control synthesis. The difficulties encountered in computing both the inverse geometric model (often called inverse kinematic model) of open chains and the direct geometric model (often called direct kinematic model) of parallel structures arise at the same time. To overcome these drawbacks, the selected parallel structure is first investigated and a new method of computation of the direct geometric model of this structure is proposed. Then, the computation of the various kinematic models required for the control of the overall structure is surveyed and subsequently applied to a particular robot manipulator possessing this structure and comprising a series of four Stewart platforms. This robot, called robot LX, has been designed and realized by the LOGABEX company.

2   Study of parallel robot manipulators

2.1   General

Parallel structures have been the focus of much attention on the part of mathematicians like Euler and Cauchy who determined the minimum number of segments ensuring rigidity as well as the maximum number required to compute the forces exerted on the segments as a function of the force and torque applied onto the mobile link relative to the fixed one (in this case the structure is said to be isostatic). Thus, six segments are needed to obtain a rigid, isostatic structure with six degrees of freedom. Numerous authors attribute the first realization of such a structure to Stewart who devised, in 1965, an aircraft simulator employing three complex segments fitted each with two electrical linear jacks. In the proposed structure, the two links are called platforms and the structure itself is sometimes referred to as Stewart's platform. Other constructions using six linear actuators could be cited in which the six linear actuator end points are connected in pairs at the corner of an equilateral triangle on at least one platform [5] [6] or in which the linear actuator end points are connected such that there are six connection points on the two platforms [7] [8] . Inoue [9] has utilized three sets of pantographs in lieu of the six linear actuators.

The advantage of this construction lies in the accuracy and rigidity capabilities. In addition, it is composed of simple modular elements based on the actuator and platform size and on the type of actuators selected, that is, electric, hydraulic or pneumatic. Combined with an effort sensor and a hybrid force-position control, it is well suited for assembly operations and may be used as a wrist on a robot [10] [7] . On the other hand, this mechanism is rather difficult to model as detailed in the following.

2.2   Modelling

Modelling difficulties of this type of structure are the inverse of those encountered with open structures. Indeed, it is the analytical computation of the *direct geometric model* which allows achievement of the location (position and orientation) of the mobile platform as a function of the actuator lengths, which is difficult whereas the analytical computation of the *inverse geometric model* which allows expression of the actuator lengths as a function of the mobile platform location, is straightforward.

2.2.1 Computation of the inverse geometric model

An orthogonal frame is associated with each platform: frame $\mathcal{R}_0(O_0, x_0, y_0, z_0)$ is linked to the lower platform and used as reference and frame $\mathcal{R}_1(O_1, x_1, y_1, z_1)$ is linked to the upper platform as shown in Fig.(1). Points O_0 and O_1 are the circle centres wherein the linear actuator end points are situated. Points A_i (B_i respectively) are linked to the lower platform (upper respectively).

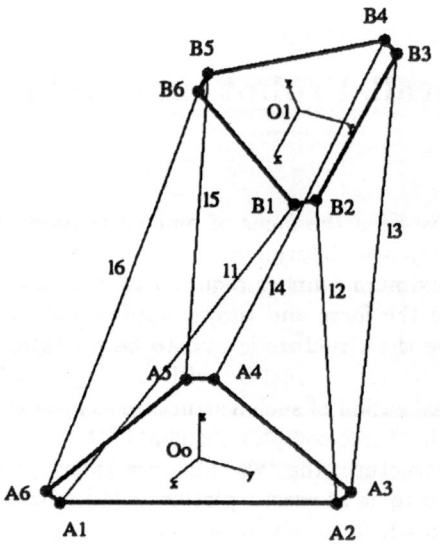

Figure 1: Placement of frames on the platforms

The calculation of the six linear actuator lengths (l_i, $i = 1$ to 6) is easily obtained from the upper platform location (position of point O_1 and orientation of frame $\mathcal{R}_1$ relative to frame $\mathcal{R}_0$):

Since:

$$\underline{A_iB_i} = -\underline{O_0A_i} + \underline{O_0O_1} + \underline{O_1B_i} \quad for \ \ i = 1 \ to \ 6$$

it is also possible to write for the components:

$$A_iB_{i(0)} = -O_0A_{i(0)} + O_0O_{1(0)} + R_{01}\,O_1B_{i(1)} \quad for \ \ i = 1 \ to \ 6$$

where R_{01} is the orientation transformation matrix from frame $\mathcal{R}_0$ to frame $\mathcal{R}_1$ and subscripts $_{(0)}$ or $_{(1)}$ denote the cartesian components in frames $\mathcal{R}_0$ or $\mathcal{R}_1$.

Then:

$$l_i = \| A_iB_{i(0)} \|$$

If X represents six independent components defining the upper platform location (obtained from $O_0O_{1(0)}$ and R_{01}) and if $L = (l_1...l_6)$, the *inverse geometric model* is represented by the transformation L=f(X) which is unique.

2.2.2 Computation of the direct geometric model

This is a complex computation. Several researchers have put forward an iterative solution involving a Jacobian matrix [8],[2]. In this paper[1], we present a solution which reduces (at least in the particular case where the linear actuator end points are connected in pairs at the corner of an equilateral triangle on at least one platform), the computation of the *direct geometric model* to that of the roots of an n order polynomial equation. This solution leads to the value $n = 8$ (asserted by Merlet [11] but has not been proved yet) and improves our preceding result leading to the value $n = 16$ [12]. Also, this solution rectifies some wrong assertions recently published in the literature [13]. Finally it may be noticed that the literal expression of the 8 order polynomial equation is explicitly obtained. Before presenting this solution, the iterative method using a Jacobian matrix is recalled.

Iterative method using a Jacobian matrix. The *direct geometric model* is computed from the explicit *inverse geometric model* L=f(X) presented in the preceding paragraph.

A classical Newton-Raphson algorithm can be used, the Jacobian matrix being the partial derivative matrix of the six linear actuator lengths relative to the six independent components defining the location of the upper platform : $J(X) = \frac{\partial L}{\partial X}(X)$. From an initial location arbitrarily chosen, the current location is modified at each iteration step in order to obtain that location which corresponds to the actuator lengths.

The method requires three or four iterations in the case where it converges and therefore involves a computation time which is a function of the number of iterations. However, this may be a drawback for real-time control. A refinement consists of computing only one constant Jacobian matrix corresponding to the mean location, i.e., when $O_0 O_{1(0)} = (0, 0, z_{mean})$ and $R_{01} = I$ (identity matrix). This increases the number of iterations but speeds up computation. However it can only be used if the final location is more or less similar to the mean location [8].

The drawback of this iterative method is that it only yields one solution associated with the initialization of the procedure, (i.e., the choice of the initial location). Usually, it does not hinder the control of these manipulators because designers intentionally limit the workspace so as to ensure method convergence [14].

Method proposed. The new solution is proposed in the case of a manipulator where the six linear actuator end points are connected in pairs at the corner of an equilateral triangle on at least one platform. However, we only deal with the simplified case depicted in Fig. (2) where there are three linear actuator end points on the two platforms [12].

In fact the problem is to find the cooordinates of points $H_0(x_0, y_0, z_0)$, $H_1(x_1, y_1, z_1)$, $H_2(x_2, y_2, z_2)$ in frame $\mathcal{R}_0$, knowing the lengths l_i (i=1 to 6).

Six relationships are associated with the actuator lengths:

$$l_1 = \| \underline{B_0 H_2} \| \quad , \quad l_2 = \| \underline{B_1 H_2} \| \quad , \quad l_3 = \| \underline{B_1 H_0} \|$$

$$l_4 = \| \underline{B_2 H_0} \| \quad , \quad l_5 = \| \underline{B_2 H_1} \| \quad , \quad l_6 = \| \underline{B_0 H_1} \|$$

---

[1]The results presented in this part differ from those revealed in Montreal during the symposium and have been considerably improved.

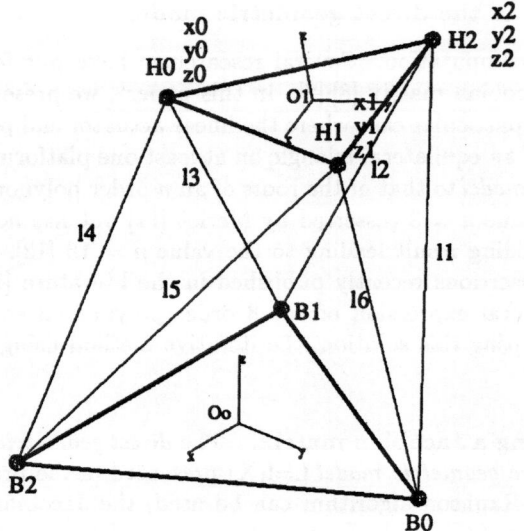

Figure 2: Simplified model

that is,

$$l_1^2 = x_2^2 + \left(y_2 - \frac{d_B}{\sqrt{3}}\right)^2 + z_2^2 \tag{1}$$

$$l_2^2 = \left(x_2 + \frac{d_B}{2}\right)^2 + \left(y_2 + \frac{d_B}{2\sqrt{3}}\right)^2 + z_2^2 \tag{2}$$

$$l_3^2 = \left(x_0 + \frac{d_B}{2}\right)^2 + \left(y_0 + \frac{d_B}{2\sqrt{3}}\right)^2 + z_0^2 \tag{3}$$

$$l_4^2 = \left(x_0 - \frac{d_B}{2}\right)^2 + \left(y_0 + \frac{d_B}{2\sqrt{3}}\right)^2 + z_0^2 \tag{4}$$

$$l_5^2 = \left(x_1 - \frac{d_B}{2}\right)^2 + \left(y_1 + \frac{d_B}{2\sqrt{3}}\right)^2 + z_1^2 \tag{5}$$

$$l_6^2 = x_1^2 + \left(y_1 - \frac{d_B}{\sqrt{3}}\right)^2 + z_1^2 \tag{6}$$

and three relations are asssociated with the geometry of the equilateral triangle H_0, H_1, H_2:

$$\parallel \underline{H_0 H_1} \parallel = \parallel \underline{H_1 H_2} \parallel = \parallel \underline{H_2 H_0} \parallel = d_H$$

that is,

$$d_H^2 = (x_2 - x_1)^2 + (y_2 - y_1)^2 + (z_2 - z_1)^2 \tag{7}$$

$$d_H^2 = (x_1 - x_0)^2 + (y_1 - y_0)^2 + (z_1 - z_0)^2 \tag{8}$$

$$d_H^2 = (x_0 - x_2)^2 + (y_0 - y_2)^2 + (z_0 - z_2)^2 \tag{9}$$

Eqs (1) and (2) are the cartesian equations of a circle in space whose axis is B_0B_1; the radius c_2 being given by:

$$c_2 = \frac{\sqrt{(l_1 + l_2)^2 - d_B^2} \sqrt{d_B^2 - (l_1 - l_2)^2}}{2\, d_B}$$

It is more convenient to represent this circle by three parametric equations using for example, as parameter, the angle θ_2 between vector $\underline{z_0}$ and plane $B_0B_1H_2$ (Fig.3) ($\theta_0 \in\ ]-\pi, +\pi]$):

$$x_2 = -\frac{\sqrt{3}}{2} c_2 \sin \theta_2 - \frac{1}{4}(a_2 + d_B) \tag{10}$$

$$y_2 = \frac{c_2}{2} \sin \theta_2 - \frac{\sqrt{3}}{4}(a_2 - \frac{d_B}{3}) \tag{11}$$

$$z_2 = c_2 \cos \theta_2 \tag{12}$$

with

$$a_2 = \frac{l_1^2 - l_2^2}{d_B}$$

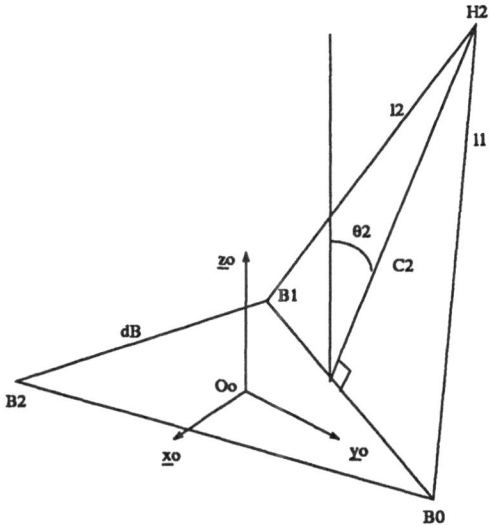

Figure 3: Definition of angle θ_2

Eqs (3) and (4) are the cartesian equations of a circle in space whose axis is B_1B_2; the radius c_0 being given by:

$$c_0 = \frac{\sqrt{(l_3 + l_4)^2 - d_B^2}\ \sqrt{d_B^2 - (l_3 - l_4)^2}}{2\,d_B}$$

This circle can be represented by three parametric equations using for example, as parameter, the angle θ_0 between vector $\underline{z_0}$ and plane $B_1 B_2 H_0$ ($\theta_0 \in\ ]-\pi, +\pi]$):

$$x_0 = \frac{a_0}{2} \tag{13}$$

$$y_0 = -c_0 \sin\theta_0 - \frac{d_B}{2\sqrt{3}} \tag{14}$$

$$z_0 = c_0 \cos\theta_0 \tag{15}$$

with

$$a_0 = \frac{l_3^2 - l_4^2}{d_B}$$

Eqs (5) and (6) are the cartesian equations of a circle in space whose axis is $B_2 B_0$; the radius c_1 being given by:

$$c_1 = \frac{\sqrt{(l_5 + l_6)^2 - d_B^2}\ \sqrt{d_B^2 - (l_5 - l_6)^2}}{2\,d_B}$$

This circle can be represented by three parametric equations using for example, as parameter, the angle θ_1 between vector $\underline{z_0}$ and plane $B_2 B_0 H_1$ ($\theta_1 \in\ ]-\pi, +\pi]$):

$$x_1 = \frac{\sqrt{3}}{2} c_1 \sin\theta_1 + \frac{1}{4}(d_B - a_1) \tag{16}$$

$$y_1 = \frac{c_1}{2} \sin\theta_1 + \frac{\sqrt{3}}{4}\left(a_1 + \frac{d_B}{3}\right) \tag{17}$$

$$z_1 = c_1 \cos\theta_1 \tag{18}$$

with

$$a_1 = \frac{l_5^2 - l_6^2}{d_B}$$

Notice that if angular sensors transmitting the value of angles $\theta_0, \theta_1, \theta_2$ are used, the computation of the direct geometric model can be effected analytically. This solution has been retained by Inoue [9] who proposed a platform differing from that of Stewart. If the values of $\theta_0, \theta_1, \theta_2$ are not known, the computation is continued as follows:

The values of coordinates $x_i, y_i, z_i (i = 0, 1, 2)$ obtained in Eqs (10) to (18) are substituted into the 3 Eqs (7), (8) and (9). A system of 3 coupled trigonometric equations with 3 unknowns $\theta_0, \theta_1, \theta_2$ is obtained:

$$\sin\theta_1 \sin\theta_2 - 2\cos\theta_1 \cos\theta_2 + q_2 \sin\theta_1 + p_1 \sin\theta_2 = R_0 \tag{19}$$

$$\sin\theta_2 \sin\theta_0 - 2\cos\theta_2 \cos\theta_0 + q_0 \sin\theta_2 + p_2 \sin\theta_0 = R_1 \tag{20}$$

$$\sin\theta_0 \sin\theta_1 - 2\cos\theta_0 \cos\theta_1 + q_1 \sin\theta_0 + p_0 \sin\theta_1 = R_2 \tag{21}$$

with

$$p_i = \frac{\sqrt{3}(d_B - a_i)}{2c_i} \quad , \quad q_i = \frac{\sqrt{3}(d_B + a_i)}{2c_i} \quad for \ i = 0, 1, 2$$

and

$$R_0 = \frac{1}{c_1 c_2}(d_H^2 + \frac{d_B^2}{4} - \frac{(l_5^2 + 3l_6^2 + 3l_1^2 + l_2^2)}{4} - \frac{(l_5^2 - l_6^2)(l_1^2 - l_2^2)}{4d_B^2})$$

$$R_1 = \frac{1}{c_2 c_0}(d_H^2 + \frac{d_B^2}{4} - \frac{(l_1^2 + 3l_2^2 + 3l_3^2 + l_4^2)}{4} - \frac{(l_1^2 - l_2^2)(l_3^2 - l_4^2)}{4d_B^2})$$

$$R_2 = \frac{1}{c_0 c_1}(d_H^2 + \frac{d_B^2}{4} - \frac{(l_3^2 + 3l_4^2 + 3l_5^2 + l_6^2)}{4} - \frac{(l_3^2 - l_4^2)(l_5^2 - l_6^2)}{4d_B^2})$$

To solve the system of Eqs (19), (20) and (21), the symmetry is suppressed and we proceed in several steps:

- From Eq (20), the value of $\cos\theta_2$ can be express as:

$$\cos\theta_2 = \frac{\sin\theta_2(\sin\theta_0 + q_0) + p_2\sin\theta_0 - R_1}{2\cos\theta_2} \tag{22}$$

and from the relationship $\cos^2\theta_2 + \sin^2\theta_2 = 1$, we deduce:

$$N_1\sin^2\theta_2 + N_2\sin\theta_2 + N_3 = 0 \tag{23}$$

in which N_1, N_2, N_3 are 2nd order polynomials in $\sin\theta_0$ only.

- The two other Eqs (19) and (21) can be written in the form of a 2nd order linear equations system with 2 unknowns $\cos\theta_1$ and $\sin\theta_1$:

$$(-2\cos\theta_0)\cos\theta_1 + (\sin\theta_0 + p_0)\sin\theta_1 = -q_1\sin\theta_0 + R_2 \tag{24}$$

$$(-2\cos\theta_0)\cos\theta_1 + (\sin\theta_2 + q_2)\sin\theta_1 = -p_1\sin\theta_2 + R_0 \tag{25}$$

whose solution is given by:

$$\cos\theta_1 = \frac{(\sin\theta_2 + q_2)(R_2 - q_1\sin\theta_0) - (\sin\theta_0 + p_0)(R_0 - p_1\sin\theta_2)}{2((\sin\theta_0 + p_0)\cos\theta_2 - (\sin\theta_2 + q_2)\cos\theta_0)} \tag{26}$$

and

$$\sin\theta_1 = \frac{(R_2 - q_1\sin\theta_0)\cos\theta_2 - (R_0 - p_1\sin\theta_2)\cos\theta_0}{(\sin\theta_0 + p_0)\cos\theta_2 - (\sin\theta_2 + q_2)\cos\theta_0} \tag{27}$$

and from the relationship $\cos^2\theta_2 + \sin^2\theta_2 = 1$, we deduce:

$$G_1\sin^2\theta_2 + G_2\cos^2\theta_2 + G_3\sin\theta_2\cos\theta_2 + G_4\sin\theta_2 + G_5\cos\theta_2 + G_6 = 0 \tag{28}$$

in which G_1, G_2, G_4, G_6 are 2nd order polynomials in $\sin\theta_0$ only, and $G_3 = \cos\theta_0 G_3'$, $G_5 = \cos\theta_0 G_5'$ with G_3', G_5' 1st order polynomials in $\sin\theta_0$ only.

Using once again $\cos^2\theta_2 + \sin^2\theta_2 = 1$, Eq (28) can be witten in the form:

$$(G_1 - G_2)\sin^2\theta_2 + G_4\sin\theta_2 + G_2 + G_6 + (G_3\sin\theta_0 + G_5)\cos\theta_2 = 0 \qquad (29)$$

and using Eq (22), Eq (29) can be written in the form:

$$\cos\theta_0(M_1\sin^2\theta_2 + M_2\sin\theta_2 + M_3) = 0 \qquad (30)$$

or since $\cos\theta_0$ is not always a solution, in the form:

$$M_1\sin^2\theta_2 + M_2\sin\theta_2 + M_3 = 0 \qquad (31)$$

with

$$M_1 = 2(G_1 - G_2) + (\sin\theta_0 + q_0)G_3'$$
$$M_2 = 2G_4 + (\sin\theta_0 + q_0)G_5' + (p_2\sin\theta_0 - R_1)G_3'$$
$$M_3 = 2(G_2 + G_6) + (p_2\sin\theta_0 - R_1)G_5'$$

M_1, M_2, M_3 are 2nd order polynomials in $\sin\theta_0$ only.

- Now, it suffices to show that there exists a common root to the Eqs (23) and (31). This can be accomplished by equating the resultant of these two equations to zero:

$$\Delta(\sin\theta_0) = \begin{vmatrix} M_1 & M_2 & M_3 & 0 \\ 0 & M_1 & M_2 & M_3 \\ N_1 & N_2 & N_3 & 0 \\ 0 & N_1 & N_2 & N_3 \end{vmatrix}$$
$$= (M_1N_3 - M_3N_1)^2 - (M_1N_2 - M_2N_1)(M_2N_3 - M_3N_2) = 0$$

Since M_i and N_i $(i = 1, 2, 3)$ are 2nd order polynomials in $\sin\theta_0$ only, $\Delta(\sin\theta_0)$ is an 8 order polynomial in $\sin\theta_0$ only.

On the one hand, this demonstrates that the maximum number of real solutions $\sin\theta_0$ in the interval $[-1, 1]$ is 8. On the other hand, it is shown in the following that this maximum number can be obtained.

Since two values of $\cos\theta_0$ can be associated with each value of $\sin\theta_0$ ($\cos\theta_0 = \sqrt{1 - \sin^2\theta_0}$ and $\cos\theta_0 = -\sqrt{1 - \sin^2\theta_0}$), the maximum number of real solutions θ_0 in the interval $]-\pi, \pi]$ is 16.

- At each value of θ_0 corresponds a unique value of θ_1 and θ_2 in the interval $]-\pi, \pi]$:

  - $\sin\theta_2 = \frac{M_3N_1 - M_1N_3}{M_1N_2 - M_2N_1}$ and $\cos\theta_2$ given by Eq (22).
  - $\cos\theta_1$ given by Eq (26) and $\sin\theta_1$ given by Eq (27).

Then the maximum number of reachable locations is 16 and the example shows hereafter that this maximum number can be obtained. Notice that the two locations corresponding to the two opposite values of $\cos\theta_0$ are symmetrical with respect to the lower platform, as shown in [16], and such that the coordinate z_0 is either positive or negative.

Bairstow's numerical method has been used to compute the roots of this polynomial equation.

Example: [2]

Computation of the platform locations based on the six actuator lengths and the module geometry.

Let:

$$l_1 = 23.0000 \quad , \quad l_2 = 23.5000 \quad , \quad l_3 = 24.5000$$

$$l_4 = 24.0000 \quad , \quad l_5 = 22.5000 \quad , \quad l_6 = 22.0000$$

$$d_H = 6\sqrt{3} \quad , \quad d_B = 13\sqrt{3}$$

Computation of constants:

```
a0 =  1.0770     a1 =  0.9882     a2 = -1.0326
c0 = 21.4729     c1 = 19.1867     c2 = 20.3374
p0 =  0.8647     p1 =  0.9717     p2 =  1.0028
q0 =  0.9516     q1 =  1.0609     q2 =  0.9149
R0 = -2.0229     R1 = -2.0737     R2 = -2.0589
```

Computation of the 8 order polynomial coefficients:

```
degree 8 :    4490.4291
degree 7 :   13549.0803
degree 6 :   16123.8510
degree 5 :    9416.4307
degree 4 :    2650.0159
degree 3 :     240.1017
degree 2 :     -32.3998
degree 1 :      -6.9443
degree 0 :      -0.2695
```

Computation of the polynomial equation roots using Bairstow's numerical method:

```
real part   imaginary part
 -0.1921        0.0000
 -0.2089        0.0000
 -0.0573        0.0000
 -0.6034        0.0000
 -0.6226        0.0000
 -0.7243        0.0000
 -0.7388        0.0000
  0.1300        0.0000
```

The 8 solutions are therefore:

---

[2]This example is derived from that of J.P. Merlet.

$\sin\theta_0$	x_0	y_0	z_0	x_1	y_1	z_1	x_2	y_2	z_2
-0.1921	0.5385	-2.3756	21.0731	-6.4733	-3.1668	13.4436	-7.0558	4.6699	20.2441
-0.2089	0.5385	-2.0139	20.9990	7.8413	5.0977	18.9754	7.6732	-3.8340	13.6653
-0.0573	0.5385	-5.2707	21.4377	5.3519	3.6604	19.1867	-4.9753	3.4687	20.3322
-0.6034	0.5385	6.4560	17.1239	-4.8294	-2.2177	15.1360	5.5217	-2.5918	15.9815
-0.6226	0.5385	6.8692	16.8033	-4.5683	-2.0670	15.3661	-9.3734	6.0079	19.8053
-0.7243	0.5385	9.0525	14.8056	9.8485	6.2566	18.4806	2.0058	-0.5619	18.4676
-0.7388	0.5385	9.3642	14.4711	9.4822	6.0451	18.5935	1.0387	-0.0036	18.9428
0.1300	0.5385	-9.2910	21.2907	-4.0665	-1.7773	15.7827	6.2410	-3.0071	15.2912

These 8 locations are illustrated in Fig. (4).

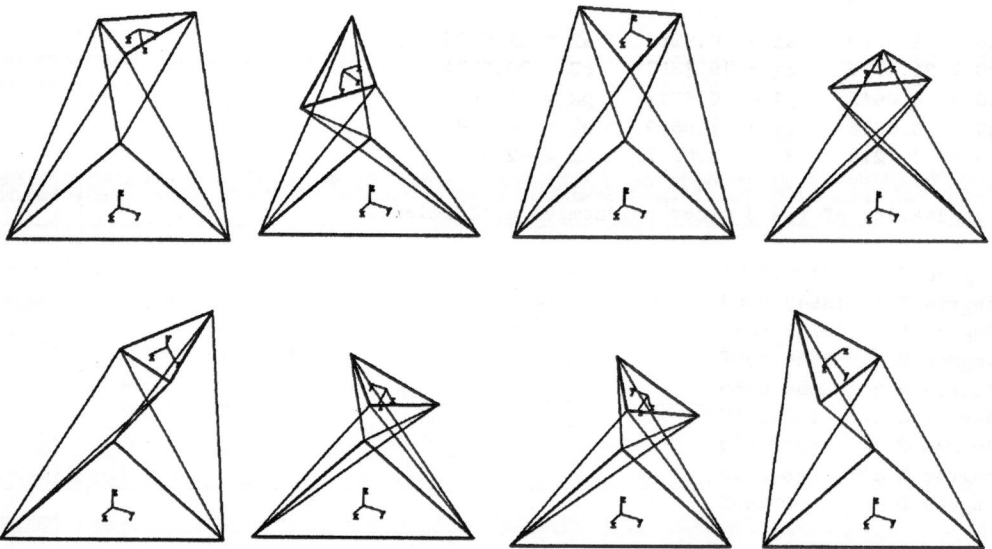

Figure 4: 8 locations corresponding to the same combination of actuator lengths (with z_0 positive)

Concluding remark on the direct geometric model computation: The approach proposed presents a great advantage: it yields all the solutions of the direct model and then deletes the convergence problems associated with iterative procedures.

2.2.3   Singularity study

The concept of parallel manipulator singularities is not so precisely defined in the literature. In fact, for this type of manipulator, *singular locations* must be defined in the same way as *singular configurations* are defined in the case of series manipulator.

A parallel manipulator possesses a singular location when the mobile platform is no longer rigid, that is, when it is unstable for a given combination of actuator lengths. This is what occurs, for example, when the upper platform is rotated 90 degrees about the z axes, both platforms being parallel; a slight screwing motion can then be performed through an external force about the z axis on the upper platform without changing the actuators lengths [5].

These singular locations can be detected by either:

- analytically or numerically cancelling the determinant of the Jacobian matrix defined previously or of the matrix of Plüker's coordinates for the six linear actuator axis [5]. The practical implementation cannot be achieved in the first case [2], it is certainly easier in the second case using Plüker's coordinates in an appropriate frame [15].

- geometrically using *Grassmann's line geometry*. The rank [16] of each linear actuator axis subset must be computed. Degeneracy cases are obtained when the rank of this subset is less than the number of axes it consists of. These cases correspond to the parallel manipulator singular locations [17].

Then, the singular configurations are obtained from the singular locations using the direct geometric model.

2.3   Workspace and control

2.3.1   Workspace characterization

The workspace is limited by several constraints:

- mechanical constraints due to linear actuator travel, joint rotation and size of the different manipulator parts;

- geometric constraints due to linear actuator end point positions on the platforms;

- geometric constraints due to singularitie avoidances.

  The geometric characterization of the workspace generated by this type of parallel manipulator is not easily obtained. The concept of *aspect* introduced by Borrel [18] for open structures, can be employed, providing it is generalized to closed structures.

  Indeed, the location workspace would then be obtained by the union of aspects. In this particular case, an aspect is a connected open set of the location workspace in which the sign of the Jacobian matrix determinant is constant. The aspects are therefore separated by a boundary which constitutes a singularity locus (determinant of the Jacobian matrix equal to zero). Thus, determining the number of aspects should permit to obtain the maximal number of direct geometric models. In other words, a singularity locus in the location space separates two distinct locations of the upper platform (one in each aspect) that correspond to the same actuator lengths [19].

  The avoidance of singularities could then be solved by limiting the location workspace to that of a single aspect.

2.3.2 Control

It may seem simple, in the first place, to carry out the control of this type of parallel robot manipulator since the *inverse geometric model* that allows computation of the actuator lengths as a function of the upper platform location is explicit. However, the analysis of this structure must be continued to highlight the existence of singularities that are awkward for the control. Thus, if the trajectory to be followed in the location workspace lies exactly in the same aspect, then the singularity problem will not exist at the level of control.

3 Application of parallel structures: robot manipulator composed of n parallel structures connected in series

3.1 Presentation

A parallel robot manipulator (referred to in the following as a module) possesses a small workspace. But if several modules are stacked in series, the manipulator will have a greater workspace and will maintain the major characteristics of a module, that is, rigidity, ease of construction and ability to lift heavy loads. These manipulators consisting of n parallel structures connected in series are then highly redundant since they possess $6n$ degrees of mobility (six per modules), this number exceeding the six degrees of freedom of the end effector as soon as at least two modules are assembled. This redundancy complicates control synthesis but nonetheless enables the robot, on the one hand, to avoid the possible obstacles in its workspace and, on the other, to reconfigure itself for continuing the task in case of the actuator becoming blocked.

These robot manipulators may be considered from two viewpoints. If it is assumed that the $6n$ generalized coordinates are the $6n$ actuator lengths, they then possess a closed chain structure and computation of their direct geometric model is highly complex. On the other hand, if the $6n$ generalized coordinates are assumed to be the $6n$ location coordinates of the n upper platforms of each module, then, they have an open chain structure - each module appearing as a complex joint - and computation of their direct geometric model is straightforward. This is why this second viewpoint has been retained in the following paragraph.

3.2 Modelling

In the following:

- *operational coordinates* refer to location coordinates of the end effector in frame $\mathcal{R}_0$ linked to the robot base. They are represented by vector $X \in R^6$.

- *generalized coordinates* stand for the location coordinates of the upper platform of each module. These generalized coordinates define the overall configuration of the robot. They are represented by vector $Q = [q_1, q_2, ..., q_n] \in R^{6n}$ where the $6n$ generalized coordinates q_{ij} $(j = 1 \ to \ 6)$ are the components of the n generalized vectors $q_i (i = 1 \ to \ n)$ and where n is the number of modules composing the robot. An orthogonal frame $\mathcal{R}_i$ is associated with each upper platform of each module i.

3.2.1 Computation of the direct geometric model

This model consists of computing the location of the robot end effector as a function of the locations of the upper platforms of each module. This computation is straightforward and can be effected by introducing the notion of homogeneous transformation matrix (one per module) [15].

The location of the end effector (regarded here as the upper platform of the last module) in frame $\mathcal{R}_0$ is given by the homogeneous matrix $T_{0,n}$:

$$T_{0,n} = T_{0,1}(q_1) \ . \ T_{1,2}(q_2) \ ... \ T_{n-1,n}(q_n)$$

where $T_{i-1,i}(q_i)$, is the homogeneous transformation matrix from frame $\mathcal{R}_{i-1}$ to frame $\mathcal{R}_i$, function of the generalized coordinates q_{ij}.

X can be obtained from $T_{0,n}$ and this model is classically denoted by X= f(Q).

3.2.2 Computation of the inverse geometric model

This model consists of computing the locations of the upper platforms of each module as a function of the end effector location. Of course by reason of redundancy there exists an infinity of solutions. Given the complexity of the problem due to the number of mobilities involved, the analytical computation seems difficult. We therefore propose an iterative numerical method relying on the computation of a Jacobian matrix.

The iterative algorithm is given in Fig. (5).

Remarks:

- The Jacobian matrix of dimension 6x6n is derived from the direct geometric model X=f(Q) in the form:

$$J(Q) = \frac{\partial X}{\partial Q}(Q)$$

- Among the set of generalized inverse matrices of J, we selected the pseudo-inverse matrix [20] given by:

$$J_{inv} = J^t (J J^t)^{-1}$$

Matrix $(J J^t)$ is square, of dimension 6 and it must be regular.

- At each iteration k, one has: $Q_k = Q_{k-1} + \partial Q_k$.

  The generalized coordinates must satisfy the mechanical constraints of each module, that is, they must belong to the workspace of a module.

Prior to dealing with robot control, the diagram depicted in Fig. (6) summarizes the modelling difficulties for such a robot.

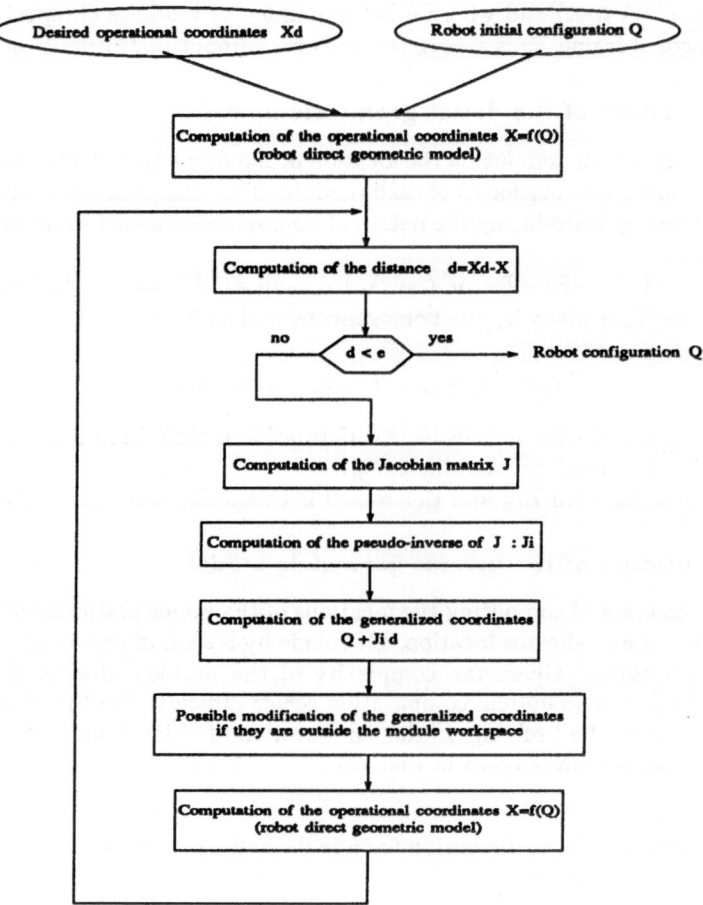

Figure 5: Iterative algorithm for the computation of the robot inverse model

3.3 Workspace and control

3.3.1 Workspace

Neither the analytical expression of the location workspace of a module nor, consequently, that of the robot are known. The latter is of course a function of the number of modules and of the workspace of each module.

3.3.2 Control

To carry out position control, the robot inverse model together with the inverse model of each module must be employed.

From Fig. (6) robot singularities may exist at different levels. In the aforementioned study of the parallel structures, we have highlighted singularities associated with the

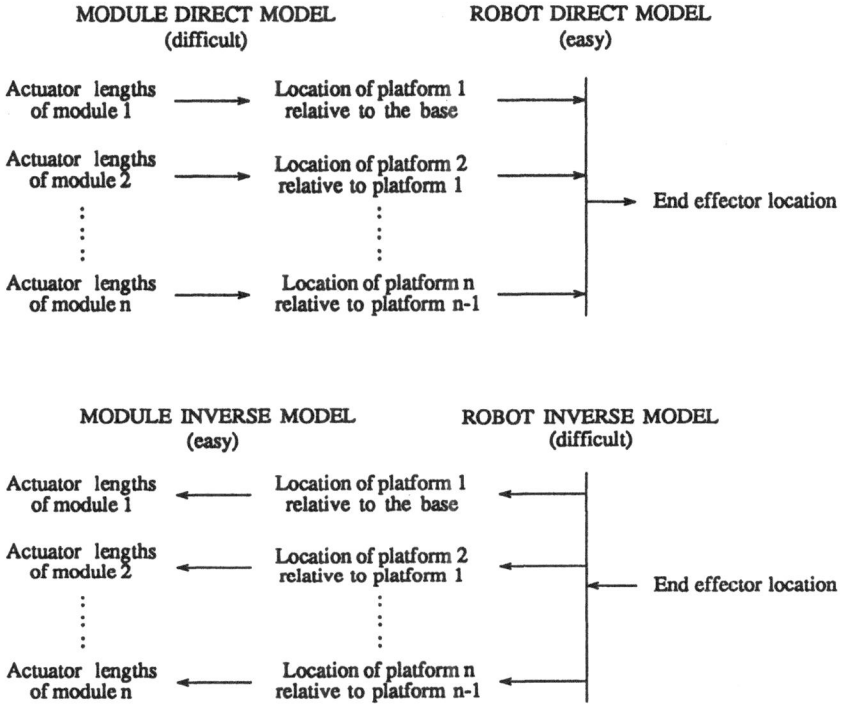

Figure 6: Different models for a robot manipulator composed of n parallel structures

geometry of the structure itself. These naturally reappear during assembly in series of several modules; the best solution is to limit the workspace of each module in order to reject these singularities outside the domain. But other singularities of a kinematic nature may occur as a result of the use of the inverse matrix.

3.4 Application: LOGABEX's robot LX

3.5 Presentation of the robot LX

LOGABEX's robot manipulator LX is depicted in Fig. (7). It consists of 4 identical modules fitted with 24 linear electrical actuators able to sustain each a 100 daN force. They are controlled by a d.c. motor, the output velocity of the actuator rod being 6 mm/s. This 120 kg robot can displace a weight of about 50 kg over the total workspace or a heavier weight in a more restricted area.

3.6 Simulation

Fig. (8) represents the cross section of the position workspace of a module.
One configuration of the robot LX is depicted in Fig (9). The program computes a

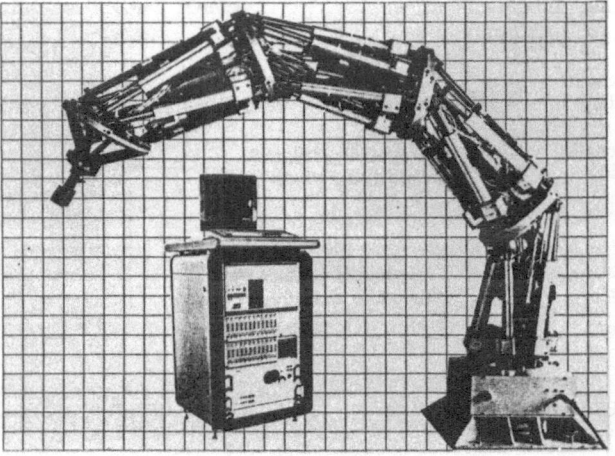

Figure 7: LOGABEX's robot LX

Figure 8: Position workspace of a module

robot configuration, that is, the generalized coordinates and the actuator lengths from the desired operational coordinates and is usually run for n modules.

3.7 Test implementation

The control program is implemented on the robot site. It operates on a multi-task, real-time, micro-computer INTEL 310.

We have controlled this robot comprising 1, 2, 3 or 4 modules and it may be noticed that the computation time for the robot inverse model increases with the number of modules. At this stage, the hardware and software facilities involve computation times that are inadequate for real-time control.

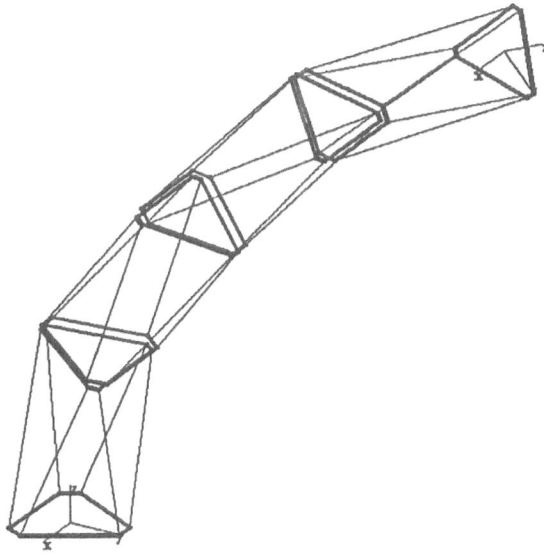

Figure 9: One configuration of the robot LX

4   Conclusion

In this paper, the major characteristics of a parallel robot manipulator have been highlighted, i.e., rigidity, ease of construction and ability to lift heavy loads. In particular, LOGABEX's modular, redundant robot manipulator LX which constitutes an original and patented application, has been presented.

The modelling of Stewart's platform as well as that of the robot LX, which is needed for control synthesis, has also been surveyed. Our contribution has primarily been concerned with a new method of computation of a module 's direct geometric model which furnishes all the solutions and the computation of the robot inverse model involving an iterative procedure utilizing a Jacobian matrix.

Nevertheless, several areas must still be investigated especially the modelling of such major domains as the characterization of the workspace and that of singularities. As to the robot, this redundant structure is interesting in order to implement and test new controls, notably in a clustered environment [21] [22].

5   Acknowledgments

The authors are grateful to Logabex [3] for providing the pictures and a movie of the robot LX.

---

[3]Logabex 3, avenue Didier Daurat 31400 Toulouse FRANCE

References

[1] I. Artobolevski. *Théorie des mécanismes et des machines*. Technique Soviétique. Editions Mir Moscou, 1977.

[2] J.P. Merlet. Parallel manipulators.part 1 : theory,design,kinematics,dynamics and control. Rapport de recherche No 646, Institut National de Recherche en Informatique et en Automatique (I.N.R.I.A.), March 1987.

[3] C. Gosselin. *Kinematic analysis, optimization and programming of parallel robotic manipulators*. PhD thesis, McGill University, Montréal, Canada, June 1988.

[4] V. Hayward and R. Kurtz. Modeling of a parallel wrist mechanism with actuator redundancy. Technical report, McGill Research Center for Intelligent Machines, McGill University, Montréal, Canada, January 1989.

[5] E.F. Fichter. A Stewart Platform-based manipulator : general theory and practical construction. *The International Journal of Robotics Research*, Vol.5(No 2):157–182, Summer 1986.

[6] H. McCallion and P.D. Truong. The analysis of a six-degree-of-freedom work station for mechanised assembly. In *Proceedings of the Fifth World Congress on Theory of Machines and Mechanisms*, pages 611–616. The American Society of Mechanical Engineers, July 1979.

[7] J.P. Merlet. Contribution à la formalisation de la commande par retour d'efforts en robotique. Application à la commande de robots parallèles. Thèse de l'Université, Paris 6, June 1986.

[8] C. Reboulet. *Techniques de la robotique. Architectures et commandes. Chapitre 8 : Modélisation des robots parallèles*. Hermès, 1988.

[9] H. Inoue, Y. Tsusaka, and T. Fukuizumi. Parallel manipulator. In *Robotics Research: The Third International Symposium, O. Faugeras and G. Giralt (Eds), MIT Press, Cambridge, Massachusetts*, pages 321–327, October 1985.

[10] C. Reboulet and A. Robert. Hybrid control of a manipulator equipped with an active compliant wrist. In *Robotics Research: The Third International Symposium, O. Faugeras and G. Giralt (Eds), MIT Press, Cambridge, Massachusetts*, October 1985.

[11] J.P. Merlet. Force-feedback control of parallel manipulators. In *IEEE International Conference on Robotics and Automation, Philadelphia (USA)*, pages 1484–1489, April 1988.

[12] S. Charentus, C. Diaz, and M. Renaud. Modular serial parallel redundant robot. In *Preprints IMACS on System Modelling and Simulation, Cetraro Italy*, pages 57–63, September 1988.

[13] P. Nanua and K.J. Waldron. Direct kinematic solution of a Stewart Platform. In *IEEE International Conference on Robotics and Automation, Scottsdale (USA)*, pages 431–437, May 1989.

[14] A. Robert. Commande hybride position-force. Mise en oeuvre et expérimentation sur un micro-manipulateur parallèle. Thèse de docteur ingénieur, Ecole Nationale Supérieure de l'Aéronautique et de l'Espace, December 1986.

[15] B. Gorla and M. Renaud. *Modèles des robots-manipulateurs. Application à leur commande.* CEPADUES Ed. Toulouse. France., 1984.

[16] K.H. Hunt. *Kinematic geometry of mechanisms.* Oxford engineering science series. Clarendon Press, Oxford, 1978.

[17] J.P. Merlet. Parallel manipulators.Part 2 : Theory.Singular configurations and Grassmann geometry. Rapport de recherche No 791, Institut National de Recherche en Informatique et en Automatique (I.N.R.I.A.), February 1988.

[18] P. Borrel. Contribution à la modélisation géométrique des robots manipulateurs. Application à la conception assistée par ordinateur. Thèse d'état, Université des Sciences et Techniques du Languedoc, Montpellier, July 1986.

[19] T. Berthomieu. Etude d'un micro-manipulateur parallèle et de son couplage avec un robot porteur. Thèse de docteur ingénieur, Ecole Nationale Supérieure de l'Aéronautique et de l'Espace, January 1989.

[20] C.A. Klein and C.H. Huang. Review of pseudoinverse control for use with kinematically redundant manipulators. *IEEE Transactions on Systems, Man, and Cybernetics*, Vol.SMC-13(No.3):245–250, march/april 1983.

[21] B. Espiau. An overview of local environment sensing in robotics applications. In *NATO advanced Research Workshop: Sensors and Sensory Systems for Advanced Robotics, Maratea (Italy)*, April 1986.

[22] T. Yoshikawa. Manipulability and redundancy control of robotic mechanisms. In *IEEE International Conference on Robotics and Automation, St Louis (USA)*, pages 1004–1009, 1985.

IDENTIFICATION AND CALIBRATION OF THE GEOMETRIC PARAMETERS OF ROBOTS

W.KHALIL* , J.L.CAENEN**, Ch.ENGUEHARD*

* E.N.S.M, Laboratoire D'automatique, URA CNRS 256, 44072 NANTES Cedex, France

** Ecole des Mines de Douai, BP 838, 59508 DOUAI Cedex, France.

Abstract

This paper presents a general method to identify the geometric parameters of serial robots. The robot location and the tool location parameters are taken into account.

Experimental results are applied on a six degree of freedom robot of cylindrical type. The measurement of the coordinates of the position of the tool is carried out by means of two theodolites.

1-Introduction

The absolute errors of robots are both nongeometric and geometric[1,2]. The former may be due to friction, compliance, gear transmission and backlash. The later may result from imprecise manufacturing of the robot links and joints, or from the deviation of the encoder offsets. It may be also due to the poor estimation of the parameters defining the location (position and orientation) of the robot with respect to the fixed reference frame, and of the poor estimation of the parameters defining the tool with respect to the terminal link.

This paper tries to increase the accuracy of the robot by identifying the errors of the geometric parameters from their nominal values. The identification model is based on the differential variation of the position of the tool as function of the differential variation of the geometric parameters.

2-Definition of the geometric parameters

In this section we define the geometric parameters needed to calculate the location of the tool with respect to a fixed reference frame. Three types of parameters have to be taken into account :

The robot parameters, the parameters defining the base of the robot with respect to a fixed frame, and the parameters defining the tool with respect to the terminal link.

2-1 Definition of the robot parameters

The system to be considered is an open loop mechanism of n joints and n+1 links. Link 0 is the base while link n is the terminal link. The coordinate frame j is assigned fixed with respect to link j. The definition of the link frames will be carried out by the modified Denavit and Hartenberg notation[3]. The z_j axis is along the axis of joint j, the x_j axis is along the common perpendicular of z_j and z_{j+1}. Frame j is defined with respect to frame j-1 by the matrix $^{j-1}T_j$ which is function of the four parameters $(\alpha_j, d_j, \theta_j, r_j)$, figure 1, such that :

$$^{j-1}T_j = Rot(x, \alpha_j)Trans(x, d_j)Rot(z, \theta_j) Trans(z, r_j)$$

$$j^{-1}T_j = \begin{bmatrix} & j^{-1}A_j & & j^{-1}P_j \\ 0 & 0 & 0 & 1 \end{bmatrix} = \begin{bmatrix} C\theta_j & -S\theta_j & 0 & d_j \\ C\alpha_jS\theta_j & C\alpha_jC\theta_j & -S\alpha_j & -r_jS\alpha_j \\ S\alpha_jS\theta_j & S\alpha_jC\theta_j & C\alpha_j & r_jC\alpha_j \\ 0 & 0 & 0 & 1 \end{bmatrix} \tag{1}$$

where:

$j^{-1}A_j$ defines the orientation of frame j with respect to frame j-1,

$j^{-1}P_j$ defines the position of the origin of frame j with respect to frame j-1.

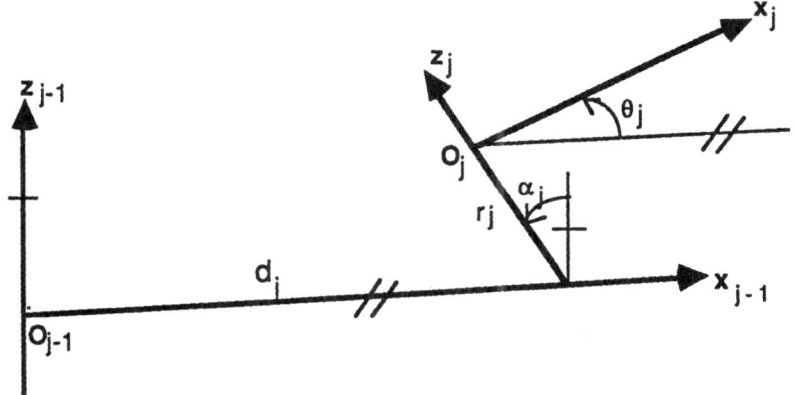

Figure 1

The joint variable j will be given as :

$$q_j = \bar{\sigma}_j \, \theta_j + \sigma_j \, r_j \tag{2}$$

where: $\sigma_j = 0$ for j rotational, $\sigma_j = 1$ for j translational, and $\bar{\sigma}_j = (1-\sigma_j)$.
We define also:

$$\bar{q}_j = \sigma_j \, \theta_j + \bar{\sigma}_j \, r_j \tag{3}$$

It is to be noted that frame 0 can be defined such that z_0 is along z_1, and x_0 is along x_1 when q_1 = 0, so $\alpha_1 = 0$, $d_1 = 0$, $\bar{q}_1 = 0$, while frame n can be defined such that $\bar{q}_n = 0$ [3,4].

2-2 Definition of the tool parameters
      Let n+1 be the tool frame, this frame can be defined arbitrary by the user, therefore its location with respect to frame n will need 6 parameters. A method to define arbitrary a frame j with respect to a frame j-1 is given in figue 2, [3,5].
Thus we can define the tool frame with respect to frame n as:

$$^nT_{n+1} = Rot(z,\gamma_t)Trans(z,b_t)Rot(x,\alpha_t)Trans(x,d_t)Rot(z,\theta_t)Trans(z,r_t) \tag{4}$$

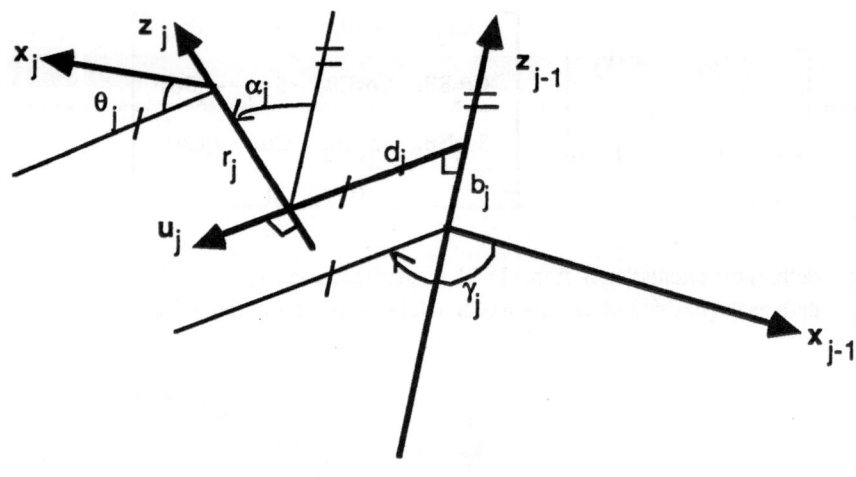

Figure 2

Concatenating this matrix with the matrix $^{n-1}T_n$ we find that:

$^{n-1}T_{n+1}=^{n-1}T_n\ ^nT_{n+1}=Rot(x, \alpha_n)Trans(x, d_n)Rot(z, \theta_n)\ Trans(z, r_n)$
$$Rot(z,\gamma_t)Trans(z,b_t)Rot(x,\alpha_t)Trans(x,d_t)Rot(z,\theta_t)Trans(z,r_t)$$

Assuming $\theta_{n'} = \theta_n + \gamma_t$, and $r_{n'} = r_n + b_t$, we deduce that

$$Rot\ (z,\theta_n)\ Trans(z,r_n)\ Rot(z,\gamma_t)Trans(z,b_t) = Rot\ (z,\theta_{n'})Tran\ (z,r_{n'})$$

So:

$^{n-1}T_{n+1}=^{n-1}T_n\ ^nT_{n+1}\ =Rot(x,\alpha_n)Tran(x,d_n)Rot(z,\theta_{n'})Trans(z,r_{n'})$
$$Rot(x,\alpha_{n+1})Tran(x,d_{n+1})Rot(z,\theta_{n+1})Trans(z,r_{n+1}) \qquad (5)$$

So by an appropriate change of variables the frame $n+1$ can be represented by 4 independent parameters similar to those defining the robot links such that $(\alpha_{n+1} = \alpha_t, d_{n+1}= d_t, \theta_{n+1}= \theta_t, r_{n+1}= r_t)$.

Remark: From this study we conclude that the concatenation of successive frames defined by six parameters, as given in Eq.4, can be reduced to the representation of four parameters as given in section 2-1. So Denavit and Hartenberg notation is not an under determined parameterization as it may seem to be [6].

2-3 Definition of the World fixed frame

The world fixed frame numbered -1 can be defined arbitrary by the user, so frame 0 will be defined with respect to frame -1 by six parameters. Using Eq.4 the matrix $^{-1}T_0$ can be defined by the parameters: $(\gamma_z, b_z, \alpha_z, d_z, \theta_z, r_z)$. Since $\alpha_1 = 0$, $d_1 = 0$, the compound transformation $^{-1}T_1$ can be given as:

$^{-1}T_1 = ^{-1}T_0\ ^0T_1 = Rot(z,\gamma_z)Trans(z,b_z)Rot(x,\alpha_z)Trans(x,d_z)Rot(z,\theta_z)Trans(z,r_z)$
$$Rot\ (z,\theta_1)\ Trans(z,r_1)$$

which gives:

$$^{-1}T_0 \ ^0T_1 = \text{Rot}(z,\gamma_z)\text{Trans}(z,b_z) \ \text{Rot}(x,\alpha_{1'})\text{Tran}(x,d_{1'})\text{Rot}(z,\theta_{1'}) \ \text{Trans}(z,r_{1'}) \tag{6}$$

with: $\alpha_{1'} = \alpha_z, d_{1'} = d_z, \theta_{1'} = \theta_1 + \theta_z$, and $r_{1'} = r_1 + r_z$

From Eq.6 we conclude that this change of variables on frame 1 parameters, will enable us to represent the fixed frame as we have done for the link frames. Such that:

$$\alpha_0 = 0, d_0 = 0, \theta_0 = \gamma_z, r_0 = b_z \tag{7}$$

As a conclusion the calculation of $^{-1}T_{n+1}$ will need $4(n+2)-2$ independent parameters.

3-The identification model

3-1 Position of the problem

The tool frame location X can be calculated by the direct (forward) geometric model given as:
$$X = {}^{-1}T_{n+1} = {}^{-1}T_0 \ ^0T_1 \ ^1T_2 \dots {}^{n-1}T_n \ ^nT_{n+1} \tag{8}$$

Owing to the errors on the geometric parameters, there will be a difference between the real location of the terminal frame and its location as calculated using the direct geometric model. The calibration problem is to adjust the geometric parameters such that this difference will be minimum.

$$X_{real} - X_{model} = \text{minimum} \tag{9}$$

As X_{model} is nonlinear function of the geometric parameters, this is a nonlinear problem. Assuming errors of first order, a linear differential model can be used [7,8,9]. If the errors cannot be considered as first order an iterative procedure based on the differential model also can be investigated using Newton-Gauss procedure.
The differential vectors defining the deviation of the tool frame from the nominal value due to the differential error in the geometric parameters can be given by:

$$\begin{bmatrix} D_{n+1} \\ \delta_{n+1} \end{bmatrix} = J_{B_{n+1}} \Delta B \tag{10}$$

where:
- D_{n+1} is the (3x1) translational differential vector of the tool frame,
- δ_{n+1} is the (3x1) rotational differential vector of the tool frame,
- ΔB defines the error in the geometric parameters,
- $J_{B_{n+1}}$ is the extended jacobian matrix.

3-2- Definition of ΔB

From section 2, the calculation of $^{-1}T_{n+1}$ will need $4(n+2)-2$ independent parameters. So the elements of ΔB are the differential variation of the corresponding parameters, but we note that the representation of equation (4) on which this modeling is based, is singular when the z axes of two successive frames are parallel. Therefore if z_{j-1} is parallel to z_j an additional differential parameter $\Delta\beta_j$ must be taken into account [9,10,11], this additional parameter will represent a rotation around the axis y_{j-1}. On the other hand we remark that the change on r_j will not be calculated because it is in the same direction of r_{j-1}, thus the maximum number of errors for each frame will remain equal to 4. On a case by case basis the errors of some other parameters may not be included in ΔB see section 3-4 of this paper.

3-3 Calculation of the jacobian matrix

Assuming the matrix defining the frame j with respect to the fixed frame as:

$$^{-1}T_j = \begin{bmatrix} ^{-1}s_j & ^{-1}n_j & ^{-1}a_j & ^{-1}P_j \\ 0 & 0 & 0 & 1 \end{bmatrix} \tag{11}$$

The calculation of the columns of the extended jacobian matrix can be calculated directly as follows [4]:

1- Since the parameter α_j represents a rotation around the x_{j-1} axis (of unit vector s_{j-1}) .The column corresponding to $\Delta\alpha_j$ in the jacobian matrix will be given as:

$$j\alpha_j = \begin{bmatrix} s_{j-1} x L_{j-1,n+1} \\ s_{j-1} \end{bmatrix} \tag{12}$$

Where :
> x denotes the vector product,
> $L_{i,n+1}$ is the (3x1) vector between the origin of frame i and the origin of frame n+1.

2- Since the parameter d_j represents a translation along the x_{j-1} . The column corresponding to Δd_j in the jacobian matrix will be given as:

$$jd_j = \begin{bmatrix} s_{j-1} \\ 0 \end{bmatrix} \tag{13}$$

3- Since the parameter θ_j represents a rotation around the z_j axis. The column corresponding to $\Delta\theta_j$ in the jacobian matrix will be given as:

$$j\theta_j = \begin{bmatrix} a_j x L_{j,n+1} \\ a_j \end{bmatrix} \tag{14}$$

4- Similarly the column corresponding to Δr_j is:

$$jr_j = \begin{bmatrix} a_j \\ 0 \end{bmatrix} \tag{15}$$

5- Since the parameter β_j represents a rotation around the y_{j-1} axis, we get:

$$j\beta_j = \begin{bmatrix} n_{j-1} x L_{j-1,n+1} \\ n_{j-1} \end{bmatrix} \tag{16}$$

All the vectors appearing in equations (12,...,16) must be referred to the measuring fixed frame. So we need to calculate the vectors s_j, n_j, a_j, $L_{j,n+1}$ referred to the fixed frame. These vectors can be obtained from the matrices $^{-1}T_j$ (j=0,...,n+1).The vector $L_{j,n+1}$ will be obtained as $= {}^{-1}P_{n+1} - {}^{-1}P_j$. These matrices can be calculated symbolically or numerically. The numerical calculation is general and preferred for the off line identification especially if ΔB cannot be considered as first order and iterative calculation is investigated to identify it. The symbolic calculation will be more efficient, from the computation cost point of view, for the correction problem.

Only the equations corresponding to the position error D_{n+1}, on sufficient number of points, are used in the identification process. The solution of the corresponding system of equation, represented by the first three equations of relation (10), is obtained by a classical least squares method.

3-4 Identifiable geometric parameters

If a column of the jacobian matrix can be obtained as linear combination of some other columns, the corresponding error parameter cannot be identified separatelly. In this case the error of this parameter will be forced to zero and the corresponding error will be calculated and calibrated through the errors on the other parameters. There is no proposed general method to determine these parameters till now, they can be obtained on a case by case basis by detecting the linear independence of the different columns of the jacobian matrix. Numerical procedure like those proposed by sheu and Walker for the inertial parameters [12] can be used.

4- Experimentation

The experimental results are carried out on a french robot TH8-ACMA, which is a six degree of freedom robot figure 3. The tool is a cylinder along the z_6 axis.

The classification of the identifiable parameters and the eliminated parameters, on which the error parameters cannot be calculated, for the TH8 robot are given in table 1. The linear relations between the columns of the jacobian matrix are given in the appendix.

4.1 Measurement of Cartesian position

4.1.1 Instrumentation

The instrumentation must be able to determine with precision the cartesian position of the tool point of the robot. So, the measurement system is based on the use of two theodolites and a cylindrical target fixed on the robot terminal link.

The angular accuracy of the theodolites is about 10^{-5} rad for the azimuth and elevation angles. The four angles given by the theodolites permit to calculate the three-dimensional position coordinates of all desired points. The theodolites are manually operated, but the data are read and the cartesian coordinates are calculated by a computer. The accuracy of measurement is about $\pm$ 0,02 mm/m for each measurement coordinates, significantly better than the repeatability of most of the robots. The point tool is materialized by a 2 mm diameter sphere.

4.1.2 Measurements

The measurements are carried out in two steps:

- In a first step, we prepare the measurements and notice a lot of the robot characteristics such as:
  - the encoder resolutions,
  - the gear transmissions,
  - the current initialization values (offsets) of the encoder ,
  - the values of the current (nominal) geometrical parameters, see table 2.
  - the robot repeatability (about 0.3 for the TH8 robot).

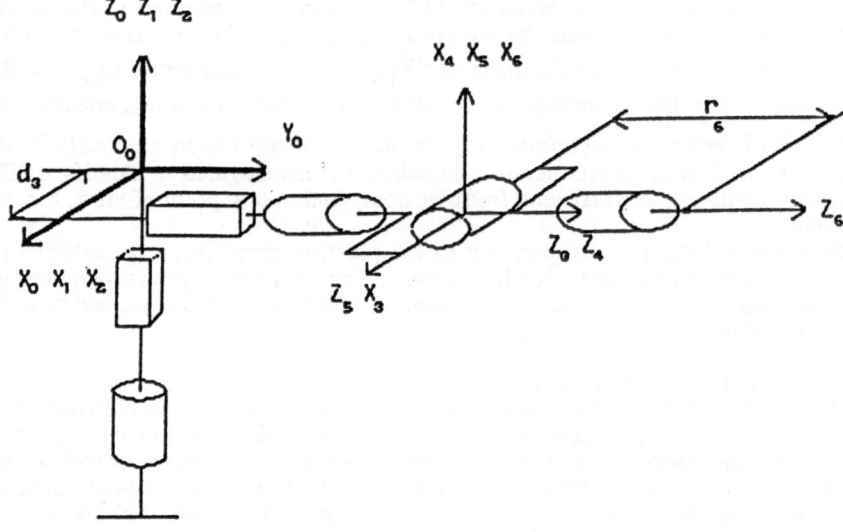

Figure 3, The TH8 Robot.

Frame	$\Delta\beta$	$\Delta\alpha$	Δd	$\Delta\theta$	Δr
0	0	0	0	Y	E
1	Y	Y	Y	Y	E
2	Y	Y	E	E	Y
3	E	Y	Y	E	Y
4	Y	Y	E	Y	E
5	E	Y	Y	Y	E*
6	E*	Y	E*	E*	Y
7	E**	E*	Y	E**	E*

* parameters eliminated owing to the given particular tool.
** parameter eliminated owing to the measure of position error only.

Table 1, Classification of the identifiable parameters (E=eliminated, Y= identifiable)

Frame	σ_j	α_j	d_j	θ_j	r_j
0		0.0	0.0	74.78	0.0
1	0	0.0	3030.2	$-163.08+\theta_1$	-415.12
2	1	0.0	0.0	0.0	r_2
3	1	-90.0	199.0	0.0	r_3
4	0	0.0	0.0	θ_4	0.0
5	0	-90.0	0.0	θ_5	0.0
6	0	90.0	0.0	θ_6	221.8
7	0	0.0	0.0	0.0	0.0

Table 2: The nominal parameters of the TH8 (distances are in mm and angles in degrees).

- In a second step, the measurements are carried out for sufficient number of robot configurations (points). For each of them, we measure the cartesian coordinates of the tool point and read the corresponding values of the joint encoders to calculate the nominal (model) position. These configurations are obtained from significant displacements of the robot axis. A total of 48 configurations covering the working space of the robot has been recorded. The duration of the measurements was about three hours for a single operator.

The measurements are defined in the frame of the theodolites considered as frame -1. This frame is chosen fixed with one of the theodolites and defined as follow:

• the z_{-1} axis is the azimuth rotation axis (vertical direction),
• the x_{-1} axis is the elevation rotation axis,

The location of frame 0 with respect to the theodolites frame ($^{-1}T_0$), can be obtaiend by identifying the circles or the lignes resulting from moving the first axes separatelly, the choice of these axes is depending on the particular structure of the robot [13]. In the case of the TH8, the definition of frame 0 is the following:
• axis z_0 is the perpendicular on the circular plane obtained by rotating joint 1,
•the origin of the frame is the center of the circle during the rotation of joint 1
• the direction of y_0 is obtained by identifying the translation of joint 3,

The accuracy of the robot with respect to frame -1 and frame 0 is given in table 3. The accuracy with respect to frame -1 includes the effect of the impricise definition of the matrix $^{-1}T_0$.

Reference frame	average (in mm)	standard deviation (in mm)
-1	3.168	1.053
0	2.764	0.877

Table 3, Absolute accuracy of the robot before calibration

4.2 Identification results

Using the same data, four identification process have been carried out:
1- Simplified error vector using measurements with respect to frame -1: We suppose that ΔB contains only the parameters which can be taken into account by the inverse geometric model. These parameters are $\Delta\theta_0, \Delta\beta_1, \Delta\alpha_1, \Delta d_1, \Delta d_3, \Delta r_6, \Delta d_7$ and the encoder offsets. The identified errors are obtained as (distances are in mm and angles in degrees):

i- constant parameters: $\Delta\theta_0 = 0.001$, $\Delta\beta_1 = 0.075$, $\Delta\alpha_1 = -0.116$, $\Delta d_1 = 0.903$, $\Delta d_3 = 0.565$,
$\Delta r_6 = -3.505$, $\Delta d_7 = -0.150$.

ii- offsets variables: $\Delta\theta_1 = -0.025$, $\Delta r_2 = 4.863$, $\Delta r_3 = 2.081$, $\Delta\theta_4 = -0.016$, $\Delta\theta_5 = -0.045$.

2- Simplified error vector using measurements with respect to frame 0. The identified errors are obtained as:

i- constant parameters: $\Delta d_3 = 1.220$, $\Delta r_6 = -3.667$, $\Delta d_7 = -0.099$

ii- variables offsets: $\Delta\theta_1 = -.037$, $\Delta r_2 = 1.651$, $\Delta r_3 = 1.897$, $\Delta\theta_4 = -0.069$, $\Delta\theta_5 = -0.083$.

3- A complete identification of all the independent elements of ΔB considering the measurements with respect to frame -1. The identified errors are given in table 4.

Frame	$\Delta\beta$	$\Delta\alpha$	Δd	$\Delta\theta$	Δr
0	0.0	0.0	0.0	0.00	0.0
1	0.137	−0.094	1.243	−0.031	0.0
2	0.020	−0.112	0.0	0.0	6.297
3	0.0	0.040	0.541	0.0	1.975
4	−0.033	0.029	0.0	−0.104	0.0
5	0.0	−0.091	−0.045	−0.012	0.0
6	0.0	0.132	0.0	0.0	−3.485
7	0.0	0.0	−0.109	0.0	0.0

Table 4: Identification of all the parameters (base and robot).

4- A complete identification of all the independent elements of ΔB considering the measurements with respect to frame 0. The identified errors are given in table 5.
The identification process has been carried out using 30 points. The convergence of the algorithm takes place by considering about 25 points.
Calibrating the direct geometric model is always possible. The results of the calibration of the robot for the different identification algorithms are given on table 6. The accuracy are given for the points used in the identification and for the points which have not been used for the identification. For this robot we see that the accuracy by calibrating all the parameters is comparable to the calibration of the simplified vector parameters. So the simplified model may be sufficient .

Frame	$\Delta\beta$	$\Delta\alpha$	Δd	$\Delta\theta$	Δr
1	0.0	0.0	0.0	0.008	0.0
2	0.005	−0.066	0.0	0.0	0.712
3	0.0	0.099	0.999	0.0	2.015
4	−0.115	0.046	0.0	−0.105	0.0
5	0.0	0.036	−0.013	−0.047	0.0
6	0.0	0.046	0.0	0.0	−3.547
7	0.0	0.0	−0.080	0.0	0.0

Table 5 : Identification of robot parameters.

Algorithm number	Points used in the identification		Points not used in the identification	
	average	standard deviation	average	standard deviation
1	0.544	0.238	0.797	0.300
2	0.695	0.339	0.831	0.249
3	0.471	0.205	0.866	0.335
4	0.606	0.227	0.919	0.363

Table 6: Accuracy of the calibrated robot

5- Calibrating the inverse geometric model

The inverse geometric model of the robot can be calibrated directly using the parameters of the simplified identification parameters. In the case of the complete identification parameters only the values appearing in the inverse geometric model can be taken into account by the robot control system. The errors of the other parameters can be compensated using a differential procedure as follows:

1) for a desired location X_d calculate the joint variables using the "simplified" calibrated robot model.

2) calculate the differential error ΔX using a completely calibrated direct geometric model or using direct differential model using Eqs.(12,...,16).

3) calculate the corresponding joint variable changes to place the robot at X_d and add to the joint solutions obtained in 1).

6-Conclusion

This paper presents a general method to identify the geometric parameters of serial robots. The robot location and the tool location parameters are treated by an approach similar to that defining the link frames. The identification model is based on the differential changes of the position of the tool as function of the error of the geometric parameters, a direct method has been presented to get this model. Experimental results are applied on a six degree of freedom robot of cylindrical type. The measurement of the coordinates of the position of the tool is carried out by means of two theodolites. More automatic and rapid measuring instrument are really needed. The problem of selecting the points satisfying the well conditionning of the identification process must pay more attention.

References

[1] D.E.Whitney,C.A. Lozinski, J.M.Rourke, "Industrial robot forward calibration method and results", ASME J. of Dynamics Systems Measurements, and Control, Vol. 108, pp.1-8, 1986.
[2] S.Roth,B.W.Mooring, B.Ravani, "An overview of robot calibration", IEEE J. Robotics and Automation Vol. RA-3, n° 5-1987.
[3] W.Khalil, J.F.Kleinfinger, "A new geometric notation for open and closed loop robots" Proc. IEEE Robotics and Automation Conf., San francisco,1986, pp.1174–1180.
[4] E.Dombre, W.Khalil, "Modélisation et commande des robots". Edition Hermès, Paris, 1988.
[5]H.W.Stone, A.C.Sanderson,C.P.Neuman, "Arm signature identification", Proc.IEEE Robotics and Automation Conf., San francisco, 1986,pp.41-47.
[6]J.Ziegert,P.Datseris,"Basic considerations for robot calibration", Proc. IEEE Robotics and Automation Conf., Philadelphia, 1988,pp.932-938.
[7] C.H.Wu, " A kinematic CAD tool for the design and control of a robot manipulator", Int. J.of robotics research, Vol.(3), 1984, pp.58-67.
[8] K.Sugimoto, T.Okada,"Compensation of positioning errors caused by geometric deviations in robot system", Proc. 2 nd Int. Symp.of robotics research, Vol.3(1), 1984, pp.58-67.
[9] D. Payannet, "Modélisation et correction des erreurs statiques des robots manipulateurs", Thèse de doctorat, Montpellier, 1986.
[10] S.A.Hayati,"Robot arm geometric link calibration", Proc. IEEE, Decision and Control Conf.,pp 798-800, 1988
[11] C.H.Wu,J.Ho,K.Y.Young,"Design of robot accuracy compensator after calibration", Proc. IEEE Robotics and Automation Conf.,pp 780-785,1988.
[12] S.Y.Sheu ,M.W.Walker,"Basis sets for manipulator inertial parameters", Proc. IEEE Robotics and Automation Conf., Scottsdale, 1989.
[13] J.L.Caenen, "Programmation hors-ligne, Etalonnage de robots: Identification des paramètres géométriques", DEA- Valenciennes, 1988.

Appendix

The linear relations between the columns of the jacobian matrix

Calculating the jacobian columns, two cases are considered:

a- relations considering position and orientation errors (the six components are taken into account):

$$jr_0 = jr_1 = jr_2 , jr_3 = jr_4 , jr_6 = jr_7 ,$$
$$j\theta_1 = j\theta_2 , j\theta_3 = j\theta_4 = j\beta_3 + d_3 jr_2 ,$$
$$j\theta_5 = j\beta_5 , j\theta_6 = j\theta_7 ,$$
$$jd_2 = jd_3 = jd_4$$

b- relations considering position errors only (first three components of each column are taken into account)):

$$j\beta_6 = 0 , j\beta_7 = 0 ,$$
$$j\alpha_6 = -r_6 \, jr_5 , j\alpha_7 = 0 ,$$
$$j\theta_5 = r_6 \, jd_6 , j\theta_6 = 0, \ j\theta_7 = 0$$

Closed-Loop Kinematic Calibration
of the Utah-MIT Hand

David J. Bennett and John M. Hollerbach
Massachusetts Institute of Technology
Artificial Intelligence Laboratory
Cambridge, MA 02139

Abstract

Determining accurate kinematic models of manipulators is often difficult and tedious, particularly for a device with the kinematic complexity of the Utah-MIT hand. Considerable effort has gone into developing automatic schemes for calibrating open-loop kinematic chains. While these schemes have been experimentally verified on many standard manipulators, they unfortunately rely on special instrumentation to track the manipulator's endpoint. Recently, it has been theoretically shown that a manipulator(s) with sufficient redundancy may form a single-loop closed kinematic chain that may self-calibrate. That is, merely by using joint angle readings and self motions, loop consistency conditions can be utilized to identify the kinematic parameters. No endpoint sensing is required. We will demonstrate that this closed-loop kinematic calibration technique is applicable to the calibration of the Utah-MIT hand.

1   Introduction

A previous paper [1] proposed a technique for calibrating closed-loop kinematic chains formed by multiple manipulators connected together. This technique differs from conventional calibration schemes[2] in that it does not require special endpoint-sensing equipment. The present work will experimentally verify this closed-loop calibration technique by calibrating the Utah-MIT hand[4][5]. We are aware of one similar unpublished experiment (using a Puma manipulator)[3].

Briefly, the closed loop kinematic calibration method is described as follows. Consider a finger of the hand system opposing the thumb, such that the fingertips are rigidly connected (Figure 1). As each finger has four degrees of freedom (DOF) an 8-DOF closed loop is formed; this loop has in general a mobility of two. Observe that fixing only two joint angles uniquely defines the configuration of this mechanism (except for the possibility of multiple inverse kinematic solutions). Thus, the other six joint angle sensors are *redundant*. This sensor redundancy may be exploited to estimate the kinematic parameters. Specifically, the loop consistency equations for a given configuration give three position and three orientation equations containing the unknown kinematic parameters. Moving

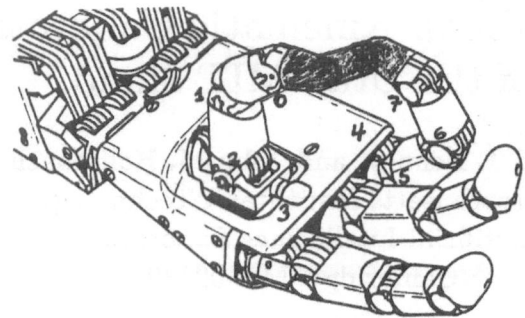

Figure 1: Hand with two fingers opposed, adapted from [4].

the fingers into different configurations while maintaining the same contact condition provides six additional equations per pose. Potentially these equations can be solved for the kinematic parameters, that is provided certain identifiability conditions are met [1].

The paper is organized as follows: (1) outline of calibration procedure in [1], (2) experimental results, and (3) discussion of future work. Actually, several modifications are made to [1]. Initial experiments revealed that the joint angles of the hand not only have joint offsets, but also joint scale factors that are difficult to determine a priori. For this reason the algorithm in [1] is augmented to include the identification of these joint scale factors. Perhaps the most difficult task in any kinematic calibration procedure is determining the initial guess at the kinematic parameters, particularly for the base and fingertip transformations. As a partial solution to this problem, the more well known parameters are fixed while the less well known ones are first estimated. Finally. singular value decomposition provides a means of dealing with parameter ambiguity, and also conveniently produces a measure of parameter observability.

2 Method

2.1 Manipulator kinematics

2.1.1 Geometric parameters

Consider two fingers of the hand connected together rigidly at their tips (i.e. the thumb opposed to a finger); see Figure 1. This closed kinematic chain has $n_f = 8$ degrees of freedom. Locate a reference (base) coordinate frame coincident with the last joint of the thumb; then number the joints proceeding from that distal joint to the palm, and then back out to the tip of the finger, as in Figure 1. Let the 4×4 homogeneous transformation A_j from link j to link $(j-1)$ be defined by the Denavit-Hartenberg (D-H) convention[6] given in Figure 2 and symbolically as:

$$A_j \;=\; Rot(z, \theta'_j) Trans(z, s_j) Trans(x, a_j) Rot(x, \alpha_j) \tag{1}$$

where the notation $Rot(x, \phi)$ indicates a rotation about an axis x by ϕ and $Trans(x, a)$ indicates a translation along an axis x by a.

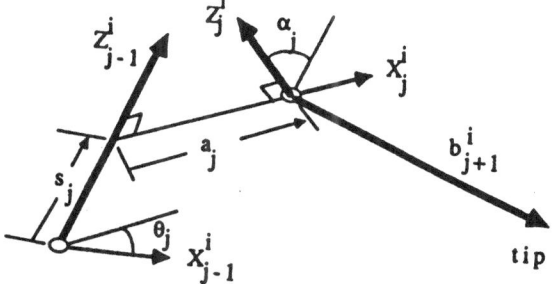

Figure 2: Denavit-Hartenberg coordinates, and tip vector $\mathbf{b}_j^i$.

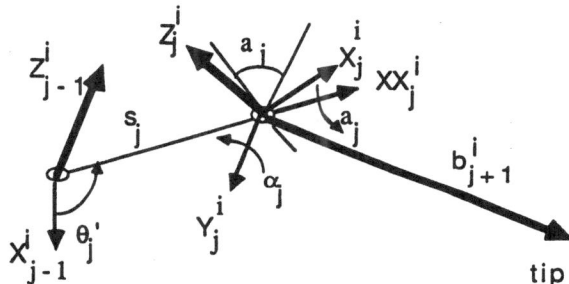

Figure 3: Hayati coordinates, and intermediate x-axis $\mathbf{xx}_j$.

Since the D-H s parameter is not uniquely defined for consecutive parallel axes the following Hayati convention[7] (see Figure 3) is employed for such axes (i.e. for joints 0, 1, 3, 5 and 6):

$$A_j = Rot(z, \theta'_j) Trans(x, s_j) Rot(x, a_j) Rot(y, \alpha_j) \tag{2}$$

The position of the last link is computed by a sequence of these homogeneous transformations:

$$T_c = A_1 A_2 ... A_{n_f} \tag{3}$$

The goal of kinematic calibration is to identify the geometric parameters s_j, α_j and a_j, and also any non-geometric parameters that may be included in the kinematic model.

2.1.2 Non-geometric parameters

The non-geometric effects on the kinematic model potentially include bearing play and joint angle sensor error. Although parametric models of both of these processes may be included we choose to only model joint angle sensor error. In particular, the D-H joint angle is presumed to be related to the sensor reading as

$$\theta'_j = k_j \theta_j + \theta_j^{off} \tag{4}$$

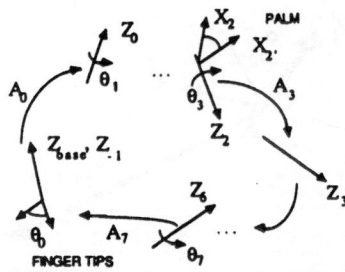

Figure 4: Coordinate assignment for loop of two fingers attached at their endpoints.

Thus, we wish to identify the constant joint angle offset θ_j^{off}, and the joint scale factor k_j. The joint scale gives the calibration factor for the measured electronic signal from the Hall effect joint sensors.

2.2 Unidentifiable and identifiable parameters

If we try to identify all the k_i there is a trivial solution $k_i = 0$, which will satisfy any set of joint angle data. To avoid this difficulty, fix a joint scale factor that can be measured independently. For the hand we choose $k_7 = 1.0$. Likewise, one link length must be known. This length defines the length scale of the closed mechanism [1]. For example, in Section 3 we set $s_0 = -1.3$.

All of the potentially identifiable kinematic parameters are placed into a single vector:

$$\underline{\varphi} \;=\; (\underline{\theta}_{off}^T, \underline{s}^T, \underline{a}^T, \underline{\alpha}^T, \underline{k}^T)^T \tag{5}$$

where $\underline{s} = (s_1, s_2, ...)^T$ etc.

2.3 Base coordinate assignment and endpoint location

The two connected fingertips may be viewed as a single effective endpoint link that connects the most distal joint of the finger to the first joint of the kinematic chain (i.e. the most distal joint of the thumb); see Figure 4. Defined as such, the endpoint always has zero orientation and position relative to the base coordinates (i.e. those coordinates that were defined to be aligned with the thumb's most distal joint axis). Notice that in addition to the robot kinematic parameters we will also identify the D-H parameters of this effective link which completes the loop.

2.4 Endpoint location error calculation

With a sufficiently good initial parameter estimate the computed endpoint location, T_c, differs only slightly from zero (the base coordinates). The endpoint error computations can therefore be simplified as follows. Let the computed position (i.e. the fourth column of T_c) be represented by $(dx_c, dy_c, dz_c)^T$. Similarly, let the calculated orientation R_c (that is, the upper left 3×3 matrix of T_c) be represented by infinitesimal xyz-Euler rotations:

$$R_x(\partial x_c) R_y(\partial y_c) R_z(\partial z_c) \approx \begin{bmatrix} 1 & -\partial z_c & \partial y_c \\ \partial z_c & 1 & -\partial x_c \\ -\partial y_c & \partial x_c & 1 \end{bmatrix} \tag{6}$$

where the right hand side of (6) is computed by directly expanding the left hand side and ignoring second order differential terms. Thus, the modeled endpoint location, evaluated at the ith joint configuration $\underline{\theta}^i$, is represented by a six vector:

$$\underline{x}_c^i = \underline{x}_c^i(\underline{\theta}^i, \underline{\varphi}) = (dx_c, dy_c, dz_c, \partial x_c, \partial y_c, \partial z_c)^T \tag{7}$$

directly computed from T_c. As stated in the previous section, the actual position and orientation of the endpoint are taken to be zero:

$$\underline{x}_a^i = (0, 0, 0, 0, 0, 0)^T \tag{8}$$

Thus, the endpoint error is:

$$\Delta \underline{x}^i = \underline{x}_a^i - \underline{x}_c^i = -\underline{x}_c^i \tag{9}$$

2.5 The loop closure equations

The six kinematic loop closure equations are therefore given by:

$$\underline{0} = \underline{x}_c^i(\underline{\theta}^i, \underline{\varphi}) + \underline{n}^i \tag{10}$$

The error term $\underline{n}^i$ has been added to emphasize that there is modeling and measurement noise.

2.6 Iterative identification

To solve for the kinematic parameters the equations (10) are linearized and iteratively identified.

2.6.1 Differential kinematics

At the ith joint configuration $\underline{\theta}^i$ the first differential of (10) is:

$$\Delta \underline{x}^i = \frac{\partial \underline{x}_c^i}{\partial \underline{\theta}} \Delta \underline{\theta} + \frac{\partial \underline{x}_c^i}{\partial \underline{s}} \Delta \underline{s} + \frac{\partial \underline{x}_c^i}{\partial \underline{a}} \Delta \underline{a} + \frac{\partial \underline{x}_c^i}{\partial \underline{\alpha}} \Delta \underline{\alpha} + \underline{n}'^i \tag{11}$$

where again $\Delta \underline{x}^i = 0 - \underline{x}_c^i$ and $\Delta \underline{s} = \underline{s} - \underline{s}^0$ etc. By denoting the combined Jacobian:

$$C^i = \frac{\partial \underline{x}_c^i}{\partial \underline{\varphi}} = \begin{bmatrix} \dfrac{\partial \underline{x}_c^i}{\partial \underline{\theta}} & \dfrac{\partial \underline{x}_c^i}{\partial \underline{\alpha}} & \dfrac{\partial \underline{x}_c^i}{\partial \underline{a}} & \dfrac{\partial \underline{x}_c^i}{\partial \underline{s}} \end{bmatrix} \tag{12}$$

equation (11) may be expressed more concisely as

$$\Delta \underline{x}^i = C^i \Delta \underline{\varphi} + \underline{n}'^i \tag{13}$$

2.6.2 Jacobian calculation

Each of the matrices in (11) are Jacobians with respect to the particular parameters. For example, $\partial \underline{x}^i_c/\partial \underline{\theta}$ is the familiar Jacobian which relates infinitesimal joint movements to end effector movements. Although the derivation of the Jacobians may be performed in a variety of ways, the following procedure[8][9] reveals that the columns of the Jacobian may be interpreted as *screw coordinates*.

Imagine the end effector variation $\Delta \underline{x}^i$ to be an instantaneous screw displacement composed of linear and angular velocity components. The combined variation in all the parameters is presumed to cause this endpoint variation. Specifically, a variation of the D-H parameter s_j along the local link z axis, z^i_{j-1}, causes a contribution to the end effector linear velocity of $\Delta s_j z^i_{j-1}$. The parameter variation $\Delta \alpha_j$ about the local link x axis, x^i_j, causes a contribution to the endpoint's angular velocity of $(\Delta \alpha_j)x^i_j = w_{\alpha_j}$, and a linear velocity contribution of $w_{\alpha_j} \times b^i_{j+1}$, where b^i_{j+1} is a vector from the jth coordinate system to the endpoint (Figure 2). The θ_j and a_j parameters are treated analogously. In total, the endpoint translation due to all of the parameter variations is given by:

$$\sum_{j=1}^n z^i_{j-1} \times b^i_j \Delta\theta_j + z^i_{j-1}\Delta s_j + x^i_j \times b^i_{j+1}\Delta\alpha_j + x^i_j \Delta a_j \tag{14}$$

and angular variation given by:

$$\sum_{j=1}^n z^i_{j-1}\Delta\theta_j + x^i_j\Delta\alpha_j \tag{15}$$

Comparing these to (11) it is seen that the columns of each of the four Jacobians are

$$\text{col}_j \frac{\partial \underline{x}^i_c}{\partial \underline{\theta}} = \left[\begin{array}{c} z^i_{j-1} \times b^i_j \\ z^i_{j-1} \end{array} \right], \quad \text{col}_j \frac{\partial \underline{x}^i_c}{\partial \underline{s}} = \left[\begin{array}{c} z^i_{j-1} \\ 0 \end{array} \right] \tag{16}$$

and

$$\text{col}_j \frac{\partial \underline{x}^i_c}{\partial \underline{a}} = \left[\begin{array}{c} x^i_j \\ 0 \end{array} \right], \quad \text{col}_j \frac{\partial \underline{x}^i_c}{\partial \underline{\alpha}} = \left[\begin{array}{c} x^i_j \times b^i_{j+1} \\ x^i_j \end{array} \right] \tag{17}$$

The Jacobian columns for parameters of the alternate Hayati convention are found analogously to be:

$$\text{col}_j \frac{\partial \underline{x}^i_c}{\partial \underline{\theta}} = \left[\begin{array}{c} z^i_{j-1} \times b^i_j \\ z^i_{j-1} \end{array} \right], \quad \text{col}_j \frac{\partial \underline{x}^i_c}{\partial \underline{s}} = \left[\begin{array}{c} xx^i_{j-1} \\ 0 \end{array} \right] \tag{18}$$

and

$$\text{col}_j \frac{\partial \underline{x}^i_c}{\partial \underline{a}} = \left[\begin{array}{c} xx^i_j \times b^i_{j+1} \\ xx^i_j \end{array} \right], \quad \text{col}_j \frac{\partial \underline{x}^i_c}{\partial \underline{\alpha}} = \left[\begin{array}{c} y^i_j \times b^i_{j+1} \\ y^i_j \end{array} \right] \tag{19}$$

where xx_j stands for the local x-axis just prior to the last rotation about the jth y-axis by α_j (see Figure 3).

In both parameter conventions the columns of the Jacobian with respect to the joint scale factors are:

$$\text{col}_j \frac{\partial \underline{x}^i_c}{\partial \underline{k}} = \theta_j^{\ i} \left[\begin{array}{c} z^i_{j-1} \times b^i_j \\ z^i_{j-1} \end{array} \right] \tag{20}$$

2.7 Data collection and parameter estimation

A series of n configurations of the actual mechanism provides n sets of joint angle measurements $\underline{\theta}^i$, and n equations of the form (13). Compactly, the equations may be written

$$\Delta \underline{X} = C\Delta\underline{\varphi} + \underline{n}' \tag{21}$$

where

$$C = \begin{bmatrix} C^1 \\ \cdot \\ C^n \end{bmatrix}, \quad \Delta\underline{X} = \begin{bmatrix} \Delta\underline{x}^1 \\ \cdot \\ \Delta\underline{x}^n \end{bmatrix} \tag{22}$$

An estimate of the parameter errors is provided by minimizing

$$LS = (\Delta\underline{X} - C\Delta\underline{\varphi})^T (\Delta\underline{X} - C\Delta\underline{\varphi}) \tag{23}$$

which yields

$$\Delta\underline{\varphi} = (C^T C)^{-1} C^T \Delta\underline{X} \tag{24}$$

Finally, the guess at the parameters is updated as

$$\underline{\varphi} = \underline{\varphi}_0 + \Delta\underline{\varphi} \tag{25}$$

and the iteration continues until $\Delta\underline{X} \to 0$.

In practice, during iteration the matrix $C^T C$ may become singular at an intermediate parameter set, even though the final parameter set does not have a singularity. To avoid the problem the least squares criteria is modified to

$$LS' = (\Delta\underline{X} - C\Delta\underline{\varphi})^T (\Delta\underline{X} - C\Delta\underline{\varphi}) + \lambda (\Delta\underline{\varphi})^T (\Delta\underline{\varphi}) \tag{26}$$

This criteria is minimized by using the singular value decomposition of C, zeroing singular values that are less that p percent of the maximum singular value, and then implementing the generalized inverse from the SVD matrices [10]. The value of p implicitly gives λ, and it is set high (e.g. $p = 5$ percent) initially and reduced once convergence occurs.

If the kinematics are over-parameterized for the collected joint angle data then C will always have zero singular values. The number of near zero singular values indicates the number of non-independent parameters.

Two other variations of the original algorithm [1] are as follows. The poorly known parameters (base and tip transformations) are first estimated by assuming that the finger parameters are correct. Then all parameters are allowed to freely vary, giving the final results. Also, it is often found useful to check that the present update to the parameters improves the endpoint positioning error. If it does not then the parameter update is repeatedly halved, until the endpoint error improves (see [10]).

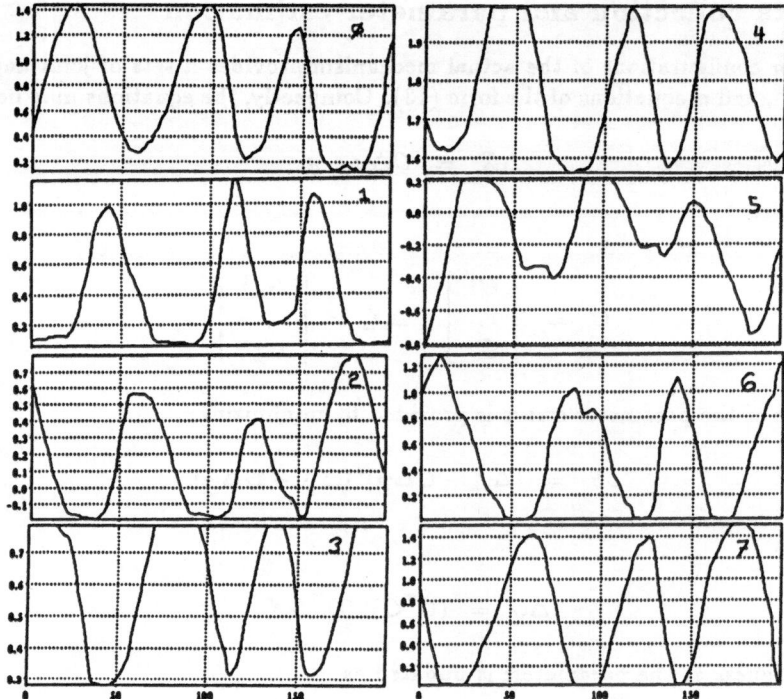

Figure 5: Joint angle(rad) vs. pose number for all 8 joints of the thumb and finger.

3   Experimental results

3.1   Calibrating a finger opposing the thumb

Finger three of the hand was opposed to the thumb (finger 0) by rigidly connecting the fingertips (screwed into a common aluminum plate). The nominal kinematic parameters for this 8 DOF closed kinematic chain were taken from [5] and are shown in Table 1. The geometric parameters (i.e. the s, a and α parameters) marked with a $\sim$ are difficult to measure independently. The joint angle offset and scale parameters are also not well known. Those geometric parameters *not* marked with a $\sim$ are likely to be more accurate than parameters identified by any identification scheme relying on the limited accuracy of the joint angle sensors.

A series of 200 configurations of the finger/thumb mechanism provided input joint angles for the above identification algorithm. These joint angles are plotted in Figure 5. The joint angles for the thumb are negated so that the identified joint axes are in the same direction as defined in [5]. Considerable care was taken to make sure that all joints moved as much as possible.

The calibration was performed with the initial parameters given in Table 1, the recorded joint angles, and the two fixed parameters $s_0 = -1.3$ and $k_7 = 1.0$. Convergence of the algorithm required $p = 0.5$ in the SVD based pseudo-inverse, and produced the parameters in Table 2.

Endpoint errors: The calibration may be assessed by measuring the error in closure

Joint	θ^{off} (rad)	s (in)	a (in)	α (rad)	k
0*	0.0000 ~	-1.3000	0.0000	0.0000	1.0000
1*	0.0000	-1.7000	0.0000	0.0000	1.0000
2	0.0000	0.0000	-0.4500	1.5708	1.0000
3*	-1.5708~	2.1200~	0.4094~	0.2080~	1.0000
4	0.0000~	2.2500~	0.6000~	-1.5708	1.0000
5*	0.0000~	1.7000	0.0000	0.0000	1.0000
6*	0.0000	1.3000	0.0000	0.0000	1.0000
7	3.1415~	-0.5000~	-1.0000~	-0.2618~	1.0000

Table 1: 8 DOF mechanism. Initial parameters. A * indicates that the parameters are those of the Hayati convention (with the respective units). The parameters marked with a ~ are difficult to measure and were either guessed (for link 7) or roughly calculated from known specifications.

Joint	θ^{off} (rad)	s (in)	a (in)	α (rad)	k
0*	0.2380	-1.3000	0.0069	0.0668	0.9508
1*	0.0850	-1.5085	-0.1515	-0.0480	0.8596
2	-0.0943	0.1604	-0.3097	1.6207	0.9336
3*	-1.7824	1.8050	0.5755	0.2890	0.9579
4	0.1389	1.7151	1.3356	-1.5836	1.0125
5*	0.5464	1.6307	0.0292	0.0737	0.8577
6*	-0.0425	1.1474	0.1745	0.1260	0.9350
7	3.7552	-0.8322	-1.4377	-0.5631	1.0000

Table 2: 8 DOF mechanism. Calibrated parameters. A * indicates that the parameters are those of the Hayati convention (with the respective units).

between the fingertips. This error is simply given by the position and orientation vector $\Delta \mathbf{x}_c^i$ for each pose i. Averaging over all poses, the pre-calibration root mean square (RMS) position error was $0.5780in$ and the RMS orientation error was $0.4645rad$. After calibration these errors are reduced to $0.0201in$ and $0.0290rad$ respectively.

Geometric parameter errors: As stated, the geometric parameters *not* marked with a $\sim$ are likely to be more accurate than the parameters identified by any identification scheme relying on the limited accuracy of the joint angle sensors. These parameter values may be used to check the validity of the calibration algorithm. Comparing these parameters in Tables 1 and 2 reasonable agreement is seen. The existing miss-match could be partly due to *unmodeled kinematics* such as joint wobble, and other factors caused by machining imprecision. Also, as discussed below, certain parameters may not be uniquely determined from the limited accuracy joint sensor readings.

Non-geometric parameter errors: It is seen from Table 2 that the joint angle offsets and scale factors are an important source of error in the kinematic model. The joint offsets marked with a $\sim$ are particularly hard to measure, and indeed show the greatest error. The non-geometric parameters that are more accessible (i.e. θ_j^{off} and k_j for $j = 0, 1, 6$) were carefully re-measured after calibration. This post hoc measurement indicated that these non-geometric parameters were identified more accurately than the initial guesses in Table 1.

3.2 Parameter identifiability

The ratio of the minimum singular value to the maximum singular value of C provides a index of parameter identifiability. This ratio is 0.003, indicating that $C^T C$ is not singular, but is not well conditioned. Thus, all the parameters are theoretically identifiable[11], but some may be sensitive to measurement noise. As stated, to obtain convergence all the singular values less than 0.5 percent of the maximum singular value were zeroed. This translated into zeroing seven singular values. Thus, for this mechanism, and for this particular joint angle data set, there are seven parameters that are close to linearly dependent upon the other 31 parameters.

It is difficult to determine the geometrical significance of these interdependent parameters. In theory, linear relations between the instantaneous parameter variations may be found from the null space of the Jacobian (the null space is determined by the span of the columns of the matrix V that correspond to (near) zero singular values, where V is part of the SVD of the Jacobian: $C = UDV^T$). It is not clear though how to interpret these first order relations to determine the source of the parameter ambiguities.

Certain parameter dependencies do have a simple explanation though. For example, consider the parameters associated with the most proximal joint of the finger and the most proximal joint of the thumb (joints angles 3 and 4). These two almost parallel joints are partially decoupled from the rest, in the sense that is possible to move them without large movements of the other joints and visa versa. From the data (Figure 5) it is seen that joints 2 and 3 move the least (only 0.4 rad). Roughly, these two joints, being perpendicular to the others, move the plane of the rest of the finger/thumb complex. The location of this plane is given by the parameters a_2 and a_4, which are measured along the common normals $\mathbf{x}_2$ and $\mathbf{x}_4$. Unfortunately, these two common normals stay almost parallel for all the data collected (mostly because of the small movement ranges of the

Joint	θ^{off} (rad)	s (in)	a (in)	α (rad)	k
-1*	0.0000	1.8050	0.5755	0.2890	N/A
0	0.1389	1.7151	1.3356	-1.5836	1.0125
1*	0.5464	1.6307	0.0292	0.0737	0.8577
2*	-0.0425	1.1474	0.1745	0.1260	0.9350
3	?	?	?	?	1.0000

Table 3: Identified finger parameters relative to a palm reference frame, where a * indicates that the parameters are those of the Hayati convention (with the respective units), and a ? means an unknown tip transformation.

joints 2 and 3). In the extreme case where they are fixed to being exactly parallel then the D-H distances a_2 and a_4 along these common normals may vary arbitrarily, as long as the difference $a_2 - a_4$ is kept the same. Though these common normals are not exactly parallel to one another, we should predict a sensitivity problem in separately identifying a_2 and a_4 from finite precision measurements. See [11] for a more general discussion of identifiability.

An alternate fingertip constraint: point contact

The identifiability of these parameters may be improved by relaxing the endpoint constraint to be a point contact (with the endpoint free to orient arbitrarily) instead of a rigid contact. In the case just discussed the axes x_2 and x_4 would then no longer be constrained to be nearly parallel. The use of a point contact constraint (or a perhaps more elaborate rolling contact constraint) has not yet been attempted because, as yet, we do not have tactile sensors that could be used to assure the point contact. We could build a ball joint to connect the fingertips, but this would not be as general, or natural way of proceeding.

3.3 Common palm reference frame conversion

Once a finger is calibrated against the thumb, as just described, it is necessary to convert the identified parameters into a reference frame located on the stationary palm of the hand, similar to [5]. This reference frame is the frame labeled 2′ in Figure 4 – that is, the frame found after the rotation of local frame 2 by the joint angle θ_3. The finger's parameters relative to this palm reference are therefore given by the parameters labeled 3 and greater (except θ_3), and are re-written in Table 3. The tip transformation is not identified by the above method. It could be identified if a point contact constraint were used instead of a fixed constraint. For practical purposes the tip D-H parameters are taken to be $\theta^{off} = 0$, $s = 0$, $a = 0.735$ and $\alpha = 0$ [5]. Finally, notice that the parameter set labeled -1 locates the first axis of the finger relative to the thumb.

The thumb parameters are found by following the kinematic chain from the palm frame 2′ backwards to the 'base' (actually, the fingertips). This simply entails reversing the sign on the respective θ, s, a and α parameters. Also, in the case of the Hayati convention parameters the order of transformations must be changed to:

$$A_j = Rot(z, \theta'_j)Rot(y, \alpha_j)Rot(x, a_j)Trans(x, s_j) \qquad (27)$$

Joint	θ^{off} (rad)	s (in)	a (in)	α (rad)	k
0	1.7824	0.0000	0.3097	-1.6207	0.9336
1'	0.0000	-0.1604	0.0000	0.0000	N/A
1*	0.0943	1.5085	0.1515	0.0480	0.8596
2*	-0.0850	1.3000	-0.0069	-0.0668	0.9508
3	?	?	?	?	1.0000

Table 4: Identified thumb parameters relative to a palm reference frame, where a *
indicates that the parameters are those of the Hayati convention (with the respective
units), and a ? means an unknown tip transformation.

The thumb parameters are thus calculated to be those shown in Table 4. Notice that an
additional s translation, 1', is required to provide the translation of -0.1604 along the
thumb joint 1 axis.

When the whole calibration procedure is repeated with other fingers opposing the
thumb then the identified palm reference frame 2' may differ from finger to finger. That
is, this frame may be arbitrarily displaced or rotated on the thumb's first joint axis
(labeled z_2 in Figures 4). One such reference frame may be defined as the 'common
reference frame' and the others related to it. The unknown translation and rotation on
z_2 (between any two palm reference frames) may be estimated by locating the thumb
in a fixed position and calculating its endpoint in both reference frames. The difference
between these endpoint locations gives the two unknown quantities.

4   Summary

Preliminary studies investigating the kinematic calibration of the Utah-MIT hand have
been presented. The approach was to use the closed loop-kinematic calibration technique
developed in [1]. Several modifications of the original algorithm in [1] were necessary.
First, it was found that joint angle scale factors had to be included as parameters to be
identified. Second, it proved to be important to identify initially the endpoint and palm
(base) transformations by assuming the initial guesses at the finger kinematic parameters
were correct. This provided improved initial estimates of these parameters, which were
otherwise tricky to measure. Once this first step was completed the full scale non-linear
parameter calibration technique was used. Finally, SVD provided a convenient way of
resolving the parameter ambiguity.

The experimental results indicate that the kinematics of the finger/thumb complex
can be identified by the proposed closed-loop kinematic calibration method. Endpoint
positioning error was improved by over an order of magnitude. Further, the identified
values of the parameters that were also accurately known a priori were in close agreement
with these a priori values.

Certain parameters were not uniquely determined from the collected joint angle data.
This parameter ambiguity, while clearly not a problem for the particular finger contact
situation studied (see RMS error), may cause troubles when the identified kinematics is

used for very different finger configurations. The parameter ambiguity is principally due to the lack of joint movement allowed by the fixed constraint (it may also be due to limited joint sensor resolution, and other unmodeled kinematics). In the future we intend to use a point contact constraint (as described in [12]). This will produce an 11 DOF mechanism with higher mobility. The natural way to implement this is to use tactile sensors to assure that the fingers stay at a point contact, while allowing arbitrary orientation of the fingertips. This experiment awaits the completion of tactile sensors for the fingers. Other more complex endpoint constraints can also be explored. In fact, the above technique may be extended to identifying the geometry of an object grasped between the fingers (see [11]).

Acknowledgments

This paper describes research done at the Artificial Intelligence Laboratory of the Massachusetts Institute of Technology. Support for the laboratory's artificial intelligence research is provided in part by the University Research Initiatives under Office of Naval Research contract N00014-86-K-0180 and in part by the Advanced Research Projects Agency of the Department of Defense under Office of Naval Research contract N00014-85-K-0124. Personal support for JMH was also provided by an NSF Presidential Young Investigator Award, and for DJB by a Fairchild fellowship and a scholarship from the Natural Sciences and Engineering Research Council of Canada. Also, thanks to Sundar Narasimhan and Bill Rockenbeck for their help with the hand system.

5 References

1. Bennett, D.J., and Hollerbach, J.M., 1988, "Self-calibration of single-loop, closed kinematic chains formed by dual or redundant manipulators," *Proc. 27th IEEE Conf. Decision and Control*, Austin, Texas, Dec. 7-9.

2. Hollerbach, J.M., 1988, "A survey of kinematic calibration," *Robotics Review*, edited by O. Khatib, J.J. Craig, and T. Lozano-Perez, Cambridge, MA, MIT Press.

3. H. Zhuang, Florida Atlantic University, personal communication.

4. Jacobsen, S.C., Iversen, E.K., Knutti, D.F., Johnson, R.T., Biggers, K.B., 1986, "Design of the Utah/MIT dextrous hand," *Proc. IEEE Int. Conf. Robotics and Automation*, San Francisco, April 7-10, pp. 1017-1023.

5. Narasimham, S., 1988, *Dextrous Robotic Hands: Kinematics and Control*, M.I.T., Masters Thesis.

6. Denavit, J., and Hartenberg, R.S., 1955, "A kinematic notation for lower pair mechanisms based on matrices," *J. Applied Mechanics*, 22, pp. 215-221.

7. Hayati, S.A., 1983, "Robot arm geometric link parameter estimation," *Proc. 22nd IEEE Conf. Decision and Control*, San Antonio, Dec. 14-16, pp. 1477-1483.

8. Whitney, D.E., 1972, "The mathematics of coordinated control of prosthetic arms and manipulators," *ASME J. Dynamic Systems, Meas., Control*, pp. 303-309.

9. Sugimoto, K., and Okada, T., 1985, "Compensation of positioning errors caused by geometric deviations in robot system," *Robotics Research: The Second International Symposium*, edited by H. Hanafusa and H. Inoue, Cambridge, Mass., MIT Press, pp. 231-236.

10. Press, W.P., Flannery, B.P., Teukolsky, S.A., and Vettering, W.T., 1988, *Numerical Recipes in C*, New York, Cambridge University Press.

11. Bennett, D.J., and Hollerbach, J.M., 1989, "Identifying the kinematics of robots and their tasks," *Proc. IEEE Int. Conf. Robotics and Automation*, Scottsdale, Arizona, May 14 - 19.

12. Bennett, D.J., and Hollerbach, J.M., 1989, "Identifying the kinematics of non-redundant serial chain manipulators by a closed-loop approach," *Proc. Fourth International Conference on Advanced Robotics*, Columbus, Ohio, June 13-15, in press.

ROBOT CALIBRATION USING LEAST-SQUARES AND POLAR-DECOMPOSITION FILTERING

Gregory Ioannides[1], Jorge Angeles[1], Randall Flanagan[2]

and David Ostry[2]

McGill University, Montréal, Canada.

Abstract

This paper reports the experimental results of a novel method to calibrate geometric errors of multi-axis robotic manipulators. The method proposed by the authors is based on a least-square estimation of the rotation matrix of a rigid body in three-dimensional Cartesian space. The error is filtered by imposing the orthogonality constraint on the rotation matrix, using the polar-decomposition theorem. The axis of rotation of the rigid body, then, is computed from the linear invariants of the rotation matrix. Finally, the wrist of the Yaskawa Motoman Robot was calibrated. The measurements of the Cartesian coordinates of points were performed using a computer vision system and LED markers on a rigid body grasped by the end-effector.

1. INTRODUCTION

It is well known that the current accuracy of robots is outside acceptable tolerance for certain manufacturing applications such as high precision assembly. Different authors have tried to give an estimate of the error involved [10],[12],[15] the need for calibrating robotic manipulators and compensating for the errors thus becoming evident. There are two kinds of errors, namely, geometric and non-geometric. The latter are caused by backlash, structure flexibility, etc. The former exist due to imprecise manufacturing of the

---

[1]Dept. of Mech. Eng., McGill Univ., 3480 University St., Montréal,Québec, H3A 2A7, Canada.
[2]Dept. of Psych., McGill Univ.,1205 Penfield Ave., Montréal, Québec, H3A 1B1, Canada.

robot links and joints. However, because the lengths can be measured easily and accurately, it is assumed that the error lies mostly in the joints and their alignment, which, in turn, causes kinematic errors.

Even though in the teach-mode programming of robotic trajectories the accurate knowledge of the parameters defining the robot architecture is not required, the off-line programming of those trajectories definitely requires the accurate knowledge of these parameters. The measurement of the aforementioned parameters is referred to as *robot geometric calibration* or, simply, *robot calibration*. Due to the unavoidable manufacturing errors mentioned above, the nominal values of the parameters defining the robot kinematic structure, e.g., the Hartenberg-Denavit parameters of the robot at hand, will be in error. Thus, axes which are nominally parallel, perpendicular or intersecting, will fail to be so in reality. This means that the positioning accuracy will be affected, unless these errors are compensated. Work in connection with robot calibration has been extensive, as reported in the literature [7],[12],[15],[9],[1]. However, as pointed out by Hayati in [7], the presence of two neighboring axes which are ideally or nearly parallel introduces serious numerical instabilities. In this paper, a method is introduced for robot calibration based on a least-square estimation of the axis of rotation of a moving rigid body, that is independent of the relative orientation of consecutive axes.

The paper is organized as follows: We begin with a short account of the problem of estimating the orientation of rigid bodies from the measurements of the Cartesian coordinates of a set of points in two finitely-separated configurations of the body. Next, we refer briefly to the least-square/polar-decomposition technique adopted for data processing, which allows us to obtain acceptable estimates of the orientation matrix. We end up with a description of the experimental procedure and discuss the results obtained.

2. THEORETICAL BACKGROUND

The determination of the orientation matrix of a rigid body from perfect measurements of the Cartesian coordinates of any three non-collinear points of the body has been the subject of extensive research [11], [3]. However, actual coordinate measurements are bound to contain noise, which can be filtered if redundant measurements, i.e., measurements of the coordinates of more than three non-collinear points, are taken. From these, an estimate of the attitude of the body, with respect to a reference configuration, is made by resorting to a simple least-square fit. However, since no constraints are imposed on the orthogonality of the estimated orientation matrix, the estimate will most likely fail to be orthogonal. A method is proposed by Higham in [4] for computing the most likely orthogonal matrix from the non-orthogonal estimate based on the *Polar-Decomposition Theorem*. Once an orthogonal estimate of the matrix defining the attitude of the body is available, the direction of the axis of the associated rotation, from the reference configuration, is determined using the concept of *vector* of a 3×3 matrix [2]. The position vector of a point of the axis of rotation is then computed using the method proposed Angeles in [3], thereby completing the calibration of one joint. The same procedure is repeated for all joints.

Studies have shown that these coordinates can be used to determine the pose—position and orientation—of a rigid body based on only three non-collinear points [3]. The work reported here is based on redundant measurements, i.e., the coordinates of $m > 3$ points are measured. Theoretically, by equating the coordinates of those points in a displaced configuration with a linear transformation of the coordinates of the same points in a reference configuration, an overdetermined system of linear algebraic equations is formed, whose least-square approximation produces the orthogonal rotation matrix involved in the coordinate transformation. However, due to unavoidable measurement and roundoff errors, rather than working with the position vectors themselves, we define a set of m vectors stemming from a common point, the centroid of the set, and pointing to each point of the set. We call the ith vector of this set $\mathbf{r}_i$, for $i = 1, \ldots, m$. Moreover the corresponding set defines, in a finitely separated configuration, a set of corresponding vectors $\mathbf{r}'_i$, for $i = 1, \ldots, m$.

Now, let the rotation carrying the rigid body from its reference to its current configuration be represented by matrix $\mathbf{Q}$. Hence, a set of m 3D vector equations linear in $\mathbf{Q}$ can be written that relate each vector $\mathbf{r}'_i$ with its counterpart $\mathbf{r}_i$. Now the entries of $\mathbf{Q}$,

q_{ij}, are to be determined. In this way, a system of m 3D vector equations that are linear in the nine unknowns q_{ij} is derived. These equations can be rewritten in a standard form by grouping the nine unknowns into the 9-dimensional vector $\mathbf{q}$, matrix $\mathbf{R}$ and vector $\mathbf{r}$ which are defined correspondingly. Clearly, $\mathbf{R}$ and $\mathbf{r}$ are known, for they are determined entirely by the components of the sets $\{\,\mathbf{r}_j\,\}_1^m$ and $\{\,\mathbf{r}'_j\,\}_1^m$, respectively. One then has:

$$\mathbf{R}\mathbf{q} = \mathbf{r} \qquad (1)$$

where $\mathbf{R}$ is a $3m \times 9$ matrix and $\mathbf{r}$ is a $3m$-dimensional vector. Since we assumed redundant measurements, $m > 3$ and hence, $3m > 9$, which renders the system of eq.(1) overdetermined. Now, let matrix $\mathbf{Q}_k$ represent the rotation carrying the rigid body from its reference to its kth current configuration, for $k = 1, \ldots, n$. This matrix contains nine unknowns and is now rewritten as the 9-dimensional vector $\mathbf{q}_k$. Since coordinate measurements of m points are made, a set of $3m$-dimensional vectors $\mathbf{v}'_k$, for $k = 1, \ldots, n$, is then formed as follows:

$$\mathbf{v}'_k \equiv [\,\mathbf{r}'^{\,T}_{1k} \quad \mathbf{r}'^{\,T}_{2k} \quad \cdots \quad \mathbf{r}'^{\,T}_{mk}\,]^T \qquad (2)$$

where $\mathbf{r}'_{jk}$, for $j = 1, \ldots, m$ and $k = 1, \ldots, n$, denote vectors $\mathbf{r}'_j$ associated with point P'_j for the kth current configuration. Furthermore, a set of n $3m \times 9$ matrices $\mathbf{R}_k$, for $k = 1, \ldots, n$, is defined for every measurement, eq.(1) thus being written n times as:

$$\mathbf{R}_k \mathbf{q}_k = \mathbf{r}_k, \quad k = 1, \ldots, n \qquad (3)$$

In this way, the matrix and the right-hand side of the algebraic system of eq.(3) are now available. The least-square approximation of this system, $\tilde{\mathbf{q}}_k$, can be expressed symbolically via the generalized inverse of $\mathbf{R}_k$, $\mathbf{R}_k{}^I$, as follows:

$$\tilde{\mathbf{q}}_k = \mathbf{R}_k{}^I \mathbf{v}'_k, \quad \mathbf{R}_k \equiv (\mathbf{R}_k{}^T \mathbf{R}_k)^{-1} \mathbf{R}_k{}^T \qquad (4)$$

The least-square estimation of vector $\mathbf{q}_k$, $\tilde{\mathbf{q}}_k$, associated with that system, is most efficiently computed using Householder reflections [5]. However, the foregoing least-square approximation provides only a first estimate, $\tilde{\mathbf{Q}}_k$, of the orthogonal matrix $\mathbf{Q}_k$. Since the measurements are noisy, the estimated rotation matrix does include some error. This error is filtered by imposing the orthogonality condition on the rotation matrix, which is done with the aid of the Polar-Decomposition Theorem (PDT) [6]. According to the PDT, a $n \times n$ matrix can be decomposed in the form :

$$\bar{Q} = OU \tag{5}$$

where O and U are both $n \times n$ matrices, the former being orthogonal, the latter, at least positive semidefinite. The polar decomposition theorem states that, if the original matrix is non singular, then matrix U is necessarily positive definite and both O and U are unique. Otherwise, U is only positive semidefinite and not unique. In our case, if neither measurement nor roundoff noise were present, O would be Q and U would be the identity matrix, 1. Since we are acknowledging the presence of errors, O is not Q, but merely an estimate of it, denoted here by $\hat{Q}$, and U is not the identity, but $1 + E$, where E is the estimation error. Note that E provides us with an estimate of the error involved, and hence, allows us to decide whether the estimate is acceptable or not.

The algorithm proposed by Higham (1986), aimed at computing the factors of the said decomposition was utilized on all $\bar{Q}_k$. Hence, the error was filtered out from the initial estimation of each rotation matrix and the filtered estimates $\hat{Q}_k$ were obtained. In this experiment, the components of the error matrix E had no more than 1% error, which was considered acceptable. Subsequently, all information about the axes was derived from the $\hat{Q}_k$ matrices, as described next.

2. COMPUTATION OF AXIS OF ROTATION FROM ORTHOGONAL MATRIX COMPONENTS

Once the estimate $\hat{Q}_k$ of the orientation matrix is available, the direction of the axis of rotation is determined from the *vector* [2] of that estimate, denoted by vect($\hat{Q}_k$). In fact, if $\hat{e}_k$ denotes the estimate of the unit vector parallel to the axis of rotation and $\hat{\phi}_k$

the estimate of the associated angle of rotation, then,

$$\text{vect}(\hat{\mathbf{Q}}_k) = \hat{\mathbf{e}}_k \sin \hat{\phi}_k \tag{6}$$

where the angle of rotation is estimated from eq.(6) and the following:

$$\text{tr}(\hat{\mathbf{Q}}_k) = 1 + 2\cos\hat{\phi}_k \tag{7}$$

Note that each measurement provides an estimate $\hat{\mathbf{e}}_k$ of the same unit vector $\mathbf{e}$ parallel to the axis under calibration. Thus, the best estimate of this vector is chosen as the normalized mean of the said estimates, i.e., for the n measurements associated with the same axis,

$$\hat{\mathbf{e}} = \frac{\mathbf{f}}{\|\mathbf{f}\|}, \quad \mathbf{f} \equiv \frac{1}{n}\sum_1^n \hat{\mathbf{e}}_k \tag{8}$$

in which $\| \cdot \|$ denotes the Euclidean norm of its vector argument. In order to completely determine the axis under calibration, all that remains is to estimate the location of one of its points. This can be, for example, the point closest to the origin, of the position vector $\mathbf{s}_0$. This vector is calculated based on the relations derived in [3], which lead to the following overdetermined linear algebraic system:

$$\mathbf{A}\mathbf{s}_0 = \mathbf{b} \tag{9}$$

where $\mathbf{A}$ is a $(3m+1) \times 3$ matrix and $\mathbf{b}$ is a $(3m+1)$-dimensional vector, both of which are given below:

$$\mathbf{A} \equiv \begin{bmatrix} \hat{\mathbf{Q}}_1 - \mathbf{1} \\ \hat{\mathbf{Q}}_2 - \mathbf{1} \\ \vdots \\ \hat{\mathbf{Q}}_n - \mathbf{1} \\ \hat{\mathbf{e}}^T \end{bmatrix}, \quad \mathbf{b} \equiv \begin{bmatrix} \mathbf{b}_1 \\ \mathbf{b}_2 \\ \vdots \\ \mathbf{b}_n \\ 0 \end{bmatrix} \tag{10a}$$

with $\mathbf{b}_k$, for $k = 1,\ldots,n$, defined as follows:

$$\mathbf{b}_k \equiv \hat{\mathbf{Q}}_k \mathbf{c} - \mathbf{c}' + \left[(\mathbf{c} - \mathbf{c}')^T \hat{\mathbf{e}}\right]\hat{\mathbf{e}}^T \tag{10b}$$

thereby completely determining the direction and the position of the axis under calibration. This procedure was applied to the calibration of the wrist axes of an industrial robot, as described in the following sections.

3. SOFTWARE TESTING

Before any experiment was performed the method was tested using the software developed for this purpose. The problem was formulated first in a constrained least-squares fashion only to find that the orthogonal rotation matrices that were obtained were unacceptable. The results of the constrained problem were far from those of the unconstrained problem. The conclusion is that the constraints were too strong and caused the solution to zero-in on an orthogonal matrix which was very different from the expected one. Hence, the unconstrained least-squares formulation was utilized.

The algorithm was written in the C programming language and it takes as input the position vectors of a number of points assumed to be on the rigid body and the same vectors after being multiplied by a specified rotation matrix. The objective of the tests was to figure out the amount of error that could be filtered out of the actual and presumably noisy measurements. For that reason, random noise was added to the input data of the program. The results showed that the least-squares computations alone performed on input contaminated with noise up to 10% yielded the required rotation matrix with an accuracy of 10^{-3}. This figure improved considerably when the polar-decomposition algorithm was used and decreased the error to 10^{-8}.

After the tests proved that it was feasible to obtain the desired robot axis with reasonable accuracy, the experiment was set up.

4. EXPERIMENTAL PROCEDURE

The objective of the experimental part of the method was to determine the pose of each link of the robot. This could be accomplished by taking measurements of the coordinates of a certain number of points on each link before and after the manipulator underwent a programmed motion. Knowing these coordinates, the matrix representing the rotation of the specific link could be evaluated. The axis of rotation of the particular link could then be computed from the linear invariants of the rotation matrix. However, because of the architecture of the robot, i.e., the limited space and the curved surfaces, it was not possible to attach a sufficient amount of markers on each link.

In order to facilitate the collection of the data, a bracket, shown in Fig. 1, was manufactured and all the light-emitting diode (LED) markers were attached to the bracket. Consequently the bracket was rigidly attached to the end-effector. This configuration allowed for better measurements since the LEDs were concentrated in one area and were not spread over the length of a link. It also has the advantage that the markers can be positioned in a configuration allowing for maximum robustness in the measurements. Another advantage is that the whole process of taking measurements was sped up by a considerable amount of time since, if the bracket is ready from the onset, all that needs to be done is to connect the bracket to the end-effector.

Even though the markers were not any more on individual links, the calibration was done now by considering motions of one joint at a time, while the others were kept locked. Hence, the locked links were considered as a single rigid body undergoing a rotation about a fixed axis, which was the robot axis to be identified.

4.1 Equipment

After researching the possibilities of equipment to be used for data acquisition, the system that is next described was chosen over laser methods and traveling microscopes because of its simplicity, suitability and availability.

The system that was used is the WATSMART (WATerloo Spatial Motion Analysis and Recording Technique) vision system [13]. It is comprised of two high resolution infrared cameras, one pre-surveyed calibration frame, several (LED) markers, along with the appropriate controllers for the cameras and the marker strobers. Finally, a computer

Fig.1: The T-bracket.

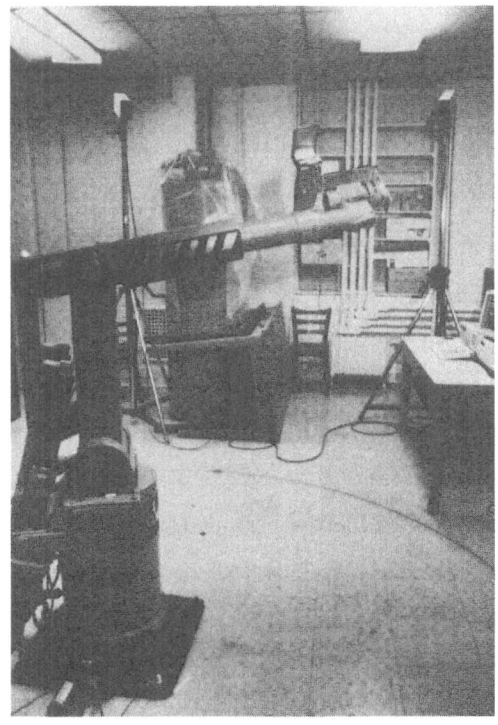

Fig.2: Experimental set up.

with an interface board to drive the camera and the strobing controllers is included in the system, as well as special software for the determination of the three-dimensional coordinates of the markers. The experimental set up is illustrated in Fig. 2.

The cameras have a resolution of 1:4000 of the viewing field. At 2m away from the cameras, where the measurements were performed, the viewing field is 1.2m wide and hence the accuracy of the cameras is 0.3mm. The calibration frame that serves as a permanent reference object is in the shape of a cube, see Fig. 3, with a side length of 21 inches. The manufactured accuracy of the frame is 0.1mm.

The robot that was calibrated is the Yaskawa Motoman AID 810 Robot. It is a six-axis manipulator designed for welding applications. The second and third axes are parallel and the wrist is supplied with an offset. A representation of the wrist is shown in Fig. 4. The AID32v controller was used to produce the required motion for the experiment and the programming language was RAIL[3].

4.2 Data Acquisition

The procedure followed to acquire the data is described next. As already mentioned the LED markers were attached to a T-shaped bracket and a part of it was bent so as to produce a three-dimensional body rather than a two-dimensional one—this has the effect of a better depth estimation. It was proved to be important to cover the bracket with a black cloth, made of colortron foam, in order to reduce reflections from the area in the proximity of the LED markers, the reason being that the cameras take the position of the marker to be the centroid of the area with the strongest light emission. The markers are strobed one at a time and hence there is no correspondence problem to solve.

Before the actual monitoring of the three-dimensional structure commenced, calibration of the vision system was needed. This was done by placing the calibration cube in the middle of the range of motion of the robot and viewing it with the cameras. The provided software, knowing the geometry of the calibration cube, calculates the regression

---

[3]Acronym for Robot Automatix Incorporated Language

Fig.3: The calibration cube.

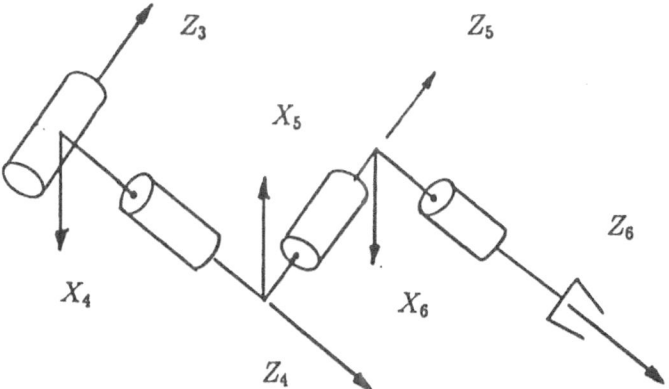

Fig.4: The wrist representation.

equations that are used to establish the three-dimensional coordinates of the markers in a reference coordinate frame, which actually is along the edges of the calibration cube. Once this procedure is completed, the reference frame is valid for all subsequent measurements and the cube is removed to allow for free motion of the robotic arm.

The next step is the data collection. The cameras see the markers on a plane perpendicular to an axis coming out of the cameras. The two-dimensional coordinates of those points are recorded separately for each camera and later combined by the software to produce the three-dimensional coordinates of the each point. The method used for this purpose is called the *direct linear transformation method* or the DLT method [14].

In this experiment, five infrared markers were used, since they are enough to render the linear algebraic system overdetermined. A small routine was written in RAIL to produce the joint-by-joint motion needed. Before any motion occurred the angle between the camera line of vision and the LEDs was optimized. The first joint of the robot arm moved in the joint coordinate space by an angle of 10°. The coordinates of the landmark points were subsequently recorded. This kind of motion was repeated six times, thereby obtaining data for an arc of 60°. The procedure was repeated for the remaining joints, always taking into account the orientation of the end-effector with respect to the cameras. Special care was taken to secure that the data had the minimum of noise, mainly due to the high reflectivity of the infrared light of the LED markers. This was achieved by covering every object in the room with black colortron foam cloth.

After the measurements were performed the data were examined using signal processing algorithms to make sure that they did not conform to any set pattern. For this purpose, a Butterworth filter in different frequencies was utilized. Since fifty measurements were taken for each position, an averaging was performed to screen out the random noise that might have affected the measurements. Having now obtained the set of coordinates of the marked points in different positions, the axes of the robot were calculated as explained in previous sections.

The experiment was repeated a second time to verify and improve the accuracy of the results. This time 12 points were marked so as to get a better least-square estimation, since that is the most critical part of the calculations. In order to increase the accuracy, the cameras were moved in closer, the number of samples was increased to 200 and the angle of rotation of the joints was reduced. However, the new plate that was manufactured to fit the 12 LEDs was flat, which resulted in loosing the good depth estimation of the cameras and therefore getting a bad estimate of the axes from the least-squares solution. We are currently manufacturing a new structure for the end-effector in the shape of a cube to eliminate this problem.

5. CALIBRATION OF THE YASKAWA MOTOMAN AID 810 ROBOT

The method described in this paper was applied on the Yaskawa Motoman Robot. The nominal Hartenberg-Denavit parameters of the robot are presented in Table 1.

Since the axis of rotation of each link is known, the angles α_i are found from the dot product of the unit vectors representing the ith and $(i + 1)$st axes. The length a_i

Table 1: Nominal H-D Parameters of the manipulator

$Joint_i$	a_i (mm)	b_i (mm)	α_i (deg)
1	0	785	90°
2	670	0	0°
3	0	0	90°
4	0	950	90°
5	0	90.0	90°
6	0	0	0°

is calculated by taking the projection on the common perpendicular of two neighboring axes of the vector joining the two known points on the same axes. The length b_i is computed as the distance between two points on the ith axis. These points are defined as the intersections of the ith axis and the two common perpendiculars between the ith and $(i - 1)$st as well as the ith and $(i + 1)$st axes. The results are listed in Table 2.

The asterisks indicate parameters that could not be calculated. The maximum error in axis location occurred between axes 5 and 6. The maximum offset error at joint 4 and

Table 2: Calibrated H-D Parameters of the manipulator

$Joint_i$	a_i (mm)	b_i (mm)	α_i (deg)
3	*	*	89.6°
4	5.9	962.1	93.7°
5	9.3	97.3	93.2°

the maximum error in the angular parameters occurred between axes 4 and 5.

6. CONCLUSIONS

The experimental aspects of a new robot calibration method were presented. The method is independent of the robot architecture. It is based on a least-square and polar-decomposition filtering of the errors in the measurement of the Cartesian coordinates of more than three non-collinear landmark points of a rigid link. Limitations in the instrument accuracy allowed the calibration of the wrist only. However, our results are promising and hence, further investigation of the method is recommended.

7. ACKNOWLEDGEMENTS

This research was possible under NSERC (Natural Sciences and Engineering Research Council of Canada) Research Grant No. A4532, and FCAR (Fonds pour la formation de chercheurs et l' aide à la recherche, du Québec) Grant No. 88AS2517.

References

[1] An, C.H., Atkeson. C.G. and Hollerbach, J.M. 1988, *"Model-Based Control of a Robot Manipulator"*, The MIT Press, ambridge, Massachusetts.

[2] Angeles, J., 1985., "On the Numerical Solution of the Inverse Kinematics Problem", *The International Journal of Robotics Research*, Vol. 4, No. 2, Summer 1985.

[3] Angeles, J., 1986, "Automatic Computation of the Screw Parameters of Rigid-Body Motions. Part I: Finitely-Separated Positions", *ASME Journal of Dynamic Systems,Measurement and Control*. Vol. 108, March 1986.

[4] Angeles, J., and Ioannides, G. 1989., "Calibration of a Robot Wrist with an Offset Axis", *ASME Winter Annual Meeting*. (To be presented.)

[5] Golub, G., and Van Loan. C., 1983. *Matrix Computations*, The Johns Hopkins University Press, Baltimore, MD.

[6] Halmos, P. R., 1974. *Finite-Dimensional Vector Spaces*, Springer-Verlag, New York.

[7] Hayati, S .A." Robot Arm Geometric Link Calibration.", *Proceedings of the 22nd IEEE Conference on Decision and Control*, pp.1477-pp.1483, December 1983.

[8] Higham, N. J., 1986, "Computing the Polar Decomposition - with Applications", *SIAM J. Sci. Stat. Comput.*, Vol. 7, No. 4, October.

[9] Kirchner H. O. K., Gurumoorthy, B., and Prinz, F. B., 1987, "A Perturbation Approach to Robot Calibration", *The International Journal of Robotics Research*, Vol. 6, No. 4, Winter.

[10] Kumar, A. and Waldron, K.J., 1981. "Numerical Plotting of Positioning Accuracy of Manipulators."*Mechanism Machine Theory* Vol. 16, No 4, pp. 361-368.

[11] Laub, A. J. and Schifflet, G. R. 1983., "A Linear Algebra Approach to the Analysis of Rigid-Body Displacement from Initial and Final Position Data", *ASME Journal of Dynamic Measurements and Control* , Vol. 105, June.

[12] Mooring, B. W., 1983, (August 7-11,Chicago)." The Effect of Joint Axis Misalignment on Robot Positioning Accuracy.",*Proc. 1983 ASME Int. Conf. on Computers in Enginnering*. New York: ASME, pp. 151-155.

[13] Nothern Digital Inc., 1986, "The WATerloo Spatial Motion Analysis and Recording Technique", *Technical Description*, August 1986.

[14] Shapiro, R., 1978. "Direct Linear Transformation Method for Three-Dimensional Cinematography", *The Research Quarterly*, Vol. 49, No. 2, pp. 197-205.

[15] Veitschegger, W. K. and Wu, C. H., 1986, "Robot Accuracy Analysis Based on Kinematics", *IEEE Journal of Robotics and Automation*, Vol. RA-2, No. 3, Fall.

[16] Veitschegger, W. K. and Wu, C. H., 1987, "A Method for Calibrating and Compensating Robot Kinematic Errors", *IEEE International Conference on Robotics and Automation*, Vol. 1, pp. 39-44.

Compliant Sliding of a Block Along a Wall

Matthew T. Mason
School of Computer Science
and Robotics Institute
Carnegie Mellon University
Pittsburgh, PA 15213

Abstract

This paper derives a compliant motion to slide a block along a wall, based on analysis of the task dynamics. We model the compliant motion by a spring with one end attached to the block, and the other end position-controlled. The attachment point of the spring on the block is the compliance center of the compliant motion strategy. We construct a strategy that produces the desired motion, independent of the actual coefficient of friction. We use a dual representation of force (Brost and Mason 1989) to analyze the dynamics of the operation, and to characterize candidate compliant motions. The same dynamic analysis has supported planning of tilting motions and pushing motions, suggesting that the task mechanics and the control strategy can be analyzed independently.

1. Introduction

One approach to robot motion planning is to hypothesize a model for the task mechanics and a model for the controller, and then to analyze the dynamic behavior of the combined system. A problem with this approach is that there are many different tasks and many different controllers, whose product gives many, many, different systems to analyze. The prospects are much better if we can analyze the task independently of the controller, vastly reducing the number of systems to be analyzed. This paper demonstrates a way to analyze the task mechanics independently of the control method.

We consider the problem of sliding a block along the side of a tray, or the wall of a room, keeping one face against the side. There are at least three ways to do it. Earlier papers (Erdmann and Mason 1988; Mason 1989) use tray-tilting motions and pushing motions to move the block. The present paper uses a compliant motion strategy. But we can plan all three using a common analysis of the task, independent of the control strategy. The task analysis results in a set of constraints on the force to be applied, which can be translated into a plan using any one of the three different control strategies.

We will model the compliant control by a spring, with one end attached to the block and the other end position-controlled. The attachment point models the compliance center of the control system. To synthesize the motion, we must choose the attachment point on the block, and then choose the motion for the other end of the spring. There is one particular choice that produces the desired motion independent of the coefficient of friction.

Throughout the paper we use the dual representation of force described in (Brost and Mason 1989). The dynamic analysis is a variation on the the techniques described by Erdmann (1984) and Rajan et al(1987), using the dual representation of force. Section 2 reviews the dual representation of force. Section 3 reviews the dynamic analysis of the block-along-wall problem. Section 4 analyzes the compliant motion strategy. Section 5 discusses and concludes.

2. Review: Representation of force by acceleration centers

This section reviews the representation of force by acceleration centers (Brost and Mason 1989), closely following (Mason 1989). The representation of force by acceleration centers is a simple graphical method to analyze planar contact problems. The approach is similar to the use of velocity centers to analyze kinematic constraints, described by Reuleaux (1876). The key observation is that the acceleration of any plane body can be described as an angular acceleration about some unaccelerated point called the *acceleration center*. Obviously this would not work for a purely translational acceleration, but these can be handled by allowing the acceleration center to range over the *projective* plane—a translational acceleration gives an acceleration center at infinity.

We can use acceleration centers to represent forces. Given some plane body with a mass m and angular inertia I, we can map any force into the resultant acceleration center. This mapping has a very useful property—the magnitude of the acceleration, and hence the magnitude of the force, is not represented. For problems involving frictional contact, this property is very useful, because the magnitudes of the contact forces are not constrained, only the lines along which the forces act.

In practice the use of acceleration centers to represent applied forces is quite simple. Some examples are shown in Figure 1. We place the center of mass at the origin, and choose a unit distance equal to the radius of gyration. Then the acceleration center lies on a perpendicular to the force through the origin. The acceleration center's distance from the origin is the inverse of the force's distance from the origin. The acceleration center has an associated sign, which is the sign of the moment of the force. We typically plot acceleration centers in a common plane, with the sign indicated next to the center. Formally, there are two planes, corresponding to positive and negative moments. The two planes have a common line at infinity, representing translational accelerations, i.e. zero moments. Topologically, the space is equivalent to a sphere. The upper hemisphere corresponds to positive moments, the lower hemisphere to negative moments, and the equator to zero moments.

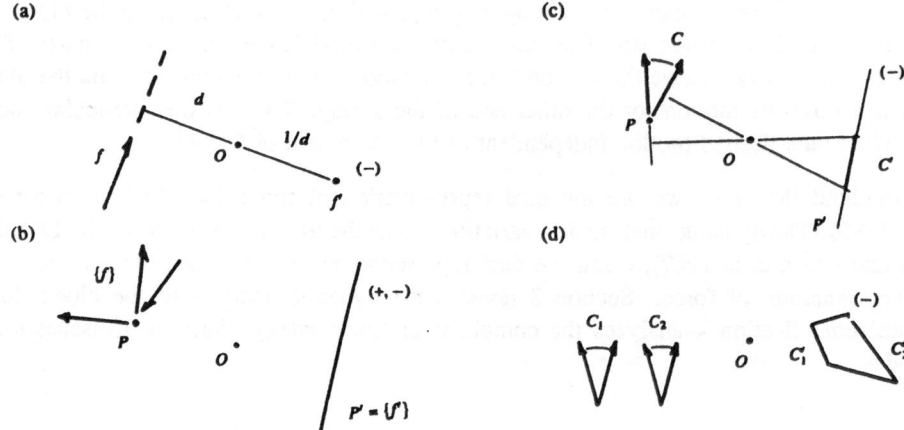

Figure 1: Some example acceleration centers, illustrating properties of the mapping of force to acceleration center. Here we have super-imposed the plane of positive moments and the plane of negative moments, using (+) and (−) to distinguish the points. (a) A line of force maps to a signed point. (b) A point, representing all forces acting through that point, maps to a line. (c) A friction cone maps to a line segment. (d) The possible resultants of several forces is obtained by taking the convex hull in the dual space. (Mason 1989)

The properties and applications of this mapping are more fully described by Brost and Mason (1989). We note two key properties:

- The mapping is nearly dual: a directed line of force maps into a point, and a point, corresponding to the set of all forces passing through that point, maps into a line, the locus of acceleration centers.

- Positive linear combinations of forces map into convex combinations of acceleration centers.

These properties are ideal for representing frictional contacts:

- The feasible motions at a point contact are represented by a linear constraint on the feasible acceleration centers.

- The feasible forces at a point contact are represented by a line segment, which is the locus of acceleration centers corresponding to a friction cone.

- The resultant of several friction cones, arising from several simultaneous contacts, lies in the positive linear combination of the forces, which defines a convex polygon in the space of acceleration centers.

When two points must be joined by a line segment, the construction is sometimes counter-intuitive: with the two planes superimposed, draw a line through the two points. Now, if the two points have the same sign, the line segment is the part of the line between the two points, as usual. But if the two points have opposite signs, take the part of the line *outside* the two points, and also include a point at infinity.

3. Dynamic analysis of the block-along-wall problem

This section closely follows the analysis of (Mason 1989). The block-along-wall problem is formulated as follows.

- A rigid rectangular body is free to move in the plane, with one edge initially against a straight wall.

- We assume Newton's laws with Coulomb friction. Gravity acts normal to the support plane. Friction occurs with the wall and with the support plane. The distribution of support forces is unknown, and may vary with time.

- The goal is to move the object forward while keeping one edge against the wall.

The purpose of the dynamic analysis is to find the mapping between applied force and body acceleration. This is complicated because the contact forces prescribed by Coulomb's law depend on the contact mode, i.e. on the relative motion at each contact. Hence the analysis focuses on identifying the mapping from applied force to contact mode.

We begin by enumerating the contact modes. Figure 2 applies Reuleaux' (1876) partitioning of the space of motion centers to determine the set of feasible contact modes. At each kinematic constraint, we construct a contact normal. To the right (left) of the normal only positive (negative) rotations are feasible. On the normal itself, either direction is feasible. We also construct the contact tangent—above the tangent positive rotations cause rightward motions and negative rotations cause leftward motions. Below the tangent, the opposite is true. The two contacts between the block and wall give rise to two normals, and a single tangent, which cut the space of acceleration centers into different sectors. Each sector, and each boundary segment, potentially corresponds to a different contact mode.

Now, for each contact mode, we want to find the set of applied forces consistent with that contact mode. For each contact mode i:

1. Construct acceleration forces A_i. Figure 2 shows the acceleration centers for each contact mode. There is only one acceleration center for the desired mode, rightward sliding, which is on the line at infinity.

2. Construct wall contact forces W_i. The possible wall forces can be represented by two point contacts, one at each corner of the block. For the desired contact mode, rightward

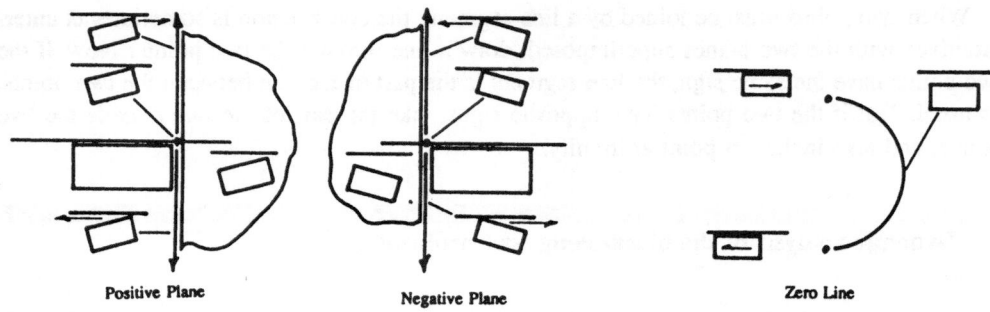

Figure 2: Acceleration centers A for each contact mode. We draw the block tipped to indicate which contacts are broken, and arrows show relative motion at any remaining contacts. The zero line is drawn as if the positive plane and negative plane were projected onto the northern and southern hemisphere, respectively, of a sphere. Then the zero line is the equator, as it would appear from the north pole. There is one contact mode not shown—rest—which corresponds to zero acceleration. (Mason 1989)

sliding, Coulomb's law constrains the direction of each force as shown in Figure 3. We construct acceleration centers for each force, and form the convex combination, to obtain the line segment shown.

3. Construct support friction forces S_i. For the desired contact mode, rightward sliding, the support frictional force reduces to a single force acting through the center of mass, as shown in Figure 3. This maps to an acceleration center at infinity.

4. Construct pushing forces $F_i = A_i \ominus W_i \ominus S_i$. The set of forces $A_i \ominus W_i \ominus S_i$ is the positive linear combination of the A_i forces, and the negations of the W_i forces and S_i forces. In the dual plane, the result is given by the convex combination of the positive plane of A_i with the negative plane of W_i and S_i, resulting in Figure 4.

The result is a mapping from each contact mode i to a set of applied forces F_i. Some of these applied force sets overlap, so that the contact mode is not always uniquely determined by the applied force. For the present case, this ambiguity is primarily due to approximations in our model of the support friction, but some ambiguities are inherent in the Coulomb model of friction.

Not every contact mode is as simple as rightward sliding. In particular, rotations are more complicated, because the floor frictional forces are unpredictable. (Mason 1989) goes through a rotating contact mode is some detail. The main idea is to apply some simple bounds on the frictional forces that can arise.

By repeating the procedure for all contact modes, we obtain the complete mapping of Figure 5. This is a multiple-valued mapping from applied force to contact mode. This mapping is an approximation; for some applied forces, some of the predicted motions cannot

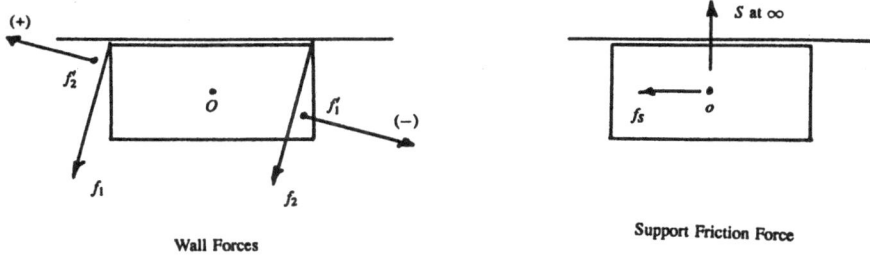

Figure 3: Wall contact forces W, and support frictional forces S for the desired motion. (Mason 1989)

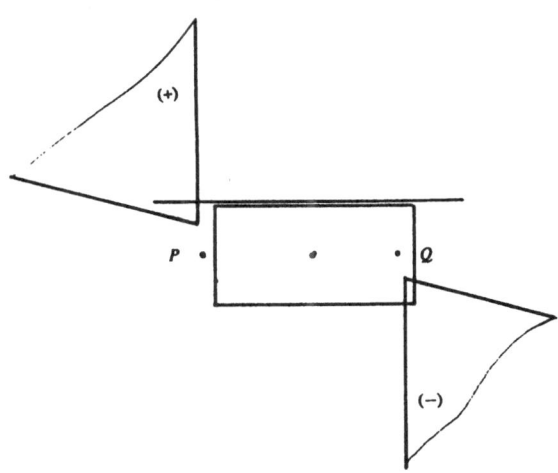

Figure 4: Applied forces F for the desired motion. The applied force must give an acceleration center in the indicated regions. An equivalent constraint is that the applied force must pass between P and Q, and make an angle greater than $\tan^{-1} \mu$ with the wall normal. (Mason 1989)

574

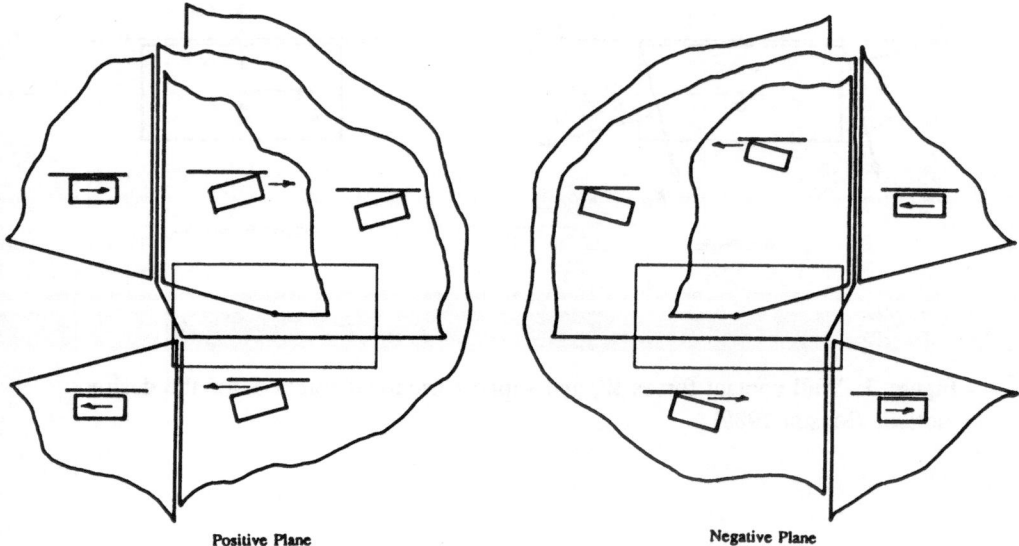

Positive Plane   Negative Plane

Figure 5: The complete mapping. Not shown is the contact mode that leaves the block motionless, which maps to the entire space of acceleration centers. The gaps in the figure correspond exactly to jamming forces, which fail to move the block no matter how large the force magnitude. Also not shown is the zero line, which in this case is a straightforward continuation of the positive plane. (Mason 1989)

really occur. But any motion that can occur will be included in the predictions. Hence, where a unique contact mode is predicted, that prediction is a correct one. Note that the desired contact mode, rightward sliding, does not overlap other modes. We can guarantee the desired motion by generating any force in the region. The problem of generating the required force is the subject of the next section.

4. Synthesis of the compliant motion

We model the compliant motion as a spring, with one end attached to the block, and the other end position-controlled. The problem is to choose the attachment point and the position trajectory so that the block moves as desired. There is a very simple solution to this problem:

1. take any point in the rightward-sliding region of Figure 5,

2. transform this point into the corresponding line of force,

3. attach the spring anywhere that the line of force crosses the block,

4. pull the other end of the spring along the desired line of force.

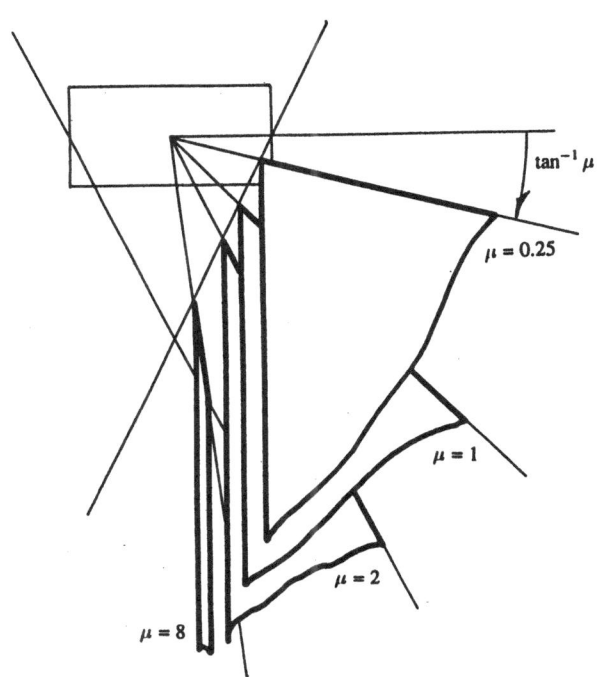

Figure 6: How the constraint on applied force varies with coefficient of friction μ. The rightward-translation region is determined by the intersections of three lines. Two of the lines are fixed in an "X" shape. The third line passes through the center of gravity at an angle $\alpha = \tan^{-1}\mu$, the friction angle. As the coefficient of friction increases, the third line pivots, and the rightward-translation region changes as shown.

With this strategy, the applied force will always be represented by the point in the desired motion region. There are actually two possible motions: for small magnitude forces, the block remains at rest, and for large magnitude forces, the block moves to the right as desired. As the robot stretches the spring, the force magnitude increases until the block begins to move.

The main problem with this simple strategy is that, as soon as the block moves, the center of gravity moves, and the acceleration center representing the spring force will move out of the desired motion region. We need to choose a strategy that works even with motion of the block. A second problem is that the coefficient of friction is rather unpredictable, and we would like to choose a strategy that is not sensitive to variations in the coefficient of friction μ. We construct the applied-force region F_i for several different values of μ in Figure 6, showing the dependence on μ.

Now, consider the dual representation of the spring force. One end of the spring is fixed to the block, at a point P. Because the spring force always passes through point P, it maps onto a line P' in the dual space. Given the path of the controlled end of the spring, we can

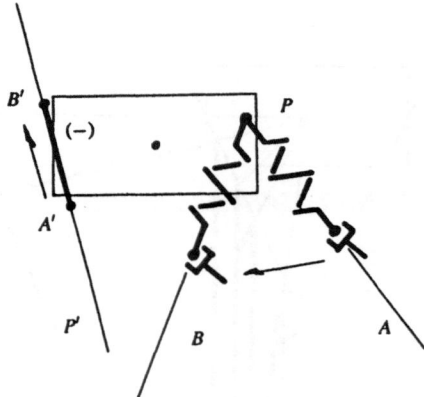

Figure 7: A compliant motion maps to a line segment in the dual space. The attachment of the spring at P determines a line P', and the path of the controlled end of the spring from A to B determines a line segment $A'B'$ in the dual space.

easily identify a segment of the line P' along which the spring force evolves as the robot makes its motion. In short, we can represent the motion strategy by a line segment in the dual space (Figure 7).

Looking again at Figure 6, we see that the line segment representing the compliant motion should be placed with one end in the center of the "X", and one end to the right of the "X" (Figure 8). The simplest choice starts at infinity to the right, and moves to the center of the "X". The spring must be attached to the center back face of the block. The path of the controlled end is not completely determined. But note that by moving parallel to the wall, the analysis remains constant even when the block begins to move. The strategy, then, agrees with intuition. We attach the spring to the back face, we pull the spring taut, pulling the block against the wall. Then we pull the spring along the wall, and the force-dual comes in from right infinity, toward the "X" center. No matter how high the coefficient of friction, eventually the force-dual point will cross into the desired contact mode. When it does, the block will move at the same velocity, so that the spring line of force remains constant from the block's viewpoint.

5. Conclusion

There are many different ways to move objects around. For each different way, there are also differences in the methods available for analysis and synthesis. Our work on the block-along-wall problems suggests that some unification may be possible. We decompose the planning problem into two parts: task dynamics, and operation planning. The task dynamics analysis can proceed in an operation-independent way, while the details of specific manipulation methods can be encapsulated in the operation planning stage. For the block-along-wall problem,

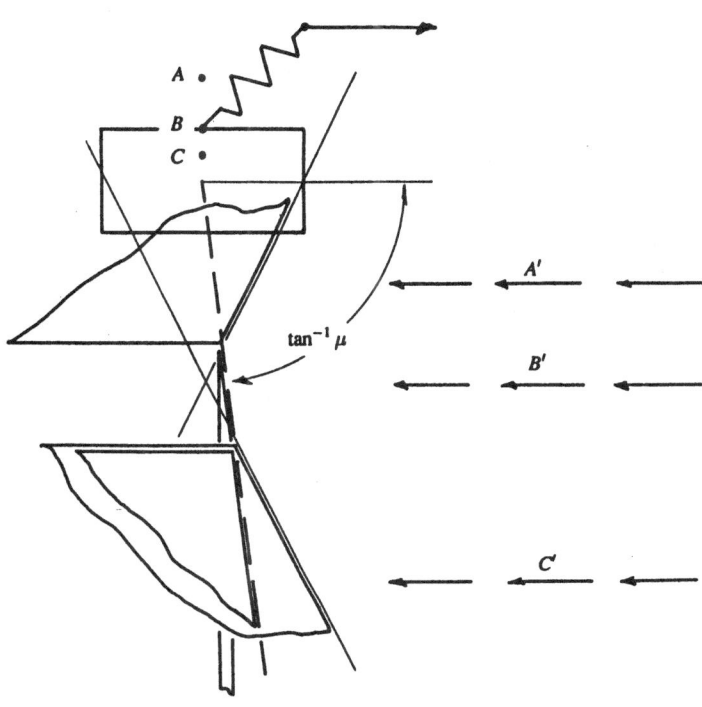

Figure 8: Choosing the compliant motion. We show the relevant contact modes in force-dual space, along with three different choices for attaching the spring. By attaching the spring at the center of the back face, and moving the controlled end to the right, we have a strategy that is insensitive to changes in the coefficient of friction, and which remains valid as the object begins to move.

the mapping of applied force to block motion is completely independent of the manipulator control strategy. After deriving the mapping, we can then turn to any of several different possible methods for producing desired forces. Another promising aspect is the method of dynamic analysis, which is amenable to automation.

Acknowledgements

I would like to thank Randy Brost who contributed to the paper in many ways, and Tom Mitchell and Alan Christiansen, who encouraged me to work on the block-along-wall problem. This work was supported by the National Science Foundation under grant DMC-8520475.

References

Brost, R. C., and Mason, M. T., "Graphical Analysis of Planar Rigid-Body Dynamics with Multiple Frictional Contacts," Fifth International Symposium on Robotics Research, Tokyo, 1989.

Erdmann, M. A., "On Motion Planning with Uncertainty," MIT Artificial Intelligence Laboratory Technical Report 810, August 1984.

Erdmann, M. A., and Mason, M. T., "An Exploration of Sensorless Manipulation," *IEEE J Robotics and Automation,* v4 n4, August 1988.

Mason, M. T., "How to push a block along a wall." *Proceedings, NASA Conference on Space Telerobotics,* Pasadena, February 1989.

Rajan, V. T., Burridge, R., and Schwartz, J. T., "Dynamics of a Rigid Body in Frictional Contact with Rigid Walls," *Proceedings, IEEE Conference on Robotics and Automation,* pp. 671-677, Raleigh, North Carolina, March 1987.

Reuleaux, F., *The Kinematics of Machinery,* Macmillan, 1876. Republished by Dover, 1963.

Simunovic, S., "Force Information in Assembly Processes," *Proceedings, 5th International Symposium on Industrial Robotics,* pp. 415-431, September 1975.

Planning Movement for Two PUMA Manipulators Holding the Same Object

R. M. C. Bodduluri, J. M. McCarthy, and J. E. Bobrow
Department of Mechanical Engineering
University of California, Irvine

Abstract

Two PUMA manipulators holding an object form a six degree of freedom closed kinematic chain. This situation occurs when heavy or long objects are to be manipulated by the robots. The constraint that the object not be dropped imposes limits on the configuration variables, which are treated as obstacles. A general program has been developed and implemented for planning paths for any multi-dimensional closed chain robot system. It is used to find a feasible path avoiding obstacles including that due to the closure of the chain. This program is used to plan a pick and place movement for two PUMA manipulators. The degrees of freedom are chosen so that both the robots can be controlled independently instead of using master-slave strategy.

1 Introduction

Two robots are said to be cooperating if they hold on to the same workpiece during their motion. The constraint that the workpiece can not be dropped by either of the robots causes the common workspace to be smaller than individual workspaces. This common workspace is the subspace of the workspaces of individual robots, see Kerr and Roth 1986, Palmquist and Duffy 1987a,b, and McCarthy et al. 1987. This workspace is further reduced by the fact that there may arise some collisions between the links of one robot with those of the other.

The fact that the workpiece can not be dropped and that there can not be any collisions between the links can be viewed as presenting obstacles in the joint space of the cooperating robots. Similarly, the real obstacles that are present in the cartesian space can be mapped into the joint space as shown in Ge and McCarthy 1989 and Lozano-Perez 1987. In this paper, we present an algorithm that plans movement in the joint space without explicit knowledge of the obstacles (including those due to closure constraint and self collisions) into the joint space.

Our approach is to represent the free workspace using one degree of freedom moves. This representation needs to determine only whether a set of one degree of freedom movements are collision free or not. This is done quickly by moving one joint of the robot at a time, or by conducting a numerical experiment. It is to be noted that earlier work by Brooks 1983, Schwartz and Sharir 1983, Gouzenes 1984, Sharir and Ariel-sheffi

1984, Lozano-Perez 1987, and Singh and Wagh 1987 establish n-dimensional free regions whereas we generate a web of one dimensional free edges, thus saving considerable amount of computation time. This representation is refined generating more free edges by a recursive subdivision of the joint space. The path thus obtained is rectilinear in the joint space which is not an optimal path in any sense. It minimizes only the number of collision detections needed to find a path. We intend to use this path as a starting point for a numerical trajectory optimization method such as Gilbert and Johnson 1985 and Bobrow 1988b.

Coordination of multiple robots has been studied by many researchers. Luh and Zheng 1987 use the joint forces to control two cooperating PUMA robot arms; Tao and Luh 1989 extend this work to redundant robots by obtaining a more uniform distribution of the forces. Other research includes control of cooperating robots using the forces in the workpiece by Schneider and Cannon 1989; control using adaptive approaches by Walker et al. 1989 and Hu and Goldenberg 1989; design of dynamic control by Tarn et al. 1987; and the force control of two planar robots by Uchiyama et al. 1987. These methods use actual measurements of the robots, so they can accommodate unexpected variation of the trajectories. While these methods guarantee locally the robot will follow the given path, they can not guarantee whether the robot will reach the goal or not. O'Donnell and Lozano-Pérez 1989 construct complete trajectories to coordinate the motion of the two robots. They first plan the paths individually (without the other robot) and then use those paths to coordinate the two. Our approach differs in a very fundamental way that the two cooperating robots are treated as one mechanical system and the paths are planned simultaneously.

The goal of this research is to develop an algorithm that will quickly plan paths for general mechanical systems. There are only two assumptions which must be satisfied to use this algorithm. The first is that the constraints are not time varying; we only consider holonomic constraint equations relating the degrees of freedom of the system. The second is that we assume a method is available to determine contact with an obstacle when one degree of freedom is moved and the others are held fixed. In this paper, this is achieved by conducting a numerical experiment, that is, by varying one degree of freedom, we determine whether there is a collision or not either with a real obstacle or among the links of the robots or if it violates the closure constraint. A more efficient algorithm (Canny 1986, Gilbert et al. 1988, Bobrow 1989) may be used to improve the performance.

In the next section, the strategy for the path planning is intuitively introduced using a simple example and then the algorithm is stated. We then present an application of the algorithm to two cooperating PUMA manipulators as an example. First the kinematic equations for the closed chain is determined. The typical computational time and memory requirements are 0.68 seconds and 64 kilobytes for a simple problem, and 163 seconds and 7.5 megabytes for a hard to find path, respectively.

2   Path Planning Strategy

In order to illustrate this algorithm, we show how it finds a path from S to G around three obstacles in a configuration space with two degrees of freedom, see Figure 1. This is intended to demonstrate the following important features of the algorithm:

- There is no preprocessing required.

- The recursive subdivision ensures that more vertices are added to the search graph only if needed.

- Only edges are to be examined. *No information regarding the interior of the region is needed.*

Let the configuration space be given by $X \times Y$, where the range of the two configuration variables is defined by $X = [l_x, u_x]$, and $Y = [l_y, u_y]$, as shown in Figure 1a. We now seek a path from the start position $S = (s_x, s_y)$ to the goal $G = (g_x, g_y)$ while avoiding collision with the three circular obstacles shown.

Initialization: The initial step is to divide the configuration space at the start point S as shown in Figure 1b. This subdivision partitions the space to form a quadtree of regions, the vertices of these regions and the edges joining them form our vertex decomposition of the configuration space. The region which contains the goal point G is further subdivided at G. Figure 1c shows the result; the configuration space is represented as fourteen vertices connected by the edges of seven regions.

Search: The vertex graph representing the space is now searched to find a path joining S and G. The search seeks to connect vertices reachable from S to neighboring vertices by testing the connecting edges for collisions with obstacles. If an edge is unobstructed then it is denoted as collision-free in the vertex graph, the darkened edges in Figure 1c. At this level of resolution, either the goal is reached or all paths from S to G are obstructed. At this stage we know all the reachable vertices, and which edges that connect them are collision-free.

Subdivision: In Figure 1c, the reachable vertices are V_1, V_2, and V_3 and the collision-free edges are darkened. Note that the points S, V_1, V_2, and V_3 are the vertices of a region bounded by the free edges. The algorithm labels this region as empty so that it is never subdivided further. Now we identify the non-empty regions which are reachable from S. These regions are subdivided at their mid-points as shown in Figure 1d and new vertices are inserted in the graph. The result is higher resolution vertex decomposition of the configuration space. The algorithm returns to the search step and, in this example, finds a path from S to G along collision-free (dark) edges.

The Algorithm

To perform the search, the following procedure is used. Assume the start and goal configurations are given by $(q_1^{st}, q_2^{st}, \ldots, q_n^{st}), (q_1^g, q_2^g, \ldots, q_n^g)$; the steps performed by the algorithm are:

1. Subdivide the root cube (n-dimensional) at the interior point $(q_1^{st}, q_2^{st}, \ldots, q_n^{st})$.

2. Determine the smallest sub-cube containing $(q_1^g, q_2^g, \ldots, q_n^g)$, and subdivide this cube at the goal position.

3. Perform an A^* search from the start to the goal position along free edges of the cubes at the current level of resolution. The Euclidean distance in configuration

582

space is used as the heuristic. The vertices visited by the search algorithm are those that are reachable from the start position. If the search algorithm reaches the vertex at the goal position, then stop.

4. If no path exists at the level of resolution of the last step, perform one uniform subdivision of all partially filled cubes[1] with at least one reachable vertex, and return to step 3.

The process of subdividing a cube consists of identifying the new sub-cubes and their vertices. Associated with each vertex are bounds on range of motion from the vertex to its neighbors which are computed. This determines whether or not an adjacent vertex can be reached.

It can be proved that this algorithm finds a path if one exists, see Bobrow and Bodduluri 1988a.

3 Kinematics of a PUMA Manipulator

A PUMA manipualtor, as shown in the Figure 2, is modeled as an RRRS open chain mechanism where the end effector is attached to the spherical joint. This allows us to view the first three joints as being used to position the end effector at any desired location in the workspace and the last spherical joint, to orient it.

The coordinate frames attached to each link are shown in the Figure 2 when all the joint angles are equal to zero. Note that the frame 0 (not shown in the figure) coincides with the frame 1 when the first joint angle is zero. The 4×4 matrices representing the transformations from one frame to the other are given by (Craig 1986)

$$
{}^{0}_{1}T = \begin{bmatrix} \cos\theta_1 & -\sin\theta_1 & 0 & 0 \\ \sin\theta_1 & \cos\theta_1 & 0 & 0 \\ 0 & 0 & 1 & 0 \\ 0 & 0 & 0 & 1 \end{bmatrix},
$$

$$
{}^{1}_{2}T = \begin{bmatrix} \cos\theta_2 & -\sin\theta_2 & 0 & 0 \\ 0 & 0 & 1 & 0 \\ -\sin\theta_2 & -\cos\theta_2 & 0 & 0 \\ 0 & 0 & 0 & 1 \end{bmatrix},
$$

$$
{}^{2}_{3}T = \begin{bmatrix} \cos\theta_3 & -\sin\theta_3 & 0 & a \\ \sin\theta_3 & \cos\theta_3 & 0 & 0 \\ 0 & 0 & 1 & 0 \\ 0 & 0 & 0 & 1 \end{bmatrix}.
$$

(1)

Given the link parameters a, d (the lengths of upper and fore arms respectively), and the joint angles θ_1, θ_2, and θ_3, the location of the end effector can be determined by the transformation equation

$$
\mathbf{X} = {}^{0}_{1}T\, {}^{1}_{2}T\, {}^{2}_{3}T\, \mathbf{x}
$$

(2)

[1]Subdivision of cubes in the order that their vertices are visited by the search algorithm will reduce many unnecessary subdivisions in the last level of subdivision required to find a path.

where $\mathbf{X} = (X, Y, Z)$ and $\mathbf{x} = (x, y, z)$ are coordinates of any point moving with the third link in the base and third link coordinate frames, respectively. If $\mathbf{x} = (0, d, 0)$, the above equation gives the coordinates of the end of the third link (also known as tool center point) in the base coordinate frame.

4 Kinematics of the Two Cooperating PUMA Manipulators

The two PUMA manipulators are placed relative to each other as shown in the Figure 3. They are oriented such that one works as a left arm and the other as right arm. The transformation matrix from the frame 3 of the robot to the fixed reference frame is given by

$$
\begin{aligned}
{}^F_3T &= {}^F_0T \, {}^0_1T \, {}^1_2T \, {}^2_3T \\
{}^F_3T' &= {}^F_0T' \, {}^0_1T' \, {}^1_2T' \, {}^2_3T'
\end{aligned}
\tag{3}
$$

where the superscript $'$ denotes the second arm (the one on the right in the figure). The transformation matrices ${}^F_0T$ and ${}^F_0T'$ base frame 0 of each robot to the fixed reference frame are given by

$$
{}^F_0T = \begin{bmatrix} 1 & 0 & 0 & t_1 \\ 0 & 1 & 0 & t_2 \\ 0 & 0 & 1 & t_3 \\ 0 & 0 & 0 & 1 \end{bmatrix}
\tag{4}
$$

and for the second robot

$$
{}^F_0T' = \begin{bmatrix} 1 & 0 & 0 & t'_1 \\ 0 & 1 & 0 & t'_2 \\ 0 & 0 & 1 & t'_3 \\ 0 & 0 & 0 & 1 \end{bmatrix}
\tag{5}
$$

Here the vectors (t_1, t_2, t_3) and (t'_1, t'_2, t'_3) are the translation vectors from F to frame 0 of the first and second robots respectively. In the configuration shown, the joint angles are $\theta_1 = 90^0$ and $\theta_2 = \theta_3 = \theta'_1 = \theta'_2 = -90^0$. In what follows it will be shown that the third joint angle of the second robot θ'_3 is dependent on the other joint angles.

The coordinates of the location of the end effector of each of the robots are given by

$$
\begin{aligned}
\mathbf{X} &= {}^F_3T\mathbf{x} \\
\mathbf{X}' &= {}^F_3T'\mathbf{x}'
\end{aligned}
\tag{6}
$$

in the fixed reference frame F. Substituting for the ${}^F_0T, {}^F_0T'$ from 1, we obtain

$$
\begin{aligned}
\mathbf{X} &= \left\{ \begin{array}{c} -dc_1s_{23} + ac_1c_2 + t_1 \\ -ds_1s_{23} + as_1c_2 + t_2 \\ -dc_{23} - as_2 + t_3 \end{array} \right\}, \\
\mathbf{X}' &= \left\{ \begin{array}{c} -d'c'_1s'_{23} + a'c'_1c'_2 + t'_1 \\ -d's'_1s'_{23} + a's'_1c'_2 + t'_2 \\ -d'c'_{23} - a's'_2 + t'_3 \end{array} \right\},
\end{aligned}
\tag{7}
$$

where $c_i = \cos(\theta_i), s_i = \sin(\theta_i), c_{23} = \cos(\theta_2 + \theta_3)$, and $s_{23} = \sin(\theta_2 + \theta_3)$ and the superscript $'$ denotes that it is for the second robot.

If these two robots are cooperating, that is, if they are holding the same object during the motion, the distance between $\mathbf{X}$ and $\mathbf{X}'$ should allways be equal to the length of the workpiece L. This condition can be expressed mathematically as

$$(\mathbf{X} - \mathbf{X}') \cdot (\mathbf{X} - \mathbf{X}') = L^2 \tag{8}$$

which is known as *closure constraint equation*. Therefore, the joint angles $\theta_1, \theta_2, \theta_3$, and $\theta_1', \theta_2', \theta_3'$ are not independent in order not to drop the workpiece. We consider the joint angles $\theta_1, \theta_2, \theta_3, \theta_1'$, and θ_2' as the independent degrees of freedom and we solve for the joint angle θ_3' from this closure equation. Though it is a six degree of freedom system, the sixth degree of freedom, the rotation of the workpiece about the axis passing through the two end effectors does not effect the configuration of the robots.

Expanding Eq. 8 using the Eq. 7, we obtain

$$Ac_{23}' + Bs_{23}' = C \tag{9}$$

where

$$\begin{aligned} A &= -a's_2' + t_3' - Z \\ B &= c_1'(a'c_1'c_2' + t_1' - X) - s_1'(a's_1'c_2' + t_2' - Y) \\ C &= (d'^2 + (a'c_1'c_2' + t_1' - X)^2 + (a's_1'c_2' + t_2' - Y)^2 + (-a's_2' + t_3' - Z)^2 - L^2)/2d' \end{aligned} \tag{10}$$

This equation can be solved for θ_3' given $\theta_1, \theta_2, \theta_3, \theta_1'$, and θ_2' as

$$\theta_3' = \arccos(\frac{C}{\sqrt{A^2 + B^2}}) + \arctan(\frac{B}{A}) - \theta_2'. \tag{11}$$

If $C^2 > A^2 + B^2$, then there doesn't exist a solution for θ_3' which implies that closure fails for the given $\theta_1, \theta_2, \theta_3, \theta_1', \theta_2'$. Otherwise, there will be two solutions in general. We consider only one solution for planning motion in this paper.

5 Examples

Two different examples are presented showing the coordinated motion of the PUMA manipulators. In both cases, the link parameters of the robots are $a = a' = 1.25$ and $d = d' = 1.5$. The translation vectors to the base joints are $(t_1, t_2, t_3) = (-1.0, 0.0, 2.25)$ and $(t_1', t_2', t_3') = (1.0, 0.0, 2.25)$. Therefore, the base separation is 2.0 whereas the length of each robot arm is atleast $a + d = 2.75$ which means that the common workspace is filled with collisions of different links. The workpiece length is chosen to be $L = 2.5$.

No real obstacles are included in this example so that the motion can be seen clearly. Though there are no real obstacles, all the collisions among the links as well as collisions with the floor which are most complicated obstacles are avoided besides satisfying the closure constraint.

A Simple Example

A set of start and goal positions are specified as shown in the Figure 4. The start position is chosen approximately vertical close to the base of the second robot and the goal position is horizontal away towards the left of the first robot. It takes 0.68 seconds to compute the path and needs 64 kilobytes of memory to store the data about the joint space on Apollo DN10000 workstation. The configuration of the robot system at different intervals during the motion is shown in the Figure 5.

A Hard Example

In this example, the goal of the "simple" example is used as the start position. A new goal position is specified that is vertical and behind the robots. This forces the robots to turn around, see Figure 6. However, if they turn towards each other, there will be collisions among the links and if one turns away from the other, then the closure constraint will not be satisfied. Therefore, they must move carefully avoiding each other to finally reach the goal Figure 7. It is very interesting to watch a simulation on the computer screen. The computational requirement for this is 163 Seconds and 7.5 megabytes.

6 Conclusions

A general algorithm has been developed for planning paths in a configuration space of n degrees-of-freedom. Applications have been given for the case of two cooperating six degree of freedom PUMA robot manipulators.

An n−dimensional web of motions along the edges of higher dimensional cubes is used to construct the search graph for an A^* algorithm. The main computational function needed to implement the algorithm is the calculation of free intervals for one configuration variable at a time, given an initial starting point. The computational requirement depends on the complexity of the problem rather than on the number of links in the system and number of the obstacles.

References

[1] J.E. Bobrow, "A Direct Minimization Approach for Obtaining the Distance Between Convex Polyhedra," *International Journal of Robotics Research,* Vol 8, No 3, June 1989.

[2] J.E. Bobrow and R.M.C. Bodduluri, "A General Robot Path Planning Algorithm," *Proceedings of the 27th IEEE Conference on Decision and Control,* Austin, Texas, Vol. 3, pp 2292-2293, December 7-9, 1988a.

[3] J.E. Bobrow, "Optimal Robot Path Planning Using The Minimum-Time Criterion," *IEEE Journal of Robotics and Automation,* Vol. 4, No. 4, pp 443-450, August 1988b.

[4] R.A. Brooks, "Planning Collision-Free Motions for Pick- and-Place Operations," *International Journal of Robotics Research* Vol. 2, No 4, pp 19-44, 1983

[5] J. Canny, "Collision Detection for Moving Polyhedra," *IEEE Transactions on Pattern Analysis and Machine Intelligence,* Vol. PAMI-8, No 2, pp 200-209, 1986.

[6] A.A. Cole, Hsu, and Sastry, "Dynamic Regrasping by Coordinated Control of Sliding for a Multifingered Hand," *Proceedings of IEEE Robotics and Automation*, Vol. 2, pp 781-786, 1989.

[7] J.J. Craig, *Introduction to Robotics: Mechanics & Control*, Addison-Wesley, Reading, MA, 1986.

[8] E.G. Gilbert and D.W. Johnson, "Distance Functions and Their Application to Robot Path Planning in the Presence of Obstacles," *IEEE Journal of Robotics and Automation*, Vol RA-1, March 1985, pp 21-30.

[9] E.G. Gilbert, D.W. Johnson, and S.S. Keerthi, "A Fast Procedure for Computing the Distance between Complex Objects in Three–Dimensional Space," *IEEE Journal of Robotics and Automation*, Vol. 4, No. 2, pp.193-203, April 1988.

[10] L. Gouzenes, "Strategies for Solving Collision-Free Trajectory Problems for Mobile and Manipulator Robots," International Journal of Robotics Research, Vol. 3, No. 4, Winter 1984.

[11] Y. Hu and A.A. Goldenberg, "An Adaptive Approach to Motion and Force Control of Multiple Coordinated Robot Arms," *Proceedings of the IEEE Robotics and Automation Conference*, Vol. 2, pp 1091-1096, 1989.

[12] J. Kerr and B. Roth, "Analysis of Multifingered Hands," *International Journal of Robotics Research*, Vol. 4, No. 4, pp.3-17, 1986.

[13] T. Lozano-Perez and M. Wesley, "An Algorithm for Planning Collision-Free Paths Among Polyhedral Obstacles," *Communications of the Association for Computer Machinery*, Vol 2, No 10, 1979, pp 560-570.

[14] T. Lozano-Perez, "Spatial Planning: A Configuration Space Approach," *IEEE Transactions on Computers,* C-32(2), pp 108-120, 1983.

[15] T. Lozano-Perez, "A Simple Motion-Planning Algorithm for General Robot Manipulators," *IEEE Journal of Robotics and Automation*, Vol RA-3, No 3, June 1987, pp 224-238.

[16] J.Y.S. Luh and Y.F. Zheng, "Constrained Relations between Two Coordinated Industrial Robots for Motion Control, " *International Journal of Robotics Research*, Vol. 6, No. 3, pp 60-70

[17] J.M. McCarthy, Q. Ge, and R.M.C. Bodduluri, "The Analysis of Cooperating Planar Robot Arms in the Image Space of the Workpiece," University of California, Irvine, Department of Mechanical Engineering Technical Memo ME-TR-1, December 1987.

[18] P.A. O'Donnell and T. Lozano-Perez, "Deadlock-Free and Collision-Free Coordination of Two Robot Manipulators," *Proceedings of IEEE Robotics and Automation*, Vol. 1, pp 484-489, 1989.

[19] R.D. Palmquist and J. Duffy, "Reachable Workspace of Two Planar 3R Cooperating Manipulators," *Proceedings of the ASME Computers in Engineering Conference*, Boston, MA, 1987.

[20] R.D. Palmquist and J. Duffy, "Dexterity of Two Planar 3R Cooperating Manipulators," *Proceedings of the ASME Computers in Engineering Conference*, Boston, MA, 1987.

[21] S.A. Schneider and R.H. Cannon, Jr., "Object Impedance Control for Cooperative Manipulation: Theory and Experimental Results,", *Proceedings of the IEEE Robotics and Automation Conference*, Vol. 2, pp 1076-1083, 1989.

[22] M. Sharir and E. Ariel-sheffi, "On the Piano Movers Problem IV. Various Decomposable Two-Dimensional Motion Planning Problems," *Communications on Pure and Applied Mathematics*, Vol 37, pp 479-493, 1984.

[23] J.T. Schwartz and M. Sharir, "On the Piano Movers Problem I. The Case of a Two-Dimensional Rigid Polygonal Body Moving Amidst Polygonal Barriers," *Communications on Pure and Applied Mathematics*, Vol 36, No. 3, pp 345-398, 1983.

[24] S. Singh, M. Wagh, "Robot Path Planning Using Intersecting Convex Shapes: Analysis and Simulation," *IEEE Journal on Robotics and Automation* Vol RA-3, No.2, pp 101-108, 1987.

[25] J.M. Tao and J.Y.S. Luh, "Coordination of Two Redundant Robots," *Proceedings of IEEE Robotics and Automation Conference*, Vol. 2, pp 425-430, 1989.

[26] T.J. Tarn, A.K. Bejczy, and X. Yun, " Design of Dynamic Control of Two Cooperating Robot Arms: Closed Chain Formulation," *Proceedings of IEEE Robotics and Automation Conference*, Raleigh, NC, pp 7-13, 1987.

[27] M. Uchiyama, N. Iwasawa, and Hakomori, "Hybrid Position/Force Control for Coordination of a Two-arm Robot," *Proceedings of IEEE Robotics and Automation Conference*, Raleigh, NC, pp 1242-1247, 1987.

[28] M.W. Walker, D. Kim, and J. Dionise, "Adaptive Coordinated Motion Control of Two Manipulator Arms," Proceedings of the IEEE Robotics and Automation Conference, Vol. 2, pp 1084-1090.

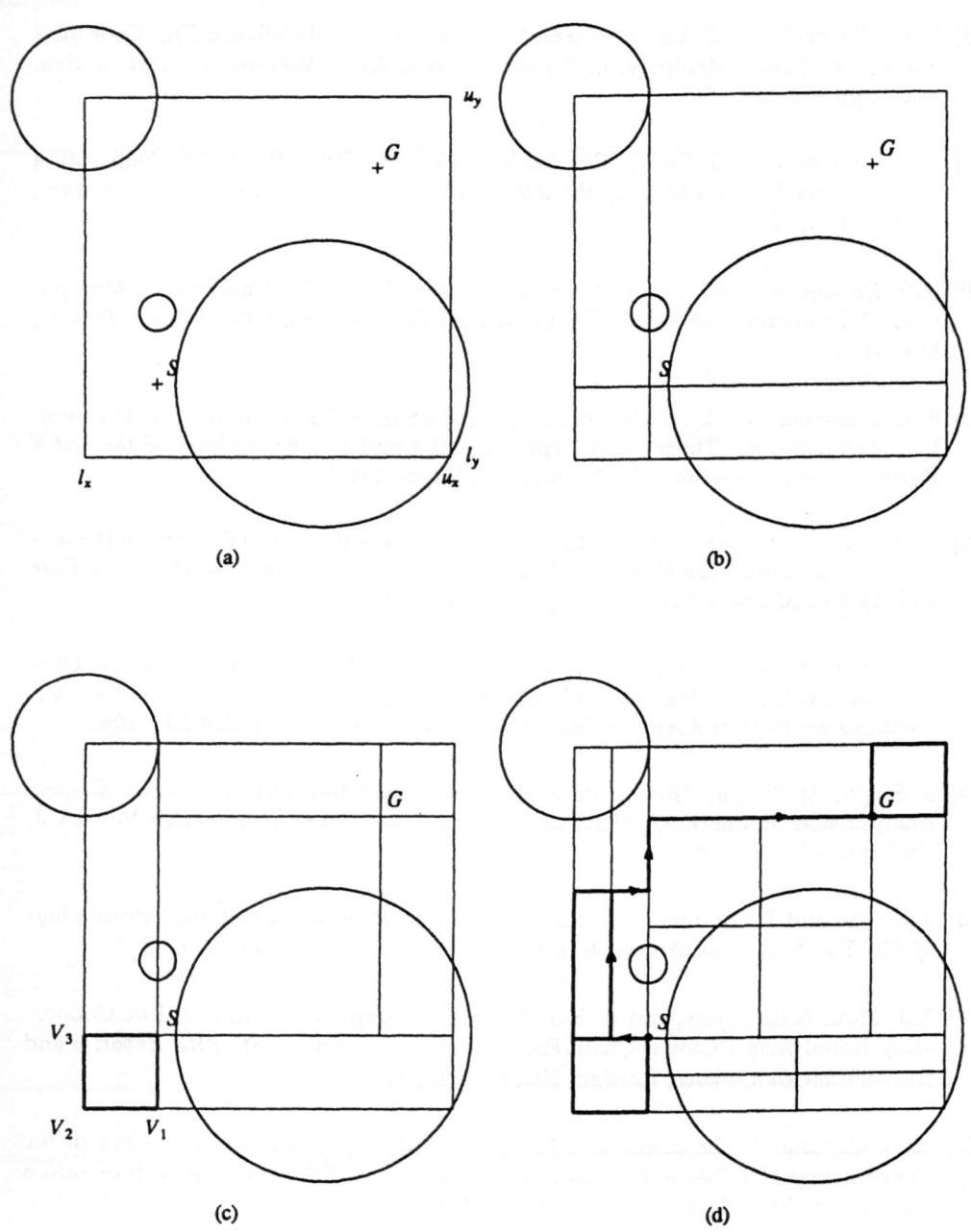

Figure 1: (a) The bounded two dimensional space to be searched for a path from S to G. (b) The space subdivided at S. (c) Further subdivision at G. (d) Subdivide all the non-empty regions with reachable vertices.

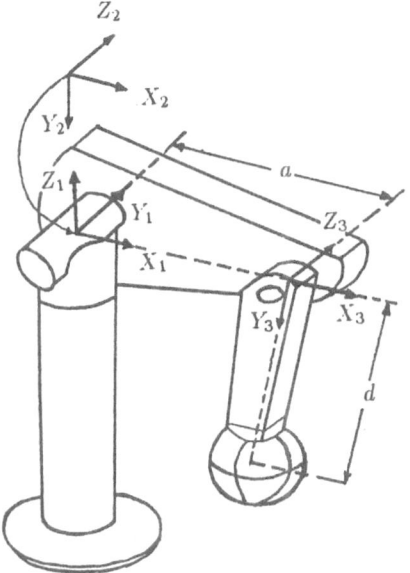

Figure 2: A PUMA robot manipulator with its coordinate frame assignment.

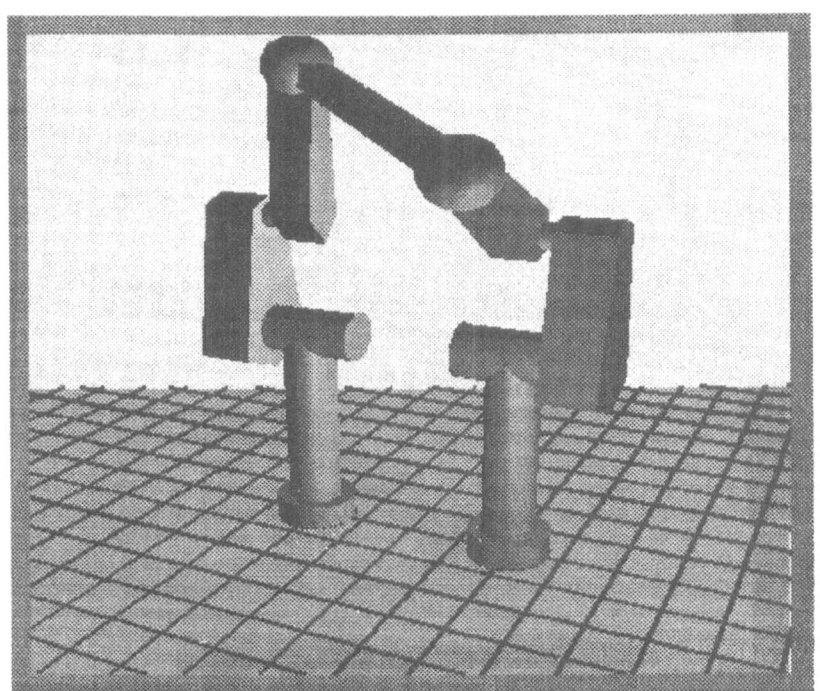

Figure 3: Two cooperating robots.

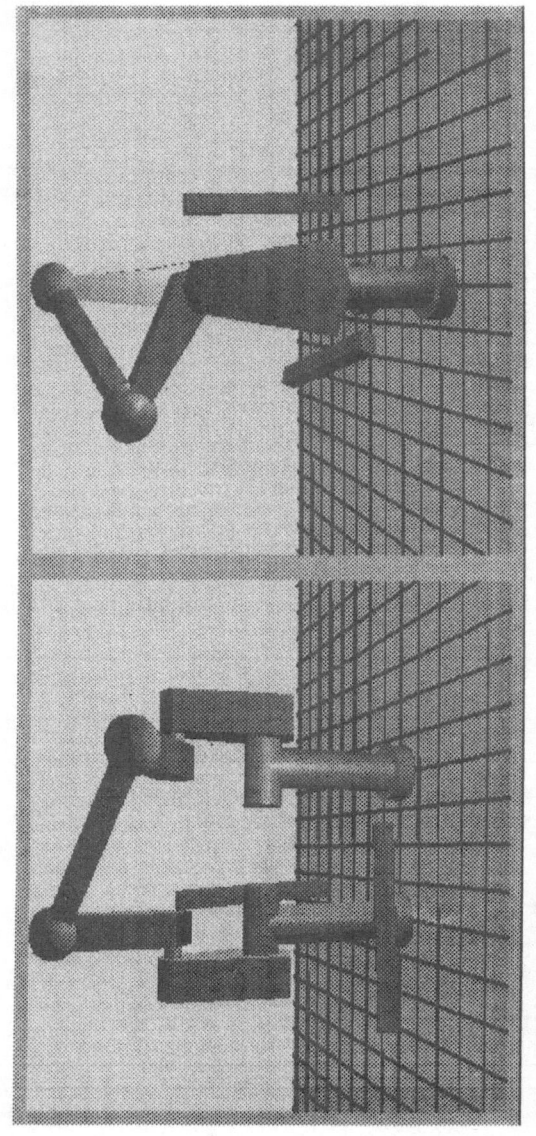

Figure 4: The Simple Problem: Start and goal positions.

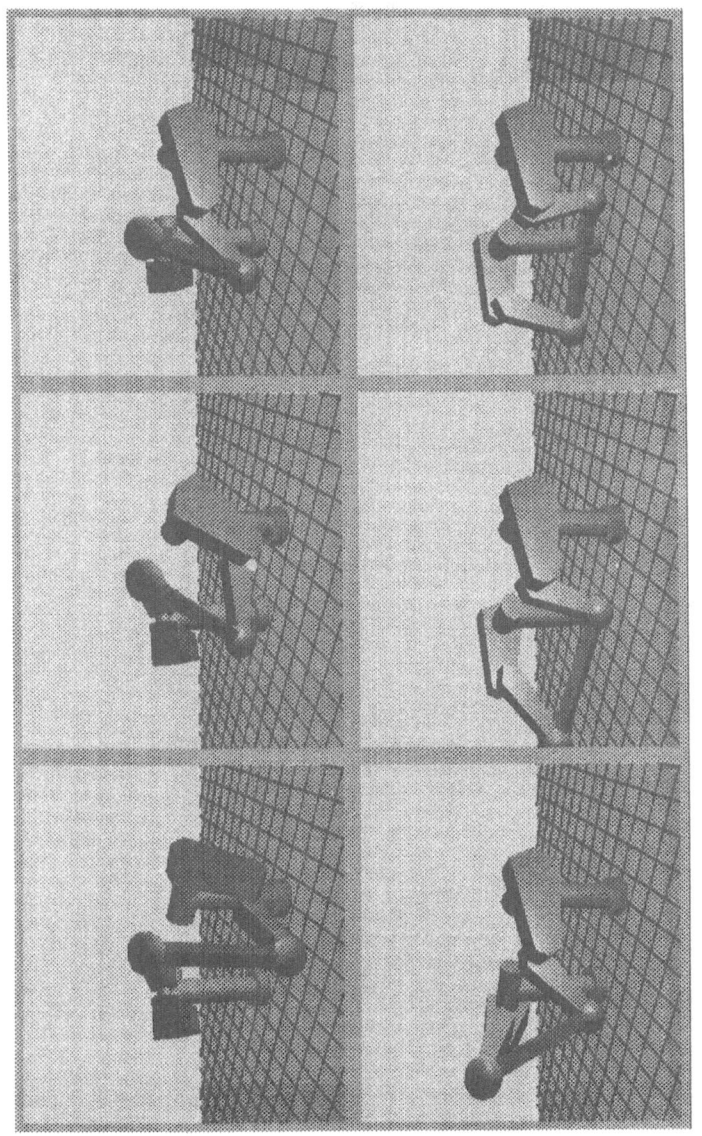

Figure 5: The Simple Problem: The path found.

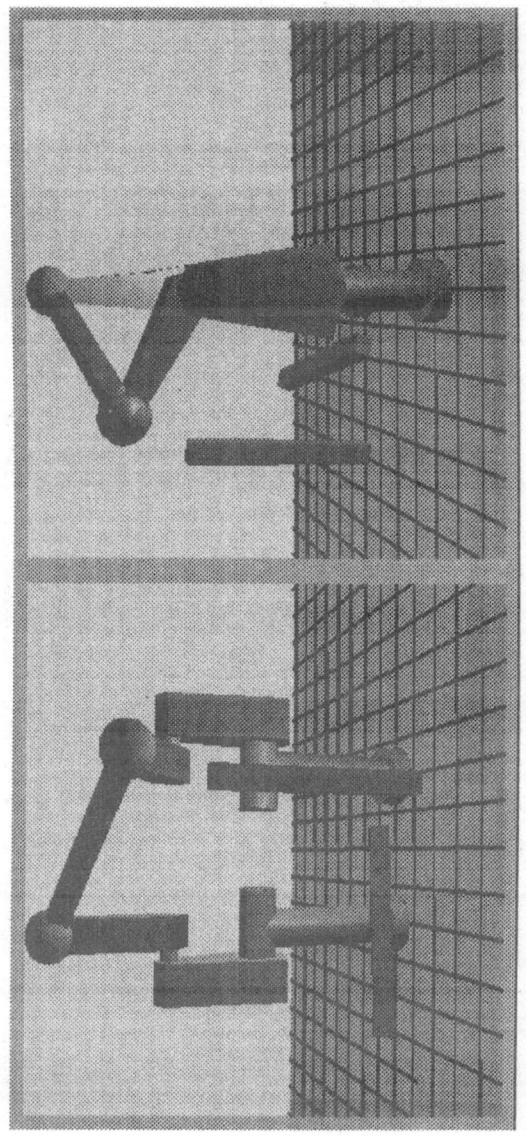

Figure 6: The Hard Problem: Start and goal positions.

593

Figure 7: The Hard Problem: Path is found. Notice how close the robots are working together.

Trajectories of Human Multi-Joint Arm Movements: Evidence of Joint Level Planning

J. Randall Flanagan and David J. Ostry

Department of Psychology, McGill University

Montreal, Quebec, Canada

Abstract

In this paper we show that staggered joint interpolation [7] with maximally smooth joint motion can account for both the hand paths and tangential velocity profiles observed in human multi-joint arm movements. Arm trajectories were recorded while subjects performed point-to-point movements in a vertical plane. The shape of the tangential velocity profile of the hand was found to be symmetrical and bell-shaped. Although this finding is predicted by the maximum-smoothness theory [6], which assumes that hand motion is maximally smooth, the theory cannot account for the curved hand paths which were often observed. The present model assumes that joint motion, rather than the motion of the hand, is maximally smooth. The results of this work suggest that multi-joint arm movements may be planned at the joint level.

1 Introduction

The problem of trajectory planning is central to the study of human motor control as well as to the field of robotics. The challenge in robotics is to develop trajectory planning methods that are both flexible and efficient. In human motor control, the challenge is to discover how a working system, which is both flexible and efficient, works.

Recently, workers in human motor control have attempted to identify the coordinates in which human multi-joint arm movement trajectories are planned. Several investigators [1][6][9] have suggested that multi-joint arm movements are planned in cartesian endpoint coordinates whereas others [7][10] have argued that these movements are planned in joint angular coordinates. There are, of course, other possibilities. For example, multi-joint arm movement trajectories may be planned in terms of muscular coordinates. Moreover, movement trajectories may not be planned at all. Indeed, it has been suggested that the central nervous system may only specify the final configuration of the arm and that the trajectory towards the final position is determined by neuromuscular dynamics and by external constraints on the movement [3][4].

2 Theory

Early evidence for endpoint planning was provided by the work of Morasso and others [1][8][9] which investigated the trajectories of point-to-point human arm movements restricted to a horizontal plane. These movements were characterized by straight line hand paths in cartesian coordinates regardless of target position and movement rate. In addition, when subjects were instructed to simply move between targets, the shape of the tangential velocity profile of the hand was found to be bell-shaped and symmetrical. Some evidence for hand level planning has also been provided by the experiments of Soechting and Lacquaniti [10] in which the hand paths of pointing movements to targets at different heights were found to form straight lines. These findings may be interpreted as evidence for hand space planning.

This work led to the formulation of the maximum-smoothness theory or minimum-jerk theory by Flash and Hogan [6]. This theory posits that movements are organized to minimize mean squared jerk at the endpoint. In agreement with the empirical findings on horizontal arm movements, the maximum smoothness theory predicts symmetrical bell-shaped tangential velocity profiles at the movement endpoint and straight line endpoint paths.

Whereas the maximum-smoothness theory can successfully describe the trajectories of human point-to-point horizontal arm movements, the theory fails to account for the trajectories of vertical movements. Atkeson and Hollerbach [2] examined the kinematics of unrestrained arm movements made between targets located in a vertical plane. In partial agreement with the minimum-jerk theory, they found that the shape of the wrist tangential velocity profiles was bell-shaped and symmetrical independent of target position, movement rate, and even hand-held load. However, in contrast to the minimum-jerk theory, they reported both straight line and curved wrist paths; the curvature of the wrist path depended on the area of the workspace in which the movement occured. This finding is not predicted by the minimum-jerk theory.

In a later paper, Hollerbach and Atkeson [7] introduced the joint level planning strategy of staggered joint interpolation to account for both the curved and straight line hand paths observed in their experiments on vertical point-to-point arm movements as well as for the hand paths reported by others. Staggered joint interpolation is a generalized form of linear joint interpolation in which all joints have the same time profiles between the start and end of movement. However, in staggered joint interpolation, unlike linear joint interpolation, the joints are not required to start (or stop) moving at the same time.

Linearly interpolated motion is generally characterized by curved hand paths. However, staggering the joint start or end times can often result in near straight paths in endpoint space [7]. It is only when straight line hand paths would require a joint reversal that near straight hand paths cannot be achieved via staggered joint interpolation. Hollerbach and Atkeson [7] have demonstrated that curved hand paths are often observed under such conditions.

On the one hand, staggered joint interpolation appears to account well for both the straight and curved endpoint paths observed in point-to-point multi-joint movements. On the other hand, the tangential velocity profile of the movement endpoint has been found to be bell-shaped and symmetrical under a wide variety of experimental conditions, as predicted by the maximum-smoothness theory. In this paper we present a version of staggered joint interpolation which can account for both the straight and curved hand paths and the bell-shaped symmetrical tangential velocity profiles that we have observed in our experiments. This version makes two assumptions: the joint angles follow minimum-jerk trajectories and the joints may be staggered at both the start and the end of the movement. To evaluate whether this version of staggered joint interpolation can account for the kinematics of point-to-point vertical arm movements, we have simulated hand paths and tangential velocity profiles and compared the results to experimental records. In the simulations we have used empirically determined initial and final joint angles and joint movement start and end times. In addition, we have examined directly the experimental joint angular velocity profiles.

2 Methods

We have examined the trajectories of point-to-point arm movements made between various targets located in a vertical plane. Movements were performed at both a preferred rate and at a faster rate. Four subjects (three females and one male) were instructed to make single smooth movements between targets and were told to not make corrective adjustments near the end of the movement.

A pegboard was used to position the targets in a plane sagittal to the subject. Targets consisted of circular disks (10 cm. in diameter) placed on the ends of pegs. Subjects were required simply to position their finger beside the target disks at the start and end of the movement.

Movements were recorded in 3-D with the WATSMART infrared imaging system. IREDs (infrared emitting diodes) were placed at the proximal and distal end of the upper and lower arms and were used to reconstruct joint angles. Each IRED was sampled at 400 Hz and digitally filtered with a Butterworth low-pass filter with a cutoff frequency of 6 Hz.

Spherical polar coordinates were used to define the position of the arm in 3-D (Figure 1). The position of the hand in space can be described by six parameters including four joint angles and the lengths of the upper and lower arms. The joint angles are upper arm (or shoulder) elevation S; upper arm yaw Y, lower arm (or elbow) elevation E; and upper arm roll R. Other joint angular coordinates can, and have, been used to describe the position of the arm [11]. However, for the purposes of the present analysis, the angles chosen were deemed appropriate since they change monotonically during point-to-point vertical arm movements; a prerequisite for joint interpolation. Although we have defined the position of the arm in 3-D, it should be noted that these movement occurred largely in the sagittal plane. Consequently, the principal angles of interest are the shoulder and elbow angles.

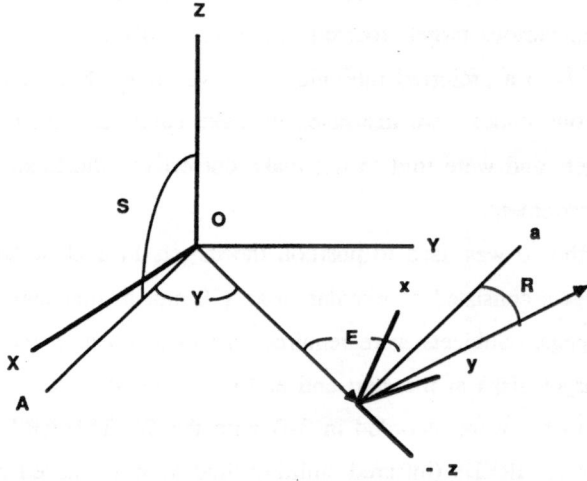

Figure 1: Arm position coordinates

3 Results

3.1 Experimental and Simulated Hand Paths

Experimental hand paths for one subject are shown in Figure 2. With the exception of the vertical movement, hand paths are unaffected by movement direction between targets. Furthermore, hand paths do not vary with movement speed. In the case of the vertical movement, the upward hand paths are more curved than the downward hand paths which are nearly straight. The hand paths observed in this study are similar to those reported by Atkeson and Hollerbach [2].

To assess how well staggered joint interpolation coupled with minimum-jerk joint trajectories could account for the experimental movement trajectories, we carried out simulations. The initial and final values for the four joint angles shown in Figure 1 and the joint movement start and end times were scored from the empirical movement trajectories. Joint start and end times were determined on the basis of 10% of peak joint angular velocity. The motions of the four joint angles were assumed to follow minimum-jerk trajectories.

Simulated hand paths are presented in Figure 3. The initial and final joint angles and joint start and end times used in these simulations were taken from the experimental hand paths shown in Figure 2 (preferred speed). We assumed that the length of the lower arm was 25% greater than the length of the upper arm in these simulations. This number was based on direct measurements of our subjects.

The simulated hand paths capture many of the features of the experimental hand paths. However, there are some discrepancies. In particular, one can see that the inward movements for paths 1 and 2 are quite curved in contrast to the experimental data. The reason for this will become clear in examining the joint velocity profiles (see Section 3.3).

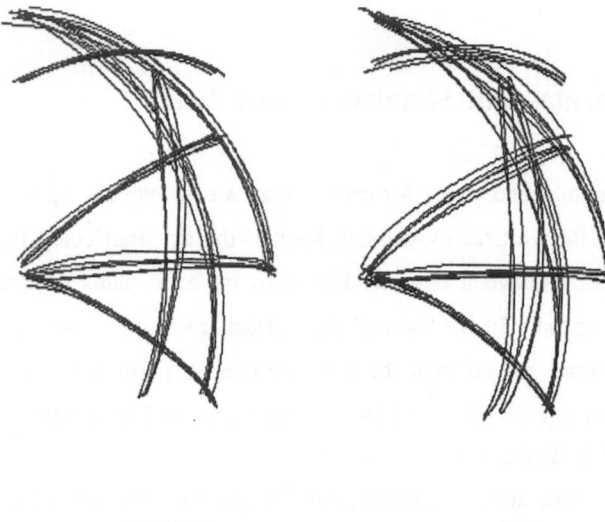

PREFERRED   FAST

Figure 2: Experimental hand paths

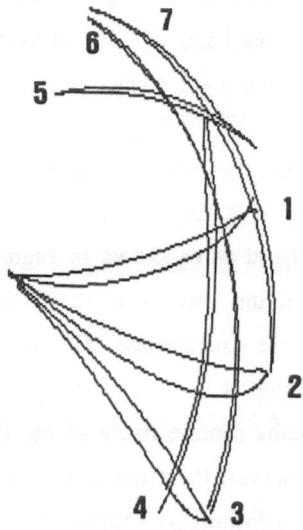

Figure 3: Simulated hand paths

3.2 Experimental and Simulated Hand Tangential Velocity Profiles

Experimental and simulated tangential velocity profiles of the hand are presented for paths 4 and 2 in Figure 4. For each path, velocity profiles are shown for both speed conditions and for both movement directions. Each set of velocity functions includes empirical (light traces) and simulated (heavy dashed trace) tangential velocity profiles together with a minimum-jerk velocity profile (heavy solid trace). Following Atkeson and Hollerbach [2], these functions have been normalized with respect to peak velocity and movement amplitude (i.e., area under the velocity curve) and shifted to minimize the area between the functions.

The experimental tangential velocity profiles shown for the vertical movements (path 4) are well approximated by both the minimum-jerk function and the simulated tangential velocity profiles based on staggered joint interpolation coupled with minimum-jerk joint trajectories and empirically determined joint movement amplitudes and durations. This holds across movement direction and rate. Similar results were found for movement paths 3, 5, 6 and 7.

For the horizontal movements (path 2) both the minimum-jerk function and the simulated tangential velocity profiles account well for the experimental velocity profiles in the case of the outwardly directed movements. However, the simulation fails to predict the experimental tangential velocity profiles in the case of the inwardly directed movements which are well described by the minimum-jerk function. The simulation also fails to account for the experimental velocity profiles observed for the inwardly directed movements of path 1.

602

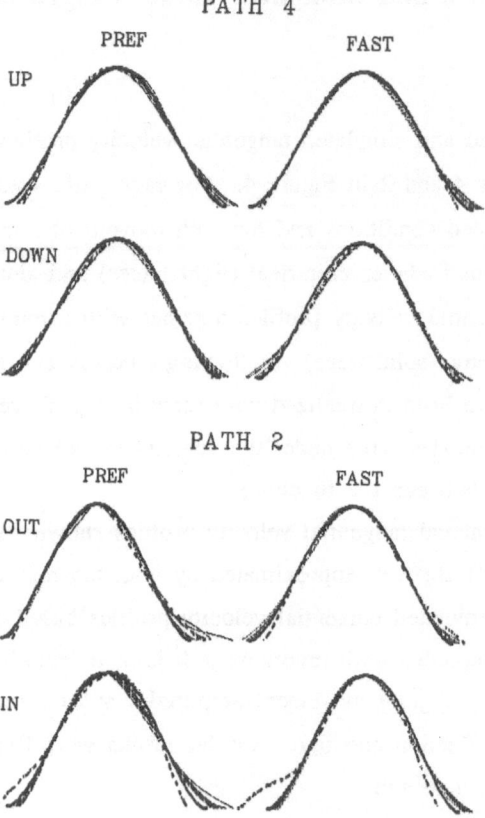

Figure 4: Tangential velocity profiles. Empirical (light), model (heavy dashed), and minimum-jerk (heavy solid) functions are shown.

In summary, the comparison of experimental and simulated hand paths and hand tangential velocity profiles demonstrates that, in general, the kinematics of point-to-point vertical arm movements are well modelled by staggered joint interpolation with the assumption of minimum-jerk joint angle trajectories. However, there are some exceptions. To further investigate these movements, we assessed the form of the joint velocity profiles directly.

3.3 Experimental Joint Velocity Profiles

Experimental shoulder and elbow velocity profiles (dotted traces) for horizontal (path 2) and vertical (path 4) movements are presented in Figure 5. Minimum-jerk velocity functions (solid traces) have been included for reference. The empirical and minimum-jerk profiles have been normalized with respect to peak velocity and movement amplitude and shifted to minimize the area between the curves [2].

For the horizontal (path 2) outward movements, the elbow velocity profiles are positively skewed. In other words, the minimum-jerk profile (used to model joint velocity profiles in the simulations) is negatively skewed relative to the empirical records. In contrast, the shoulder velocity profiles are bell-shaped and symmetrical. Importantly, the amplitude (and peak velocity) of the elbow movements is about twice as great as the shoulder movements (see Figure 6, upper panel) and consequently the form of the velocity profile of the elbow has a greater influence on the shape of the tangential velocity profile. Since the minimum-jerk function is negatively skewed relative to the elbow velocity profile, it is not suprising that the simulated tangential velocity profile is also negatively skewed relative to the experimental data.

A similar but opposite pattern is observed for the inward horizontal (path 2) movement. Once again, the amplitude of the elbow is double the amplitude of the shoulder (see Figure 6, middle panel). However, as shown in Figure 5, the elbow velocity profiles are now bell-shaped and symmetrical whereas the shoulder profiles are positively skewed. Since the elbow velocity profile is similar to the minimum-jerk function, the tangential velocity profile of the hand is well approximated by the simulation.

As noted above, the vertical (path 4) upward movement features a curved hand path whereas the hand path of the vertical downward movement

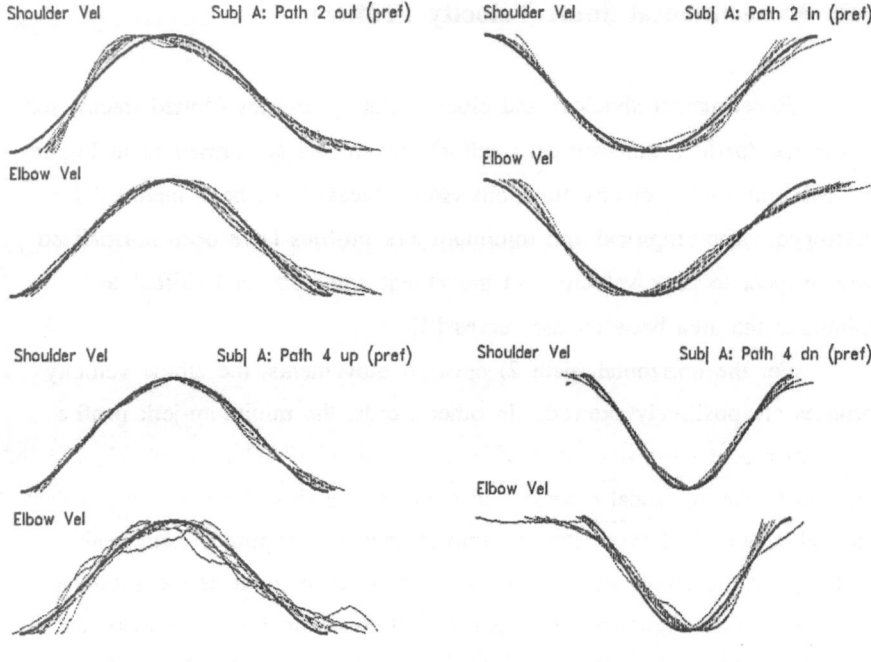

Figure 5: Joint angular velocity profiles. Empirical (dotted traces) and minimum-jerk (solid trace) functions are shown.

is closer to a straight line. The elbow and shoulder velocity profiles for these movements are presented in Figure 5. For both the upward and downward movements, the shoulder velocity profile is bell-shaped and symmetrical. The elbow velocity profile for the downward movement is also bell-shaped and symmetrical. In constrast, the elbow is essentially still

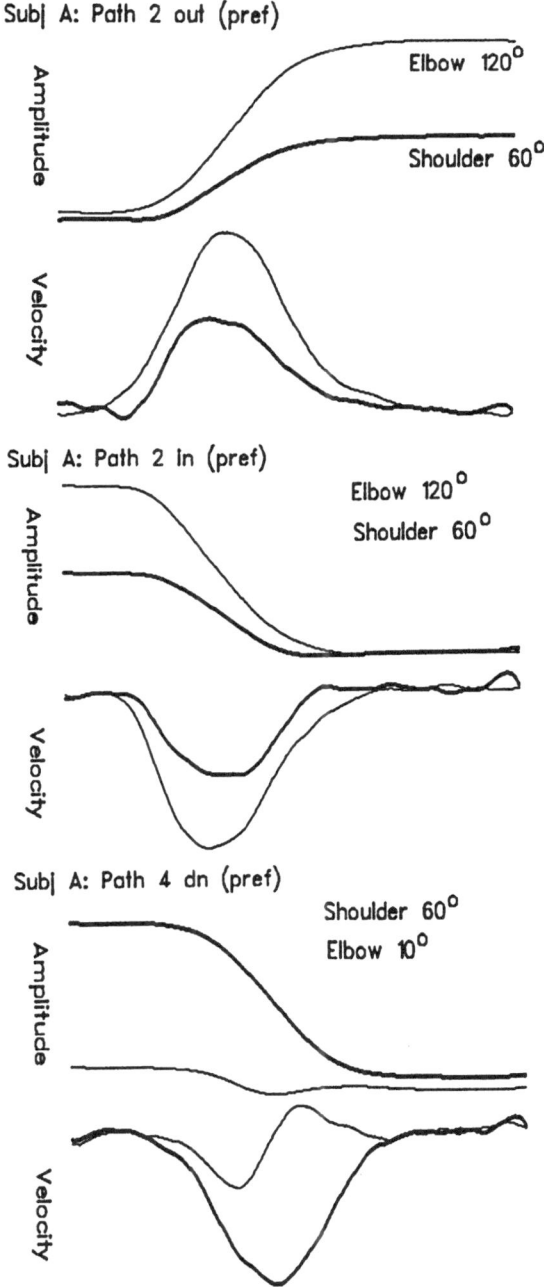

Figure 6: Elbow and shoulder amplitude and velocity

during the upward movement and consequently, the hand path is curved. A typical record of the vertical downward movement is shown in Figure 6 (lower panel). The small elbow flexion at the beginning of the movement accounts for the near straight hand path. This example shows how staggered joint motion can often produce near straight hand paths where curved paths would be seen given linear joint interpolation.

3.4 Interaction of Joint and Hand Tangential Trajectories

In linear joint interpolation, the path (but not the tangential velocity profile) of the endpoint is independent of the form of the joint velocity profiles. However, in staggered joint interpolation, both the path and the tangential velocity profile of the endpoint depend, in part, on the form of the angular velocity profiles at the joints. In order to investigate the relationship between the shape of the joint angular velocity profiles and the trajectory of the hand, we carried out simulations in which the arm was modelled as a two-joint planar manipulator with equal link lengths.

Figure 7 shows the joint and tangential velocity profiles for a linearly joint interpolated movement. The joints follow minimum-jerk trajectories and are shown in the same scale; the amplitude of the elbow is double that of the shoulder. The tangential velocity profile is shown in the middle panel. The lower panel shows a scored tangential velocity profile (movement start and end determined on the basis of 10% peak velocity) with a minimum-jerk profile overlaid. The tangential velocity profile was scored to simulate data analysis used in this and other studies. These profiles have been normalized with respect to peak velocity and amplitude and shifted to maximize the overlap between the curves (see [2]).

Although the joint velocity profiles shown in Figure 7 are bell-shaped and symmetrical, the resulting tangential velocity profile is skewed positively. However, when the tangential velocity profile is scored and normalized, it is

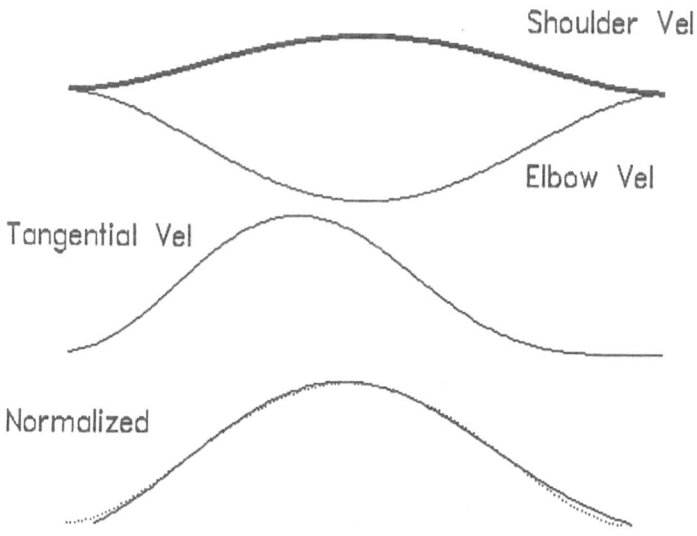

Figure 7: Simulated joint and tangential velocity profiles.
Maximally smooth joint motion with linear joint interpolation.

well approximated by a minimum-jerk function; this is a general
characteristic of two-joint planar motion under linear joint interpolation.

Figure 8 illustrates the joint and tangential velocity profiles for joint
interpolated movements in which elbow motion onset has been delayed by 0,
10 and 20% of movement duration. The amplitude of shoulder flexion is
equal to the amplitude of elbow extension in these movements.
Consequently, under linear joint interpolation, the tangential velocity profile
will have the same shape as the joint profiles. The movements illustrated in
Figure 8 have minimum-jerk joint trajectories. The hand paths for these
movements are shown in Figure 9 (upper panel). Whereas the joint onset
staggering markedly effects the path of the hand, the effect on the tangential
velocity profile of the hand is less clear. Only when the elbow onset is
delayed by 20% of movement duration can a clear difference between the
scored and normalized tangential velocity profile and the minimum-jerk

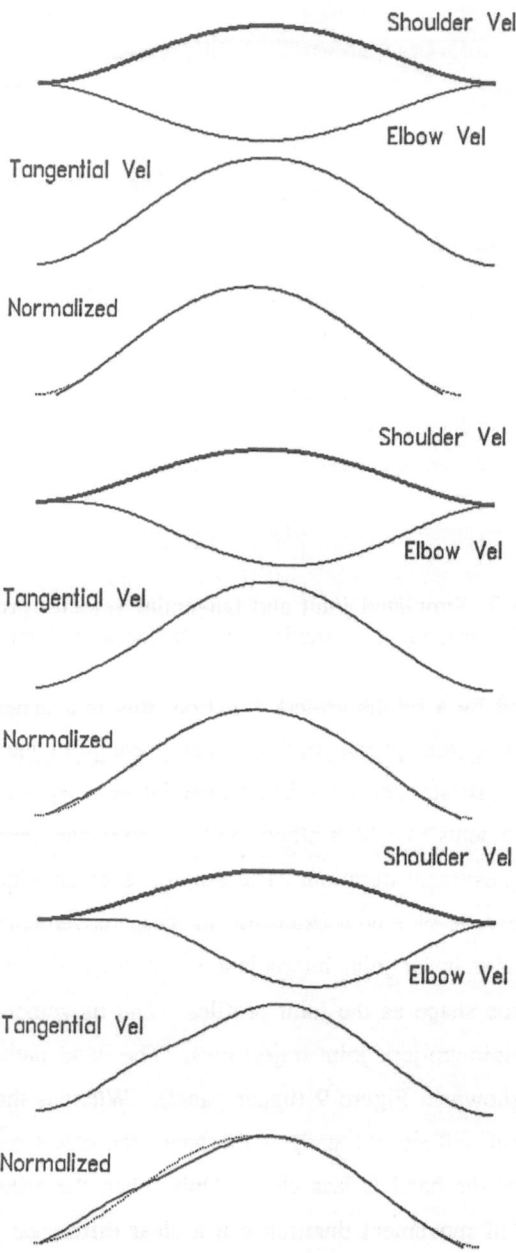

Figure 8: Simulated joint and tangential velocity profiles.
Maximally smooth joint motion with 0% (upper 3 curves), 10%
(middle 3 curves), and 20% (lower 3 curves) initial elbow stagger.

function be observed. However, even in this instance, the departure from the minimum-jerk profile is not great.

Joint interpolated movements in which the joint velocity profiles are positively skewed are presented in Figure 10. A beta density function was used to simulate the skewed joint velocity profiles. The figure shows movements in which elbow onset has been delayed by 10 and 20% of

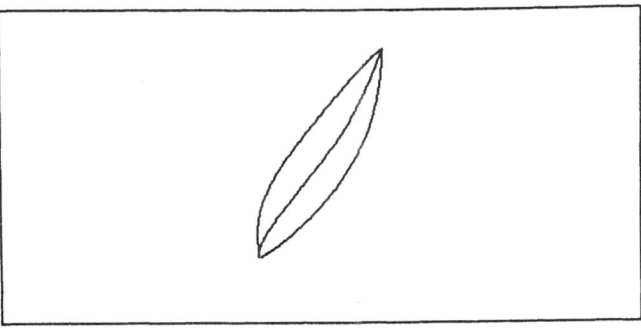

1/5, 1/10, 0 Elbow Stagger (Left -> Right)

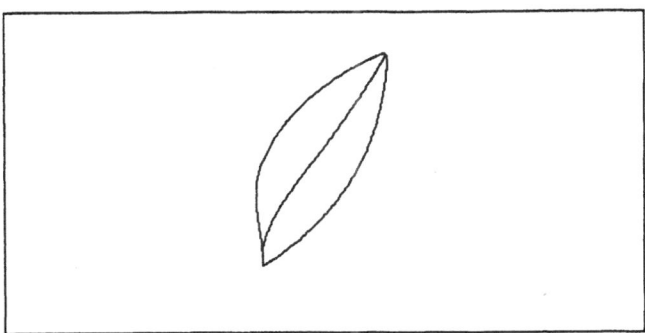

1/5, 1/10, 0 Elbow Stagger; Skewed JVPs

Figure 9: Simulated hand paths. Symmetrical (upper panel) and skewed (lower panel) joint velocity profiles with 20, 10 and 0% (left to right) initial elbow stagger.

movement duration. Once again, the path of the hand is strongly effected by the degree of elbow stagger (see Figure 9, lower panel). Comparison of the upper and lower panels of Figure 9 demonstrates the dependence of the hand path on the form of the joint velocity profiles. With 10% elbow stagger, the shape of the scored and normalized tangential velocity profile is reasonably well accounted for by the minimum-jerk function. However, when the

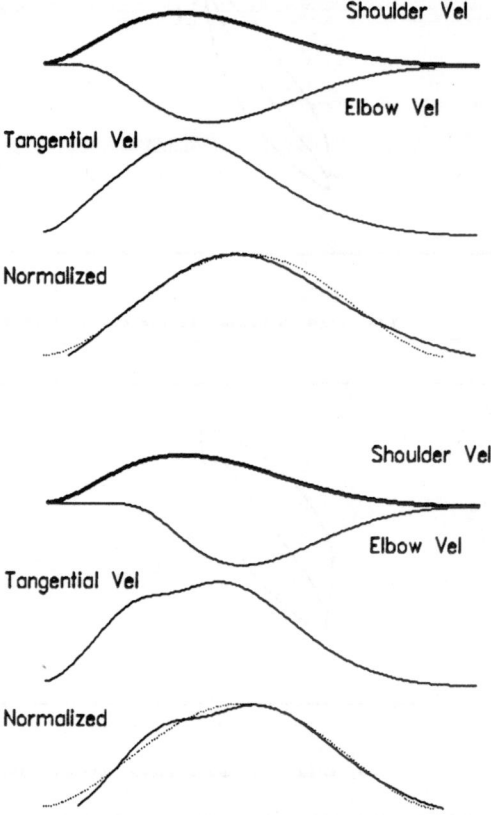

Figure 10: Simulated velocity profiles of skewed joint motion with 10% (upper panel) and 20% (lower panel) initial elbow stagger.

skewed joint velocity profiles are combined with 20% stagger, the tangential velocity profile of the hand is clearly not bell-shaped or symmetrical.

5 Conclusions

We have examined the trajectories of point-to-point multi-joint arm movements made between targets in a vertical plane. In agreement with previous work [2] the shape of the tangential velocity profile of the hand was found to be bell-shaped and symmetrical and the curvature of hand path was found to depend on target location in the workspace. We have shown that, in general, both the form of the hand path and the shape of the tangential velocity profile can be modelled with a version of staggered joint interpolation in which the joints are assumed to follow maximally smooth trajectories.

To test this model, we carried out simulations in which the initial and final joint angles and the joint start and end times were taken from experimental records and the joints were made to follow minimum-jerk trajectories. In most cases, the simulated hand paths and tangential velocity profiles agreed well with the empirical data. However, the model failed to account for the hand paths and tangential velocity profiles observed for inwardly directed horizontal movements. On examination of the experimental joint angular velocity profiles of these movements, it was found that the elbow velocity profile was positively skewed. Consequently, the assumption of bell-shaped and symmetrical (i.e., minimum-jerk) joint trajectories made in the model was inappropriate for these movements.

The finding that minimum-jerk joint trajectories generally correspond to tangential velocity profiles which are similar in form seems suprising since the mapping between joint velocity and hand velocity is non-linear. However, simulations showed that, under linear joint interpolation with minimum-jerk joint velocity profiles, tangential velocity profiles remain approximately bell-shaped and symmetrical when the tangential velocity profiles are scored using standard techniques (see Section 3.4). Even when joint staggering is introduced, the tangential velocity profiles are often reasonably bell-shaped and symmetrical. However, if the joint velocity profiles are skewed then the tangential velocity profile will tend not to be bell-shaped and symmetrical.

In summary, the kinematics of the vertical arm movements examined in this study can be well accounted for by a model combining staggered joint interpolation and minimum-jerk joint trajectories. This finding suggests that human arm movements are planned, at least in part, in joint coordinates.

References

[1] Abend, W., Bizzi, E. and Morasso, P., 1982. Human arm trajectory formation. *Brain*, 105, 331-348.

[2] Atkeson, C.G. and Hollerbach, J.M., 1985. Kinematic features of unrestrained vertical arm movements. *J. Neurosci.*, 9, 2318-2330.

[3] Berkinblit, M.B., Feldman, A.G. and Fukson, O.I., 1986. Adaptability of innate motor patterns and motor control mechanisms. *Behav. Brain Sci.*, 9, 585-638.

[4] Bizzi, E., Chapple, W. and Hogan, N., 1982. Mechanical properties of muscles: implications for motor control. *Trends Neurosci.*, 5, 395-398.

[5] Flanagan, J.R. and Ostry, D.J., 1988. Kinematics of two and three link sagittal arm and arm with pointer movements. *Soc. Neurosci. Abstr.*, 14, 951.

[6] Flash, T. and Hogan, N., 1985. The coordination of arm movements: An experimentally confirmed mathematical model. *J. Neurosci.*, 7, 1688-1703.

[7] Hollerbach, J.M. and Atkeson, C.G., 1987. Deducing planning variables from experimental arm trajectories: Pitfalls and possibilities. *Biol. Cybern.*, 56, 279-292.

[8] Hollerbach, J.M. and Flash, T., 1982. Dynamic interactions between limb segments between during planar arm movement. *Biol. Cybern.*, 44, 67-77.

[9] Morasso, P., 1981. Spatial control of arm movements. *Exp. Brain Res.*, 42, 223-227.

[10] Soechting, J.F. and Lacquaniti, F., 1981. Invariant characteristics of a pointing movement in man. *J. Neurosci.*, 1, 710-720.

[11] Soechting, J.F. and Ross, B., 1984. Psychophysical determination of coordinate representation of human arm orientation. *Neurosci.*, 13, 595-604.

Lecture Notes in Control and Information Sciences

Edited by M. Thoma and A. Wyner

Lecture Notes in Control and Information Sciences

Edited by M. Thoma and A. Wyner

Lecture Notes in Control and Information Sciences

Edited by M. Thoma and A. Wyner

Vol. 137: S. L. Shah, G. Dumont (Eds.)
Adaptive Control Strategies for
Industrial Use
Proceedings of a Workshop
Kananaskis, Canada, 1988
VI, 360 pages. 1989

Vol. 138: D. C. McFarlane, K. Glover
Robust Controller Design Using
Normalized Coprime Factor
Plant Descriptions
X, 206 pages. 1990

Vol. 139: V. Hayward, O. Khatib (Eds.)
Experimental Robotics I
The First International Symposium
Montreal, June 19 - 21, 1989
XVII, 613 pages. 1990